Arbitrarily Close

An Introduction to Real Analysis

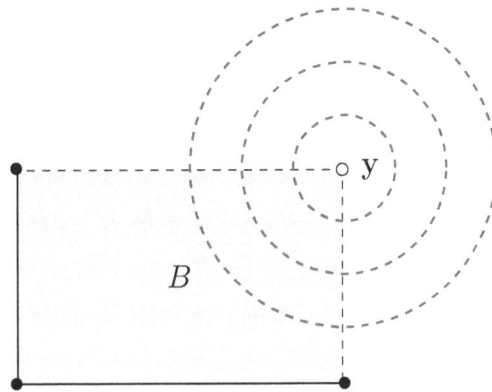

John A. Rock

First edition, 2025

Hardcover ISBN 978-1-958469-31-6

Paperback ISBN 978-1-958469-30-9

Published by 619 Wreath

Contents

Chapter 0

Preface

0.1   Dedication

This book is dedicated to $\mathbf{y}$ and B.

0.2   Thanks

Dylan Alvarenga	Jacob Henn	Jeffrey Robbins
Ryan Aniceto	Casey Hernandez	Rachel Robbins
Grace Bajar	Stephen Hutchins	John Rodriguez
Kendra Calman	Rasha Issa	Felix Rojas
Herman Carrillo	Michelle Lin	Rolando de Santiago
Jose Castillo	Hugo Marquez	Erin Sewell
Bryan Chao	Wendy Marquez	
Justin Chrien	Grace Meigs	Kyle Teats
Maricarmen Chavez	Eugene Monforte	Ciara Thomas
Sara Elakesh	Jorge Montes-Guzman	Jamie Thomas
Samya Faraj	Sandra Moreno	Michael Thompson
Ian Farish	Philip Nicoll	Weizhong Wang
Berit Givens	Lekha Patil	Anna Yee
Latimer Harris-Ward	Noah Reef	

0.3   Content overview

I think of analysis as the mathematics of estimation. When do we know our estimates are arbitrarily precise? And what are we estimating? For instance, difference quotients estimate derivatives. To me, the idea that brings it all together—the kernel of analysis—is a formal definition for the phrase *arbitrarily close*. This line of inquiry often takes the form of studying linear combinations: When we take the set of all points arbitrarily close to a set of linear combinations of some nice objects, what do we get? For instance, polynomials are linear combinations of monomials, and linear combinations of heights and widths (areas of rectangles) are arbitrarily close to the areas under curves defined by (nearly) continuous functions. These ideas and more are made precise and studied throughout.

This book is not a novel, it's a resource for people who would like an introduction to real analysis supported with a lot of detail. Please make use of it as you see fit.

For readers who have never taken a course on real analysis: Take your time, read as much as you can, and take notes as you go. Familiarity with inequalities, basic algebra, trigonometry, some set theory, and writing mathematical proofs is assumed.

Lots of details and figures are included to help you along the way. However, there is much more to real analysis than is covered here. The choice of content is based on the course MAT 3140

Introduction to Real Analysis I at Cal Poly Pomona. In particular, the material up to Chapter 4 and part of Chapter 5 deals with basics topics in analysis of Euclidean spaces, but thereafter the focus is limited to the real line.

It's easy to lose sight of a big picture if there's too much focus on details. Please skip around when it makes sense for you. The details are here for when you want to see them, but they don't tell the story.

For readers with more experience including instructors, please skip around as you see fit. However, I recommend you do not skip Sections 1.1, 1.3, and 1.5. These are vital to the development of the entire book since they are where the kernel of analysis—*arbitrarily close*—is defined for both the real line and Euclidean spaces, respectively.

To all my readers, please let me know what you think! If you spot any errors or have recommendations for ways to improve any aspect of what you find here, please reach out to me via email: jarock@cpp.edu.

Remark 0.3.1: Unconventional terminology and notation

Informal use of the phrase "arbitrarily close" can be found across the literature on analysis and topology. However, this book takes the unconventional approach of using technical definitions to drive the development of material. See Definitions 1.1.8 and 1.5.1. Additionally, since "arbitrarily close" appears frequently throughout the book, the shorthand notation "acl" is used to streamline arguments and computations much in the way "lim" shortens work done with limits. See Definitions 2.2.1 and 5.1.2.

Another break from convention appears with the definitions for *supremum* and *infimum* in Definition 1.1.14 where they are given in terms arbitrarily close. These are equivalent but distinct characterizations of the more conventional definitions of *least upper bound* (Definition 1.3.9) and *greatest lower bound* (Definition 1.4.2). Equivalence is established with Theorems 1.3.10 and 1.4.3, respectively.

0.4   Mistakes, play, and learning

Failure is the amuse-bouche of learning.

- Chef Aurie Taamu, *Tears of the Kingdom*

The exercises are there to play with! Do scratch work, draw stuff, and make mistakes—make *lots* of mistakes—before worrying about writing proofs. Mistakes are an unavoidable and essential part of learning and writing mathematics. You will undoubtedly find some typos and even mathematical mistakes as you read this book. And you know what, *that's a good thing*. I encourage you to read Dr. Francis Su's wonderful book titled "Mathematics for Human Flourishing" [12].

If you find mistakes I've made, you're learning. If you try something that ends up not working, you're learning. If an idea works for a bit and you get stuck, put the work aside and do something else. You're still learning. Your brain is amazing, it'll keep churning on the math even if you're not actively thinking about it.

For people who play video games, do you ever read an entire manual or guide before playing a new game? Probably not. Even if you do, do you find that reading the manual made you into an expert right away? I seriously doubt it. My guess is you just grabbed a controller and went for it. Why not take the same approach with mathematics?

Play with the ideas! Playing and learning go hand-in-hand. Grab the controller, crash your character into a wall, get a "Game Over", and try again. When you figure something out, share it with your friends, classmates, and instructors. When you get stuck, ask them what they did or would do. Communicating your ideas (poorly formed, haphazard, and sometimes silly) and thinking through the ideas shared by your friends, classmates, and instructors will get you through.

And how about driving a car? You can watch me do it, I can explain to you how I go about the way I drive, you can read about it. But none of that compares to how much you will learn by getting behind the wheel yourself.

A point I'm trying to make is that I can explain how I see the mathematics of real analysis, but it's more important for you to discover your own perspective on the material.

0.5   Color scheme

Colors are used to parse the types of content found in the book.

Definition 0.5.1: A unique color for definitions

Technical definitions of mathematical ideas can be found in boxes like this one.

Example 0.5.2: A unique color for examples

Examples contain varying degrees of detail.

Theorem 0.5.3: Lemmas, Theorems, and Corollaries have to share

Technical results appear in boxes like this one. Most are accompanied by Scratch Work and a proof.

Scratch Work 0.5.4: Play around with ideas

Mathematics can be messy before we figure out nice ways of explaining what we have in mind. As a result, the Scratch Work found in this book comes in variety of flavors with no definitive structure. Some have lots of detail that are eventually used in a proof, some provide a summary of an upcoming proof, and others are somewhere in between. The idea of Scratch Work is really to motivate you to do your own. Play around with the ideas. Let your own scratch work be messy until you figure things out. Jay Cummings' wonderful book *Real Analysis: A Long-Form Mathematics Textbook* also features a lot of scratch work to go along with excellent dialogue around the technical material. (Nice work, Jay!)

Proofs look like this. Not every technical statement found is this book is proven here, but most are. Many examples are accompanied by Scratch Work and proofs. For the most part, proofs are intentionally heavy on details and references (hyperlinks), as can be seen with Theorem 2.3.1. By the way, Theorem 2.3.1 happens to be the fundamental result connecting the notion of arbitrarily close (Definition 1.5.1) with convergence of sequences in Euclidean spaces (Definition 2.2.1).

The amount of technical detail found in the proofs is motivated by the lack thereof in many mathematical textbooks on real analysis. This is especially true of the classic—to some, *the definitive classic*—textbook by Walter Rudin [10], colloquially referred to as "Baby Rudin". Don't get me wrong, I learned to love "Baby Rudin" and credit it as a reason to choose analysis for my own field of study. The amount of detail there leaves a lot for readers to discern for themselves, which is not necessarily a bad thing. □

Remark 0.5.5: Notation, Remarks, and Problems have to share

Other blocks of text look like this. They are generally not as detailed, but something about them is worth highlighting.

0.6   Walkthroughs

You are encouraged to write up *walkthroughs* as you read. Here are some activities to consider, in no particular order:

- Rewrite statements in your own words.

- Come up with examples and nonexamples.

- Draw figures to accompany or supplement the given material.

- Create activities using Desmos, GeoGebra, WolframAlpha, or other freely available resources.

- Come up with scratch work that may or may not fit a given proof.

- Write proofs that are more thorough.

- Write proofs that are more concise.

- Do any activity you can think of that will help you understand what's going on.

Ultimately, walkthroughs are like Scratch Work in that they are what you make of them. Do what makes sense for you. I believe the more you supplement your reading with walkthroughs, whatever a walkthrough means to you, the more the mathematics will make sense and the more beautiful it will become.

0.7   Themes

The book has a number of recurring themes.

Remark 0.7.1: Arbitrarily close

A simple question drove the development of this book.

> How close can a point be to a set?

The following answer redefined my perspective on analysis and topology.

> So close there's no distance between them: *arbitrarily close.*

Once I defined arbitrarily close in terms of comparing a point to a set, as opposed to comparing one point and another, I felt like I was really onto something. See Definitions 1.1.8 and 1.5.1 along with the title of Chapter 1, and of course, the title of the book.

Here are some of the ways *arbitrarily close* appears throughout the book.

- The supremum of a set of real numbers is the upper bound arbitrarily close to the set (Definition 1.1.14). Similarly, The infimum of a set of real numbers is the lower bound arbitrarily close to the set.

- Zero is the only real number arbitrarily close to the sets of positive and negative real numbers (Lemma 1.5.14).

- A limit of a sequence is arbitrarily close to its sequence (Theorem 2.3.1). In fact, the precise connection between the concepts of limits for sequences and points arbitrarily close to the range of a sequence is quite deep and forms a foundation for many of the results presented in the book .

- A closed set comprises the points arbitrarily close to the set. See Definition 3.1.1 and Theorem 3.2.4.

- A connected set comprises partitions with a point in one set which is arbitrarily close to another. See Definitions 3.3.1 and 3.3.4.

- A function preserves closeness at a point if and only if it is continuous at the point. See Definitions 4.2.5 and 4.3.2 as well as Theorem 4.4.7.

- A limit of a function is arbitrarily close to the range of the function. See Theorem 5.1.14.

- A derivative is arbitrarily close to the set of difference quotients. See Definition 5.3.1 and Corollary 5.3.14.

- An integral is arbitrarily close to sets of linear combinations defining sums of areas of rectangles. See Definition 6.1.6.

- A continuous function on a closed and bounded interval is arbitrarily close the set of polynomials. See the Weierstrass Approximation Theorem 7.4.7.

- The sum of a series is arbitrarily close to the set of partial sums. See Definitions 8.1.1 and 8.3.1.

The setting for arbitrarily close is limited to Euclidean spaces throughout the book, with additional restrictions to the real line when warranted somehow. See Definitions 1.1.8 and 1.5.1.

Results stated for Euclidean spaces hold for the real line (as well as complete metric spaces in many cases). The intuition of distance between points provided by metric spaces is at the heart of the concept of arbitrarily close, but the definition itself is topological in nature.

Chapter 1 develops the main topic: A formal definition for the phrase *arbitrarily close* and fundamental results on analysis built from there.

Remark 0.7.2: Linearity

Properties that distribute across addition and allow constants to factor out exhibit *linearity*, a concept so pervasive in analysis that it is taken for granted. An attempt is made to reverse this trend throughout the book by explicitly stating results on linearity and mentioning them in scratch work and proofs.

Here are some of the ways linearity appears in analysis.

- Linearity of arbitrarily close, Theorem 1.6.12.

- Linearity of sequential limits, Theorem 2.3.9 and Corollary 2.3.13.

- Linearity of continuity, Theorem 4.5.5 and Corollary 4.5.7.

- Linearity of functional limits, Theorem 5.2.6 and Corollary 5.2.8.

- Linearity of differentiation, Theorem 5.4.1 and Corollary 5.4.3.

- Linearity of integration, Theorem 6.3.6 and Corollary 6.3.9.

- Linearity of uniform convergence, Theorem 7.2.8 and Corollary 7.2.10.

Remark 0.7.3: Thresholds and convergence

Thresholds take concepts of arbitrarily close a step further by telling us convergence is ensured in some way. Roughly speaking:

> Outputs are within a chosen distance whenever inputs are beyond or within a suitable threshold.

As a result:

> Thresholds control inputs to ensure outputs are as close as we like.

Thresholds help us define a variety of concepts, including some notions of local behavior:

- threshold (continuity), 260
- threshold (functional limit), 314
- threshold (local maximum), 349
- threshold (local minimum), 349
- threshold (locally linear), 337
- threshold (pointwise series), 479

- threshold (pointwise), 423
- threshold (sequential limit), 91
- threshold (uniform continuity), 305
- threshold (uniform convergence), 425
- threshold (uniform series), 479

Remark 0.7.4: Cauchy criteria

French mathematician and engineer Augustin-Louis Cauchy had such a tremendous impact on the world of mathematics that his last name is used as an adjective. In the context of real analysis, the adjective *Cauchy* captures convergence without a candidate for the limit in mind. For instance, see Definition 2.6.1. This freedom is used in many proofs throughout the book.

Below is a list of *Cauchy criteria* with ties to convergence.

- Cauchy criterion for sequences, Theorem 2.6.5.
- Cauchy criterion for integrability, part (vi) of Theorem 6.2.12.
- Cauchy criterion for uniform convergence, Theorem 7.2.11.
- Cauchy criterion for series, Theorem 8.2.2.

0.8   QR codes

Links to a variety of free online computational activities on Desmos and GeoGebra appear throughout the book to provide dynamic supplements to static figures and exercises. The links are given as QR codes and hyperlinks. For example, Figure 0.8.1 provides a QR code for the Desmos activity "y acl B" that accompanies the figure on the title page, Figure 1.5.1, and Definition 1.5.1. Play around with these activities and make your own!

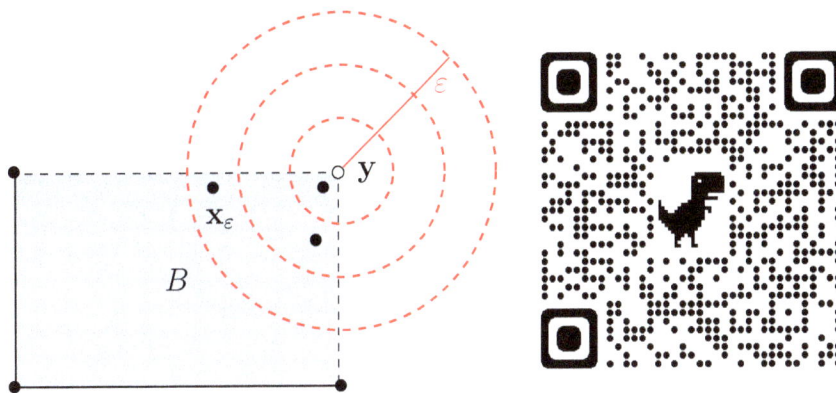

Figure 0.8.1: QR code for the Desmos activity "y acl B" to accompany the figure on the title page, Figure 1.5.1, and Definition 1.5.1. https://desmos.com/calculator/nfbdjs8pdh

Chapter 1

Kernel of Analysis

> A point is *arbitrarily close* to a set when every neighborhood of the point intersects the set.

This notion of *arbitrarily close* establishes the theme of the book, serving as a foundation for classic results in real analysis on *closure, limits and convergence, connectedness, continuity, differentiation,* and *integration*.

1.1   Arbitrarily close

Let's start with a couple of intervals and see if either has a largest element. The set of real numbers is denoted by $\mathbb{R}$ throughout the book, and the notation "$x \in \mathbb{R}$" means x is a real number.

Problem 1.1.1: Two intervals

Consider the closed interval F and open interval G in Figure 1.1.1 given by

$$F = (-\infty, 3140] = \{x \in \mathbb{R} : x \leq 3140\} \quad \text{and} \tag{1.1.1}$$
$$G = (-\infty, 3140) = \{x \in \mathbb{R} : x < 3140\}. \tag{1.1.2}$$

Which interval has a largest element? That is, which has a *maximum*?

Do you know what is meant by "largest element", exactly? How is "large" used in this context? The following definitions of *upper bound, maximum, lower bound,* and *minimum* codify the context and allow us to prove results in a rigorous way.

Definition 1.1.2: Maximum of a set of real numbers

A real number b is an *upper bound* for a set of real numbers S if for every x in S we have $x \leq b$. In this case, we say S is *bounded above*. A real number q is the *maximum* of S if q is an upper bound for S and q is an element of S. That is,

(i) for every x in S we have $x \leq q$, and

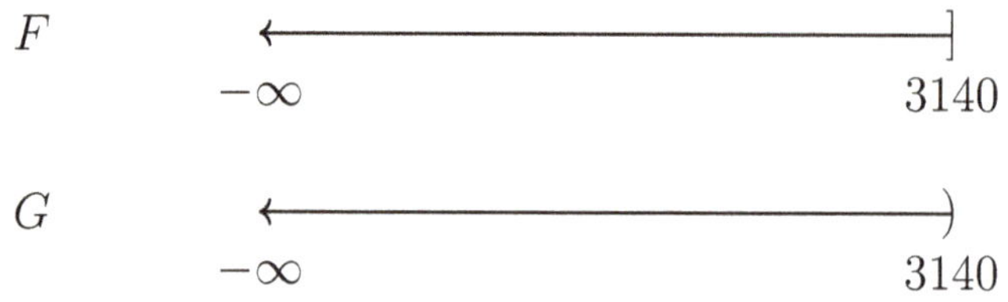

Figure 1.1.1: The intervals $F = (-\infty, 3140]$ and $G = (-\infty, 3140)$ from Problem 1.1.1.

(ii) q is in S.

In this case, we write $q = \max S$.

Similarly, a real number a is a *lower bound* for a set of real numbers S if for every x in S we have $a \leq x$. In this case, we say S is *bounded below*. A real number v is the *minimum* of S if v is a lower bound for S and v is an element of S. That is,

(iii) for every x in S we have $v \leq x$, and

(iv) v is in S.

In this case, we write $v = \min S$.

Let's use Definition 1.1.2 to prove a couple facts about F and G.

Example 1.1.3: Closed interval with a maximum

The interval $F = (-\infty, 3140]$ has a largest element. In other words, $\max F$ exists.

Proof for Example 1.1.3. By checking the properties in Definition 1.1.2, since (i) $x \leq 3140$ for every x in F (thus 3140 is an upper bound for F) and (ii) 3140 is in F, the interval F has a maximum—its largest element—given by $\max F = 3140$. □

What about an interval like $G = (-\infty, 3140)$? My intuition suggests the maximum might be 3140. The problem is no real number is both (i) an upper bound for G and (ii) an element of G.

Example 1.1.4: Open interval with no maximum

The interval $G = (-\infty, 3140)$ has no largest element. In other words, $\max G$ does not exist.

Proof for Example 1.1.4. Any real number greater than or equal to 3140 is not in G, so (ii) fails in this case. Also, any real number in G is strictly less than 3140, so there's always a larger number in G. For instance, the midpoint between 3140 and the given real number in G is also in G. Thus, no element of G is an upper bound for G, so (i) fails in this case.

Since no real number satisfies both parts (i) and (ii) of Definition 1.1.2 with respect to G, the open interval G has no maximum. In other words, *G has no largest element.* □

Even though 3140 is not the maximum of G, it plays a special role: The point 3140 is an upper bound for the set G which is as close to G as possible without actually being in G. In other words, 3140 is both an upper bound for G and so close to G there is no distance between them. But what does that mean, exactly? And how can we prove it?

In order to prove the results we can visualize and believe to be true, mathematical definitions capture our intuition and allow us to control the context of the problems we're trying to solve. Definitions allow us to be specific in addressing intuitive ideas like how close a point can be to a set or having no distance between a point and a set.

Remark 1.1.5: A first question

What does it mean for a point to be *arbitrarily close* to a set? To ensure we have a sound mathematical foundation, let's define a notion for distance between points in the real line so we can eventually be precise about how close points and sets can be to each other. The distance we use relies on a formal definition for the absolute value of a real number.

Definition 1.1.6: Absolute value and distance in the real line $\mathbb{R}$

For every x in $\mathbb{R}$, the *absolute value* of x, denoted by $|x|$, is given by

$$|x| = \begin{cases} x, & \text{if } x \geq 0, \\ -x, & \text{if } x < 0. \end{cases} \tag{1.1.3}$$

For every x and y in $\mathbb{R}$, the *distance* between x and y, denoted by $d_{\mathbb{R}}(x,y)$, is defined by

$$d_{\mathbb{R}}(x,y) = |x - y|. \tag{1.1.4}$$

Remark 1.1.7: Absolute value and negative numbers

It may be odd to see absolute value defined piecewise as it is in Definition 1.1.6, but doing so gives a rigorous way to work with absolute values when writing proofs.

Still, the "$-x$" part may look weird. I think of it as multiplying a negative real number x by -1 rather than just "dropping the negative sign" since the latter sometimes fails to capture the process of finding the absolute value of a negative number. For instance, $x = 2 - \sqrt{5}$ is a negative real number with no clear negative sign to "drop". On the other hand,

$$x = 2 - \sqrt{5} < 0 \quad \implies \quad |x| = |2 - \sqrt{5}| = (-1)(2 - \sqrt{5}) = \sqrt{5} - 2,$$

where the symbol " $\implies$ " is read as "implies".

Again, by thinking of taking the absolute value of a negative real as multiplying the negative number by -1, we have a rigorous mathematical process we can rely on for proofs involving absolute value.

The definition for distance in Definition 1.1.6 is based on a comparison between two points in

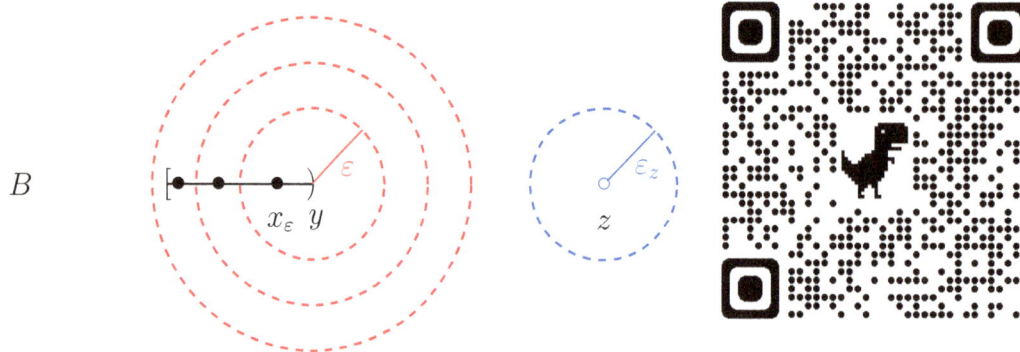

Figure 1.1.2: Examples of real numbers y and z along with a set B where $y \operatorname{acl} B$ and $z \operatorname{awf} B$ as in Definition 1.1.8. Also, a QR code for the Desmos activity "y acl B interval" to accompany this figure. https://www.desmos.com/calculator/ehogaxpv8v

the real line, not a point and a set as mentioned in Remark 1.1.5. Even so, it gives us something to work with: Given a point and a set, we can consider *any* amount of distance around the point and see if there are points from the set within that distance.

Here is a definition for *arbitrarily close* in the real line, what I consider to be the kernel of analysis and the fundamental theme of the book. See Figure 1.1.2. (A more general definition in the context of Euclidean spaces is presented in Definition 1.5.1.)

Definition 1.1.8: Arbitrarily close and away from (real line)

Let y be a real number and let B be a set of real numbers. The point y is said to be *arbitrarily close* to the set B, and we write $y \operatorname{acl} B$, if for every $\varepsilon > 0$ there is some x_ε in B such that

$$d_{\mathbb{R}}(x_\varepsilon, y) = |x_\varepsilon - y| < \varepsilon. \tag{1.1.5}$$

The phrase "B is *arbitrarily close* to y" is also defined and denoted in the same way. That is, $B \operatorname{acl} y$ is taken to mean the same thing as $y \operatorname{acl} B$.[a]

If some real number z is not arbitrarily close to B, then there is some $\varepsilon_z > 0$ such that for every x in B we have

$$d_{\mathbb{R}}(x, z) = |x - z| \geq \varepsilon_z > 0. \tag{1.1.6}$$

In this case, we say z is *away from* B and write $z \operatorname{awf} B$. The phrase "B is *away from* z" is also defined and denoted in the same way.

---
[a]Thanks to Berit Givens for suggesting the notation "$y \operatorname{acl} B$" to represent the phrase "y is arbitrarily close to B".

In Figure 1.1.2, the set B is an interval that does not contain its right endpoint y. The point z

Figure 1.1.3: The open interval $G = (-\infty, 3140)$ along with the real number $z = 4710$ and distance $\varepsilon_z = 785$ from Example 1.1.11.

is away from B since the blue dashed circle centered at z has a positive radius ε_z with no points of B inside. On the other hand, y is arbitrarily close to B since circles centered at y of any positive radius ε have a point x_ε from the set B within them. Three such circles are dashed in red, and each has at least one such x_ε within its radius indicated by one of the three •. In order to keep things from getting too cluttered, only one red ε appears in the figure.

Remark 1.1.9: As small as we like

In Definition 1.1.8, it may help to think of the positive real number ε as the amount of error or "wiggle room" we'd like to allow, the idea being that we can allow *any* amount of error, no matter how small. In this way, $y \operatorname{acl} B$ means B gets as close to y as we like, no matter how close that may be. So, $y \operatorname{acl} B$ is exactly what it means when there is no distance between a point y and a set B.

Remark 1.1.10: Subscript or not

How can we prove $y \operatorname{acl} B$ and $z \operatorname{awf} B$ based on the definitions? To prove $y \operatorname{acl} B$, we should respond to *each* distance $\varepsilon > 0$ with a point x_ε in B that is within ε of y. To prove $z \operatorname{awf} B$, we need just *one* distance $\varepsilon_z > 0$ that separates z from all the points in B by a distance of ε_z or more.

The notation for the variables in Definition 1.1.8 play subtle roles. The x_ε and ε_z both have a subscript indicating something special is going on. Specifically, x_ε is a particular real number in B that is found in response to $\varepsilon > 0$, and $\varepsilon_z > 0$ is a particular positive number found in response to the real number z and its relationship to B. On the other hand, ε represents *any* positive real number and x represents *any* real number in B.

Example 1.1.11: Points arbitrarily close and away from a set

For the open interval $G = (-\infty, 3140)$, I claim the real number $y = 3140$ is arbitrarily close to G while $z = 4710$ is away from G. To prove $4710 \operatorname{awf} G$, we can find just *one* positive distance ε_z to separate 4710 from all the points in G. Let's do that first.

To align with Figure 1.1.3, the radius of the dashed blue circle is half the distance between 4710 and 3140, which looks good enough to keep every point in G at least that far from

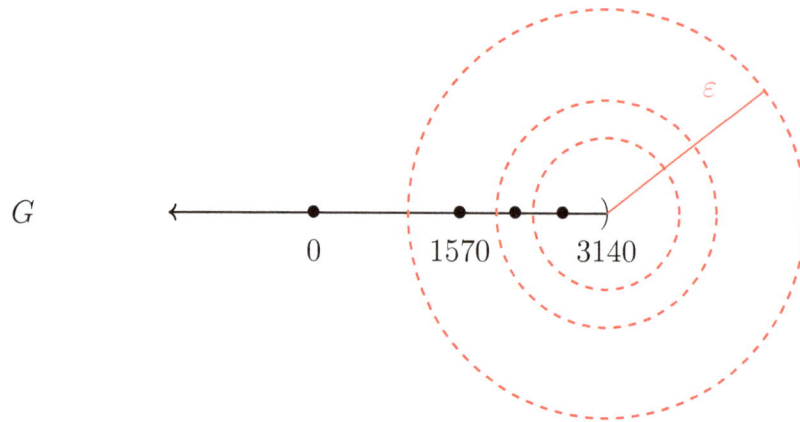

Figure 1.1.4: The open interval G along with multiple distances (in red) and corresponding points (the •) as in Scratch Work 1.1.12. The real number 1570 is within the largest circle whose radius is ε (in red), but not the smaller circles.

4710^{a}. So we can take

$$\varepsilon_z = \frac{|4710 - 3140|}{2} = \frac{1570}{2} = 785 > 0. \tag{1.1.7}$$

---

^{a}The real number $z = 4710$ ensures the figure is to scale.

Proof of 4710 awf G in Example 1.1.11. Consider the real number $z = 4710$ and let x be any real number in $G = (-\infty, 3140)$. Then we have

$$x < 3140 < 4710. \tag{1.1.8}$$

Now let $\varepsilon_z = 785 > 0$. Since $x - 4710 < 0$ and $x < 3140$ according to (1.1.8), by the definition of absolute value (Definition 1.1.6) we have

$$|x - 4710| = 4710 - x > 4710 - 3140 = 1570 > 785 = \varepsilon_z. \tag{1.1.9}$$

Therefore, every point in G is more than $\varepsilon_z = 785 > 0$ away from 4710, and so 4710 awf G. See Figure 1.1.3.   □

  Proving 3140 is arbitrarily close to G takes more effort. It's not good enough to consider just one distance. To prove 3140 acl G, we should respond to *every* positive distance or "error" $\varepsilon > 0$ with its own point x_ε that's both in G and within ε of 3140. See Figure 1.1.4.

Scratch Work 1.1.12: A point arbitrarily close to a set

Consider drawing figures whenever you do scratch work, such as Figure 1.1.4. The largest circle has radius $\varepsilon > 0$. For that ε and any larger radius, it suffices to respond with

$$x_\varepsilon = 1570 \tag{1.1.10}$$

which is good enough since 1570 is both in G and within ε of 3140. The other two circles in Figure 1.1.4 have smaller radii and leave out 1570, but each has at least one point (one of the ●) that is in G and also within the corresponding distance of 3140. To verify *every* $\varepsilon > 0$ comes with at least one point x_ε that's both in G and within ε of 3140, the variable ε is taken to represent all positive distances at the same time and we respond to each with a point x_ε defined by a function of ε.

On to the proof.

Proof of $3140 \operatorname{acl} G$ *in Example 1.1.11.* Let $\varepsilon > 0$. (By not specifying a particular value of ε, we are accounting for *all* positive distances, or "errors".) Choose

$$x_\varepsilon = 3140 - \frac{\varepsilon}{2}. \tag{1.1.11}$$

Since

$$x_\varepsilon = 3140 - \frac{\varepsilon}{2} < 3140, \tag{1.1.12}$$

we have x_ε is in $G = (-\infty, 3140)$ and $x_\varepsilon - 3140 < 0$. Hence,

$$|x_\varepsilon - 3140| = -(x_\varepsilon - 3140) = -\left(3140 - \frac{\varepsilon}{2} - 3140\right) = \frac{\varepsilon}{2} < \varepsilon. \tag{1.1.13}$$

Therefore, $3140 \operatorname{acl} G$.   □

Remark 1.1.13: Does scratch work help?

Hang on. What just happened? Did every step make sense? Does Scratch Work 1.1.12 help you see how I came up with this proof? Take your time reasoning through my proof and consider writing up a walkthrough to help you find your own understanding.

After laying out Scratch Work 1.1.12, I reorganized and rewrote stuff to produce the proof showing 3140 is arbitrarily close to G for Example 1.1.4. How would you have done the scratch work and proof?

Moving forward, scratch work typically appears before a corresponding proof. The amount of detail varies, but scratch work usually includes motivation for the steps in proofs. Scratch work should entail anything that helps you figure stuff out.

With the definition for arbitrarily close we now have enough to define *supremum*. A supremum is a lot like a maximum in that both are upper bounds for a given set, but the supremum is not necessarily in the set like the maximum has to be. An analogous statement holds for the definition of *infimum* which is defined in terms of a suitable lower bound and is a lot like minimum.[1]

Definition 1.1.14: Supremum and infimum

A real number u is the *supremum* of a nonempty set of real numbers S, and we write $u = \sup S$, if u is an upper bound for S and arbitrarily close to S. That is, $u = \sup S$ if

---

[1]The definitions for *supremum* and *infimum* in Definition 1.1.14 are not the classic versions, but they are equivalent. See Theorems 1.3.10 and 1.4.3.

(i) for every $x \in S$ we have $x \leq u$, and

(ii) $u \operatorname{acl} S$.

Similarly, a real number ℓ is the *infimum* of S, and we write $\ell = \inf S$, if ℓ is a lower bound for S and arbitrarily close to S. That is, $\ell = \inf S$ if

(iii) for every $x \in S$ we have $\ell \leq x$, and

(iv) $\ell \operatorname{acl} S$.

Have you noticed how similar the definitions for maximum and supremum are to one another? Compare Definitions 1.1.2 and 1.1.14.

Example 1.1.15: Revisiting an open interval

As seen in Example 1.1.11, the real number 3140 is arbitrarily close to the open interval $G = (-\infty, 3140)$. Since $x < 3140$ for each x in G, we have 3140 is also an upper bound for G. Therefore, $\sup G = 3140$ and yet $\max G$ does not exist (as pointed out in Example 1.1.3). Since G has no lower bounds, neither $\min G$ nor $\inf G$ exist.

Let's consider one more example to see what our results give us regarding the open interval G from Problem 1.1.1 as well as Examples 1.1.4 and 1.1.15, this time with specific values of $\varepsilon > 0$ in mind.

Example 1.1.16: Checking our reasoning

For the interval $G = (-\infty, 3140)$, we proved $3140 \operatorname{acl} G$ with a key step of choosing

$$x_\varepsilon = 3140 - \frac{\varepsilon}{2} \qquad (1.1.14)$$

in response to any given $\varepsilon > 0$.

What does this formula give us for the specific values $\varepsilon_1 = 1$, $\varepsilon_2 = 1/2$, and $\varepsilon_3 = 1/100$? Are the values of x_ε really in G?

First, we can show x_1, x_2, and x_3 are in G by verifying they satisfy the inequality defining G in line (1.1.2). Respectively, we have[a]

$$x_1 = 3140 - \frac{\varepsilon_1}{2} = 3140 - \frac{1}{2} < 3140, \qquad (1.1.15)$$

$$x_2 = 3140 - \frac{\varepsilon_2}{2} = 3140 - \frac{1}{4} < 3140, \quad \text{and} \qquad (1.1.16)$$

$$x_3 = 3140 - \frac{\varepsilon_3}{2} = 3140 - \frac{1}{200} < 3140. \qquad (1.1.17)$$

Hence, x_1, x_2 and x_3 are in $G = (-\infty, 3140)$.

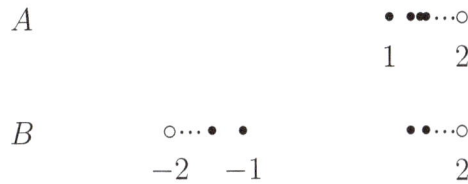

Figure 1.1.5: The sets A and B from Example 1.1.17. Neither set contains 2, indicated by one of the ∘ in the figure. But how close do they get? What about -2? Can you prove your answers?

Now, are these x_k for $k = 1, 2, 3$ close enough to 3140? Yes, but each one only needs to be within their corresponding ε_k to get their jobs done. Since $x_k < 3140$ implies $x_\varepsilon - 3140 < 0$ for each $k = 1, 2, 3$, by the definition of absolute value (Definition 1.1.6) we have

$$|x_1 - 3140| = \left| -\frac{1}{2} \right| = \frac{1}{2} < 1 = \varepsilon_1, \tag{1.1.18}$$

$$|x_2 - 3140| = \left| -\frac{1}{4} \right| = \frac{1}{4} < \frac{1}{2} = \varepsilon_2, \quad \text{and} \tag{1.1.19}$$

$$|x_3 - 3140| = \left| -\frac{1}{200} \right| = \frac{1}{200} < \frac{1}{100} = \varepsilon_3. \tag{1.1.20}$$

We're good!

---

^aThe notation is changed from x_{ε_1} to x_1 since the smaller subscript is so tiny.

Let's play around with a pair of sets that are *not* intervals.

Example 1.1.17: Two countable sets

Consider the following sets of real numbers A and B where $\mathbb{N}$ denotes the set of positive integers:

$$A = \left\{ a_n = 2 - \frac{1}{\sqrt{n}} : n \in \mathbb{N} \right\} \quad \text{and} \tag{1.1.21}$$

$$B = \left\{ b_n = \left(2 - \frac{1}{\sqrt{n}} \right)(-1)^n : n \in \mathbb{N} \right\}. \tag{1.1.22}$$

See Figure 1.1.5. I claim $2 \operatorname{acl} A$, $2 \operatorname{acl} B$, and $-2 \operatorname{acl} B$ while $-2 \operatorname{awf} A$.

Before attempting proofs, consider some scratch work for the set A along with Figure 1.1.6.

Scratch Work 1.1.18: Two countable sets

The "away from" part looks like it might be easiest to prove since a radius of $\varepsilon_z = 1$ is enough distance to keep $z = -2$ away from all the points in A.

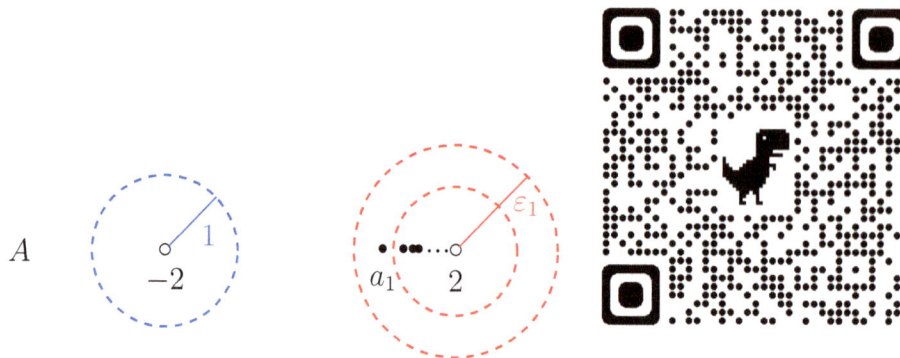

Figure 1.1.6: A figure to accompany Scratch Work 1.1.18 featuring the set A along with a distance 1 around -2 and a distance $\varepsilon_1 > 1$ around -2 along with a QR code for the Desmos activity "2 acl A" to accompany this figure. https://www.desmos.com/calculator/ceczc717wa

The "arbitrarily close" part—$2\,\mathrm{acl}\,A$—deserves more scrutiny. See Figure 1.1.6. For the largest red circle around 2 (the center $\circ$ on the right) whose radius ε_1 is greater than 1, the real number $x_{\varepsilon_1} = a_1$ is in A and close enough to 2. But a_1 is not close enough for the smaller circle whose radius is less than 1. That's okay, though. For each radius $\varepsilon > 0$, we need just one point x_ε that's in A and within ε of 2, and each ε can have its own x_ε.

So, if we can find a way to take a given but arbitrary $\varepsilon > 0$ and respond to it with a suitably defined x_ε, we're good.

Remark 1.1.19: Careful choice

Setting

$$x_\varepsilon = 2 - \frac{\varepsilon}{2} \tag{1.1.23}$$

will not be good enough to show $2\,\mathrm{acl}\,A$, even though a similar choice was made for x_ε and the interval G in Example 1.1.4. To see why, temporarily set $\varepsilon = 3/2$. Since $0 < \varepsilon = 3/2 < 2$, we have

$$x_\varepsilon = 2 - \frac{3/2}{2} = 2 - \frac{3}{4} = \frac{5}{4}. \tag{1.1.24}$$

Since $5/4 < 2$, the definition of absolute value (Definition 1.1.6) yields

$$d_{\mathbb{R}}(x_\varepsilon, 2) = |x_\varepsilon - 2| = 2 - \frac{5}{4} = \frac{3}{4} < \frac{3}{2} = \varepsilon, \tag{1.1.25}$$

which means $x_\varepsilon = 5/4$ is close enough to 2. The problem is, $5/4$ *is not in* A. For $x_\varepsilon = 5/4$ to be an element of A, there must be a positive integer n_ε where

$$x_\varepsilon = 2 - \frac{1}{\sqrt{n_\varepsilon}} = 5/4. \tag{1.1.26}$$

But solving for n_ε yields $n_\varepsilon = 16/9$, which is not a positive integer. So, we need to be more careful when choosing x_ε in response to ε.

Let's start over with some new scratch work.

Scratch Work 1.1.20: Start at the end

With any scratch work, consider starting at the end. To show $2 \operatorname{acl} A$, we want to consider each $\varepsilon > 0$ and end up with a point x_ε which is both in A and within ε of 2. Since A is defined in terms of the set of positive integers $\mathbb{N}$, work with

$$x_\varepsilon = a_{n_\varepsilon} = 2 - \frac{1}{\sqrt{n_\varepsilon}} \tag{1.1.27}$$

where n_ε is in $\mathbb{N}$. Such x_ε is in A, but we also need x_ε to be within ε of 2, like this:

$$|x_\varepsilon - 2| = \left| \left(2 - \frac{1}{\sqrt{n_\varepsilon}} \right) - 2 \right| = \frac{1}{\sqrt{n_\varepsilon}} < \varepsilon. \tag{1.1.28}$$

This amounts to solving the inequality for n_ε and making sure we choose n_ε to be a positive integer. We get

$$\frac{1}{\sqrt{n_\varepsilon}} < \varepsilon \quad \Longleftrightarrow \quad n_\varepsilon > \frac{1}{\varepsilon^2}. \tag{1.1.29}$$

So, choosing a *positive integer* n_ε large enough to satisfy $n_\varepsilon > 1/\varepsilon^2$ ensures the choice

$$x_\varepsilon = a_{n_\varepsilon} = 2 - \frac{1}{\sqrt{n_\varepsilon}}, \tag{1.1.30}$$

which is both in A and within ε of 2.

Time for a proof. While Scratch Work 1.1.20 is where I really figured things out, the proof amounts to a careful reorganization of the scratch work where the details are put in an appropriate order.

Proof of $-2 \operatorname{awf} A$ *and* $2 \operatorname{acl} A$ *in Example 1.1.17.* First, to show $-2 \operatorname{awf} A$, note that for every n in $\mathbb{N}$ we have

$$|a_n - (-2)| = \left| \left(2 - \frac{1}{\sqrt{n}} \right) - (-2) \right| = 4 - \frac{1}{\sqrt{n}} \geq 3 > 1 > 0. \tag{1.1.31}$$

Hence, every real number a_n in A is a distance of at least 1 away from -2.

Next, to show $2 \operatorname{acl} A$, let $\varepsilon > 0$. (By not specifying a particular value of ε, we are accounting for *all* positive distances, or radii or "errors", at the same time.) Choose a positive integer n_ε large enough to satisfy $n_\varepsilon > 1/\varepsilon^2$. We have

$$n_\varepsilon > \frac{1}{\varepsilon^2} \quad \Longleftrightarrow \quad \frac{1}{\sqrt{n_\varepsilon}} < \varepsilon. \tag{1.1.32}$$

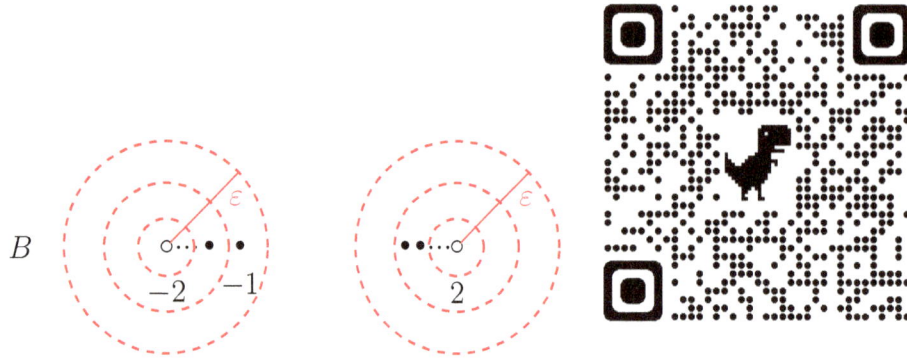

Figure 1.1.7: The set B from Example 1.1.17 along with various distances around 2 and -2 as well as a QR code for the Desmos activity "2 acl B" to accompany this figure. https://www.desmos.com/calculator/ysaphndtqh

Now consider $a_{n_\varepsilon} = 2 - (1/\sqrt{n_\varepsilon})$. Then a_{n_ε} is in A and

$$d_{\mathbb{R}}(a_{n_\varepsilon}, 2) = |a_{n_\varepsilon} - 2| = \left|\left(2 - \frac{1}{\sqrt{n_\varepsilon}}\right) - 2\right| = \frac{1}{\sqrt{n_\varepsilon}} < \varepsilon. \tag{1.1.33}$$

Therefore, $2 \operatorname{acl} A$.   □

Remark 1.1.21: Different indices

In the proofs of $-2 \operatorname{awf} A$ and $2 \operatorname{acl} A$ as stated in Example 1.1.17, I used the variables n and n_ε in subtly different ways, much like the differences between ε versus ε_z and x versus x_ε. Both n and n_ε represent positive integers, but n represents *any* positive integer while n_ε is a particular positive integer chosen in response to $\varepsilon > 0$ (hence the subscript) and how the scratch work turned out based on the definition of the set A.

Proving $-2 \operatorname{acl} B$ and $2 \operatorname{acl} B$ from Example 1.1.17 is similar to proving $2 \operatorname{acl} A$, but there is a significant difference. When proving $2 \operatorname{acl} A$, it was enough to find a positive integer n_ε in response to ε. But when proving $-2 \operatorname{acl} B$ and $2 \operatorname{acl} B$, the parity of n_ε—whether n_ε is even or odd—changes whether the corresponding point in B is close to 2 or -2. Such a point in B might be close enough to either 2 or -2, but maybe not both.

For instance, Figure 1.1.7, $b_1 = -1$ is in B and within the largest red circle on the left, so it is within the corresponding radius ε of -2. Hence, $b_1 = -1$ is close enough to -2 for that particular ε. But -1 is not close enough to 2 since it is outside the largest red circle on the right which has the same radius ε. Hence, for the $\varepsilon > 0$ that gives the radius of the two largest red circles, $n_\varepsilon = 1$ produces $b_1 = -1$ which is close enough to -2 but not close enough to 2. On the other hand, $n_\varepsilon + 1 = 2$ produces $b_2 = 2 - (1/\sqrt{2})$, which is close enough to 2.

Loosely speaking, the b_n with even indices are near 2 while those with odd indices are near -2. In turn, parity of positive integers affects the proof. And again, by not specifying a particular value of ε, we are accounting for *all* positive distances, or "errors", around both -2 and 2 simultaneously.

Proof of $-2 \operatorname{acl} B$ and $2 \operatorname{acl} B$ in Example 1.1.17. Let $\varepsilon > 0$. Choose an *odd* positive integer j_ε

large enough to satisfy $j_\varepsilon > 1/\varepsilon^2$. We have

$$j_\varepsilon > \frac{1}{\varepsilon^2} \quad \Longleftrightarrow \quad \frac{1}{\sqrt{j_\varepsilon}} < \varepsilon. \tag{1.1.34}$$

Now, since j_ε is odd, we have $(-1)^{j_\varepsilon} = -1$ and so $b_{j_\varepsilon} = -2 + (1/\sqrt{j_\varepsilon})$. Then b_{j_ε} is in B and

$$d_\mathbb{R}(b_{j_\varepsilon}, -2) = \left| \left(-2 + \frac{1}{\sqrt{j_\varepsilon}} \right) - (-2) \right| = \frac{1}{\sqrt{j_\varepsilon}} < \varepsilon. \tag{1.1.35}$$

Hence $-2 \operatorname{acl} B$.

Next, choose an *even* positive integer k_ε large enough to satisfy $k_\varepsilon > 1/\varepsilon^2$. We have

$$k_\varepsilon > \frac{1}{\varepsilon^2} \quad \Longleftrightarrow \quad \frac{1}{\sqrt{k_\varepsilon}} < \varepsilon. \tag{1.1.36}$$

Now, since k_ε is even, we have $(-1)^{k_\varepsilon} = 1$ and so $b_{k_\varepsilon} = 2 - (1/\sqrt{k_\varepsilon})$. Then b_{k_ε} is in B and

$$d_\mathbb{R}(b_{k_\varepsilon}, 2) = \left| \left(2 - \frac{1}{\sqrt{k_\varepsilon}} \right) - 2 \right| = \frac{1}{\sqrt{k_\varepsilon}} < \varepsilon. \tag{1.1.37}$$

Hence $2 \operatorname{acl} B$ as well. □

Before going any deeper, Section 1.2 lays out some of the assumptions I've made about the mathematical knowledge and experiences readers are expected to have. In some ways, it may have been better to put these assumptions first, but I wanted to start with a notion for arbitrarily close in the real line that serves as the foundation for the entire book. This comes at the cost of using some facts before proving them; a small price to pay for a first section on such a difficult topic. Section 1.2 also establishes more notation and background material.

Exercises

Once again, here are the sets discussed throughout Section 1.1:

$$F = (-\infty, 3140] = \{x \in \mathbb{R} : x \le 3140\}, \tag{1.1.38}$$
$$G = (-\infty, 3140) = \{x \in \mathbb{R} : x < 3140\}, \tag{1.1.39}$$
$$A = \{2 - (1/\sqrt{n}) : n \in \mathbb{N}\}, \quad \text{and} \tag{1.1.40}$$
$$B = \{[2 - (1/\sqrt{n})](-1)^n : n \in \mathbb{N}\}. \tag{1.1.41}$$

1.1.1. For the sets F, G, A, and B given above and visualized in Figure 1.1.8, determine whether each set has a maximum, minimum, supremum, and infimum and prove your results. (Some cases have already been discussed and proven in Section 1.1, so don't prove those again unless you want to. You can treat those proofs like templates, carefully adjusting what I've done to fit a similar situation accordingly.)

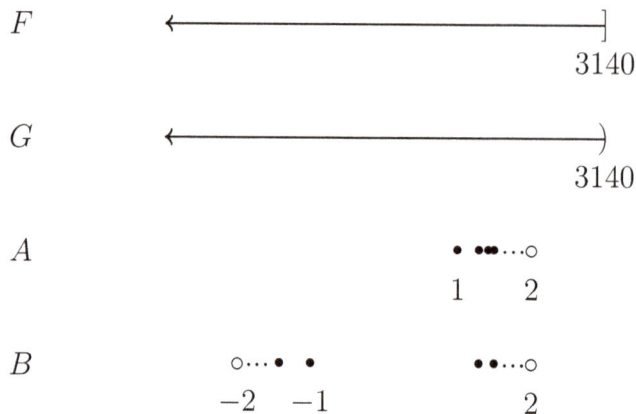

Figure 1.1.8: The sets F, G, A, and B are explored further in the following exercises.

Figure 1.1.9: The set of real numbers T from Exercise 1.1.3.

1.1.2. The following statement is *false*: For any set of real numbers S that is bounded above we have

$$u = \max S \qquad \text{if and only if} \qquad u = \sup S. \tag{1.1.42}$$

 (i) Find a *counterexample* for this statement and prove your result. That is, find, describe, and draw a set of real numbers S that is bounded above but for which the "if and only if" part of line (1.1.42) does not hold.

 (ii) Revise the above false statement to create an implication which is true and prove your result.

(iii) Write a similar implication involving minimum and infimum, and prove it too.

1.1.3. Consider the set of real numbers T given by

$$T = [0, 2) \cup \{4 - (1/n) : n \in \mathbb{N}\}. \tag{1.1.43}$$

See Figure 1.1.9. That is, every real number x in T satisfies either $0 \le x < 2$ or $x = 4 - (1/n)$ for some positive integer n. The symbol $\cup$ stands for *union* and is discussed in the next section.

 (i) Prove $\min T = \inf T = 0$, $\sup T = 4$, and $\max T$ does not exist.

 (ii) What can you say and prove about the real numbers 2 and 3 here? Are either of them arbitrarily close to T? Away from T? Can you prove your answers?

1.1.4. Find examples of sets of real numbers with the following properties:

 (i) A set U where neither $\sup U$ nor $\inf U$ exist and $U \ne \mathbb{R}$.

(ii) A set V where $\min V$ and $\max V$ exist with $\min V < \max V$, but V is not an interval.

(iii) A set W where $\inf W$ exists, $\sup W$ exists, and we have $\inf W = \sup W$.

Be sure to draw figures for each of your sets and justify your results. What other properties do your examples have?

1.2 Preliminary concepts and background material

To set the stage for the more technical aspects of the book, this section provides some notation and terminology. I am assuming a fair amount of familiarity with inequalities, basic algebra, trigonometry, set theory, and writing mathematical proofs.

Notation 1.2.1: Conventions used throughout

(i) Lower case letters like $a, b, c, f, g, x, y, z, \varepsilon$, and δ typically denote real numbers or functions.

(ii) Capital letters like B, S, and V typically denote sets.

(iii) Boldface lower case letters like $\mathbf{y}$ and $\mathbf{w}$ typically denote points (or vectors) in some Euclidean space $\mathbb{R}^m$.

(iv) Along these lines, notation that starts with lower case letters such as "max", "sup", and "lim" represent real numbers or points while "Coda" and "Slim" represent sets.

(v) The notation " $\Longrightarrow$ " reads "implies".

(vi) The notation "$\Longleftrightarrow$" reads "if and only if".

Notation 1.2.2: Sets and points

If S is any set comprising any kind of elements we like, we write $x \in S$ when x is in—or x is an element of, or x belongs to—the set S. If z is not in S, we write $z \notin S$. Throughout the book, elements are also referred to as *points*. The set with no elements is called the *empty set*, denoted by $\varnothing$. If a set has one or more elements, it is called *nonempty*.

Notation 1.2.3: Subsets and equality of sets

If A and B are sets where every element of A is also an element of B, we say A is a *subset* of B or B *contains* A and write $A \subseteq B$. If A is a subset of B but there is an element of B that is not in A, we say A is a *proper* subset of B and write $A \subsetneq B$. In the case where we have both $A \subseteq B$ and $B \subseteq A$, we say A and B are *equal* and write $A = B$. Equivalently,

$A = B$ if and only if A and B have the exact same elements.

Notation 1.2.4: Set builder

Sets can be described in many different ways, all of which are useful and provide a variety of perspectives. For instance, with $\mathbb{Z}$ denoting the set of integers, consider:

(i) Let $S = \{-3, -2, -1, 0, 1, 2, 3\}$.

(ii) Let S be the set of integers between -3.5 and 3.999.

(iii) Let $S = \{z \in \mathbb{Z} : |z| \leq 3\}$.

Items (i), (ii), and (iii) define the same exact set S in three different ways, each of which is perfectly valid for the purposes of writing proofs.

Another way to describe a set is to draw a figure. Many figures are provided throughout the text because I think they can be incredibly helpful for building intuition and exploring technical ideas. In fact, I regularly encourage you to draw figures to supplement proofs, examples, and pretty much everything in the book. However, figures don't suffice for proofs (...but just barely).

Unions and intersections of sets play a significant role.

Definition 1.2.5: More set theory

For sets A and B, the *union* of A and B is denoted by $A \cup B$ and defined by

$$A \cup B = \{x : x \in A \text{ or } x \in B\}. \tag{1.2.1}$$

The "or" within the definition of $A \cup B$ is the "inclusive or". That is, $A \cup B$ is the set of elements in A, in B, or in both A and B. The *intersection* of A and B is denoted by $A \cap B$ and defined by

$$A \cap B = \{x : x \in A \text{ and } x \in B\}. \tag{1.2.2}$$

The notation $A \backslash B$ denotes the set of points in A that are not in B. That is,

$$A \backslash B = \{x : x \in A \text{ and } x \notin B\}. \tag{1.2.3}$$

The variables x_ε, ε_z, and n_ε from Section 1.1 show us that indexing variables can be helpful. Similarly, we sometimes want to consider intersections and unions of collections of sets. Also, variables and sets can be indexed by other sets like $\mathbb{N}$, $\mathbb{Z}$, or the set of positive real numbers.

Definition 1.2.6: Indexed sets

Given a set B, a nonempty set Λ is an *index set* for B if for each $\lambda \in \Lambda$ there is a subset S_λ of B. The collection of these sets is called an *indexed family* of sets and is denoted by $\{S_\lambda\}_{\lambda \in \Lambda}$.

Definition 1.2.7: Operations with indexed sets

Given an indexed family of sets $\{S_\lambda\}_{\lambda \in \Lambda}$, the *union* and *intersection* of all of the sets in the indexed family are defined by

$$\bigcup_{\lambda \in \Lambda} S_\lambda = \{x : x \in S_\lambda \text{ for at least one } \lambda \in \Lambda\} \quad \text{and} \tag{1.2.4}$$

$$\bigcap_{\lambda \in \Lambda} S_\lambda = \{x : x \in S_\lambda \text{ for all } \lambda \in \Lambda\}, \quad \text{respectively.} \tag{1.2.5}$$

In other words and slightly different notation, we have $x \in \cup_{\lambda \in \Lambda} S_\lambda$ whenever x is in some of the S_λ. We have $x \in \cap_{\lambda \in \Lambda} S_\lambda$ only when x is in *every* S_λ. In the special case where the index set is the set of positive integers $\mathbb{N}$, we write

$$\bigcup_{n=1}^{\infty} S_n \quad \text{and} \quad \bigcap_{n=1}^{\infty} S_n, \quad \text{respectively.} \tag{1.2.6}$$

Notation 1.2.8: Important sets

The following sets appear throughout the book.[a]

$$
\begin{aligned}
\textit{positive integers:} \quad & \mathbb{N} = \{1, 2, 3, \ldots\}. \\
\textit{integers:} \quad & \mathbb{Z} = \{\ldots, -2, -1, 0, 1, 2, 3, \ldots\}. \\
\textit{rational numbers:} \quad & \mathbb{Q} = \{m/n : m, n \in \mathbb{Z} \text{ and } n \neq 0\}. \\
\textit{real numbers:} \quad & \mathbb{R} = \{x : x \in \mathbb{Q} \text{ or } x \text{ is a gap near } \mathbb{Q}\}.
\end{aligned}
$$

---

[a]The set of positive integers is also known as the set of natural numbers, among other things.

Intervals are used throughout as well.

Definition 1.2.9: Intervals

For $a, b \in \mathbb{R}$ with $a < b$, each of the following sets is an *interval*.

(i) $(-\infty, \infty) = \mathbb{R}$

(ii) $(-\infty, b) = \{x \in \mathbb{R} : x < b\}$

(iii) $(a, \infty) = \{x \in \mathbb{R} : a < x\}$

(iv) $(-\infty, b] = \{x \in \mathbb{R} : x \leq b\}$

(v) $[a, \infty) = \{x \in \mathbb{R} : a \leq x\}$

(vi) $(a, b) = \{x \in \mathbb{R} : a < x < b\}$

(vii) $[a, b] = \{x \in \mathbb{R} : a \leq x \leq b\}$

(viii) $(a, b] = \{x \in \mathbb{R} : a < x \leq b\}$

(ix) $[a, b) = \{x \in \mathbb{R} : a \leq x < b\}$

See Figure 1.2.1. Furthermore, we have:

- $(a, b), (-\infty, b), (a, \infty)$, and $(-\infty, \infty)$ are *open intervals*;

- $[a, b], (-\infty, b], [a, \infty)$, and $(-\infty, \infty)$ are *closed intervals*;

(a, b)

$[a, b]$

$(a, b]$

$[a, b)$

$(-\infty, b)$

(a, ∞)

$(-\infty, b]$

$[a, \infty)$

$(-\infty, \infty)$

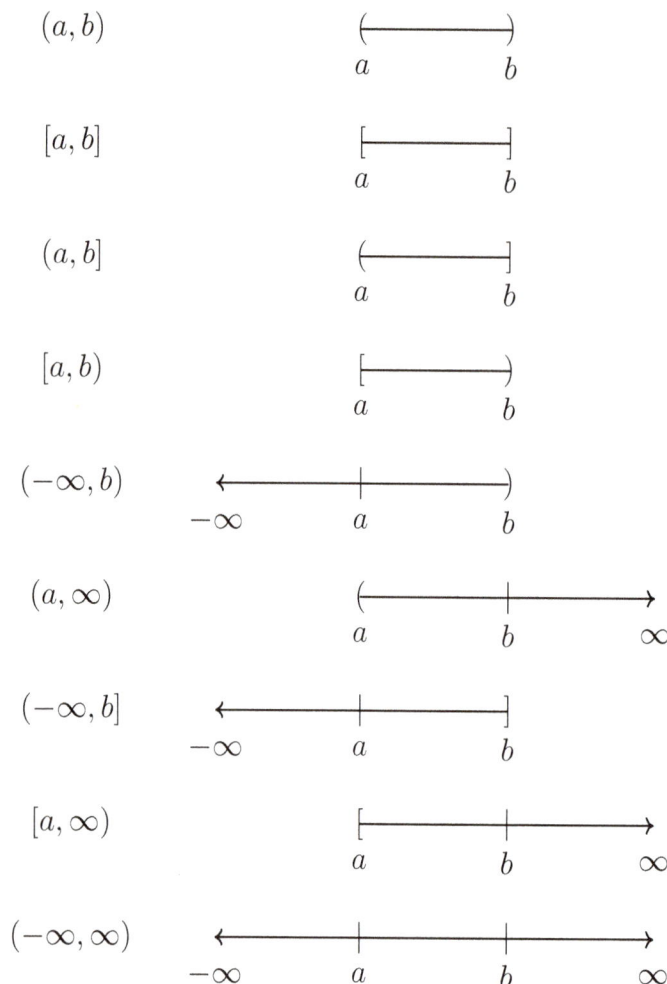

Figure 1.2.1: Plots of all nine types of intervals. See Definition 1.2.9.

- $(-\infty, b)$ and $(-\infty, b]$ are *bounded above*, but *unbounded*;

- (a, ∞) and $[a, \infty)$ are *bounded below*, but *unbounded*;

- $(a, b), [a, b], (a, b]$, and $[a, b)$ are *bounded*, meaning they are both bounded above and bounded below.

The real line $\mathbb{R} = (-\infty, \infty)$ is both an open interval and a closed interval. If this doesn't seem right to you, you're not alone. However, this choice of terminology is conventional and so prevalent I will not try to replace it. The study of open, closed, and other types of sets—*topology*—is the focus of Chapter 3.

Other sets playing a prominent role in the book are the set of *irrational numbers* denoted by

$\mathbb{R}\backslash\mathbb{Q}$ and *Euclidean spaces* denoted by $\mathbb{R}^m$ where m is a positive integer. We have

$$\mathbb{R}\backslash\mathbb{Q} = \{x : x \in \mathbb{R} \text{ and } x \notin \mathbb{Q}\} \quad \text{and} \tag{1.2.7}$$

$$\mathbb{R}^m = \left\{ \mathbf{x} = \begin{bmatrix} x_1 \\ x_2 \\ \vdots \\ x_m \end{bmatrix} : m \in \mathbb{N} \text{ and } x_1, x_2, \ldots, x_m \in \mathbb{R} \right\}, \tag{1.2.8}$$

where the real numbers $x_1, x_2, \ldots, x_m$ are the *coordinates* (or *components* or *entries*) of the *point* (or *vector*) $\mathbf{x} \in \mathbb{R}^m$. Similarly, the coordinates of a point $\mathbf{y} \in \mathbb{R}^m$ are denoted by $y_1, y_2, \ldots, y_m$. In every Euclidean space $\mathbb{R}^m$, the special *zero vector* $\mathbf{0}$ is the vector whose coordinates all are 0.

There are many deep relationships between the sets described so far. For instance, we have

$$\mathbb{N} \subseteq \mathbb{Z} \subseteq \mathbb{Q} \subseteq \mathbb{R}, \tag{1.2.9}$$

and each of these relationships represents a proper subset. With that said, there are some important differences between these sets that influence the way I choose to state theorems and write proofs. For instance, the set of rational numbers $\mathbb{Q}$ has addition, multiplication, and inequalities that all play nicely together, but the set has gaps. The real line $\mathbb{R}$ has addition, multiplication, and inequalities that all play nicely together while having no gaps. (We will look at this much more closely in the next section.)

Remark 1.2.10: Euclidean spaces provide the setting

Many of the results presented throughout are in the setting of an arbitrary Euclidean space $\mathbb{R}^m$ where the positive integer m is left unspecified. This allows us to discuss and prove results in all of these spaces simultaneously. In particular, for $m = 1$ we have $\mathbb{R}^1 = \mathbb{R}$ (the real line) and for $m = 2$ we have $\mathbb{R}^2$ (the plane).

Remark 1.2.11: Important vector spaces

We will often use the fact that $\mathbb{R}$ and $\mathbb{R}^m$ for any $m \in \mathbb{N}$ are *vector spaces*. Basically, this means their points can be multiplied by constants (called *scalars*) and their points can added together, both processes creating new points as follows: For any *scalar* $\alpha \in \mathbb{R}$ and any $\mathbf{x}, \mathbf{y} \in \mathbb{R}^m$ we have

$$\alpha\mathbf{x} = \begin{bmatrix} \alpha x_1 \\ \alpha x_2 \\ \vdots \\ \alpha x_m \end{bmatrix} \quad \text{and} \quad \mathbf{x} + \mathbf{y} = \begin{bmatrix} x_1 + y_1 \\ x_2 + y_2 \\ \vdots \\ x_m + y_m \end{bmatrix}. \tag{1.2.10}$$

Functions are an integral part of mathematics which allow us to talk about the various ways we can transform points and sets. A fairly formal definition for functions is provided here along with related terminology and properties.

Definition 1.2.12: Function

A *function* f is a relation between two sets A and B where every element of A is associated with exactly one element of B. In this case we write

$$f : A \to B \tag{1.2.11}$$

and say f is a *function from A to B* or *f maps A to B*.

Furthermore, each element $x \in A$ is called an *input* and is associated with an element $y \in B$ called an *output*. When f associates an input x with an output y, we write

$$f(x) = y. \tag{1.2.12}$$

For functions that map one Euclidean space to another—not just the real line to the real line—boldface variables like $\mathbf{x}$ and $\mathbf{y}$ are used to represent inputs and outputs.

Definition 1.2.13: Domain, codomain, and range

Given a function $f : A \to B$, the set of inputs A is called the *domain* of f, the set B is called the *codomain* of f, and the set of outputs is called the *range* of f. Furthermore, the range of f is denoted by $f(A)$ and given by

$$f(A) = \{y \in B : f(x) = y \text{ for some } x \in A\}. \tag{1.2.13}$$

Functions can be thought of as transforming elements of the domain into elements of the range. Similarly, functions transform subsets of the range into subsets of the range. We capture these ideas with the notion of *images*.

Definition 1.2.14: Image

When $f : A \to B$ and given an input $x \in A$, the *image* of x is its output $f(x) \in B$. Given a subset S of the domain A, the *image* of S is the subset of B whose elements are outputs of at least one element in S. In this case, the image of S is denoted by $f(S)$ and we have

$$f(S) = \{y \in B : f(x) = y \text{ for some } x \in S\} \subseteq f(A). \tag{1.2.14}$$

Example 1.2.15: Range is not codomain

In general, the range is a subset of the codomain, but they are not necessarily the same set. For instance, when $f : \mathbb{R} \to \mathbb{R}$ is given by $f(x) = x^2$, we have codomain $B = \mathbb{R}$ with range $f(\mathbb{R}) = [0, \infty)$. So, $-1 \in B$ but there is no input $x \in A = \mathbb{R}$ where $f(x) = -1$. Hence, $-1 \notin f(\mathbb{R})$.

Definition 1.2.16: Onto

A function $f : A \to B$ is *onto* if every element in the codomain B is an output of f. That is, f is onto if

$$B \subseteq f(A), \tag{1.2.15}$$

meaning for every point $y \in B$ there is an input $x \in A$ where $f(x) = y$.

Remark 1.2.17: Onto

Again, whenever $f : A \to B$, the range is a subset of the codomain. That is,

$$f(A) \subseteq B. \tag{1.2.16}$$

So, f is onto if and only if the range equals the codomain and we have

$$f(A) = B. \tag{1.2.17}$$

Definition 1.2.18: One-to-one

A function $f : A \to B$ is *one-to-one* (also written 1-1) if distinct inputs yield distinct outputs. That is, f is one-to-one if for every $x, y \in A$ where $x \neq y$ we have $f(x) \neq f(y)$.

Functions that are both one-to-one and onto play a special role throughout mathematics.

Definition 1.2.19: Bijection

A function is a *bijection* if it is both one-to-one and onto.

Remark 1.2.20: Injective and surjective

The terms one-to-one and onto are essentially adjectives, but there are other equivalent adjectives used by mathematical community. Basically:

$$\text{one-to-one} \quad \longleftrightarrow \quad \text{injective} \tag{1.2.18}$$
$$\text{onto} \quad \longleftrightarrow \quad \text{surjective} \tag{1.2.19}$$
$$\text{bijective} \quad \longleftrightarrow \quad \text{injective and surjective} \tag{1.2.20}$$

Similarly, the word bijection is a noun and we have:

$$f \text{ is 1-1} \quad \longleftrightarrow \quad f \text{ is an injection} \tag{1.2.21}$$
$$f \text{ is onto} \quad \longleftrightarrow \quad f \text{ is a surjection} \tag{1.2.22}$$

Remark 1.2.21: Clarifying our notion of graphs

The *graphs* of functions provided throughout the text—along with their domains and ranges—are designed to help fuel our discussion and drive us towards a formal definition for *continuity* (Definition 4.3.2).

By the *graph* of a function from the real line to the real line, I mean a single plot that combines the domain and range, technically plotted in the plane. I believe there is value at looking at all three types of figures for a given function, namely its graph, domain, and range.

The concepts of finite and infinite sets will appear in various ways throughout the book.

Definition 1.2.22: Finite and infinite sets

A set is *finite* if it is empty or if it has n_0 elements for some positive integer n_0. On the other hand, a set is *infinite* if it not finite.

Factorials and *binomial coefficients* play a role in defining some important sequences and functions.

Notation 1.2.23: Factorial and binomial coefficient

Given a nonnegative integer $n \in \mathbb{N} \cup \{0\}$ we have $0! = 0$ and

$$n! = n(n-1) \cdots (2)(1) \tag{1.2.23}$$

for $n \in \mathbb{N}$. Also, the notation $n!$ is pronounced "*n factorial*".

For $n \in \mathbb{N} \cup \{0\}$ and $k = 0, 1, \ldots, n$, the *binomial coefficient* is given by

$$\binom{n}{k} = \frac{n!}{k!(n-k)!}. \tag{1.2.24}$$

From the perspective of combinatorics, we have that for any set S with n elements, the binomial coefficient $\binom{n}{k}$ is the number of subsets of S with k elements. As a result, they provide a classic way to expand binomials.

The proof of the following theorem is interesting but omitted for the sake of brevity and exposition.

Theorem 1.2.24: Binomial Theorem

For any pair $x, y \in \mathbb{R}$ and any $n \in \mathbb{N} \cup \{0\}$ we have

$$(x+y)^n = \sum_{k=0}^{n} \binom{n}{k} x^k y^{n-k} \tag{1.2.25}$$

where we use the convention $a^0 = 1$ for all $a \in \mathbb{R}$.

Distance is a recurring theme in analysis. Basically, the notions for distance we use in this book are functions which assign a nonnegative real number to every pair of points in some set.

Notation 1.2.25: Distance and norm

The types of distance we consider are limited to the standard notions in the real line $\mathbb{R}$ and Euclidean spaces of the form $\mathbb{R}^m$. In all cases, these distances stem from the standard *norms* and *metrics* on Euclidean spaces $\mathbb{R}^m$, denoted by $\|\cdot\|_m$ and d_m, respectively.[a]

Given a point $\mathbf{x} \in \mathbb{R}^m$, we have the *norm* of $\mathbf{x}$ given by

$$\|\mathbf{x}\|_m = \sqrt{x_1^2 + x_2^2 + \cdots + x_m^2}. \tag{1.2.26}$$

Given a pair of points $\mathbf{x}, \mathbf{y} \in \mathbb{R}^m$, the *distance* between $\mathbf{x}$ and $\mathbf{y}$ is given by the classic Pythagorean distance formula:

$$\begin{aligned} d_m(\mathbf{x}, \mathbf{y}) &= \|\mathbf{x} - \mathbf{y}\|_m \\ &= \sqrt{(x_1 - y_1)^2 + (x_2 - y_2)^2 + \cdots + (x_m - y_m)^2}. \end{aligned} \tag{1.2.27}$$

The notation for points, norms, and distances in $\mathbb{R}^m$ represents *any* Euclidean space, including the real line $\mathbb{R}^1 = \mathbb{R}$ and the plane $\mathbb{R}^2$ as special cases. When working in the real line $\mathbb{R}$, we use absolute value of the difference for distance:

$$d_{\mathbb{R}}(x, y) = |x - y|. \tag{1.2.28}$$

---

[a]The concept of arbitrarily close can be explored in the much more general settings of metric spaces (where a wide variety of distances are taken into consideration) and even topological spaces (where notions of distance are not necessarily in play).

The distances $d_{\mathbb{R}}$ and d_m are *metrics*: They satisfy a special set of properties which will be used throughout the book.

Definition 1.2.26: Metric and triangle inequality

Suppose S is a nonempty set. A function d is a *metric* on S if it satisfies all of the following properties for any x, y, and z in S:

$$d(x, y) \geq 0; \tag{1.2.29}$$
$$d(x, y) = 0 \iff x = y; \tag{1.2.30}$$
$$d(x, y) = d(y, x); \quad \text{and} \tag{1.2.31}$$
$$d(x, y) \leq d(x, z) + d(z, y). \tag{1.2.32}$$

Inequality (1.2.32) is called the *triangle inequality*.

Remark 1.2.27: Recap of metrics

To rephrase the defining properties of a metric:

- (1.2.29) Distances given by a metric are nonnegative.

- (1.2.30) The distance between two points is zero if and only if the points are the same.

- (1.2.31) Metrics are symmetric: The order doesn't matter.

- (1.2.32) The distance between two points is the same or shorter than the total distance when an additional point is considered.

Our notions of distance are metrics, but the proof of this fact is left as an exercise. The establishment of the triangle inequality is the key step.

Theorem 1.2.28: Distances are metrics

The functions $d_{\mathbb{R}}$ and d_m (for any positive integer m) are metrics on $\mathbb{R}$ and $\mathbb{R}^m$, respectively.

Remark 1.2.29: An important difference

Instances of the distances $\|\mathbf{x} - \mathbf{c}\|_m$ in Euclidean spaces and $|x - c|$ in the real line appear throughout the textbook.

The following corollary stems from Theorem 1.2.28 and, likewise, the proof is omitted.

Corollary 1.2.30: Norms, distances, and inequalities

For any scalar $\alpha \in \mathbb{R}$ and any three points $\mathbf{x}, \mathbf{y}, \mathbf{c} \in \mathbb{R}^m$, the triangle inequality (1.2.32) combined with properties of vector spaces, norms, and metrics yield:

$$\|\alpha \mathbf{x}\|_m = |\alpha| \|\mathbf{x}\|_m; \tag{1.2.33}$$

$$\|\mathbf{x} - \mathbf{c}\|_m = \|\mathbf{x} \underbrace{-\mathbf{y} + \mathbf{y}}_{\text{add } \mathbf{0}} -\mathbf{c}\|_m \leq \|\mathbf{x} - \mathbf{y}\|_m + \|\mathbf{y} - \mathbf{c}\|_m; \tag{1.2.34}$$

$$\|\mathbf{x} + \mathbf{c}\|_m \leq \|\mathbf{x}\|_m + \|\mathbf{c}\|_m; \tag{1.2.35}$$

$$\|\mathbf{x} - \mathbf{c}\|_m \leq \|\mathbf{x}\|_m + \|\mathbf{c}\|_m; \quad \text{and both} \tag{1.2.36}$$

$$\|\mathbf{x}\|_m - \|\mathbf{c}\|_m \leq \|\mathbf{x} - \mathbf{c}\|_m \quad \text{and} \quad \|\mathbf{c}\|_m - \|\mathbf{x}\|_m \leq \|\mathbf{x} - \mathbf{c}\|_m. \tag{1.2.37}$$

Remark 1.2.31: Triangle inequalities

The property given by (1.2.33) is called the *homogeneity* of the Euclidean norm. Inequality (1.2.34) is another version of the triangle inequality (1.2.32) which is appropriately named thanks to figures like Figure 1.2.2. The intermediate step in (1.2.34) amounts to adding **0** (the vector whose coordinates are all 0) inside the norm before applying the triangle inequality (1.2.32). We'll use little techniques like this often when writing proofs. Adding

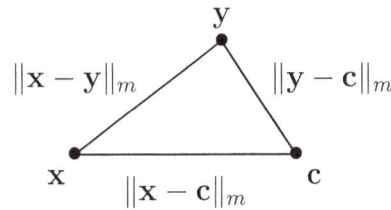

Figure 1.2.2: The version of the triangle inequality found in line (1.2.34).

a nice version of **0** in conjunction with the triangle inequality is particularly common.

Inequalities (1.2.35) and (1.2.36) are also particular instances of the triangle inequality (1.2.32). Inequalities (1.2.37) are each a version of the *reverse triangle inequality*. Combined, they are equivalent to

$$\big|\,\|\mathbf{x}\|_m - \|\mathbf{c}\|_m\big| \le \|\mathbf{x} - \mathbf{c}\|_m. \tag{1.2.38}$$

One last fact to conclude the section. The proof is left as an exercise.

Lemma 1.2.32: Important factorizations

Suppose $x, c \in \mathbb{R}$, $n \in \mathbb{N}$, and the convention $a^0 = 1$ is used as appropriate. Then each of the following holds:

$$x^2 - c^2 = (x - c)(x + c), \tag{1.2.39}$$
$$x^3 - c^3 = (x - c)(x^2 + cx + c^2), \tag{1.2.40}$$
$$\vdots \tag{1.2.41}$$
$$x^n - c^n = (x - c) \sum_{j=0}^{n-1} c^j x^{n-1-j} \tag{1.2.42}$$
$$= (x - c)(x^{n-1} + cx^{n-2} + \cdots + c^{n-2}x + c^{n-1}). \tag{1.2.43}$$

The following section focuses on introducing properties of the real line $\mathbb{R}$, especially the subtle notion of *completeness*.

Exercises

1.2.1. Prove that if x is irrational and q is rational where $q \neq 0$, then both $q + x$ and qx are irrational.

1.2.2. Determine which of the following statements are true and which are false regarding sets A, B, and C. Find counterexamples for the false ones.

(i) $A \cap (B \cup C) = (A \cap B) \cup C$.

(ii) $A \cap (B \cap C) = (A \cap B) \cap C$.

(iii) $A \cap (B \cup C) = (A \cap B) \cup (A \cap C)$.

1.2.3. Find a collection of infinite sets $A_1, A_2, A_3, \ldots$ (that is, each set is infinite) where

$$A_j \cap A_k = \varnothing \text{ when } j \neq k \qquad \text{and yet} \qquad \bigcup_{n=1}^{\infty} A_n = \mathbb{Z}. \qquad (1.2.44)$$

1.2.4. For sets $S_1 \supseteq S_2 \supseteq S_3 \ldots$, determine which of the following statements are true and which are false. Find counterexamples for the false ones.

(i) If S_n is infinite for each $n \in \mathbb{N}$, then the union $\cup_{n=1}^{\infty} S_n$ is infinite.

(ii) If S_n is infinite for each $n \in \mathbb{N}$, then the intersection $\cap_{n=1}^{\infty} S_n$ is infinite.

(iii) If S_n is nonempty and finite for each $n \in \mathbb{N}$, then the union $\cup_{n=1}^{\infty} S_n$ is nonempty and finite.

(iv) If S_n is nonempty and finite for each $n \in \mathbb{N}$, then the intersection $\cap_{n=1}^{\infty} S_n$ is nonempty and finite.

1.2.5. For sets $T_1 \subseteq T_2 \subseteq T_3 \ldots$, determine which of the following statements are true and which are false. Find counterexamples for the false ones.

(i) If T_n is infinite for each $n \in \mathbb{N}$, then the union $\cup_{n=1}^{\infty} T_n$ is infinite.

(ii) If T_n is infinite for each $n \in \mathbb{N}$, then the intersection $\cap_{n=1}^{\infty} T_n$ is infinite.

(iii) If T_n is nonempty and finite for each $n \in \mathbb{N}$, then the union $\cup_{n=1}^{\infty} T_n$ is nonempty and finite.

(iv) If T_n is nonempty and finite for each $n \in \mathbb{N}$, then the intersection $\cap_{n=1}^{\infty} T_n$ is nonempty and finite.

1.2.6. Consider the square function $s : \mathbb{R} \to \mathbb{R}$ given by $s(x) = x^2$, and Let $I = [0, 3]$ and $J = [1, 4]$. Find and compare the images $s(I), s(J), s(I \cap J), s(I) \cap s(J), s(I \cup J)$, and $s(I) \cup s(J)$. Draw stuff.

1.2.7. Prove that for any function g, we have the image of an intersection is a subset of the intersection of images, as follows: For any two subsets A and B of the domain, we have

$$g(A \cap B) \subseteq g(A) \cap g(B). \qquad (1.2.45)$$

Also, state a prove a similar statement regarding the image of a union and the union of images.

1.2.8. Infinite sets are weird: They have bijections with proper subsets of themselves.

(i) Find a bijection f where $f : \mathbb{N} \to \mathbb{Z}$.

(ii) Find a bijection g where $g : (0, \infty) \to \mathbb{R}$.

(iii) Prove a set S is infinite if and only if there is a bijection between S and a proper subset of S.

1.2.9. Prove Theorem 1.2.28. Draw figures for each statement, too.

1.2.10. Prove Corollary 1.2.30. Draw figures for each statement, too.

1.2.11. Prove the *Parallelogram Law*: For all $\mathbf{x}, \mathbf{y} \in \mathbb{R}^m$, we have

$$\|\mathbf{x} + \mathbf{y}\|_m^2 + \|\mathbf{x} - \mathbf{y}\|_m^2 = 2\|\mathbf{x}\|_m^2 + 2\|\mathbf{y}\|_m^2. \tag{1.2.46}$$

To get a geometric sense of how this equation earned its name, draw figures in the case of the plane $\mathbb{R}^2$.

1.2.12. Prove that if $\mathbf{x}_1, \mathbf{x}_2, \ldots, \mathbf{x}_k \in \mathbb{R}^m$, then

$$\|\mathbf{x}_1 + \mathbf{x}_2 + \cdots + \mathbf{x}_k\|_m = \left\|\sum_{n=1}^{k} \mathbf{x}_n\right\|_m \leq \sum_{n=1}^{k} \|\mathbf{x}_n\|_m = \|\mathbf{x}_1\|_m + \|\mathbf{x}_2\|_m + \cdots + \|\mathbf{x}_k\|_m. \tag{1.2.47}$$

1.2.13. Prove Lemma 1.2.32.

1.3 The real line $\mathbb{R}$ is a complete ordered field

To help motivate a rigorous investigation into real analysis, we will work with the following underlying assumption:

> The real line $\mathbb{R}$ is a *complete ordered field*.

But what does this mean?

The goal of this section is to define each part of the phrase "complete ordered field". Loosely speaking: a *field* is a set of mathematical objects where both addition and muliplication are defined and play nicely together; a set is *ordered* if inequalities make sense in a concrete and (hopefully) familiar way; and a set is *complete* if it knows its limits in some specific sense. The assumption above amounts to the existence of a set with these three properties, and we call this set the real line $\mathbb{R}$.

A formal description of what makes $\mathbb{R}$ an ordered field is provided by Axiom 1.3.1 below. I am assuming properties therein are familiar and will work them without explicitly citing them.

But what about completeness? Among the concepts of fields, order, and completeness, the latter is very much at the heart of real analysis but it is possibly the least familiar. We'll get to it in a bit.

Axiom 1.3.1: The real line is an ordered field

There exists a set called the *real line* or the *set of real numbers* which is denoted by $\mathbb{R}$ and has the following properties involving addition, multiplication, and inequalities for any

$x, y, z \in \mathbb{R}$:

(i) *Commutativity*:
   $x + y = y + x$ and $xy = yx$.

(ii) *Associativity*:
   $(x + y) + z = x + (y + z)$ and $(xy)z = x(yz)$.

(iii) *Distributive property*:
   $x(y + z) = xy + xz$.

(iv) *Additive identity*:
   There is a unique $0 \in \mathbb{R}$ such that
   for every $x \in \mathbb{R}$ we have $x + 0 = x$.

(v) *Additive inverse*:
   For each $x \in \mathbb{R}$, there is a unique $y \in \mathbb{R}$ where $x + y = 0$.
   (We write $y = -x$.)

(vi) *Multiplicative identity*:
   There is a unique $1 \in \mathbb{R}$ such that $1 \neq 0$ and
   for every $x \in \mathbb{R}$ we have $x(1) = x$.

(vii) *Multiplicative inverse*:
   For each $x \in \mathbb{R}$ where $x \neq 0$, there is a unique $y \in \mathbb{R}$ where $xy = 1$.
   (We write $y = 1/x$ or $y = x^{-1}$.)

(viii) *Translation invariance*:
   If $x < y$, then $x + z < y + z$.

(ix) *Transitivity*:
   If $x < y$ and $y < z$, then $x < z$.

(x) *Trichotomy*:
   For any $x, y \in \mathbb{R}$, exactly one of the following is true:
   $x = y$, $x < y$, or $x > y$.

(xi) *Multiplication inequality*:
   If $x < y$ and $z > 0$, then $xz < yz$.

Here are some consequences of the assumption that $\mathbb{R}$ is an ordered field (Axiom 1.3.1). I believe these ideas are familiar from calculus and other courses, so the proof is left as an exercise for those who would like to explore the details.

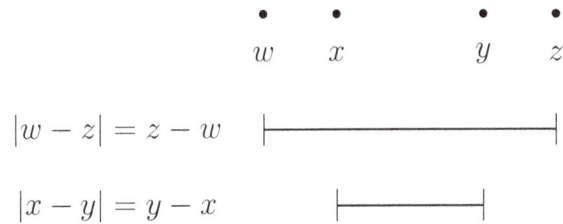

Figure 1.3.1: A quartet of real numbers $w < x < y < z$ and a comparison of the distance between the outer endpoints w and z to the distance between the inner endpoints x and y. See part (vi) of Theorem 1.3.2.

Theorem 1.3.2: Properties of inequalities

For any $w, x, y, z \in \mathbb{R}$ we have:

(i) If $x < y$, then $-y < -x$.

(ii) $0 < 1$.

(iii) If $0 < x < y$, then $0 < 1/y < 1/x$.

(iv) If $x < y$ and $z < 0$, then $xz > yz$.

(v) $x^2 \geq 0$.

(vi) If $w \leq x \leq y \leq z$, then

$$0 \leq |x - y| = y - x \leq z - w = |w - z|. \tag{1.3.1}$$

To get into what it means to say the real line $\mathbb{R}$ is *complete*, first consider my perspective on an important and subtle feature of the set of rational numbers $\mathbb{Q}$. I hope it helps you find your own perspective:

The set of rational numbers $\mathbb{Q}$ has arbitrarily small gaps.

For instance, no rational number is the square root of 2. And yet there are rational numbers that are as close together as we like whose squares are less than and greater than 2, respectively. To make this pair of assertions more concrete, consider the following theorem and problem.

Theorem 1.3.3: No rational number is the square root of two

If r is a rational number, then $r^2 \neq 2$.

Let's prove this with a classic contradiction argument.

Proof of Theorem 1.3.3. First, if $r = 0$, then $r^2 = 0 \neq 2$. Now, to set up a contradiction, suppose r is a nonzero rational number where $r^2 = 2$. We can write $r = m/n$ where m and n are integers that are not both even and neither is zero. We then have $r^2 = m^2/n^2 = 2$ and so $m^2 = 2n^2$, thus m^2 is even. This means m itself is even as well since the square of an odd integer is odd.

As an even number, we can write $m = 2k$ for some integer k. So now we have

$$m^2 = 2n^2 = 4k^2. \tag{1.3.2}$$

The right-hand side simplifies to

$$n^2 = 2k^2. \tag{1.3.3}$$

This means n^2 is even, which implies n is even, too.

We have arrived at our contradiction: We assumed m and n are not both even, yet both must be even when $r^2 = m^2/n^2 = 2$. Therefore, 2 is not the square of any rational number.   □

Loosely speaking, Theorem 1.3.3 says that the set of rational numbers $\mathbb{Q}$ has a gap at the square root of 2. And yet, the size of that gap is arbitrarily small.

Problem 1.3.4: Arbitrarily close to the square root of two

Consider the pair of rational numbers

$$a_1 = 1.5 = 15/10 \quad \text{and} \quad b_1 = 1.4 = 14/10. \tag{1.3.4}$$

We have

$$|a_1 - b_1| = |1.5 - 1.4| = 0.1 = 1/10 \tag{1.3.5}$$

while

$$b_1^2 = 1.96 < 2 < 2.25 = a_1^2. \tag{1.3.6}$$

Next, consider $a_2 = 1.42 = 142/100$ and $b_2 = 1.41 = 141/100$. Then

$$|a_2 - b_2| = |1.42 - 1.41| = 0.01 = 1/100 \tag{1.3.7}$$

while

$$b_2^2 = 1.9881 < 2 < 2.0164 = a_2^2. \tag{1.3.8}$$

The process results in something like Figure 1.3.2. Try to describe how to continue finding pairs of rational numbers a_n and b_n where, for each positive integer n, we have

$$|a_n - b_n| = 1/10^n \quad \text{and} \quad b_n^2 < 2 < a_n^2. \tag{1.3.9}$$

There's no need to find formulas for a_n and b_n, but you should try to describe any process you use.

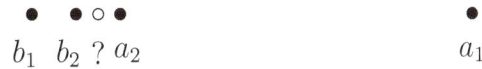

Figure 1.3.2: The rational numbers $a_1, a_2, b_1,$ and b_2 from Problem 1.3.4 giving initial approximations to the square root of two.

Remark 1.3.5: Arbitrarily small

Intuitively, $1/10^n$ can be made as small as we like by taking a positive integer n to be as large as we need. So the inclusion of the bound $1/10^n$ in Problem 1.3.4 is my way of indicating the size of the gap in the rationals at the square root of 2 is indeed arbitrarily small. We can and will prove this concretely once we have more mathematical tools—definitions, theorems, etc.—at our disposal.

Between any two rational numbers there is another, despite gaps like the square root of 2.

Lemma 1.3.6: Rationals between rationals

Suppose p and q are rational numbers where $p < q$. Then there is some rational number r where $p < r < q$.

Taking r to be the average of p and q gives us the result.

Proof of Lemma 1.3.6. Suppose $p = m/n$ and $q = s/t$ where $m, n, s,$ and t are integers where n and t are nonzero. Let

$$r = \frac{p+q}{2}. \tag{1.3.10}$$

With $p < q$ and after applying some algebra we have

$$p = \frac{p+p}{2} < \frac{p+q}{2} < \frac{q+q}{2} = q, \tag{1.3.11}$$

and so $p < r < q$. Also, by finding a common denominator we get

$$r = \frac{p+q}{2} = \frac{mt+ns}{2nt}. \tag{1.3.12}$$

Since $mt + ns$ and $2nt$ are integers and $2nt$ is nonzero, we have r is rational. $\square$

To recap, the set of rational numbers $\mathbb{Q}$ has arbitrarily small gaps and yet between any two there are always more. So what about the set of real numbers $\mathbb{R}$?

Remark 1.3.7: Rationals with gaps

The assumption that the real line $\mathbb{R}$ is complete is a way to ensure, from the start, that it has no gaps. In my opinion:

> The real line $\mathbb{R}$ is the set of rational numbers and the arbitrarily small gaps between them.

$$S \quad\quad \longlongrightarrow) \;\circ\;\bullet\;\bullet\cdots\circ$$
$$\quad\quad\quad\quad\quad\quad\quad\quad\quad\quad y \quad\quad x_y\; b$$

Figure 1.3.3: As in Definition 1.3.9, when b is the least upper bound of a set S and $y < b$, y is not an upper bound for S. So, there must be a point x_y in S where $y < x_y$.

There are a number of interesting, robust, and beautiful ways to construct the real line $\mathbb{R}$ from the set of rational numbers $\mathbb{Q}$. But to me they all amount to starting with the rational numbers then identifying the gaps from different perspectives.[a] Identifying an arbitrarily small gap in the rational numbers amounts to identifying an *irrational* number.

---

[a]Dedekind cuts use sets to partition $\mathbb{Q}$ in nice ways to identify the gaps. Equivalence classes of certain sequences (called Cauchy sequences) of rational numbers can be used to identify the gaps in a very different way.

Finally, here is a formal description of what it means for the real line $\mathbb{R}$ to be complete.

Axiom 1.3.8: Axiom of Completeness

Every nonempty subset of the real line that is bounded above has a supremum.

The Axiom of Completeness 1.3.8 formally describes what it means that the real line $\mathbb{R}$ has no arbitrarily small gaps (unlike the set of rational numbers $\mathbb{Q}$, see Problem 1.3.4). However, you're not expected to see why this means the real line has no gaps just yet. This perspective is a goal to be approached gradually as we explore the structure of the real line.

Also, the Axiom of Completeness 1.3.8 ensures every rational number and every arbitrarily small gap in the rational numbers can be identified as a real number: Each is the supremum of a set of rational numbers that is bounded above.

A more conventional way to set up an Axiom of Completeness is to use the notion of a *least upper bound* to define supremum (unlike, but equivalent to, Definition 1.1.14). Convention is often and deliberately broken throughout the book.

Definition 1.3.9: Least upper bound

Suppose S is a nonempty subset of the real line $\mathbb{R}$. A real number b is the *least upper bound* of S if

(i) for every $x \in S$ we have $x \leq b$, and

(ii) if $y < b$, then y is not an upper bound for S.

Statement (i) says b is an upper bound for S while (ii) says no real number smaller than b is an upper bound for S. When a real number y is not an upper bound for S, there is some x_y in S where y is less than x_y. See Figure 1.3.3.

On the other hand, any real number that's greater than b is another upper bound for S. Thus,

Figure 1.3.4: A set of real numbers S along with its supremum u which is also its least upper bound. See Theorem 1.3.10.

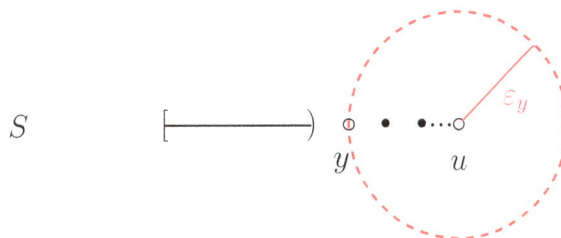

Figure 1.3.5: As in the proof of Theorem 1.3.10, any real number y strictly less than the supremum u of a set S creates a positive distance $\varepsilon_y = u - y$.

(i) and (ii) combine to identify b as the *unique* least upper bound for S. *Try drawing stuff!*

The concept of supremum as presented in Definition 1.1.14 is equivalent to that of least upper bound in Definition 1.3.9, so the Axiom of Completeness 1.3.8 is equivalent to more conventional notions found in other texts such as [1] and [10]. The following theorem codifies this equivalence.

Theorem 1.3.10: Supremum is the least upper bound

Suppose $u \in \mathbb{R}$ is an upper bound for a set $S \subseteq \mathbb{R}$. Then $u = \sup S$ if and only if u is the least upper bound of S.

See Figure 1.3.4 for a set of real numbers S, which is neither an interval nor a sequence, along with its supremum u.

Scratch Work 1.3.11: A thorough application of definitions

The proof of Theorem 1.3.10 uses the definitions of absolute value, arbitrarily close, supremum, and least upper bound (Definitions 1.1.6, 1.1.8, 1.1.14, and 1.3.9, respectively). For this particular proof, my goal is to be thorough and indicate where some definitions are used as clearly as I can, though this makes the proof longer than it would otherwise need to be. That's okay, especially this early in the book.

Proof of Theorem 1.3.10. First, assume u is an upper bound for S and $u = \sup S$. Next, suppose $y < u$ with the goal to show that y is not an upper bound for S. Then $u - y > 0$ and we let $\varepsilon_y = u - y$, thinking of this as a specific distance or "error" we want to allow. See Figure 1.3.5.

Since $u \, \text{acl} \, S$ ((ii) in Definition 1.1.14), by the definition of arbitrarily close (Definition 1.1.8) there must be some x_y in S where

$$d_{\mathbb{R}}(x_y, u) = |x_y - u| < \varepsilon_y = u - y. \tag{1.3.13}$$

See Figure 1.3.6.

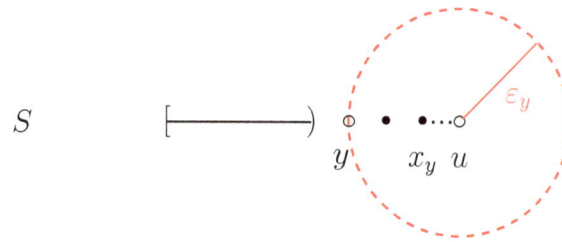

Figure 1.3.6: As in the proof of Theorem 1.3.10, since $u \operatorname{acl} S$, there must be a point x_y in S within $\varepsilon_y = u - y$ of $u = \sup S$.

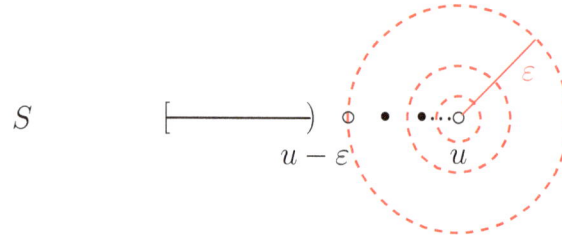

Figure 1.3.7: As in the proof of Theorem 1.3.10, the least upper bound u is within every positive distance ε of the set S. Only one ε is drawn to keep things from getting too cluttered.

Since u is also an upper bound for S, we have $x \le u$ for every x in S. So $x_y - u \le 0$, and by the definition of absolute value (Definition 1.1.6) we have

$$|x_y - u| = -(x_y - u) = u - x_y < u - y. \tag{1.3.14}$$

By subtracting u from the right hand side of the inequality above then multiplying the result by -1, we get $y < x_y$. (This might seem clear from Figure 1.3.6, but the figure falls just short for the purposes of this proof.) Thus, as we wanted to show, y is not an upper bound for S. That is, no real number smaller than u is an upper bound for S. So, u satisfies both parts of Definition 1.3.9 and is the least upper bound of S.

Next, for the other direction of the proof, assume u is an upper bound for S that is also the least upper bound. Now let $\varepsilon > 0$. (We let $\varepsilon > 0$ to allow for any amount of "error" and set up a verification of the definition of arbitrarily close, Definition 1.1.8). Then we have $u - \varepsilon < u$. See Figure 1.3.7.

Since u is the *least* upper bound of S, $u - \varepsilon$ is not an upper bound for S. This means there is a real number x_ε in S where

$$u - \varepsilon < x_\varepsilon. \tag{1.3.15}$$

See Figure 1.3.8.

Rearranging inequality (1.3.15) slightly yields

$$u - x_\varepsilon < \varepsilon. \tag{1.3.16}$$

Since u is an upper bound for S and x_ε is in S we have $x_\varepsilon \le u$, and so $u - x_\varepsilon \ge 0$. By the definition of absolute value (Definition 1.1.6), we have

$$|x_\varepsilon - u| = -(x_\varepsilon - u) = u - x_\varepsilon < \varepsilon. \tag{1.3.17}$$

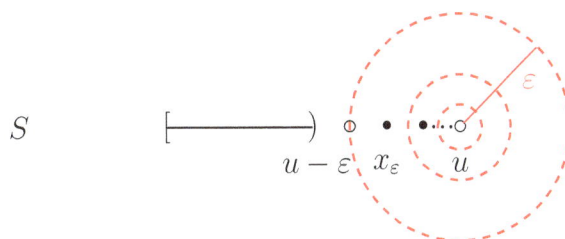

Figure 1.3.8: As in the proof of Theorem 1.3.10, x_ε is only close enough to the least upper bound u for the largest value of ε, as shown. It is not close enough for smaller values of ε (the smaller radii), but the proof only requires us to find one x_ε in the set S for each ε distance, separately.

We have shown u is an upper bound for S and arbitrarily close to S. Therefore, $u = \sup S$ according to Definition 1.1.14. □

To recap the above proof: When u is the supremum of S, no smaller number can be an upper bound for S, so u is the least upper bound. Also, when u is the least upper bound for S, it must be arbitrarily close to S since otherwise some smaller number would be an upper bound for S.

Remark 1.3.12: Walkthrough

Did the proof of Theorem 1.3.10 make sense to you? *Every line of it?* At first, reading and understanding proofs in analysis can take *a long time*. Please be patient. There are often so many details to parse and double-check, or so many steps left out, that it can be difficult to get a feeling for what's going on. But this is okay, and to me learning how to deal with parsing and understanding proofs is part of the development of every mathematician.

When you're confronted with a proof you find hard to follow, I suggest you write up a *walkthrough*. This can and should be whatever you want it to be, but basically a walkthrough should reflect your own thoughts and perspectives on the proof you've been given. Maybe drawing a figure will help, or writing out some algebraic steps that were left out, or writing out the definition or statement of a theorem when it is cited in the proof, or just rewriting the proof in your own words. If you can't see why a certain step works, don't hesitate to ask a friend, a professor, or even me what's going on.

Exercises

1.3.1. Prove that if $x < y$, then $x < \frac{x+y}{2} < y$.

1.3.2. This exercise compares the *arithmetic mean* $\frac{a+b}{2}$ to the *geometric mean* $\sqrt{ab}$. Suppose $a \geq 0$ and $b \geq 0$. Prove

$$\sqrt{ab} \leq \frac{a+b}{2}. \tag{1.3.18}$$

1.3.3. Suppose $0 < x < y$.

(i) Prove $0 < x^2 < y^2$. (The square function is increasing for positive inputs.)

(ii) Prove $0 < \sqrt{x} < \sqrt{y}$. (The square root function is increasing.)

1.3.4. Prove $\sqrt{3}$ is irrational by showing no rational number has square equal to 3.

1.3.5. Prove $\sqrt{6}$ is irrational by showing no rational number has square equal to 6.

1.3.6. Determine which of the following statements regarding two real numbers a and b is true statements and which is false. Write a proof for the true statement and find a counterexample for the false one.

(i) $a < b$ if and only if $a < b + \varepsilon$ for every $\varepsilon > 0$.

(ii) $a \le b$ if and only if $a < b + \varepsilon$ for every $\varepsilon > 0$.

1.3.7. Consider the real numbers $a_1, a_2, a_3 \ldots$ defined recursively by

$$a_1 = 6 \quad \text{and} \quad a_{n+1} = \frac{2a_n - 6}{3} \text{ for each } n \in \mathbb{N}. \tag{1.3.19}$$

(See Exercise 1.2.12 of [1].)

(i) Use induction to prove $a_n > -6$ for all $n \in \mathbb{N}$.

(ii) Use induction again to show $a_{n+1} \le a_n$ for all $n \in \mathbb{N}$. (This shows the sequence (a_n) is decreasing, see Chapter 2.)

1.3.8. Prove Theorem 1.3.2. Hint: Use Axiom 1.3.1.

1.3.9. Use Axiom 1.3.1 to prove the following statements about real numbers x, y, and z when $x \ne 0$.

(i) If $x \ne 0$ and $xy = xz$, then $y = z$.

(ii) If $x \ne 0$ and $xy = x$, then $y = 1$.

(iii) If $x \ne 0$ and $xy = 1$, then $y = 1/x$.

1.3.10. Find an example of sets $A, B \subseteq \mathbb{R}$ where $\sup A \le \sup B$ and yet no element of B is an upper bound for A.

1.3.11. Suprema have interesting relationships with unions of sets.

(i) Suppose $S, T \subseteq \mathbb{R}$ where both S and T are bounded above. Prove

$$\sup(S \cup T) = \max\{\sup S, \sup T\}. \tag{1.3.20}$$

(ii) Suppose $S_1, S_2, S_3, \ldots$ be an infinite collection of sets where, for each $n \in \mathbb{N}$, S_n is a set of real numbers which is bounded above. Prove that for any $k \in \mathbb{N}$ we have

$$\sup\left(\cup_{n=1}^{k} S_n\right) = \max\{\sup S_1, \sup S_2, \ldots, \sup S_k\}. \tag{1.3.21}$$

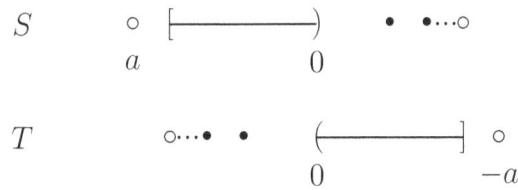

Figure 1.4.1: A set of real numbers S along with a lower bound a and their "mirror images" T and $-a$. See the proof of Theorem 1.4.1.

(iii) Find an infinite collection of sets $T_1, T_2, T_3, \ldots$ where, for each $n \in \mathbb{N}$, T_n is a set of real numbers which is bounded above and yet

$$\sup (\cup_{n=1}^{\infty} T_n) \neq \max\{\sup T_1, \sup T_2, \ldots\}. \tag{1.3.22}$$

1.4 Implications of completeness

What does completeness do for us? Generally speaking, it provides a way for us to ensure the *existence* of real numbers: They could be rational numbers, positive integers, or something else, but the big idea is for us to be able to assert their existence when the time is right.

This section explores a collection of results stemming from the Axiom of Completeness 1.3.8, including its mirror image in terms of infimum.

Theorem 1.4.1: Bounded below means the infimum exists

Every nonempty set of real numbers that is bounded below has an infimum.

Multiplication of real numbers by -1 flips them around 0 and reverses order. So, the idea behind the proof of Theorem 1.4.1 is to identify the infimum of one set as the negative of the supremum of another.

Proof of Theorem 1.4.1. Suppose S is a set of real numbers that is bounded below by a. Let $T = \{-s : s \in S\}$. Loosely speaking, T is the "mirror image" of S comprising the negatives of every point in S. See Figure 1.4.1.

Since $a \leq s$ for every $s \in S$, by part (i) of Theorem 1.3.2 we have $-s \leq -a$, so $-a$ is an upper bound for T. By the Axiom of Completeness 1.3.8, T has a supremum, so let $u = \sup T$.

For the final step we show $-u = \inf S$. Since u is an upper bound for T, we have $-s \leq u$ for every s in S. Hence, $-u \leq s$ for every s in S, so $-u$ is a lower bound for S. See Figure 1.4.2. We also have $u \operatorname{acl} T$, so for every $\varepsilon > 0$ there is some x_ε in S where

$$|u - (-x_\varepsilon)| = |(-u) - x_\varepsilon| < \varepsilon. \tag{1.4.1}$$

Hence, $-u \operatorname{acl} S$ and we have $-u = \inf S$. $\qquad \square$

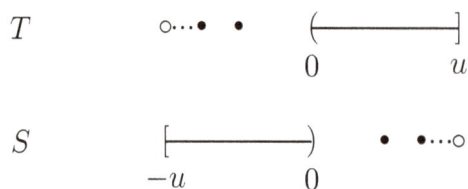

Figure 1.4.2: The set of real numbers T along with its supremum u and their "mirror images" S and $-u$, where $-u$ is the infimum of S. See the proof of Theorem 1.4.1.

Theorem 1.3.10 has a mirror image as well, this one is in terms of infimum and *greatest lower bound*.

Definition 1.4.2: Greatest lower bound

Suppose S is a nonempty subset of the real line $\mathbb{R}$. A real number w is the *greatest lower bound* of S if

(i) for every $x \in S$ we have $w \le x$, and

(ii) if $w < y$, then y is not a lower bound for S.

Theorem 1.4.3: Infimum is the greatest lower bound

Suppose $\ell \in \mathbb{R}$ is a lower bound for a set $S \subseteq \mathbb{R}$. Then $\ell = \inf S$ if and only if ℓ is the greatest lower bound of S.

Theorem 1.4.3 is so similar to Theorem 1.3.10 even their proofs are mirror images of each other. So the proof of Theorem 1.4.3 is left as an exercise. A careful consideration and modification of one of the proofs can lead to a proof of the other, making for some good practice.

The proof of the following corollary makes use of Theorems 1.3.10 and 1.4.3 which allow us to interpret supremum and infimum in the classic sense as the least upper bound and greatest lower bound, respectively. The corollary itself plays a key role in the development of properties of *integrals* in Chapter 6.

Corollary 1.4.4: Order properties of suprema and infima

Suppose A and B are nonempty bounded sets of real numbers. Then

$$A \subseteq B \quad \implies \quad \sup A \le \sup B \quad and \quad \inf B \le \inf A. \qquad (1.4.2)$$

Scratch Work 1.4.5: Draw stuff!

Draw some pairs of sets A and B that fit the hypothesis of Corollary 1.4.4. When I do this, I generally see that the smaller set A does not reach as far to the left nor as far to the right as the larger set B. This means B has a potentially smaller infimum or larger supremum than A.

Proof of Corollary 1.4.4. Suppose A and B are nonempty bounded sets of real numbers.

Case (i): Suppose $A = B$. Then $\sup A = \sup B$ and $\inf B = \inf A$.

Case (ii): Suppose $A \subsetneq B$ and there is some real number w where $w \in B$, $w \notin A$, and $x < w$ for all $x \in A$. Then w is *an* upper bound for A. By Theorem 1.3.10, $\sup A$ is *the least* upper bound for A, making it less than or equal w. Therefore, since $w \in B$ and $\sup B$ is an upper bound for B, we have

$$\sup A \leq w \leq \sup B. \tag{1.4.3}$$

Case (iii): Suppose $A \subsetneq B$ and there is some real number z where $z \in B$, $z \notin A$, and $z < x$ for all $x \in A$. Then z is *a* lower bound for A. By Theorem 1.4.3, $\inf A$ is *the greatest* lower bound for A, making it greater than or equal to z. Therefore, since $z \in B$ and $\inf B$ is a lower bound for B, we have

$$\inf B \leq z \leq \inf A. \tag{1.4.4}$$

$\square$

An intuitive idea is that given any real number, there is a larger positive integer. Actually, we already used this idea in the proofs for Example 1.1.17, but we are now in position to formally prove it.

Theorem 1.4.6: Archimedean Property

Given a real number x there is some positive integer n_x where $x < n_x$.

Proof of the Archimedean Property (Theorem 1.4.6). To set up a contradiction, let $x \in \mathbb{R}$ and assume the set of positive integers $\mathbb{N}$ is bounded above by x so that $n \leq x$ for every n in $\mathbb{N}$. By the Axiom of Completeness 1.3.8, $\mathbb{N}$ has a supremum $u = \sup \mathbb{N}$. Since $7 > 0$ and u is both an upper bound for $\mathbb{N}$ and arbitrarily close to $\mathbb{N}$, there must be some $n \in \mathbb{N}$ where

$$|u - n| = u - n < 7. \tag{1.4.5}$$

Rearranging the inequality yields $u < n + 7$, which implies u is *not* an upper bound for $\mathbb{N}$ since $n + 7$ is also a positive integer. This contradicts the assertion that $u = \sup \mathbb{N}$, meaning x cannot be an upper bound for $\mathbb{N}$. Therefore, there must be some n_x in $\mathbb{N}$ where $x < n_x$. $\square$

Remark 1.4.7: Why 7?

Did 7 play a special role in the proof of the Archimedean Property 1.4.6 somehow? No, not really. It was good enough and *that's all I needed*. While preparing the scratch work for the previous proof, there was a choice I was free to make. I needed a positive number to play the role of ε from the definition of arbitrarily close (Definition 1.1.8) so the inequality in line (1.4.5) is valid, and I needed this positive number to be an integer so that $n + 7$ is also a positive integer. So 7 was good enough, but any positive integer could have played

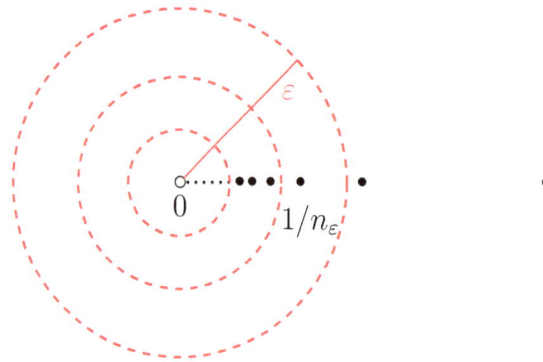

Figure 1.4.3: Corollary 1.4.8 shows there is no smallest real number. In particular, no matter how small a positive real number ε is, there's a rational number of the form $1/n_\varepsilon$ that's less than ε.

the role just as well.

There is no smallest positive real number. This subtle and perhaps surprising fact follows immediately from the Archimedean Property 1.4.6. A formal statement is provided by the next corollary which also gives us a way to guarantee there is a positive integer large enough that its reciprocal (a rational number) is as small as we like. See Figure 1.4.3.

Corollary 1.4.8: A consequence of the Archimedean Property

Given $\varepsilon > 0$, there is a positive integer n_ε where

$$0 < \frac{1}{n_\varepsilon} < \varepsilon. \tag{1.4.6}$$

Proof of Corollary 1.4.8. Suppose $\varepsilon > 0$. Then $1/\varepsilon$ is a real number and we can apply the Archimedean Property 1.4.6. So there is a positive integer n_ε such that $1/\varepsilon < n_\varepsilon$. Using some algebraic properties of inequalities and noting n_ε is positive yields

$$0 < \frac{1}{n_\varepsilon} < \varepsilon. \tag{1.4.7}$$

See Figure 1.4.3.   □

Yet another intuitive idea we can formally prove is the notion that every real number is between two consecutive integers. See Figure 1.4.4.

Corollary 1.4.9: Between consecutive integers

For every $x \in \mathbb{R}$ there is some $m_x \in \mathbb{Z}$ such that

$$m_x \leq x < m_x + 1. \tag{1.4.8}$$

The proof addresses a subtlety: The set to which we want to apply the Axiom of Completeness 1.3.8 may or may not be empty.

$(-\infty, \infty)$

$[m_x, m_x + 1)$

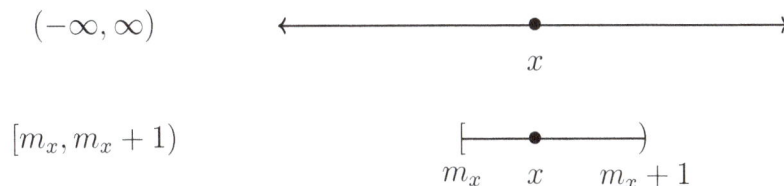

Figure 1.4.4: Each given real number x lies between a pair of consecutive integers m_x and $m_x + 1$. See Corollary 1.4.9.

Proof of Corollary 1.4.9. Let $x \in \mathbb{R}$ and let $S_x = \{z \in \mathbb{Z} : z \leq x\}$. Since $-x \in \mathbb{R}$, by the Archimedean Property 1.4.6, there is an $n_{-x} \in \mathbb{N}$ where $-x < n_{-x}$. Hence, $-n_{-x} < x$ and $-n_{-x} \in \mathbb{Z}$. Thus, $-n_{-x} \in S_x$, so S_x is nonempty and bounded above by x.

Now, since S_x is nonempty and bounded above, by the Axiom of Completeness 1.3.8 we have $u = \sup S_x$ exists. By Theorem 1.3.10, u is the least upper bound for S_x, which implies $u - 1$ is not an upper bound for S_x. So, there is some $m_x \in S_x \subseteq \mathbb{Z}$ where $u - 1 < m_x$. This implies $u < m_x + 1$, and since u is an upper bound for S_x, we have $m_x + 1$ is an integer that is not in S_x. Hence, $x < m_x + 1$. Therefore, m_x is an integer where $m_x \leq x < m_x + 1$. □

No matter how close two distinct real numbers are, there is always a rational number between them.

Theorem 1.4.10: Density of the rationals in the reals

Let $x, y \in \mathbb{R}$ where $x < y$. Then there is some $r \in \mathbb{Q}$ where $x < r < y$.

The proof below has some steps which may not make sense on a first reading. So, before that, let me share some of my scratch work with you.

Scratch Work 1.4.11: Density of the rationals in the reals

The goal is to come up with an integer m and a positive integer n where

$$x < r = \frac{m}{n} < y. \tag{1.4.9}$$

How can we prove this in a mathematically rigorous way? By that I mean, can we get the result we want by relying on the Axioms, Definitions, Theorems, Corollaries, and other technical statements we've come across so far?

We can use Corollary 1.4.8 to find an $n \in \mathbb{N}$ where $1/n$ is small enough to be less than the distance between x and y as in Figure 1.4.5. That is

$$1/n < |x - y| = y - x. \tag{1.4.10}$$

But we still need a numerator $m \in \mathbb{Z}$ to produce the inequalities in line (1.4.9). We can aim for m to specifically give us

$$\frac{m-1}{n} \leq x < \frac{m}{n} < y. \tag{1.4.11}$$

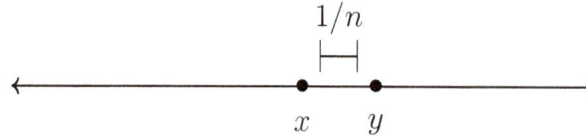

Figure 1.4.5: The distance between any two real numbers x and y is larger than the reciprocal $1/n$ of some positive integer n. See line (1.4.10) in Scratch Work 1.4.11.

Figure 1.4.6: As in line (1.4.13) from the proof of Theorem 1.4.10, the real number $nx + 1$ is between two consecutive integers m and $m + 1$.

Such a numerator m will be furnished by Corollary 1.4.9, stemming from a real number that lines up nicely for the conclusion of the proof. But which real number exactly? Solving for m in the leftmost inequality in (1.4.11) yields $m \leq nx + 1$. Now for the proof.

Proof of Theorem 1.4.10. Suppose $x, y \in \mathbb{R}$ where $x < y$. Then $y - x > 0$, so Corollary 1.4.8 guarantees there is an $n \in \mathbb{N}$ where $1/n < y - x$. We can rearrange the inequality to get

$$nx + 1 < ny. \tag{1.4.12}$$

Since $nx + 1$ is a real number, it is between consecutive integers. See Figure 1.4.6. By Corollary 1.4.9, there is an $m \in \mathbb{Z}$ where

$$m \leq nx + 1 < m + 1. \tag{1.4.13}$$

The right-hand side of (1.4.13) implies $nx < m$, so $x < m/n$. In fact,

$$\frac{m - 1}{n} \leq x < \frac{m}{n}. \tag{1.4.14}$$

See Figure 1.4.7. Also, combining the left-hand side of (1.4.13) with (1.4.12) yields

$$m \leq nx + 1 < ny. \tag{1.4.15}$$

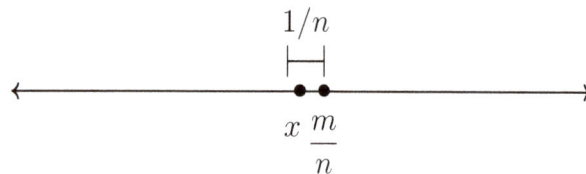

Figure 1.4.7: As in line (1.4.14) from the proof of Theorem 1.4.10, the rational number m/n is less than $1/n$ away from the real number x.

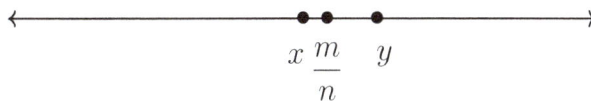

Figure 1.4.8: As in line (1.4.16) at the end of the proof of Theorem 1.4.10, the rational number m/n is between the real numbers x and y.

Hence, $m < ny$ and so $m/n < y$. Therefore, as in Figure 1.4.8,

$$x < \frac{m}{n} < y. \tag{1.4.16}$$

$\square$

Theorem 1.3.3 shows that no rational number is the square root of 2, but that doesn't immediately imply the square root of 2 is a real number. This is something we can prove using the completeness of the real line and its consequences.

Theorem 1.4.12: The square root of two is a real number

There is an $x \in \mathbb{R}$ where $x^2 = 2$.

The proof makes use of a carefully chosen set of real numbers that's bounded above whose supremum plays a key role along with a pair of contradictions. The supremum of this set exists by the Axiom of Completeness 1.3.8 and the way the set is chosen allows us to show, using a pair of similar contradictions, that the square of its supremum can be neither larger nor smaller than 2. So by the trichotomy property of $\mathbb{R}$, the supremum must be 2 (see Axiom 1.3.1). However, only one of the contradictions is derived here.

Half of the proof of Theorem 1.4.12. Let S be the set of real numbers whose squares are less than 2. That is, $S = \{y \in \mathbb{R} : y^2 < 2\}$.

Since $1^2 = 1 < 2$ we have $1 \in S$, so S is nonempty. Since $1 < 7 < y$ implies $2 < 49 = 7^2 < y^2$, we have that 7 is an upper bound for S. By the Axiom of Completeness 1.3.8, there is a real number u where $u = \sup S$. Since u is an upper bound for S, we have $0 < 1 \le u$. In particular, u is positive.

To establish a contradiction, suppose $u^2 < 2$. The goal is to find an element in S that is greater than u. Since $u^2 < 2$, we have $2 - u^2 > 0$. Since $u > 0$, we have $2u + 1 > 0$ as well. Thus,

$$\frac{2 - u^2}{2u + 1} > 0, \tag{1.4.17}$$

which is enough wiggle room to work with. By Corollary 1.4.8, there is a positive integer n large enough where both

$$0 < \frac{1}{n} < \frac{2 - u^2}{2u + 1} \quad \text{and} \quad n > 1. \tag{1.4.18}$$

Thus, after some algebraic manipulation we have

$$\frac{2u + 1}{n} < 2 - u^2, \tag{1.4.19}$$

and by choosing $n > 1$ we have $1/n^2 < 1/n$. Combining these results we have

$$\left(u + \frac{1}{n}\right)^2 = u^2 + \frac{2u}{n} + \frac{1}{n^2} \tag{1.4.20}$$

$$< u^2 + \frac{2u}{n} + \frac{1}{n} \tag{1.4.21}$$

$$= u^2 + \frac{2u + 1}{n} \tag{1.4.22}$$

$$< u^2 + 2 - u^2. \tag{1.4.23}$$

This implies $(u + (1/n))^2 < 2$, therefore $u + (1/n)$ is in S and greater than u. But this contradicts the assertion that u is an upper bound for S, so we must have $u^2 \geq 2$.

The case where $u^2 > 2$ is assumed leads to another contradiction following from a similar argument. So, the completion of the proof of Theorem 1.3.3 is left as an exercise. $\qquad\square$

Theorem 1.4.12 tells us that there is a positive real number whose square is 2 and this number is denoted by $\sqrt{2}$. Theorem 1.3.3 tells us $\sqrt{2}$ is not a rational number, so it must be *irrational*. That is, $\sqrt{2} \in \mathbb{R}\backslash\mathbb{Q}$. We can use this to prove that between any two real numbers, there is an irrational number.

> **Corollary 1.4.13: Density of irrationals in the reals**
>
> Let $x, y \in \mathbb{R}$ where $x < y$. Then there is some $v \in \mathbb{R}\backslash\mathbb{Q}$ where $x < v < y$.

Proof of Corollary 1.4.13. Suppose $x < y$ so that $x - \sqrt{2} < y - \sqrt{2}$. By the Density of $\mathbb{Q}$ in $\mathbb{R}$ (Theorem 1.4.10), there is a rational number r such that

$$x - \sqrt{2} < r < y - \sqrt{2}. \tag{1.4.24}$$

Therefore, $x < r + \sqrt{2} < y$.

Now let, $v = r + \sqrt{2}$. To complete the proof, we should show v is irrational. This can be done by establishing a contradiction. We have $r \in \mathbb{Q}$, so suppose $v \in \mathbb{Q}$ as well. Then there are integers m and s along with positive integers n and t where $r = m/n$ and $v = s/t$. Then we have

$$\sqrt{2} = v - r = \frac{ns - mt}{nt}. \tag{1.4.25}$$

Since $ns - mt$ is an integer and nt is a positive integer, we have that $\sqrt{2}$ is rational, establishing a contradiction of Theorem 1.3.3. $\qquad\square$

The Axiom of Completeness 1.3.8 allows us to prove many results about the structure and properties of the real line. Some of which you may have seen or worked with before, others may be new. Still, more questions remain to be asked and answered: In addition to $\sqrt{2}$, what other kinds of real numbers are irrational? How many irrational numbers are there? What other properties does the real line have in store for us?

Exercises

1.4.1. Prove Theorem 1.4.3: Suppose $\ell \in \mathbb{R}$ is a lower bound for a set $S \subseteq \mathbb{R}$. Then $\ell = \inf S$ if and only if ℓ is the greatest lower bound of S. Hint: Carefully modify the proof of Theorem 1.3.10, which connects supremum and least upper bound.

1.4.2. Prove for any prime number p and every rational number r that $r^2 \neq p$. (So $\sqrt{p}$ is irrational). Modify proof of Theorem 1.3.3.

1.4.3. Complete the proof of Theorem 1.4.12 for the case where $u = \sup S$ and it is assumed $u^2 > 2$. Hint: Carefully modify the half of the proof already provided.

1.4.4. This exercise justifies the definitions and properties of rational and irrational powers of real numbers, under meaningful conditions. (See Exercise 1.6 of [10].)

Fix $b > 1$.

(i) Suppose $m, p \in \mathbb{Z}$ and $p, q \in \mathbb{N}$ where $r = m/n = p/q$. Show that

$$(b^m)^{1/n} = (b^p)^{1/q}. \tag{1.4.26}$$

(This allows us to define $b^r = (b^m)^{1/n}$ for all $r \in \mathbb{Q}$.)

(ii) Prove that if $r, s \in \mathbb{Q}$, then $b^{r+s} = b^r b^s$.

(iii) Suppose $x \in \mathbb{R}$ and define $T(x)$ by

$$T(x) = \{b^t : t \in \mathbb{Q} \text{ and } t \leq x\}. \tag{1.4.27}$$

Prove that when $r \in \mathbb{Q}$, we have

$$b^r = \sup T(r). \tag{1.4.28}$$

(So, it makes sense to define $b^x = \sup T(x)$ for each $x \in \mathbb{R}$.)

(iv) Prove $b^{x+y} = b^x b^y$ for all $x, y \in \mathbb{R}$.

1.4.5. This exercise justifies the definition and properties of logarithms. (See Exercise 1.7 of [10].)

Fix $b > 1$ and $y > 0$, and assume the results of Exercise 1.4.4 hold. Complete the following steps to prove there is a unique $x \in \mathbb{R}$ such that $b^x = y$. Here, x is called the *logarithm of y* with respect to the *base b*.

(i) For any positive integer $n \in \mathbb{N}$, we have $b^n - 1 \geq n(b-1)$.

(ii) By (i), show $b - 1 \geq n(b^{1/n} - 1)$.

(iii) Prove that if $t > 1$ and $n > (b-1)/(t-1)$, then $b^{1/n} < t$.

(iv) Prove that if $z \in \mathbb{R}$ is such that $b^z < y$, then $b^{z+(1/n)} < y$ for sufficiently large n by applying (iii) with $t = y \cdot b^{-z}$.

(v) Prove that if z is as in (iv), $b^z > y$, then $b^{z-(1/n)} > y$ for sufficiently large $n \in \mathbb{N}$.

(vi) Let $B_y = \{w \in \mathbb{R} : b^w < y\}$. Prove $x = \sup B_y$ satisfies $b^x = y$.

(vii) Prove that the value of x in (vi) is unique.

1.5 Arbitrarily close in Euclidean spaces

In the more general setting of points and sets in a Euclidean space $\mathbb{R}^m$, we have the following definition for *arbitrarily close*. To replace the notion of distance in the real line given by a difference in absolute values, here we use the more general Pythagorean distance

$$d_m(\mathbf{x}, \mathbf{y}) = \|\mathbf{x} - \mathbf{y}\|_m$$
$$= \sqrt{(x_1 - y_1)^2 + (x_2 - y_2)^2 + \cdots + (x_m - y_m)^2}. \tag{1.5.1}$$

Definition 1.5.1: Arbitrarily close

Let $\mathbf{y}$ be a point in $\mathbb{R}^m$ and let $B \subseteq \mathbb{R}^m$. The point $\mathbf{y}$ is said to be *arbitrarily close* to the set B, and we write $\mathbf{y} \, \mathrm{acl} \, B$, if for every $\varepsilon > 0$ there is a point $\mathbf{x}_\varepsilon$ in B such that

$$d_m(\mathbf{x}_\varepsilon, \mathbf{y}) = \|\mathbf{x}_\varepsilon - \mathbf{y}\|_m < \varepsilon. \tag{1.5.2}$$

The phrase "B is *arbitrarily close* to $\mathbf{y}$" is defined in the same way.

Example 1.5.2: A point arbitrarily close to a set

In Figure 1.5.1, B is a solid rectangle containing three of its corners and two of its sides. The point $\mathbf{y}$ is the corner of the rectangle that is not in B but is arbitrarily close to B.

Remark 1.5.3: As small as we like, revisited

As in the definition for arbitrarily close in the real line (Defintion 1.1.8), we can think of the positive real number ε as the amount of error or "wiggle room" we'd like to allow. In Figure 1.5.1, just one radius ε is drawn to keep things from getting too cluttered. However, since Definitions 1.1.8 and 1.5.1 allow for *any* $\varepsilon > 0$, we can take the error to be as small as we like. So, $\mathbf{y} \, \mathrm{acl} \, B$ means B gets as close to $\mathbf{y}$ as we like, no matter how close. In order to prove $\mathbf{y} \, \mathrm{acl} \, B$, it suffices can respond to each $\varepsilon > 0$ with a point $\mathbf{x}_\varepsilon$ which is in B and within ε of $\mathbf{y}$.

As a first result, points in a set are arbitrarily close to the set.

Lemma 1.5.4: A point in a set is arbitrarily close

Let $\mathbf{y} \in \mathbb{R}^m$ and let $B \subseteq \mathbb{R}^m$. If $\mathbf{y} \in B$, then $\mathbf{y} \, \mathrm{acl} \, B$.

The idea of the proof is to choose the same point $\mathbf{x}_\varepsilon = \mathbf{y}$ for each distance $\varepsilon > 0$.

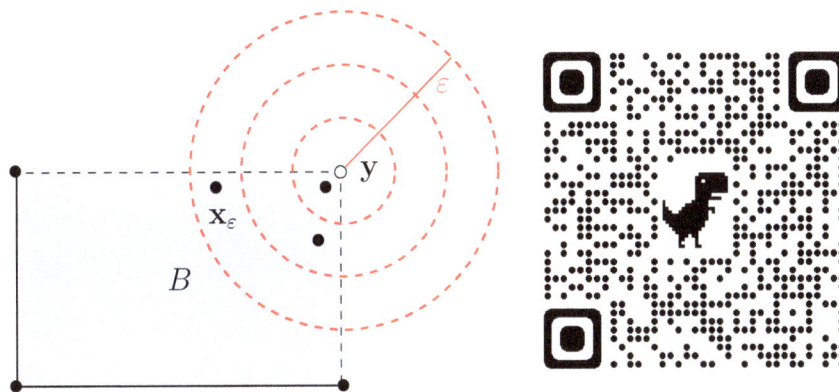

Figure 1.5.1: A point $\mathbf{y}$ arbitrarily close to a set B in the plane $\mathbb{R}^2$. The point $\mathbf{x}_\varepsilon$ is in B and within a distance of ε from $\mathbf{y}$. That is, $\mathbf{x}_\varepsilon \in B \cap V_\varepsilon(\mathbf{y})$. See Definition 1.5.1. Also, a QR code for the Desmos activity "y acl B" to accompany this figure and the one on the title page. https://www.desmos.com/calculator/nfbdjs8pdh

Proof of Lemma 1.5.4. Assume $\mathbf{y} \in B$ and let $\varepsilon > 0$. Choosing $\mathbf{x}_\varepsilon = \mathbf{y}$ yields

$$d_m(\mathbf{x}_\varepsilon, \mathbf{y}) = \|\mathbf{x}_\varepsilon - \mathbf{y}\|_m = \|\mathbf{y} - \mathbf{y}\|_m = 0 < \varepsilon. \tag{1.5.3}$$

Hence, $\mathbf{y} \operatorname{acl} B$. $\qquad\square$

A natural question students have asked is: "What does it mean when two points are arbitrarily close?" Since the definition of arbitrarily close (Definition 1.5.1) compares a point to a set, we need to be creative in order to properly to use it to answer this question. We can replace one of the points with a singleton.

Lemma 1.5.5: Equal points are arbitrarily close

Let $\mathbf{x}, \mathbf{y} \in \mathbb{R}^m$. We have $\mathbf{x} = \mathbf{y}$ if and only if $\mathbf{x} \operatorname{acl} \{\mathbf{y}\}$.

Proof of Lemma 1.5.5. Suppose $\mathbf{x} = \mathbf{y}$. Then for every $\varepsilon > 0$ we have $\|\mathbf{x} - \mathbf{y}\|_m = 0 < \varepsilon$. Therefore, $\mathbf{x} \operatorname{acl} \{\mathbf{y}\}$.

Now, suppose $\mathbf{x} \neq \mathbf{y}$. Then $0 < \|\mathbf{x} - \mathbf{y}\|_m/2 \leq \|\mathbf{x} - \mathbf{y}\|_m$. Therefore, $\mathbf{x}$ is not arbitrarily close to $\{\mathbf{y}\}$. $\qquad\square$

Remark 1.5.6: Two points arbitrarily close are equal

Lemma 1.5.5 says that in any Euclidean space $\mathbb{R}^m$ two points are arbitrarily close if and only if they are equal. *But this is not necessarily the case when we extend the definition of arbitrarily close to topological spaces!* If you're not familiar with topology yet, don't worry. I'm just pointing out that cool things happen in topology when other beautiful structures aside from Euclidean spaces are in play.

Problem 1.5.7: Drawing figures for a lemma

Draw some figures to go with the proofs of Lemmas 1.5.4 and 1.5.5. Playing with examples in the plane $\mathbb{R}^2$ isn't a bad idea.

To provide another perspective for notions of closeness, consider the terminology of *neighborhoods* found in topology.

Definition 1.5.8: Neighborhood

Let $\varepsilon > 0$ and $\mathbf{c} \in \mathbb{R}^m$. The *$\varepsilon$-neighborhood* of $\mathbf{c}$, denoted by $V_\varepsilon(\mathbf{c})$, is the set of points within ε of $\mathbf{c}$. That is,

$$V_\varepsilon(\mathbf{c}) = \{\mathbf{x} \in \mathbb{R}^m : d_m(\mathbf{x}, \mathbf{c}) = \|\mathbf{x} - \mathbf{c}\|_m < \varepsilon\}. \tag{1.5.4}$$

The value of ε used here can still be thought of as an "error" or as a bound for the distance from $\mathbf{c}$ we want to allow. Also, the word *neighborhood* means the same thing as an "ε-neighborhood of $\mathbf{c}$" and is used when ε and $\mathbf{c}$ need not be specified.

Given a fixed $m \in \mathbb{N}$ and its Euclidean space $\mathbb{R}^m$, what do ε-neighborhoods look like? See Figure 1.5.2 for ε-neighborhoods of some fixed $\varepsilon > 0$ centered at $\mathbf{c}_3$ in $\mathbb{R}^3$, $\mathbf{c}_2$ in $\mathbb{R}^2$, and c_1 in $\mathbb{R}$, respectively.

- $V_\varepsilon(\mathbf{c}_3)$ is the open sphere of radius ε centered at $\mathbf{c}_3$ and does not include points on its surface exactly ε away from $\mathbf{c}_3$.

- $V_\varepsilon(\mathbf{c}_2)$ is the open disk of radius ε centered at $\mathbf{c}_2$ and does not include points on the circle exactly ε away from $\mathbf{c}_2$.

- $V_\varepsilon(c_1)$ is the open interval of length 2ε centered at c_1 and does not include the endpoints $c_1 - \varepsilon$ and $c_1 + \varepsilon$.

Remark 1.5.9: Arbitrarily close via neighbhorhoods

As suggested at the start of this chapter, neighborhoods provide an important perspective for defining notions of arbitrarily close. For a point $\mathbf{y}$ and a set B in a Euclidean space $\mathbb{R}^m$, we have $\mathbf{y} \operatorname{acl} B$ if and only if every ε-neighborhood of $\mathbf{y}$ intersects B. That is, $\mathbf{y} \operatorname{acl} B$ if and only if for every $\varepsilon > 0$ we have

$$V_\varepsilon(\mathbf{y}) \cap B \neq \varnothing, \tag{1.5.5}$$

where $\varnothing$ denotes the empty set. See Definitions 1.5.1 and 1.5.8.

The following lemma highlights a special property of neighborhoods in the real line that turns out to be quite useful. In particular, the real line is ordered, so any neighborhood of a real number has only two directions to go in. This effect is established by the Trichotomy Property of the real line (see Axiom 1.3.1) and the definition of absolute value (Definition 1.1.6). Specifically, every pair real numbers x and y satisfy exactly one of the following: $x < y$, $y < x$, or $x = y$. The proof of the lemma is left to Exercise 1.5.3.

$$V_\varepsilon(\mathbf{c}_3) \subseteq \mathbb{R}^3$$

$$V_\varepsilon(\mathbf{c}_2) \subseteq \mathbb{R}^2$$

$$V_\varepsilon(c_1) \subseteq \mathbb{R}$$

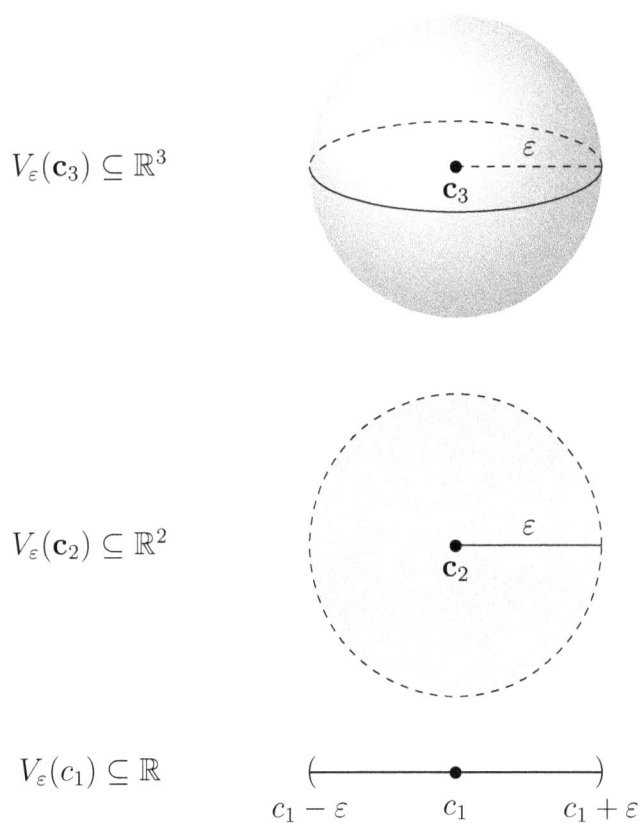

Figure 1.5.2: Some ε-neighborhoods centered at $\mathbf{c}_3$ in $\mathbb{R}^3$, $\mathbf{c}_2$ in $\mathbb{R}^2$, and c_1 in $\mathbb{R}$, respectively, for some fixed $\varepsilon > 0$.

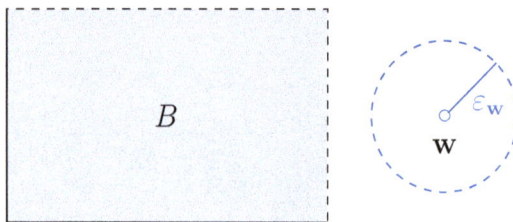

Figure 1.5.3: A point $\mathbf{w}$ away from a set B where all points in B are more than some positive distance $\varepsilon_{\mathbf{w}}$ away from $\mathbf{w}$. Thus, $V_{\varepsilon_{\mathbf{w}}}(\mathbf{w}) \subseteq \mathbb{R}^m \backslash B$. See Definition 1.5.11.

Lemma 1.5.10: Breakdown of neighborhoods in the real line

Let c and x be real numbers. For each $\varepsilon > 0$, the following are equivalent:

$$x \in V_\varepsilon(c) \iff |x - c| < \varepsilon \tag{1.5.6}$$
$$\iff -\varepsilon < x - c < \varepsilon \tag{1.5.7}$$
$$\iff c - \varepsilon < x < c + \varepsilon. \tag{1.5.8}$$

Let's try turning things around. What if some point $\mathbf{w} \in \mathbb{R}^m$ is not arbitrarily close to B? Negation leads immediately to the following definition.

Definition 1.5.11: Away from

We say $\mathbf{w}$ is *away from* B and write $\mathbf{w}$ awf B if there is some $\varepsilon_{\mathbf{w}} > 0$ such that for every $\mathbf{x}$ in B we have

$$d_m(\mathbf{x}, \mathbf{w}) = \|\mathbf{x} - \mathbf{w}\|_m \geq \varepsilon_{\mathbf{w}}. \tag{1.5.9}$$

Remark 1.5.12: Distance between a set and a point

To recap, we have $\mathbf{y}$ acl B if there is no distance between $\mathbf{y}$ and B, while $\mathbf{w}$ awf B if there is some distance between $\mathbf{w}$ and B. Equivalently, $\mathbf{w}$ awf B if there is some $\varepsilon_{\mathbf{w}}$-neighborhood $V_{\varepsilon_{\mathbf{w}}}(\mathbf{w})$ where we have $V_{\varepsilon_{\mathbf{w}}}(\mathbf{w}) \subseteq \mathbb{R}^m \backslash B$. So in order to prove $\mathbf{w}$ awf B, all we need is one fixed distance $\varepsilon_{\mathbf{w}} > 0$ that keeps the all points in B that far from $\mathbf{w}$ or more. See Figure 1.5.3.

Theorem 1.4.10, Corollary 1.4.13, and Lemma 1.5.10 combine to produce a nice result on the relationships between rational numbers, irrational numbers, and ε-neighborhoods in the real line.

Theorem 1.5.13: All neighborhoods in the real line contain rationals and irrationals

Let $c \in \mathbb{R}$ and $\varepsilon > 0$. There is a rational number and an irrational number in the ε-neighborhood $V_\varepsilon(c) = (c - \varepsilon, c + \varepsilon)$.

Proof of Theorem 1.5.13. Let $c \in \mathbb{R}$ and $\varepsilon > 0$. Then $c - \varepsilon < c + \varepsilon$, so by Theorems 1.4.10 and Corollary 1.4.13, there is an $r \in \mathbb{Q}$ and a $v \in \mathbb{R} \backslash \mathbb{Q}$ where

$$c - \varepsilon < r < c + \varepsilon \quad \text{and} \quad c - \varepsilon < v < c + \varepsilon. \tag{1.5.10}$$

By Lemma 1.5.10 we have $r \in (c - \varepsilon, c + \varepsilon)$ and $v \in (c - \varepsilon, c + \varepsilon)$. $\qquad\square$

Another property of real numbers that might seem intuitive is the notion that the only real number arbitrarily close to both positive and negative real numbers is zero. To this end, let $\mathbb{R}^+ = (0, \infty)$ and $\mathbb{R}^- = (-\infty, 0)$.

Lemma 1.5.14: Zero, postive, and negative

A real number ℓ is equal to 0 if and only if $\ell \operatorname{acl} \mathbb{R}^+$ and $\ell \operatorname{acl} \mathbb{R}^-$.

Proof of Lemma 1.5.14. First, assume $\ell = 0$ and let $\varepsilon > 0$. Then $\varepsilon/2 \in \mathbb{R}^+$, $-\varepsilon/2 \in \mathbb{R}^-$, and

$$-\varepsilon < -\frac{\varepsilon}{2} < \ell = 0 < \frac{\varepsilon}{2} < \varepsilon. \tag{1.5.11}$$

By Lemma 1.5.10, $0 \operatorname{acl} \mathbb{R}^+$ and $0 \operatorname{acl} \mathbb{R}^-$.

Next, assume $\ell > 0$ and let $y < 0$. Then

$$y < 0 < \frac{\ell}{2} < \ell. \tag{1.5.12}$$

Since $\varepsilon_0 = \ell/2 > 0$, it follows that $|\ell - y| > \ell/2$. Therefore $\ell \operatorname{awf} \mathbb{R}^-$. Similarly, no negative number is arbitrarily close to $\mathbb{R}^+$. $\qquad\square$

To close out this chapter, we will often want to consider the set of points arbitrarily close to a given set.

Definition 1.5.15: Closure

Let $S \subseteq \mathbb{R}^m$. The *closure* of S, denoted by $\overline{S}$, is the set of points arbitrarily close to S. Thus,

$$\overline{S} = \{\mathbf{y} \in \mathbb{R}^m : \mathbf{y} \operatorname{acl} S\}. \tag{1.5.13}$$

Example 1.5.16: Arbitarily close and closure

In Figure 1.5.4, $\overline{B}$ is the closure of the rectangle B. $\overline{B}$ is a solid rectangle that contains its corners and sides, thus the corner $\mathbf{y}$ is in $\overline{B}$.

Example 1.5.17: Closed neighborhoods

There are subtle differences between ε-neighborhoods and their closures. Can you spot the differences between Figures 1.5.2 and 1.5.5?

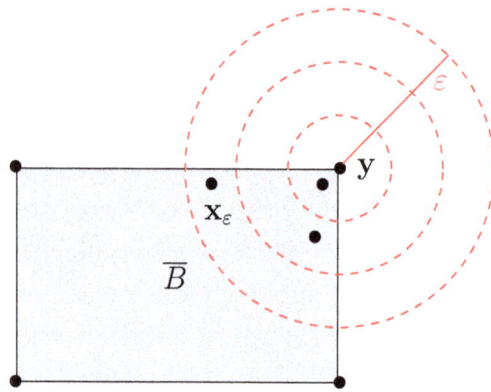

Figure 1.5.4: The closure $\overline{B}$ contains all points in and arbitrarily close to the rectangle B, including the corner $\mathbf{y}$ and the sides.

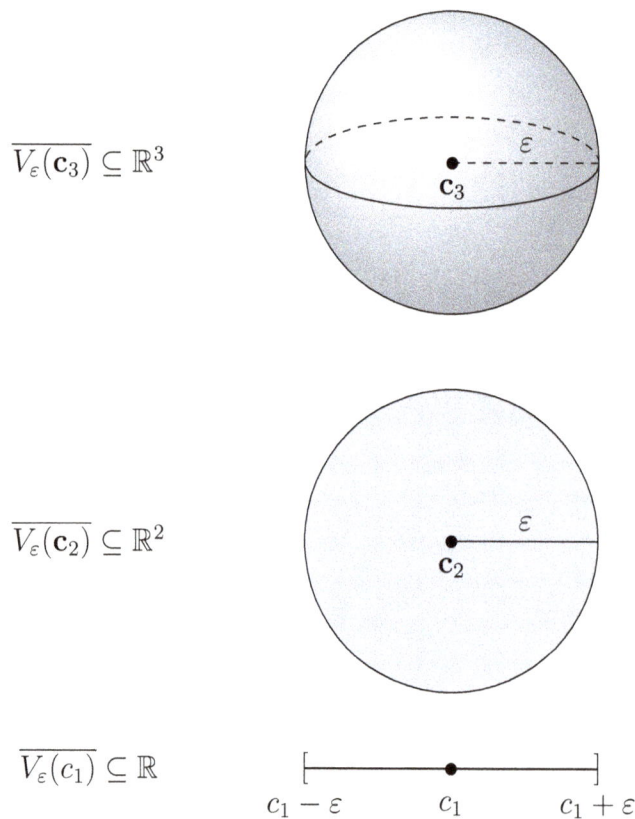

Figure 1.5.5: Closures of ε-neighborhoods centered at $\mathbf{c}_3$ in $\mathbb{R}^3$, $\mathbf{c}_2$ in $\mathbb{R}^2$, and c_1 in $\mathbb{R}$, respectively, for some fixed $\varepsilon > 0$. Which points are in these sets that are not in their counterparts in Figure 1.5.2?

- $\overline{V_\varepsilon(\mathbf{c}_3)}$ is the closed sphere of radius ε centered $\mathbf{c}_3$, including the points on the surface exactly ε away from $\mathbf{c}_3$.

- $\overline{V_\varepsilon(\mathbf{c}_2)}$ is the closed disk of radius ε centered $\mathbf{c}_2$, including the points on the circle exactly ε away from $\mathbf{c}_2$.

- $\overline{V_\varepsilon(c_1)}$ is the closed interval of length 2ε centered c_1, including the endpoints $c_1 - \varepsilon$ and $c_1 + \varepsilon$.

Remark 1.5.18: Convential notion of closure

Some of you may have seen a definition for closure in a class on real analysis, topology, or some other topic. Definition 1.5.15 is equivalent to more conventional definitions of closure, but the justification is left to Chapter 3. For now, closures give us a powerful tool for exploring properties of sets before the boss fight in Chapter 2: Defining limits and convergence for *sequences*.

Example 1.5.19: Closures of some sets

Recall the sets A, B, F, and G from previous Examples 1.1.4 and 1.1.17 from Section 1.1. What are their closures? We have:

$$
\begin{aligned}
A &= \{2 - (1/\sqrt{n}) : n \in \mathbb{N}\} &\implies& \quad \overline{A} = A \cup \{2\}, \\
B &= \{[2 - (1/\sqrt{n})](-1)^n : n \in \mathbb{N}\} &\implies& \quad \overline{B} = B \cup \{-2, 2\}, \\
F &= (-\infty, 3140] = \{x \in \mathbb{R} : x \le 3140\} &\implies& \quad \overline{F} = (-\infty, 3140] = F, \\
G &= (-\infty, 3140) = \{x \in \mathbb{R} : x < 3140\} &\implies& \quad \overline{G} = (-\infty, 3140] = F.
\end{aligned}
$$

As usual, I think figures help. See Figure 1.5.6.

Each of the sets in Example 1.5.19 is bounded above. See Figure 1.5.6. In Euclidean spaces that are not the real line, the notions of bounded above and bounded below lose meaning since there are infinitely many directions to move around in. Even so, *boundedness* is a property sets in Euclidean space can have. The idea is to bound the norms.

Definition 1.5.20: Bounded set

A set $S \subseteq \mathbb{R}^m$ is *bounded* if there is a $b \ge 0$ such that for all $\mathbf{x}$ in S we have

$$\|\mathbf{x}\|_m \le b. \tag{1.5.14}$$

In this case, b is called a *bound* for S and we say S is *bounded by b*.

Remark 1.5.21: Bounded above and below

In the real line $\mathbb{R}$, a set S is bounded if and only if it is bounded above and bounded below.

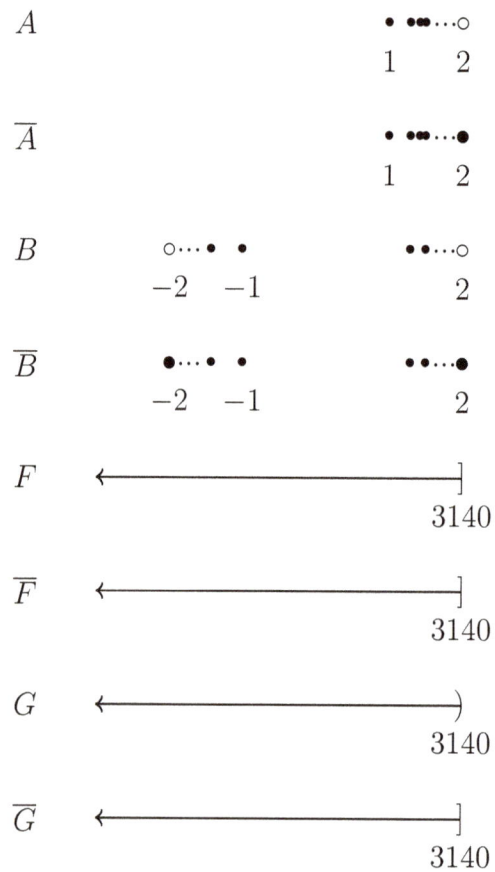

Figure 1.5.6: The sets of real numbers A, B, F, and G from Examples 1.1.4 and 1.1.17, along with their closures. In each case, the closure contains the points in the original set along with points arbitrarily close to the original set.

Figure 1.5.7: As in Lemma 1.5.23, here is a set of real numbers S along with a lower bound a, an upper bound b, and a point y where $y \operatorname{acl} S$.

This section wraps up with a trio of results about bounded sets. Lemmas 1.5.22 and 1.5.23 stem from the order inherent to the real line, while Lemma 1.5.24 pertains to bounded sets in Euclidean space. All of these results make use of arbitrarily close in some way (Definitions 1.1.8 and 1.5.1).

Lemma 1.5.22: Sided arbitrarily close

Suppose x, a, and b are real numbers. Then

(i) $x \leq b$ if and only if for every $\varepsilon > 0$ we have $x < b + \varepsilon$.

(ii) $x \geq a$ if and only if for every $\varepsilon > 0$ we have $x > a - \varepsilon$.

The proofs of statements (i) and (ii) in Lemma 1.5.22 are similar enough that we will only work on (i) here and leave (ii) as an exercise. Also, since $b < b + \varepsilon$ for every $\varepsilon > 0$, one direction of (i) has a short proof. For the other direction, a contraposition argument yields the result.

Proof of (i) in Lemma 1.5.22. First, suppose $x \leq b$ and let $\varepsilon > 0$. Then

$$x \leq b < b + \varepsilon. \tag{1.5.15}$$

For the other direction, assume $x > b$ to set up a contraposition argument. Then $x - b > 0$ and so

$$\varepsilon_0 = \frac{x - b}{2} > 0. \tag{1.5.16}$$

Hence,

$$b + \varepsilon_0 = b + \frac{x - b}{2} = \frac{x + b}{2} < \frac{x + x}{2} = x. \tag{1.5.17}$$

Therefore, there is some $\varepsilon_0 > 0$ where $x > b + \varepsilon_0$. $\square$

Lemma 1.5.23: Order and arbitrarily close

Suppose $S \subseteq \mathbb{R}$ and $y \operatorname{acl} S$.

(i) If b is an upper bound for S, then $y \leq b$.

(ii) If a is a lower bound for S, then $a \leq y$.

The statements (i) and (ii) in Lemma 1.5.23 are similar enough that we will only work on (i) here and leave (ii) as an exercise. See Figure 1.5.7.

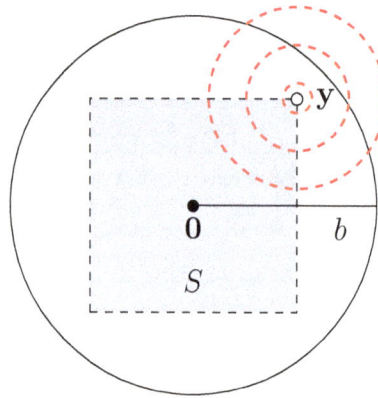

Figure 1.5.8: As in Lemma 1.5.24 and Lekha Patil's proof, a set S in the plane $\mathbb{R}^2$ bounded by a positive real number b along with a point $\mathbf{y}$ in $\mathbb{R}^2$ where $\mathbf{y}\,\mathrm{acl}\,S$ and $\|\mathbf{y}\|_m \leq b$.

Proof of (i) in Lemma 1.5.23. To set up a contraposition argument, suppose $y\,\mathrm{acl}\,S$ and $y > b$. Then $\varepsilon_b = y - b > 0$. By the definition of arbitrarily close (Definition 1.5.1), there is some x_b in S where

$$|x_b - y| < \varepsilon_b = y - b. \tag{1.5.18}$$

Therefore,

$$y - x_b \leq |x_b - y| < y - b. \tag{1.5.19}$$

So, by subtracting y then multiplying through by -1 in (1.5.19) we have

$$x_b > b. \tag{1.5.20}$$

Since x_b is in S, b is not an upper bound for S. □

The points in the closure of a bounded set are so close to the original set that there is no distance between them. As a result, a bounded set and its closure respect the same bounds. See Figure 1.5.8.

Lemma 1.5.24: A bound is a bound for the closure

Suppose $S \subseteq \mathbb{R}^m$ is bounded by b. Then $\overline{S}$ is bounded by b as well. That is, if $\mathbf{y}\,\mathrm{acl}\,S$, then

$$\|\mathbf{y}\|_m \leq b. \tag{1.5.21}$$

Lemma 1.5.24 was presented as an exercise to my undergraduate real analysis class in Summer 2021. One student, Lekha Patil, came up with an elegant proof. Her proof is provided below along with Figure 1.5.8. An alternate proof is provided thereafter.

Lekha Patil's proof of Lemma 1.5.24. Suppose $\mathbf{y}\,\mathrm{acl}\,S$ and b is a bound for S. Then for every

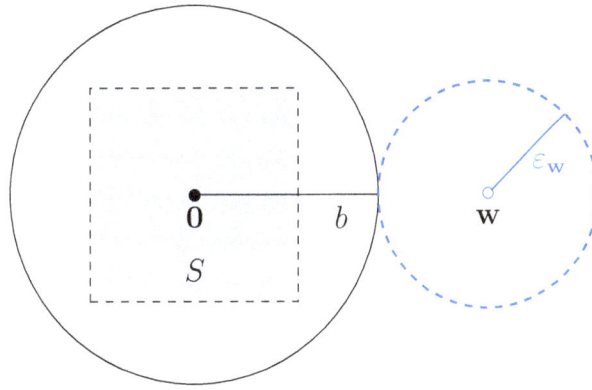

Figure 1.5.9: In the second proof of Lemma 1.5.24, a set S in the plane $\mathbb{R}^2$ bounded by a positive real number b is away from any point $\mathbf{w}$ in $\mathbb{R}^2$ where $\|\mathbf{w}\|_m > b$.

point $\mathbf{x}$ in S we have

$$\|\mathbf{y}\|_m = \|\mathbf{y} - \mathbf{x} + \mathbf{x}\|_m \tag{1.5.22}$$
$$\leq \|\mathbf{y} - \mathbf{x}\|_m + \|\mathbf{x}\|_m \quad \text{(triangle inequality (1.2.32))} \tag{1.5.23}$$
$$\leq \|\mathbf{y} - \mathbf{x}\|_m + b. \quad \text{(Definition 1.5.20)} \tag{1.5.24}$$

Now let $\varepsilon > 0$. Since $\mathbf{y}$ acl S, there is some $\mathbf{x}_\varepsilon$ in S where

$$\|\mathbf{y}\|_m \leq \|\mathbf{y} - \mathbf{x}_\varepsilon\|_m + b < \varepsilon + b. \tag{1.5.25}$$

Since $\varepsilon > 0$ is arbitrary (standing for *any* positive number, so Lemma 1.5.22 applies), we have

$$\|\mathbf{y}\|_m \leq b. \tag{1.5.26}$$

Hence, b is a bound for $\overline{S}$ as well. □

The line "Since $\varepsilon > 0$ is arbitrary" highlights a key intuitive idea behind the definition of arbitrarily close, much like Lemma 1.5.5: If the difference between two objects is less than *every* $\varepsilon > 0$, then there's no difference at all. To me, this intuition is reinforced by Lemmas 1.5.22 and 1.5.24.

An alternative proof of Lemma 1.5.24 makes use of a contraposition argument and the definition of *away from* (Definition 1.5.11). Basically, a point whose norm is greater than a bound for a set is away from the set. See Figure 1.5.9.

Alternate proof of Lemma 1.5.24. Suppose b is a bound for S and $\|\mathbf{w}\|_m > b$. For every $\mathbf{x}$ in S we have $\|\mathbf{x}\|_m \leq b$, so $-b \leq -\|\mathbf{x}\|_m$. Define

$$\varepsilon_{\mathbf{w}} = \|\mathbf{w}\|_m - b > 0. \tag{1.5.27}$$

Then we have

$$0 < \varepsilon_{\mathbf{w}} = \|\mathbf{w}\|_m - b \tag{1.5.28}$$
$$\leq \|\mathbf{w}\|_m - \|\mathbf{x}\|_m \tag{1.5.29}$$
$$\leq \|\mathbf{w} - \mathbf{x}\|_m \quad \text{(reverse tri. ineq. (1.2.37))} \tag{1.5.30}$$

Hence, $\mathbf{w}$ awf S and $\mathbf{w}$ is not in $\overline{S}$. □

The following chapter explores properties of sequences through the lens of arbitrarily close (Definition 1.5.1). In particular, a central tenet of this book is the way arbitrarily close allows us to parse the definitions of *convergence* and *limits* (Definition 2.2.1).

Exercises

1.5.1. Prove the claims about the closures in Example hold 1.5.19 hold. See Figure 1.5.6.

1.5.2. Determine the closures of the following sets.

(i) $A = (0,1)$

(ii) $B = [0,1]$

(iii) $C = \mathbb{Z}$

(iv) $D = \mathbb{Q}$

(v) $E = \{(-1)^n : n \in \mathbb{N}\}$

(vi) $F = \{(-2)^n : n \in \mathbb{N}\}$

(vii) $G = \{(-1)^n/n : n \in \mathbb{N}\}$

(viii) $H = \{(-1)^n + (1/n) : n \in \mathbb{N}\}$

1.5.3. Prove Lemma 1.5.10. Hint: Use the definition of absolute value (Definition 1.1.6) and Axiom 1.3.1, especially the Trichotomy property.

1.5.4. Prove that if $x \in \mathbb{R}$, then $x \operatorname{acl} \mathbb{Q}$ and $x \operatorname{acl} \mathbb{R}\backslash\mathbb{Q}$. Thus, both the set of rational numbers $\mathbb{Q}$ and the set of irrational numbers $\mathbb{R}\backslash\mathbb{Q}$ are *dense* in $\mathbb{R}$. (See Definition 3.6.13.)

1.5.5. Prove part (ii) of Lemma 1.5.22. The proof follows from a careful consideration and modification of the proof of part (i).

1.5.6. Prove the closure of a set is closed. That is, given $A \subseteq \mathbb{R}^m$, prove $\overline{\overline{A}} = \overline{A}$.

1.5.7. Pick a nonempty set S in the plane $\mathbb{R}^2$. Draw S and, depending on what you drew, try to draw three points as follows: One that is arbitrarily close to both S and its complement $\mathbb{R}^2\backslash S$, one that is away from S, and one that is away from $\mathbb{R}^2\backslash S$. Next, draw the three following sets: The set of all points arbitrarily close to both S and its complement $\mathbb{R}^2\backslash S$, the set of all points away from S, and the set of all points away from $\mathbb{R}^2\backslash S$. Are there any points away from both S and $\mathbb{R}^2\backslash S$?

1.6   Linearity and arbitrarily close

Linearity is a recurring theme in analysis. This section lays out some of the ways linearity appears throughout the textbook, mainly as the pair of properties called *additivity* and *homogeneity*. Most of the proofs are omitted.

Vector spaces provide the general context for linearity whose objects are called *vectors*. The following definitions may be familiar from coursework on linear algebra.

Definition 1.6.1: Vector space

A *vector space* over the real line $\mathbb{R}$ is a nonempty set V with two operations, called *addition* and *scalar multiplication*, acting on objects called *vectors* and satisfying the following ten properties for all vectors $\mathbf{u}, \mathbf{v}, \mathbf{w} \in V$ and all *scalars* $a, b \in \mathbb{R}$.

(i) $\mathbf{u} + \mathbf{v} \in V$ (*additivity*).

(ii) $a\mathbf{u} \in V$ (*homogeneity*).

(iii) There is a vector $\mathbf{0} \in V$ called the *zero vector* such that for all $\mathbf{u} \in V$ we have $\mathbf{u} + \mathbf{0} = \mathbf{u}$ (*additive identity*).

(iv) $\mathbf{u} + \mathbf{v} = \mathbf{v} + \mathbf{u}$ (*commutativity of vector addition*).

(v) $(\mathbf{u} + \mathbf{v}) + \mathbf{w} = \mathbf{u} + (\mathbf{v} + \mathbf{w})$ (*associativity of vector addition*).

(vi) $a(\mathbf{u} + \mathbf{v}) = a\mathbf{u} + a\mathbf{v}$ (*distributivity of scalars*).

(vii) $(a + b)\mathbf{u} = a\mathbf{u} + b\mathbf{v}$ (*distributivity of vectors*).

(viii) $a(b\mathbf{u}) = (ab)\mathbf{u}$ (*associativity of scalar multiplication*).

(ix) For every $\mathbf{u} \in V$ there is vector $-\mathbf{u} \in V$ called the *negative of* $\mathbf{u}$ where $\mathbf{u} + (-\mathbf{u}) = \mathbf{0}$ (*additive inverse*).

(x) $1\mathbf{u} = \mathbf{u}$ (*scalar identity*).

In the context of analysis, *linear combinations* of certain vectors are rich enough to be arbitrarily close to all objects in a given vector space while being malleable enough to work with.

Definition 1.6.2: Linear combination and span

A *linear combination* is a finite sum of scaled vectors. That is, linear combinations are of the form

$$\sum_{j=1}^{m} a_j\mathbf{v}_j = a_1\mathbf{v}_1 + a_2\mathbf{v} + \cdots + a_m\mathbf{v}_m \tag{1.6.1}$$

where $m \in \mathbb{N}$, $a_j \in \mathbb{R}$, and $\mathbf{v}_j \in V$ where V is a vector space.

The *span* of a set of vectors $A \subseteq V$, denoted by Span (A), is the set of all linear combinations of vectors in A. That is,

$$\text{Span}\,(A) = \left\{ \sum_{j=1}^{m} a_j\mathbf{v}_j : m \in \mathbb{N}, \mathbf{v}_j \in A \right\}. \tag{1.6.2}$$

Vector spaces contain all linear combinations of their vectors. The next lemma formalizes this result, but its proof is omitted since it follows from an induction argument using parts (i) and (ii) of Definition 1.6.1.

> **Lemma 1.6.3: Vector spaces contain linear combinations**
>
> If V is a vector space, $m \in \mathbb{N}$, and $\mathbf{v}_j \in V$ for each $j = 1, \ldots, m$, then the linear combination $\sum_{j=1}^{m} a_j \mathbf{v}_j$ is also in V.

Euclidean spaces are fundamental examples of vector spaces.

> **Example 1.6.4: Euclidean spaces and vector spaces**
>
> The following sets are vector spaces.
>
> - The real line $\mathbb{R}$ is a vector space where each real number is both a vector and a scalar.
>
> - More generally, every Euclidean space $\mathbb{R}^m$ is a vector space.
>
> - The set of sequences of real numbers denoted by $\mathbb{R}^{\mathbb{N}}$ is a vector space where vector addition amounts to term-by-term addition.
>
> - More generally, the set of sequences of vectors in a Euclidean space is a vector space.
>
> - Given a nonempty set A, the set of all functions from the common domain A to a Euclidean space $\mathbb{R}^m$ is a vector space. The special case of functions from A to the real line $\mathbb{R}$ is denoted by $\mathbb{R}^A$.

Remark 1.6.5: Vectors are functions

Although it may not look like it, every vector space in Example 1.6.4 is a special case of a set of functions. For instance, we have:

$$\mathbb{R}^m = \left\{ \mathbf{x} = \begin{bmatrix} x_1 \\ x_2 \\ \vdots \\ x_m \end{bmatrix} : x_1, x_2, \ldots, x_m \in \mathbb{R} \right\} \tag{1.6.3}$$

$$= \{ f : f : \{1, \ldots, m\} \to \mathbb{R} \} \tag{1.6.4}$$

and

$$\mathbb{R}^{\mathbb{N}} = \{ (x_n) : (x_n) \subseteq \mathbb{R} \} \tag{1.6.5}$$

$$= \{ f : f : \mathbb{N} \to \mathbb{R} \}. \tag{1.6.6}$$

Subspaces are subsets of vector spaces which are themselves vector spaces. Lots of them pop up in analysis.

Definition 1.6.6: Subspace

A subset S of a vector space V is a *subspace* of V if S is a vector space whose addition and scalar multiplication are given by those of V.

Given a subset of a vector space, there is a relatively simple way to determine if the subset is a subspace. The following lemma codifies this notion, but the proof is omitted.

Lemma 1.6.7: Determining subspaces

A subset S of a vector space V is a subspace of V if S satisfies each of the following conditions.

(i) S is *closed under vector addition*: If $\mathbf{u}, \mathbf{v} \in S$, then $\mathbf{u} + \mathbf{v} \in S$.

(ii) S is *closed under scalar multiplication*: If $a \in \mathbb{R}$ and $\mathbf{u} \in S$, then $a\mathbf{u} \in S$.

(iii) S contains $\mathbf{0}$, the zero vector of V: $\mathbf{0} \in S$.

Polynomials are linear combinations of *monomials*.

Definition 1.6.8: Polynomials and monomials

A *monomial* is a function $f_n : \mathbb{R} \to \mathbb{R}$ given by

$$f_n(x) = \begin{cases} 1, & \text{if } n = 0, \\ x^n, & \text{if } n \in \mathbb{N}. \end{cases} \tag{1.6.7}$$

A *polynomial* is a linear combination of monomials. That is, a polynomial is a function $p : \mathbb{R} \to \mathbb{R}$ given by

$$p(x) = \sum_{j=0}^{n} a_j x^j = a_0 + a_1 x + a_2 x^2 + \cdots + a_{n-1} x^{n-1} + a_n x^n \tag{1.6.8}$$

where $n \in \mathbb{N} \cup \{0\}$, the coefficients $a_0, a_1, a_2, \ldots, a_{n-1}, a_n$ are real numbers, and the convention $x^0 = 1$ is used.

Remark 1.6.9: The set of polynomials is a span

The set of all polynomials over the real line $\mathbb{R}$ is the span of the set of monomials. That is, we have

$$p(x) \in \text{Span}\{x^k : k \in \mathbb{N}\} \qquad \Longleftrightarrow \qquad p(x) = \sum_{j=0}^{n} a_j x^j \tag{1.6.9}$$

where the coefficients $a_0, a_1, a_2, \ldots, a_{n-1}, a_n$ are real numbers.

Vector spaces of functions are plentiful and appear throughout the textbook. Given $a, b \in \mathbb{R}$

where $a < b$, the set of functions from the closed and bounded interval $[a, b]$ to the real line $\mathbb{R}$, denoted by $\mathbb{R}^{[a,b]}$, is a vector space with a wealth of interrelated subspaces. Relationships between these subspaces are explored throughout the textbook.

Example 1.6.10: Subspaces of $\mathbb{R}^{[a,b]}$

Each of the following sets is a subspace of the vector space of real-valued functions $\mathbb{R}^{[a,b]}$ with a common domain $[a, b]$:

- $C[a, b]$, the set of real-valued continuous functions on $[a, b]$ (see Definition 4.3.2).

- $D[a, b]$, the set of real-valued differentiable functions on $[a, b]$ (see Definition 5.3.1).

- $C^n[a, b]$ where $n \in \mathbb{N} \cup \{0\}$, the set of real-valued functions whose nth derivative exists and is continuous on $[a, b]$. By convention, $C^0[a, b] = C[a, b]$.

- $C^\infty[a, b]$, the set of real-valued functions whose nth derivative exists for every $n \in \mathbb{N}$.

- $R[a, b]$, the set of integrable functions over $[a, b]$ (see Definition 6.1.6).

Also, these subspaces satisfy the following string of strict containments: For all $n \in \mathbb{N}$, we have

$$C^\infty(a, b) \subsetneq C^n[a, b] \subsetneq D[a, b] \subsetneq C[a, b] \subsetneq R[a, b] \subsetneq \mathbb{R}^{[a,b]}. \qquad (1.6.10)$$

Connections between linearity, linear combinations, span, subspaces, and arbitrarily close permeate analysis. One unconventional connection that seems to belong in this section is the notion of arbitrarily close exhibiting its own version of linearity when it comes to sums of sets. This idea also sets the stage for many results stated later on.

Definition 1.6.11: Linear combination of sets

Suppose V is a vector space over the real line, $A, B \subseteq V$, and $c \in \mathbb{R}$. The *sum of sets* A and B, denoted by $A + B$, is defined by

$$A + B = \{\mathbf{x} + \mathbf{y} : \mathbf{x} \in A \text{ and } \mathbf{y} \in B\}. \qquad (1.6.11)$$

The *scaled set* denoted by cA is defined by

$$cA = \{c\mathbf{x} : \mathbf{x} \in A\}. \qquad (1.6.12)$$

Finally, a *linear combination of sets* is a finite sum of scaled sets. That is, given $n \in \mathbb{N}$, sets $A_1, \ldots, A_n \subseteq V$, and scalars $\{c_1, \ldots, c_n\} \subseteq \mathbb{R}$, the set given by

$$\sum_{j=1}^{n} c_j A_j = c_1 A_1 + c_2 A_2 + \cdots + c_{n-1} A_{n-1} + c_n A_n \qquad (1.6.13)$$

is a linear combination of sets.

Next up is a look at the interplay between linearity and arbitrarily close. These results are not terribly important in the sense that they do not appear elsewhere in the textbook, but the scratch work and arguments for their proofs provide an amuse-bouche for similar results throughout analysis.

Theorem 1.6.12: Linearity of arbitrarily close

Suppose $\mathbf{x}, \mathbf{y} \in \mathbb{R}^m$, and $A, B \subseteq \mathbb{R}^m$.

  (i) If $\mathbf{x}\,\mathrm{acl}\,A$ and $\mathbf{y}\,\mathrm{acl}\,B$, then $(\mathbf{x} + \mathbf{y})\,\mathrm{acl}\,(A + B)$ (*additivity*).

  (ii) If $c \in \mathbb{R}$ and $\mathbf{x}\,\mathrm{acl}\,A$, then $(c\mathbf{x})\,\mathrm{acl}\,(cA)$ (*homogeneity*).

Remark 1.6.13: Recurring techniques

To prove each statement in Theorem 1.6.12, we can verify the definition of arbitrarily close (Definition 1.5.1) holds by considering an arbitrary $\varepsilon > 0$ and finding a suitable point from the set under consideration within the distance ε of the given point. Recurring techniques used whenever linearity is to be shown in some way include starting scratch work at the end, splitting the distance ε into smaller chunks or scaling it by some positive amount, applying definitions that yield distances each within some chunk or scaling of ε, then using some form of the triangle inequality (1.2.32) to bring the argument together.

Scratch Work 1.6.14: Additivity of arbitrarily close

To derive some scratch work for directly showing

$$(\mathbf{x} + \mathbf{y})\,\mathrm{acl}\,(A + B), \tag{1.6.14}$$

let's start at the end. Given $\varepsilon > 0$, we want to end up with $\mathbf{a} \in A$ and $\mathbf{b} \in B$ where

$$\|(\mathbf{a} + \mathbf{b}) - (\mathbf{x} + \mathbf{y})\|_m < \varepsilon. \tag{1.6.15}$$

We can assume

$$\mathbf{x}\,\mathrm{acl}\,A \quad \text{and} \quad \mathbf{y}\,\mathrm{acl}\,B, \tag{1.6.16}$$

so by Definition 1.5.1 applied twice, given any $\varepsilon > 0$ there are points $\mathbf{a}_\varepsilon \in A$ and $\mathbf{b}_\varepsilon \in B$ where

$$\|\mathbf{a}_\varepsilon - \mathbf{x}\|_m < \varepsilon \quad \text{and} \quad \|\mathbf{b}_\varepsilon - \mathbf{y}\|_m < \varepsilon. \tag{1.6.17}$$

Combining both inequalities in (1.6.17) and applying triangle inequality (1.2.35) gives us

$$\begin{aligned}
\|(\mathbf{a}_\varepsilon + \mathbf{b}_\varepsilon) - (\mathbf{x} + \mathbf{y})\|_m &= \|(\mathbf{a}_\varepsilon - \mathbf{x}) + (\mathbf{b}_\varepsilon - \mathbf{y})\|_m & (1.6.18) \\
&\leq \|\mathbf{a}_\varepsilon - \mathbf{x}\|_m + \|\mathbf{b}_\varepsilon - \mathbf{y})\|_m & (1.6.19) \\
&< \varepsilon + \varepsilon & (1.6.20) \\
&= 2\varepsilon. & (1.6.21)
\end{aligned}$$

This doesn't quite get us to our goal (1.6.15), but we can adapt: The definition of arbitrarily close (Definition 1.5.1) ensures that we can respond to *any positive distance* we like with suitable points from the set. So, split ε in half and consider $\varepsilon/2 > 0$ and apply the definition of arbitrarily close twice, yielding $\mathbf{a}_{\varepsilon/2} \in A$ and $\mathbf{b}_{\varepsilon/2} \in B$ where

$$\|\mathbf{a}_{\varepsilon/2} - \mathbf{x}\|_m < \frac{\varepsilon}{2} \qquad \text{and} \qquad \|\mathbf{b}_{\varepsilon/2} - \mathbf{y}\|_m < \frac{\varepsilon}{2}. \tag{1.6.22}$$

From there, the sum $\mathbf{a}_{\varepsilon/2} + \mathbf{b}_{\varepsilon/2}$ is both in $A + B$ and, thanks to the triangle inequality (1.2.35), within ε of $\mathbf{x} + \mathbf{y}$.

Proof of (i) in Theorem 1.6.12. Suppose $\mathbf{x}, \mathbf{y} \in \mathbb{R}^m$ and $A, B \subseteq \mathbb{R}^m$ where $\mathbf{x} \operatorname{acl} A$ and $\mathbf{y} \operatorname{acl} B$. Let $\varepsilon > 0$. Since $\varepsilon/2 > 0$, the definition for arbitrarily close (Definition 1.5.1) applies and there exist $\mathbf{a}_{\varepsilon/2} \in A$ and $\mathbf{b}_{\varepsilon/2} \in B$ where

$$\|\mathbf{a}_{\varepsilon/2} - \mathbf{x}\|_m < \frac{\varepsilon}{2} \qquad \text{and} \qquad \|\mathbf{b}_{\varepsilon/2} - \mathbf{y}\|_m < \frac{\varepsilon}{2}. \tag{1.6.23}$$

Then by the definition of the sum of sets (Definition 1.6.11) and the triangle inequality (1.2.35), we have $(\mathbf{a}_{\varepsilon/2} + \mathbf{b}_{\varepsilon/2}) \in (A + B)$ and

$$\|(\mathbf{a}_{\varepsilon/2} + \mathbf{b}_{\varepsilon/2}) - (\mathbf{x} + \mathbf{y})\|_m = \|(\mathbf{a}_{\varepsilon/2} - \mathbf{x}) + (\mathbf{b}_{\varepsilon/2} - \mathbf{y})\|_m \tag{1.6.24}$$
$$\leq \|\mathbf{a}_{\varepsilon/2} - \mathbf{x}\|_m + \|\mathbf{b}_{\varepsilon/2} - \mathbf{y})\|_m \tag{1.6.25}$$
$$< \frac{\varepsilon}{2} + \frac{\varepsilon}{2} \tag{1.6.26}$$
$$= \varepsilon. \tag{1.6.27}$$

Therefore, $(\mathbf{x} + \mathbf{y}) \operatorname{acl} (A + B)$. $\square$

Next, let's prove the homogeneity of arbitrarily close, part (ii) of Theorem 1.6.12.

Scratch Work 1.6.15: Homogeneity of arbitrarily close

Once again, let's start at the end. Given a distance $\varepsilon > 0$ and a scalar $c \in \mathbb{R}$, we want to end up a point $\mathbf{a} \in A$ where

$$\|c\mathbf{a} - c\mathbf{x}\|_m < \varepsilon. \tag{1.6.28}$$

By the homogeneity of the Euclidean norm (1.2.33), we have

$$\|c\mathbf{a} - c\mathbf{x}\|_m = |c|\|\mathbf{a} - \mathbf{x}\|_m < \varepsilon. \tag{1.6.29}$$

So if $c \neq 0$, then $|c| \neq 0$ as well and we can divide both sides of the rightmost inequality in (1.6.29) to get

$$\|\mathbf{a} - \mathbf{x}\|_m < \frac{\varepsilon}{|c|}. \tag{1.6.30}$$

The definition for arbitrarily close (Definition 1.5.1) ensures a point $\mathbf{a}_{\varepsilon/|c|} \in A$ can be found in response to the positive distance $\varepsilon/|c|$ that will suffice, as long as $c \neq 0$.

Proof of part (ii) in Theorem 1.6.12. This proof has two cases: (i) $c = 0$ and (ii) $c \neq 0$.

Case (i), $c = 0$: Suppose $c = 0$, $\mathbf{x}\operatorname{acl} A$, and $\varepsilon > 0$. Then for every $\mathbf{a} \in A$ (or $\mathbb{R}^m$ for that matter), we have

$$cA = 0A = \{\mathbf{0}\} \qquad \text{and} \qquad \|c\,\mathbf{a} - c\,\mathbf{x}\|_m = \|\mathbf{0} - \mathbf{0}\|_m = 0 < \varepsilon. \tag{1.6.31}$$

Since $\mathbf{0}\operatorname{acl}\{\mathbf{0}\}$ by Lemma 1.5.4, we have $(c\mathbf{x})\operatorname{acl}(cA)$.

Case (ii), $c \neq 0$: Suppose $c \neq 0$, $\mathbf{x}\operatorname{acl} A$, and $\varepsilon > 0$. Then $\varepsilon/|c| > 0$ and by the definition of arbitrarily close (Definition 1.5.1), there is a point $\mathbf{a}_{\varepsilon/|c|} \in A$ such that

$$\|\mathbf{a}_{\varepsilon/|c|} - \mathbf{x}\|_m < \frac{\varepsilon}{|c|}. \tag{1.6.32}$$

So, by the definition for scaled sets (Definition 1.6.11) and the homogeneity of the Euclidean norm (1.2.33), we have $c\,\mathbf{a}_{\varepsilon/|c|} \in cA$ and

$$\|c\,\mathbf{a}_{\varepsilon/|c|} - c\,\mathbf{x}\|_m = |c|\|\mathbf{a}_{\varepsilon/|c|} - \mathbf{x}\|_m < |c|\frac{\varepsilon}{|c|} = \varepsilon. \tag{1.6.33}$$

Therefore, $(c\,\mathbf{x})\operatorname{acl}(cA)$. $\qquad\square$

The linearity of arbitrarily close established in Theorem 1.6.12 extends to linear combinations via induction. Actually, the same is true for the numerous results on linearity throughout the textbook.

Corollary 1.6.16: Arbitrarily close and linear combinations of sets

Suppose $A_j \subseteq \mathbb{R}^m$ for each $j = 1,\dots,n$ where $n \in \mathbb{N}$. If $\mathbf{x}_j\operatorname{acl} A_j$ and $c_j \in \mathbb{R}$ for each $j = 1,\dots,n$, then

$$\left(\sum_{j=1}^n c_j\mathbf{x}_j\right)\operatorname{acl}\left(\sum_{j=1}^n c_jA_j\right). \tag{1.6.34}$$

Scratch Work 1.6.17: Induction extends linearity to linear combinations

An induction argument allows us to extend the linearity of arbitrarily close to linear combinations of sets. This extension applies to all the other notions of linearity explored later in the textbook.

Proof of Corollary 1.6.16. The base case follows from the linearity of arbitrarily close (Theorem 1.6.12) and the inductive case follows from the base case.

Base case: Suppose $r, s \in \mathbb{R}$, $A, B \subseteq \mathbb{R}^m$, $\mathbf{x}\operatorname{acl} A$, and $\mathbf{y}\operatorname{acl} B$. By the homogeneity of arbitrarily close (part (ii) of Theorem 1.6.12), we have $(r\mathbf{x})\operatorname{acl}(rA)$ and $(s\mathbf{y})\operatorname{acl}(sB)$. So by the additivity of arbitrarily close (part (i) of Theorem 1.6.12), we have

$$(r\mathbf{x} + s\mathbf{y})\operatorname{acl}(rA + sB). \tag{1.6.35}$$

Inductive case: Suppose $c_j \in \mathbb{R}$, $A_j \subseteq \mathbb{R}^m$, and $\mathbf{x}_j \operatorname{acl} A_j$ for each $j = 1, \ldots, k+1$. For the inductive hypothesis, suppose

$$\left(\sum_{j=1}^{k} c_j \mathbf{x}_j \right) \operatorname{acl} \left(\sum_{j=1}^{k} c_j A_j \right) \qquad \text{and} \qquad \mathbf{x}_{k+1} \operatorname{acl} A_{k+1}. \tag{1.6.36}$$

Set $r = 1, s = c_{k+1}, \mathbf{x} = \sum_{j=1}^{k} c_j \mathbf{x}_j, \mathbf{y} = \mathbf{x}_{k+1}, A = \sum_{j=1}^{k} c_j A_j$, and $B = A_{k+1}$. Then

$$r\mathbf{x} + s\mathbf{y} = \sum_{j=1}^{k+1} c_j \mathbf{x}_j \qquad \text{and} \qquad rA + sB = \sum_{j=1}^{k+1} c_j A_j. \tag{1.6.37}$$

So, by the inductive hypothesis and the base case we have

$$\left(\sum_{j=1}^{k+1} c_j \mathbf{x}_j \right) \operatorname{acl} \left(\sum_{j=1}^{k+1} c_j A_j \right). \tag{1.6.38}$$

Therefore, for every $n \in \mathbb{N}$, if $\mathbf{x}_j \operatorname{acl} A_j$ and $c_j \in \mathbb{R}$ for each $j = 1, \ldots, n$, then

$$\left(\sum_{j=1}^{n} c_j \mathbf{x}_j \right) \operatorname{acl} \left(\sum_{j=1}^{n} c_j A_j \right). \tag{1.6.39}$$

$\square$

Remark 1.6.18: Linearity throughout analysis

Notions of linearity like those for arbitrarily close in Theorem 1.6.12 and its extension to linear combinations in Corollary 1.6.16 hold for all kinds of concepts in analysis. As in the preface (Chapter 0), here's a list of some of the results found throughout the textbook.

- Linearity of sequential limits, Theorem 2.3.9 and Corollary 2.3.13.

- Linearity of continuity, Theorem 4.5.5 and Corollary 4.5.7.

- Linearity of functional limits, Theorem 5.2.6 and Corollary 5.2.8.

- Linearity of differentiation, Theorem 5.4.1 and Corollary 5.4.3.

- Linearity of integration, Theorem 6.3.6 and Corollary 6.3.9.

- Linearity of uniform convergence, Theorem 7.2.8 and Corollary 7.2.10.

For readers familiar with the fruitful combination of linear algebra and analysis in the context of Hilbert spaces, this chapter concludes with a definition for *orthonormal basis* in terms of orthonormal vectors, span, and arbitrarily close.

Definition 1.6.19: Orthonormal basis

An *orthonormal basis* of a Hilbert space is a set of orthonormal vectors whose span is arbitrarily close to every vector in the space.

Chapter 2

Sequences, Tails, and Limits

The technical definitions of *limit* and *convergence* (Definition 2.2.1) along with *continuity* (Definition 4.3.2) are among the most difficult in undergraduate mathematics. As explored in this chapter, the definition of arbitrarily close (Definitions 1.1.8 and 1.5.1) provides a stepping stone towards understanding these challenging concepts.

If you're interested, you might like to take a look at [3, 4, 6, 9, 11] for various approaches to address this difficulty. Also see [5, 9] for a thorough discussion of the challenges that come with teaching convergence and limits.

This chapter explores limits and convergence for sequences, highlighted by Definition 2.2.1. We start with a formal definition of a sequence (Definition 2.1.1) and an investigation of what it means for points to be arbitrarily close to sequences.

2.1   Sequences and tails

One way to think of a sequence is as an unending list of objects. More specifically, a sequence imbues two structures on a collection of objects: (i) The objects are listed in ordered (first, second, third...); and (ii) the list is infinitely long, even if some objects are repeated.

Formally, a *sequence* is a function from the set of positive integers $\mathbb{N}$ into some nonempty set. The ordering of the domain $\mathbb{N}$ becomes an ordering on the range of the function, generating the *terms* of the sequence. This ordering pairs with the notion of *tails* to expand the notion of arbitrarily close to *limits* and *convergence*.

Definition 2.1.1: Sequence, index, term, and range

A *sequence* $(\mathbf{x}_n)$ is a function whose domain is $\mathbb{N}$. For a nonempty set $B \subseteq \mathbb{R}^m$, a *sequence of points in B* is a function $f : \mathbb{N} \to B$. Each input $n \in \mathbb{N}$ is called an *index* and has a corresponding output called a *term* written

$$\mathbf{x}_n = f(n). \tag{2.1.1}$$

The *range* of a sequence is the set of terms given by

$$f(\mathbb{N}) = \{\mathbf{x}_n \in B : n \in \mathbb{N} \text{ and } \mathbf{x}_n = f(n)\}. \tag{2.1.2}$$

Remark 2.1.2: Conventional notation for sequences

Following convention, sequences in this book are written $(\mathbf{x}_n)$ or (x_n) when the terms are real numbers. We also write

$$(\mathbf{x}_n) = (\mathbf{x}_1, \mathbf{x}_2, \mathbf{x}_3, \ldots) \tag{2.1.3}$$

to help us expand the terms and identify patterns.

The conventional notation $(\mathbf{x}_n)$ suppresses the fact that a sequence represents a function but keeps the ordering inherited from $\mathbb{N}$ intact. Technically, the range $f(\mathbb{N})$ does not repeat elements and does not have any particular ordering on its own.

Throughout the book, figures for sequences of real numbers are sometimes graphs with both inputs (positive integers) on a horizontal axis and outputs (terms) on a vertical axis. Other times, figures for sequences are just plots of the ranges.

Notation 2.1.3: Sequence arbitrarily close to a point

Following another convention which is a mild abuse of notation, for a sequence $(\mathbf{x}_n)$ defined by $f : \mathbb{N} \to B$ we take the phrase "$\mathbf{y}$ is arbitrarily close to the sequence $(\mathbf{x}_n)$" to mean the point $\mathbf{y}$ is arbitrarily close to the range $f(\mathbb{N})$. So,

$$\mathbf{y} \operatorname{acl} (\mathbf{x}_n) \quad \text{means} \quad \mathbf{y} \operatorname{acl} f(\mathbb{N}). \tag{2.1.4}$$

Similarly,

$$(\mathbf{x}_n) \subseteq \mathbb{R}^m \quad \text{means} \quad f(\mathbb{N}) \subseteq \mathbb{R}^m. \tag{2.1.5}$$

The following examples give us a basis for comparison throughout this section. They will help us build a bridge from *arbitrarily close* to *limits* for sequences with a keystone highlighted by their differences with their *tails*.

Example 2.1.4: Two sequences of real numbers

Consider the sequences of real numbers (a_n) and (b_n) defined for each $n \in \mathbb{N}$ by

$$a_n = 2 - \frac{1}{\sqrt{n}} \quad \text{and} \quad b_n = \left(2 - \frac{1}{\sqrt{n}}\right)(-1)^n. \tag{2.1.6}$$

See Figure 2.1.1 which is nearly indentical to Figure 1.1.5 from Example 1.1.17. I claim $2 \operatorname{acl} (a_n)$ and $2 \operatorname{acl} (b_n)$.

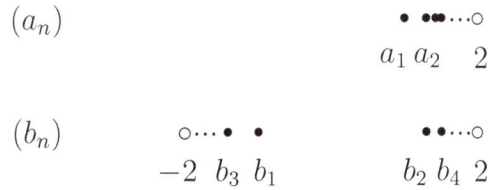

Figure 2.1.1: The ranges of the sequences (a_n) and (b_n) from Example 2.1.4 are the sets A and B, respectively, from Example 1.1.17. The black dots, like •, represent the terms of the sequences while the circles, like ∘, represent points arbitrarily close to the sequence but are not terms.

In Example 1.1.17, we proved $2 \operatorname{acl} A$ and $2 \operatorname{acl} B$. Since A is the range of (a_n) and B is the range of (b_n), we can already conclude $2 \operatorname{acl}(a_n)$ and $2 \operatorname{acl}(b_n)$. However, revisiting some of the scratch work and proofs will help us parse the definition of convergence using arbitrarily close and tails.

Scratch Work 2.1.5: Start at the end

To show $2 \operatorname{acl}(a_n)$, we should consider an unspecified $\varepsilon > 0$ and end up with a term $a_n = 2 - (1/\sqrt{n})$ where

$$|a_n - 2| = \left|\left(2 - \frac{1}{\sqrt{n}}\right) - 2\right| = \frac{1}{\sqrt{n}} < \varepsilon. \tag{2.1.7}$$

See Figure 2.1.2. We can find a suitable term by solving the rightmost inequality for the index n. We get

$$\frac{1}{\sqrt{n}} < \varepsilon \quad \Longleftrightarrow \quad n > \frac{1}{\varepsilon^2}. \tag{2.1.8}$$

From there, we can choose our index n_ε to be any positive integer large enough to satisfy $n_\varepsilon > 1/\varepsilon^2$ to give us a suitable term $a_{n_\varepsilon} = 2 - (1/\sqrt{n_\varepsilon})$.

Proof for $2 \operatorname{acl}(a_n)$ in Example 2.1.4. Let $\varepsilon > 0$. Choose an index $n_\varepsilon \in \mathbb{N}$ large enough to satisfy $n_\varepsilon > 1/\varepsilon^2$. We have

$$n_\varepsilon > \frac{1}{\varepsilon^2} \quad \Longleftrightarrow \quad \frac{1}{\sqrt{n_\varepsilon}} < \varepsilon. \tag{2.1.9}$$

Hence, the term $a_{n_\varepsilon} = 2 - (1/\sqrt{n_\varepsilon})$ is within ε of 2:

$$d_{\mathbb{R}}(a_{n_\varepsilon}, 2) = |a_{n_\varepsilon} - 2| = \left|\left(2 - \frac{1}{\sqrt{n_\varepsilon}}\right) - 2\right| = \frac{1}{\sqrt{n_\varepsilon}} < \varepsilon. \tag{2.1.10}$$

Therefore, $2 \operatorname{acl}(a_n)$. $\square$

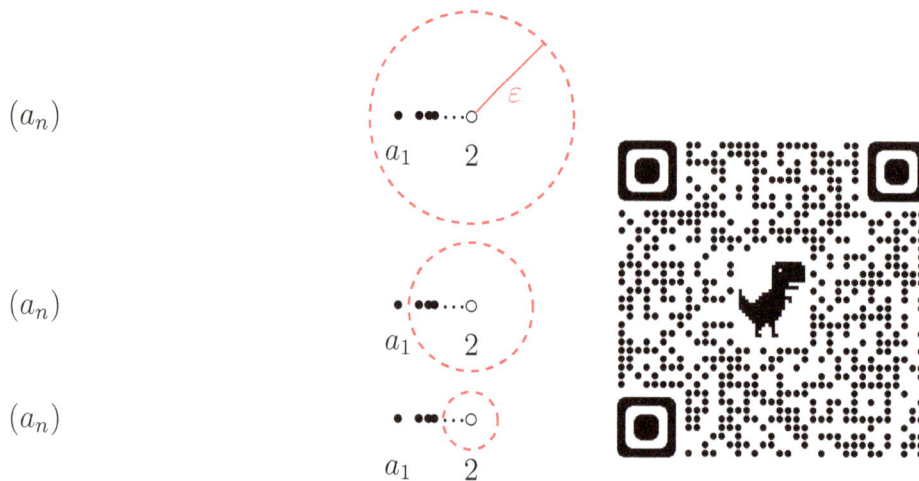

Figure 2.1.2: The real number 2 is arbitrarily close to the sequence (a_n) from Example 2.1.4. For every distance $\varepsilon > 0$, an infinite number of the a_n terms are within ε of 2 *and* only a finite number of the terms are more than ε away from 2. So, every ε-neighborhood $V_\varepsilon(2)$ contains a tail of (a_n). Also, use the QR code to play around with the Desmos activity "2 acl A". https://www.desmos.com/calculator/ceczc717wa

Figure 2.1.3: The real number 2 is arbitrarily close to the sequence (b_n) from Example 2.1.4. For every distance $\varepsilon > 0$, an infinite number of the b_n terms— with even indices—are within ε of 2. Unlike the sequence (a_n), an infinite number of the b_n terms—with odd indices—are more than ε away from 2 when ε is relatively small. For these small ε, no ε-neighborhood $V_\varepsilon(2)$ contains a tail of (b_n). Also, use the QR code to play around the Desmos activity "2 acl B". https://www.desmos.com/calculator/ysaphndtqh

The scratch work and proof showing $2\,\mathrm{acl}\,(b_n)$ are similar to those for $2\,\mathrm{acl}\,(a_n)$, except the parity of the index $n \in \mathbb{N}$ as either an even or odd integer plays a key role. See the scratch work and proof showing $2\,\mathrm{acl}\,B$ in Example 1.1.17, if you'd like a refresher.

Scratch Work 2.1.6: Start at the end

To show $2\,\mathrm{acl}\,(b_n)$, we should consider an unspecified $\varepsilon > 0$ and end up with a term $b_n = [2 - (1/\sqrt{n})](-1)^n$ where

$$|b_n - 2| < \varepsilon. \tag{2.1.11}$$

See Figure 2.1.3.

In this example, the term b_n and distance $|b_n - 2|$ do not simplify readily since b_n depends on whether the index n is even or odd. More specifically,

$$(-1)^n = \begin{cases} -1, & \text{when } n \text{ is odd,} \\ 1, & \text{when } n \text{ is even,} \end{cases} \tag{2.1.12}$$

which implies

$$b_n = \left(2 - \frac{1}{\sqrt{n}}\right)(-1)^n = \begin{cases} -2 + \dfrac{1}{\sqrt{n}}, & \text{when } n \text{ is odd,} \\[2mm] 2 - \dfrac{1}{\sqrt{n}}, & \text{when } n \text{ is even.} \end{cases} \tag{2.1.13}$$

Hence, the distance $|b_n - 2|$ depends on the parity of n as well:

$$|b_n - 2| = \left|\left(2 - \frac{1}{\sqrt{n}}\right)(-1)^n - 2\right| = \begin{cases} 4 - \dfrac{1}{\sqrt{n}}, & \text{when } n \text{ is odd,} \\[2mm] \dfrac{1}{\sqrt{n}}, & \text{when } n \text{ is even.} \end{cases} \tag{2.1.14}$$

So, in order to ensure $|b_n - 2| < \varepsilon$ for *any* given distance $\varepsilon > 0$, we should consider only terms b_n whose indices are even since we can make the distance $|b_n - 2| = 1/\sqrt{n}$ as small as we like. Therefore, we should choose an index n which is both even *and* large enough to give us a term b_n within ε of 2.

When n is even we have $|b_n - 2| = 1/\sqrt{n}$, so we can determine how large the index n needs to be by solving the following inequality for n:

$$|b_n - 2| = \left|\left(2 - \frac{1}{\sqrt{n}}\right)(-1)^n - 2\right| = \frac{1}{\sqrt{n}} < \varepsilon \quad \Longleftrightarrow \quad n > \frac{1}{\varepsilon^2}. \tag{2.1.15}$$

From there, we should choose our index n_ε to be an *even* positive integer large enough to satisfy $n_\varepsilon > 1/\varepsilon^2$. This index n_ε gives us a suitable term

$$b_{n_\varepsilon} = \left(2 - \frac{1}{\sqrt{n_\varepsilon}}\right)(-1)^{n_\varepsilon} = 2 - \frac{1}{\sqrt{n_\varepsilon}}. \tag{2.1.16}$$

Proof for $2 \operatorname{acl}(b_n)$ *in Example 2.1.4.* Let $\varepsilon > 0$. Choose an *even* index n_ε large enough to satisfy $n_\varepsilon > 1/\varepsilon^2$. We have

$$n_\varepsilon > \frac{1}{\varepsilon^2} \quad \Longleftrightarrow \quad \frac{1}{\sqrt{n_\varepsilon}} < \varepsilon. \tag{2.1.17}$$

Since $(-1)^{n_\varepsilon} = 1$, the corresponding term b_{n_ε} satisfies

$$b_{n_\varepsilon} = \left(2 - \frac{1}{\sqrt{n_\varepsilon}}\right)(-1)^{n_\varepsilon} = 2 - \frac{1}{\sqrt{n_\varepsilon}}. \tag{2.1.18}$$

Hence, the term b_{n_ε} is within ε of 2:

$$d_{\mathbb{R}}(b_{n_\varepsilon}, 2) = |b_{n_\varepsilon} - 2| = \left|\left(2 - \frac{1}{\sqrt{n_\varepsilon}}\right) - 2\right| = \frac{1}{\sqrt{n_\varepsilon}} < \varepsilon. \tag{2.1.19}$$

Therefore, $2 \operatorname{acl}(b_n)$. $\qquad\qquad\qquad\qquad\qquad\qquad\qquad\qquad\qquad\qquad\qquad\qquad\quad \square$

Compare Figures 2.1.2 and 2.1.3 featuring the sequences (a_n) and (b_n) from Example 2.1.4 along with a few ε-neighborhoods of 2. Can you describe the differences you see? Does the behavior of each sequence seem to depend on the parity of the indices? Better yet, can you define what you're seeing?

In Figure 2.1.2, every ε-neighborhood of 2 captures all of the terms of (a_n) except for a finite number which are more than ε away from 2. In Figure 2.1.3, the given ε-neighborhoods are too small to contain the terms of (b_n) with odd indices. So, sometimes an infinite number of the terms of (b_n) are more than ε away from 2.

The notion of capturing all but a finite number of the terms of a sequence is codified by the *tails* of a sequence.

Definition 2.1.7: Tail of a sequence

A *tail* of a sequence $(\mathbf{x}_n)$ is a sequence of the form

$$(\mathbf{x}_{n \geq k}) = (\mathbf{x}_k, \mathbf{x}_{k+1}, \mathbf{x}_{k+2}, \ldots) \tag{2.1.20}$$

which is a copy of $(\mathbf{x}_n)$ starting at the kth term $\mathbf{x}_k$ for some positive integer k. When we want to be more specific, we refer to $(\mathbf{x}_{n \geq k})$ as the *k-tail.*

Example 2.1.8: Tails and not a tail

Once again, consider the sequences (a_n) and (b_n) from Example 2.1.4 which are defined for each $n \in \mathbb{N}$ by

$$a_n = 2 - \frac{1}{\sqrt{n}} \quad \text{and} \quad b_n = \left(2 - \frac{1}{\sqrt{n}}\right)(-1)^n. \tag{2.1.21}$$

Which of the following are tails of (a_n)? Of (b_n)? Each figure displays a graph of the corresponding sequence where the points in the plane stem from indices along the horizontal axis and terms along the vertical axis.

terms

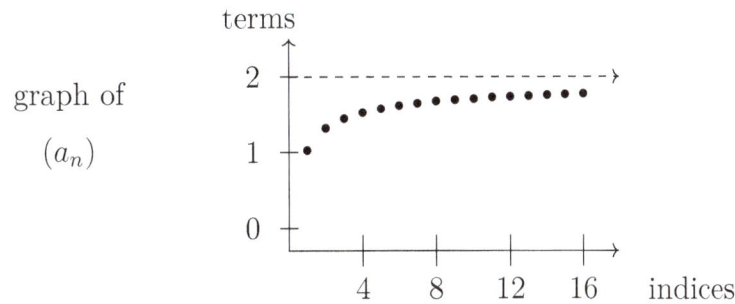

Figure 2.1.4: A graph of the sequence (a_n) from Examples 2.1.4 and 2.1.8. Technically, the black dots are *not* the terms of the sequence. The terms are represented by heights which would lie on the vertical axis, but they are not plotted here.

- Figure 2.1.4: $(a_n) = (a_1, a_2, a_3, \ldots)$

- No figure: $(a_{n \geq 4}) = (a_4, a_5, a_6, \ldots)$

- Figure 2.1.5: $(a_{n \geq 8}) = (a_8, a_9, a_{10}, \ldots)$

- Figure 2.1.6: $(b_n) = (b_1, b_2, b_3, \ldots)$

- Figure 2.1.7: $(b_{n \geq 8}) = (b_8, b_9, b_{10}, \ldots)$

- Figure 2.1.8: $(b_{2k}) = (b_2, b_4, b_6, \ldots)$

Every sequence is its own 1-tail, so we have $(a_n) = (a_{n \geq 1})$ and $(b_n) = (b_{n \geq 1})$. The sequence $(a_{n \geq 4})$ is the 4-tail of (a_n) whose first term is a_4 while $(a_{n \geq 8})$ is the 8-tail whose first term is a_8. The sequence $(b_{n \geq 8})$ is the 8-tail of (b_n) whose first term is b_8.

On the other hand, the sequence (b_{2k}) comprising the terms of (b_n) with even indices, is not a tail of (b_n). Every tail must contain all terms of some original sequence after a particular initial term, whether the indices are even or odd. The sequence (b_{2k}) is not a tail of (b_n) since (b_{2k}) contains no terms with odd indices.

Remark 2.1.9: Subsequences are discussed later

While they are not tails of (b_n), the sequences (b_{2k}) and (b_{2k-1}) comprising the terms with even and odd indices, respectively, are *subsequences* of (b_n). See Figure 2.1.8. The formal definition for subsequences and their properties are explored in Section 2.9.

The purpose of this section is to put us in position to explore the definition of *convergence* and *limits* for sequences in the next section by considering arbitrarily close and tails. One more look at the sequences (a_n) and (b_n) from Examples 2.1.4 and 2.1.8 will help us prepare.

terms

graph of

$(a_{n \geq 8})$

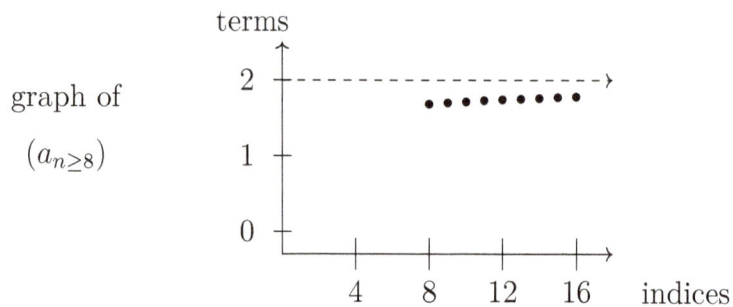

Figure 2.1.5: A graph of the 8-tail $(a_{n \geq 8})$ from Example 2.1.8. The first seven terms (heights) of the original (a_n) are discarded while all remaining terms with index $n \geq 8$ are included.

graph of

(b_n)

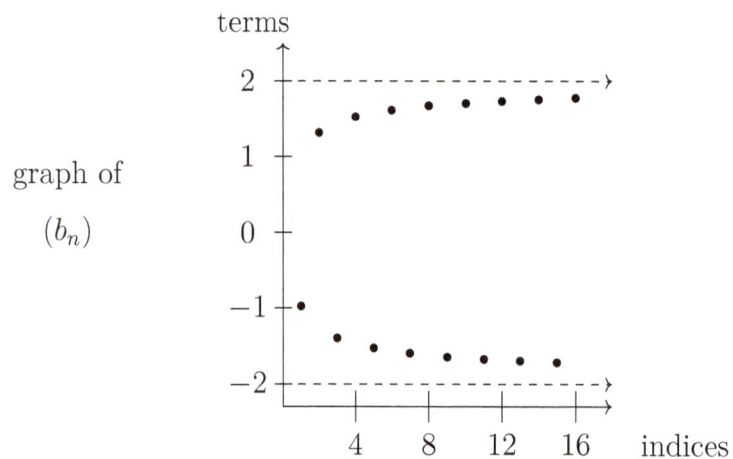

Figure 2.1.6: A graph of the sequence (b_n) from Examples 2.1.4 and 2.1.8. Note the terms (heights) alternate between being close to 2 or -2 as the indices increase.

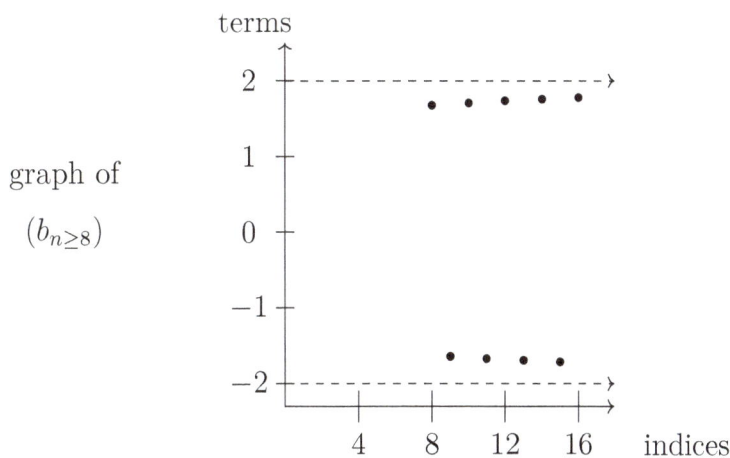

Figure 2.1.7: A graph of the 8-tail sequence $(b_{n \geq 8})$ from Example 2.1.8. The first seven terms of (b_n) are discarded while all remaining terms with index $n \geq 8$ are included. The terms (heights) of the 8-tail $(b_{n \geq 8})$ alternate between being close to 2 or -2 as the indices increase.

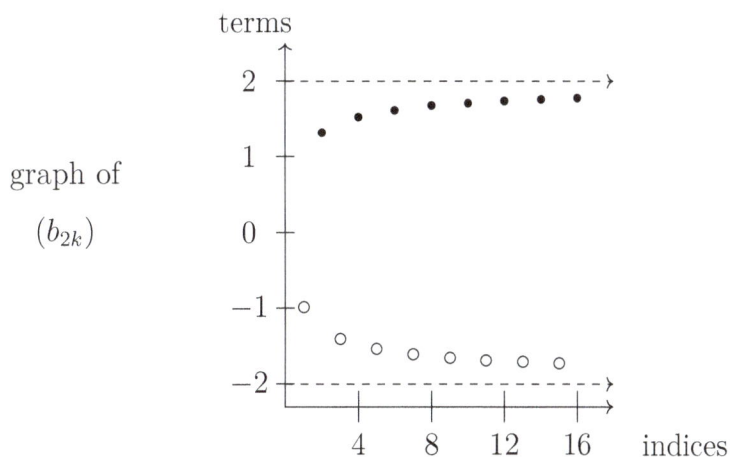

Figure 2.1.8: A graph of the subsequence (b_{2k}) featuring the terms (heights) of (b_n) whose indices are even. See Example 2.1.8. The black dots, like $\bullet$, represent the sequence (b_{2k}) while the circles, like $\circ$, represent the subsequence (b_{2k-1}) with odd indices which are not part of (b_{2k}). Neither (b_{2k}) nor (b_{2k-1}) is a tail of the original (b_n).

Example 2.1.10: Neighborhoods and tails

Once again, consider the sequences (a_n) and (b_n) defined for each index $n \in \mathbb{N}$ by

$$a_n = 2 - \frac{1}{\sqrt{n}} \qquad \text{and} \qquad b_n = \left(2 - \frac{1}{\sqrt{n}}\right)(-1)^n. \tag{2.1.22}$$

Even though $2 \operatorname{acl}(a_n)$ and $2 \operatorname{acl}(b_n)$ as proven for Example 2.1.4, every ε-neighborhood of 2 contains a tail (a_n), but some ε-neighborhoods of 2 do not contain a tail of (b_n). See Figures 2.1.2, 2.1.3, and 2.1.9.

Remark 2.1.11: Scratch work for arbitrarily close versus convergence

A benefit of exploring points arbitrarily close to sequences before convergence is the scratch work. In order to show all ε-neighborhoods of a point contain a tail of a sequence, we can build on the scratch work developed to prove the point is arbitrarily close to the sequence. This process allows us to prove such a sequence converges to the point. See Definition 2.2.1.

Scratch Work 2.1.12: From arbitrarily close to convergence

To prove every ε-neighborhood of 2 contains a tail of (a_n), we can build on Scratch Work 2.1.5 by showing that not only do we have a *term* a_{n_ε} within a given distance ε of 2, but a whole *tail* $(a_{n \geq n_\varepsilon})$ is also within ε of 2. Equivalently, this shows $(a_{n \geq n_\varepsilon}) \subseteq V_\varepsilon(2)$. The proof itself considers the terms a_n with indices large enough to satisfy $n \geq n_\varepsilon$ where the index n_ε is found in scratch work.

Proof that every neighborhood of 2 contains a tail of (a_n) in Example 2.1.10. Let $\varepsilon > 0$. Choose an index $n_\varepsilon \in \mathbb{N}$ large enough to satisfy $n_\varepsilon > 1/\varepsilon^2$, as motivated by Scratch Work 2.1.5. We have

$$n_\varepsilon > \frac{1}{\varepsilon^2} \quad \Longleftrightarrow \quad \frac{1}{\sqrt{n_\varepsilon}} < \varepsilon. \tag{2.1.23}$$

Now consider every index $n \in \mathbb{N}$ large enough to have $n \geq n_\varepsilon$. Then, since the square root function is increasing, we have

$$n \geq n_\varepsilon > \frac{1}{\varepsilon^2} \quad \Longrightarrow \quad \frac{1}{n} \leq \frac{1}{n_\varepsilon} < \varepsilon^2 \quad \Longrightarrow \quad \frac{1}{\sqrt{n}} \leq \frac{1}{\sqrt{n_\varepsilon}} < \varepsilon. \tag{2.1.24}$$

Hence, every term $a_n = 2 - (1/\sqrt{n})$ where $n \geq n_\varepsilon$ is in the n_ε-tail $(a_{n \geq n_\varepsilon})$ and within ε of 2:

$$d_{\mathbb{R}}(a_n, 2) = |a_n - 2| = \left|\left(2 - \frac{1}{\sqrt{n}}\right) - 2\right| = \frac{1}{\sqrt{n}} \leq \frac{1}{\sqrt{n_\varepsilon}} < \varepsilon. \tag{2.1.25}$$

Therefore, every ε-neighborhood of 2 contains its n_ε-tail $(a_{n \geq n_\varepsilon})$. $\qquad \square$

Now to show an ε_0-neighborhood of 2 contains no tails of (b_n).

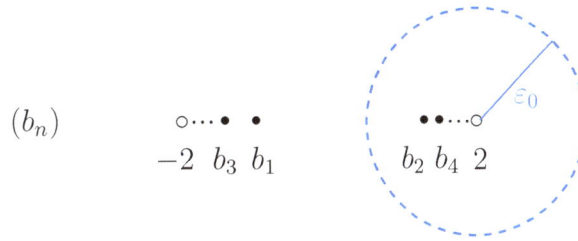

Figure 2.1.9: The distance $\varepsilon_0 = 3/2$ defines a neighborhood around 2 which contains no tails of the sequence (b_n) from Examples 2.1.4 and 2.1.8. That is, the ε_0-neighborhood $V_{\varepsilon_0}(2)$ does not contain a tail of (b_n).

Scratch Work 2.1.13: Arbitrarily close but not convergence

See Figure 2.1.9 where the distance $\varepsilon_0 = 3/2$ yields a neighborhood $V_{\varepsilon_0}(2)$ which does not seem to contain any tails of (b_n). The proof entails showing all terms with odd indices are more than $\varepsilon_0 = 3/2$ away from 2.

Proof regarding tails of (b_n) in Example 2.1.10. Consider the distance $\varepsilon_0 = 3/2 > 0$. Following Scratch Work 2.1.6, for every *odd* index $n \in \mathbb{N}$ we have $(-1)^n = -1$ and so

$$b_n = \left(2 - \frac{1}{\sqrt{n}}\right)(-1)^n = -2 + \frac{1}{\sqrt{n}}. \tag{2.1.26}$$

Hence, the distance $|b_n - 2|$ yields

$$d_{\mathbb{R}}(b_n, 2) = |b_n - 2| = \left|-2 + \frac{1}{\sqrt{n}} - 2\right| = 4 - \frac{1}{\sqrt{n}}. \tag{2.1.27}$$

Therefore, the distance $|b_n - 2|$ is more than $\varepsilon_0 = 3/2$ when n is odd:

$$|b_n - 2| = 4 - \frac{1}{\sqrt{n}} \geq 3 > \frac{3}{2} = \varepsilon_0. \tag{2.1.28}$$

Finally, since every tail of (b_n) must contain terms with even and odd indices, the ε_0-neighborhood of 2 contains no tails of (b_n). $\square$

The remainder of this section considers a sequence in the plane $\mathbb{R}^2$.

$\circ$ **v**

$\bullet$ $\mathbf{z}_4$

$(\mathbf{z}_n)$ $\bullet$ $\mathbf{z}_2$

$\circ \cdots \bullet$ $\bullet$

u $\mathbf{z}_3$ $\mathbf{z}_1$

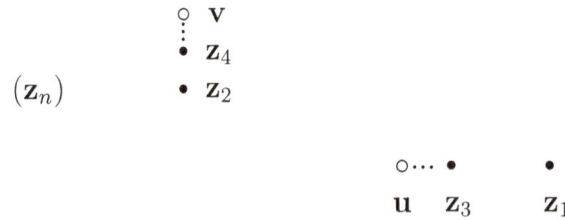

Figure 2.1.10: The terms $\mathbf{z}_n$ where n is odd seem to be bunching together near **u** while the terms where n is even are bunching together near **v**. See Example 2.1.14.

Example 2.1.14: A sequence arbitrarily close to two points in the plane

Consider the sequence in the plane $\mathbb{R}^2$ given by

$$\mathbf{z}_n = \begin{cases} \begin{bmatrix} 2 + (2/n) \\ 1 \end{bmatrix}, & \text{if } n \text{ is odd}, \\[2em] \begin{bmatrix} -1 \\ 3 - (2/n) \end{bmatrix}, & \text{if } n \text{ is even}. \end{cases} \tag{2.1.29}$$

Also, consider the points **u** and **v** given by

$$\mathbf{u} = \begin{bmatrix} 2 \\ 1 \end{bmatrix} \quad \text{and} \quad \mathbf{v} = \begin{bmatrix} -1 \\ 3 \end{bmatrix}. \tag{2.1.30}$$

See Figure 2.1.10 for a plot of the sequence $(\mathbf{z}_n)$ along with the points **u** and **v**. It turns out we have both $\mathbf{u} \, \mathrm{acl}(\mathbf{z}_n)$ and $\mathbf{v} \, \mathrm{acl}(\mathbf{z}_n)$.

Proof for Example 2.1.14. Let $\varepsilon > 0$. By the Corollary of the Archimedean Property 1.4.8, there is an *odd* integer j_ε that's large enough to give us[1]

$$d_2(\mathbf{z}_{j_\varepsilon}, \mathbf{u}) = \|\mathbf{z}_{j_\varepsilon} - \mathbf{u}\|_2 = \frac{2}{j_\varepsilon} < \varepsilon. \tag{2.1.31}$$

Again by the Corollary of the Archimedean Property 1.4.8, there is also an *even* integer k_ε that's large enough to give us

$$d_2(\mathbf{z}_{k_\varepsilon}, \mathbf{v}) = \|\mathbf{z}_{k_\varepsilon} - \mathbf{v}\|_2 = \frac{2}{k_\varepsilon} < \varepsilon. \tag{2.1.32}$$

Hence, every ε-neighborhood of **u** and every ε-neighborhood of **v** contains a term of $(\mathbf{z}_n)$. So, $\mathbf{u} \, \mathrm{acl}(\mathbf{z}_n)$ and $\mathbf{v} \, \mathrm{acl}(\mathbf{z}_n)$. $\square$

---

[1]Don't see how these computations work out? If so, that's okay. Some steps were skipped. Fill them in!

In Definition 2.2.1 below, we build on the notions of arbitrarily close and tails to define limits and convergence for sequences, also called *sequential limits and convergence*. This definition is notoriously difficult to understand. See the works [3, 4, 6, 9, 11] for other approaches to teaching real analysis at the undergraduate level which are, in part, designed to address this difficulty. Also see [5, 9] for a thorough discussion of the challenges that come with the pedagogy of convergence and limits.

Exercises

2.1.1. Consider the sequence of real numbers (x_n) defined for each index $n \in \mathbb{N}$ by

$$x_n = \frac{4(-1)^n}{\sqrt{n+1}}. \tag{2.1.33}$$

(i) Prove $0 \, \text{acl} \, (x_n)$.

(ii) Prove every ε-neighborhood of 0 contains a tail of (x_n).

2.1.2. Consider the sequence of real numbers (z_n) defined for each index $n \in \mathbb{N}$ by

$$z_n = \frac{4(-1)^n}{\sqrt{n+1}} + 2(-1)^n. \tag{2.1.34}$$

(a) Prove $2 \, \text{acl} \, (z_n)$.

(b) Find an ε_0-neighborhood of 2 which contains no tails of (z_n) and prove your result.

2.1.3. Determine the set of points arbitrarily close to the range of each sequence defined below. Draw stuff, and note $n \in \mathbb{N}$.

(i) $a_n = n$.

(ii) $b_n = 8 - \dfrac{(-1)^n}{n}$.

(iii) $c_n = 3(-1)^n + \dfrac{1}{n}$.

(iv) $d_n = 1 - \dfrac{1}{10^n}$.

(v) $e_n = \begin{cases} 0, & n \text{ is odd} \\ n/2, & n \text{ is even}. \end{cases}$

Note we have $(d_n) = (.9, .99, .999, \ldots)$ and $(e_n) = (0, 1, 0, 2, 0, 3, \ldots)$.

2.1.4. Let $\varepsilon_0 = 1/101 > 0$ and $n \in \mathbb{N}$.

(i) For $a_n = n$, prove $V_{1/101}(3140)$ contains no tails of (a_n).

(ii) For $b_n = 8 - \dfrac{(-1)^n}{n}$, prove $V_{1/101}(8)$ contains a tail of (b_n).

(iii) For $c_n = 3(-1)^n + \dfrac{1}{n}$, prove $V_{1/101}(3)$ contains no tails of (c_n).

(iv) For $d_n = 1 - \dfrac{1}{10^n}$, prove $V_{1/101}(1)$ contains a tail of (d_n).

(v) For $e_n = \begin{cases} 0, & n \text{ is odd,} \\ n/2, & n \text{ is even,} \end{cases}$ prove $V_{1/101}(0)$ contains no tail of (e_n).

Note we have $(d_n) = (.9, .99, .999, \ldots)$ and $(e_n) = (0, 1, 0, 2, 0, 3, \ldots)$.

2.1.5. This exercise explores an example of a sequence defined as the sum of two other sequences.

(i) Consider the sequences (x_n) and (y_n) defined for each index $n \in \mathbb{N}$ by

$$x_n = 100 + \frac{1}{n^2} \qquad \text{and} \qquad y_n = \pi + \frac{1}{\sqrt{n}}. \tag{2.1.35}$$

Given the distance $\varepsilon_0 = 1/3140$, find indices $s_0, t_0 \in \mathbb{N}$ where

$$|x_{s_0} - 100| < \frac{1}{3140} \qquad \text{and} \qquad |y_{t_0} - \pi| < \frac{1}{3140}. \tag{2.1.36}$$

(ii) Consider the sequence (z_n) defined for each index $n \in \mathbb{N}$ by

$$z_n = x_n + y_n = 100 + \frac{1}{n^2} + \pi + \frac{1}{\sqrt{n}}. \tag{2.1.37}$$

Given the distance $\varepsilon_0 = 1/3140$, find an index $m_0 \in \mathbb{N}$ where

$$|z_{m_0} - (100 + \pi)| < \frac{1}{3140}. \tag{2.1.38}$$

(iii) Once again, consider the sequence (z_n) from part (ii). Given *an arbitrary* positive distance $\varepsilon > 0$, prove there is an index $n_\varepsilon \in \mathbb{N}$ where

$$|z_{n_\varepsilon} - (100 + \pi)| < \varepsilon. \tag{2.1.39}$$

(iv) One more time, consider the sequence (z_n) from part (ii). Prove every ε-neighborhood of $100 + \pi$ contains a tail of (z_n).

2.1.6. This exercise explores another example of a sequence defined as the sum of two other sequences.

(i) Consider the sequences (a_n) and (b_n) defined for each $n \in \mathbb{N}$ by

$$a_n = 100(-1)^n + \frac{1}{n^2} \qquad \text{and} \qquad b_n = \pi(-1)^{n+1} + \frac{1}{\sqrt{n}}. \tag{2.1.40}$$

Given the distance $\varepsilon_0 = 1/3140$, find indices $s_0, t_0 \in \mathbb{N}$ where

$$|a_{s_0} - 100| < \frac{1}{3140} \qquad \text{and} \qquad |b_{t_0} - \pi| < \frac{1}{3140}. \tag{2.1.41}$$

(ii) Consider the sequence (c_n) defined for each $n \in \mathbb{N}$ by

$$c_n = a_n + b_n = 100(-1)^n + \frac{1}{n^2} + \pi(-1)^{n+1} + \frac{1}{\sqrt{n}}. \tag{2.1.42}$$

Given the distance $\varepsilon_0 = 1/3140$, explain why there is no index $n \in \mathbb{N}$ where

$$|c_n - (100 + \pi)| < \frac{1}{3140}. \tag{2.1.43}$$

2.1.7. Given a real number y, construct a sequence of real numbers (x_n) where y is the only point arbitrarily close to every tail of (x_n) and yet no neighbhorhood of y contains a tail of (x_n). Note that neither (a_n) nor (b_n) from Example 2.1.4 satisfies this condition.

2.2   Limit of a sequence

The *convergence* of a sequence can be summarized as follows:

> A sequence *converges* to a point if every neighborhood of the point contains a tail of the sequence.

A purpose of formal definitions is to capture our intuition about concepts such as convergence. As we delve into the formal definitions, please keep in mind whatever properties you expect limits and converging sequences to have. The technical versions of these concepts take time to fully understand, so please be patient. Play around with the ideas and let the understanding come to you.

Definition 2.2.1: Convergence, threshold, and limit of a sequence

Let $(\mathbf{x}_n)$ be a sequence of points in $\mathbb{R}^m$ and let $\mathbf{y}$ be a point in $\mathbb{R}^m$. The sequence $(\mathbf{x}_n)$ *converges to* $\mathbf{y}$ if for every distance $\varepsilon > 0$ there is an index $n_\varepsilon \in \mathbb{N}$ such that for every index $n \in \mathbb{N}$ we have

$$n \geq n_\varepsilon \quad \Longrightarrow \quad d_m(\mathbf{x}_n, \mathbf{y}) = \|\mathbf{x}_n - \mathbf{y}\|_m < \varepsilon. \tag{2.2.1}$$

In this case, the index n_ε is called a *threshold*, $\mathbf{y}$ is called the *limit* of the sequence $(\mathbf{x}_n)$, and we write

$$\lim_{n \to \infty} \mathbf{x}_n = \mathbf{y} \quad \text{or} \quad \lim_{n \to \infty} \mathbf{x}_n = \mathbf{y}. \tag{2.2.2}$$

A sequence *diverges* if it does not converge to any point.

Remark 2.2.2: Thresholds define tails

The definition of convergence and limit for a sequence in Definition 2.2.1 is equivalent to saying every neighborhood of the point contains a tail of the sequence. In particular, the implication (2.2.1) is equivalent to

$$(\mathbf{x}_{n \geq n_\varepsilon}) = (x_{n_\varepsilon}, x_{n_\varepsilon+1}, x_{n_\varepsilon+2}, \ldots) \subseteq V_\varepsilon(\mathbf{y}). \tag{2.2.3}$$

So when $(\mathbf{x}_n)$ converges to $\mathbf{y}$, a threshold n_ε defines an n_ε-tail of $(\mathbf{x}_n)$ contained in the neighborhood $V_\varepsilon(\mathbf{y})$.

Remark 2.2.3: Arbitrarily close and convergence are not equivalent

The concepts arbitrarily close and convergence are deeply related but they are not equivalent. A key difference is the role played by the index n_ε and whether it is a threshold for convergence or not. Every index defines both a term and a tail of a given sequence. When n_ε produces a term within ε of a point, it leads to the point being arbitrarily close to the sequence. When a whole n_ε-tail is within ε of a point, the point is the limit of the sequence.

By first working with a formal definition for arbitrarily close in the context of sequences, we can parse convergence and limits using arbitrarily close and tails. You may find that finding a suitable threshold n_ε—which we do for both arbitrarily close and convergence—is often the most difficult part about proving convergence.

Definition 1.5.1: $\mathbf{y}$ is arbitrarily close to $(\mathbf{x}_n)$	Definition 2.2.1: $\mathbf{y}$ is the limit of $(\mathbf{x}_n)$
$\mathbf{y} \, \mathrm{acl} \, (\mathbf{x}_n)$	$\mathbf{y} = \lim_{n \to \infty} (\mathbf{x}_n)$
Every neighborhood of $\mathbf{y}$ contains a *term* of $(\mathbf{x}_n)$.	Every neighborhood of $\mathbf{y}$ contains a *tail* of $(\mathbf{x}_n)$.
$\forall \, \varepsilon > 0,$ $\exists \, n_\varepsilon \in \mathbb{N}$ such that $\mathbf{x}_{n_\varepsilon} \in V_\varepsilon(\mathbf{y}).$	$\forall \, \varepsilon > 0,$ $\exists \, n_\varepsilon \in \mathbb{N}$ such that $(\mathbf{x}_{n \geq n_\varepsilon}) \subseteq V_\varepsilon(\mathbf{y}).$
$\forall \, \varepsilon > 0,$ $\exists \, n_\varepsilon \in \mathbb{N}$ such that $\|\mathbf{x}_{n_\varepsilon} - \mathbf{y}\|_m < \varepsilon.$	$\forall \, \varepsilon > 0,$ $\exists \, n_\varepsilon \in \mathbb{N}$ such that $n \geq n_\varepsilon \implies \|\mathbf{x}_n - \mathbf{y}\|_m < \varepsilon.$

The table above compares the technical definitions of arbitrarily close and convergence for sequences. The quantifier "$\forall$" means "for all" while "$\exists$" means "there exists". With both arbitrarily close and convergence, $\varepsilon > 0$ tells us how close we would like the terms of the sequence $(\mathbf{x}_n)$ to be to $\mathbf{y}$. Again, the key difference is role played by the index $n_\varepsilon \in \mathbb{N}$. In both cases, n_ε depends on the given positive distance or "error" ε and the corresponding ε-neighborhood $V_\varepsilon(\mathbf{y})$. With arbitrarily close, n_ε ensures at least one term $\mathbf{x}_{n_\varepsilon}$ is within ε of $\mathbf{y}$, so $\mathbf{x}_{n_\varepsilon}$ is in the neighborhood $V_\varepsilon(\mathbf{y})$. With convergence, n_ε is a threshold ensuring *all* terms in the tail $(\mathbf{x}_{n \geq n_\varepsilon})$ are within ε of $\mathbf{y}$, thus $V_\varepsilon(\mathbf{y})$ contains the n_ε-tail of $(\mathbf{x}_n)$. See the following table and Figure 2.2.1.

It turns out the scratch work for showing a point is arbitrarily close to a sequence often leads to a suitable threshold to prove the point is actually the limit of the sequence. See Example 2.1.10 and the scratch work for the related proofs. Scratch work is an important part of developing any proof, but it is vital to proving some point is the limit of a sequence (equivalently, a sequence converges).

Remark 2.2.4: Guide for sequential limit proofs

Below is a guide for developing scratch work and proving $\lim_{n \to \infty} \mathbf{x}_n = \mathbf{y}$ via Definition 2.2.1.

Scratch work for $\lim_{n \to \infty} \mathbf{x}_n = \mathbf{y}$:

- Consider the inequality you want to end up with, typically:

$$d_m(\mathbf{x}_n, \mathbf{y}) = \|\mathbf{x}_n - \mathbf{y}\|_m < \varepsilon. \tag{2.2.4}$$

- *Key step*: Work backwards from this inequality to find a formula for a potential threshold n_ε, likely involving an inequality with ε. The triangle inequality as it appears in (1.2.32) and (1.2.34) is used quite often in scratch work and proofs involving limits and convergence.

- Consider all indices $n \geq n_\varepsilon$ and their distances $d_m(\mathbf{x}_n, \mathbf{y}) = \|\mathbf{x}_n - \mathbf{y}\|_m$ in order to ensure n_ε is a threshold for the convergence of $(\mathbf{x}_n)$ to its limit $\mathbf{y}$. Ultimately, we want to verify the neighborhood $V_\varepsilon(\mathbf{y})$ contains the n_ε-tail of $(\mathbf{x}_n)$.

- Draw a figure with the sequence $(\mathbf{x}_n)$ and the point $\mathbf{y}$ which serves as a candidate for the limit.

Proving $\lim_{n \to \infty} \mathbf{x}_n = \mathbf{y}$:

- Early in the proof, perhaps the first step, write "Let $\varepsilon > 0$" or something similar, indicating you are accounting for *all* positive distances at the same time.

- Define or choose an index n_ε based on your scratch work using a formula or expression that depends on ε.

- Verify n_ε is truly a threshold and $\mathbf{y}$ is the limit by showing

$$n \geq n_\varepsilon \quad \Longrightarrow \quad d_m(\mathbf{x}_n, \mathbf{y}) = \|\mathbf{x}_n - \mathbf{y}\|_m < \varepsilon. \tag{2.2.5}$$

This shows the neighborhood $V_\varepsilon(\mathbf{y})$ contains the n_ε-tail of $(\mathbf{x}_n)$ and $\lim_{n \to \infty} \mathbf{x}_n = \mathbf{y}$.

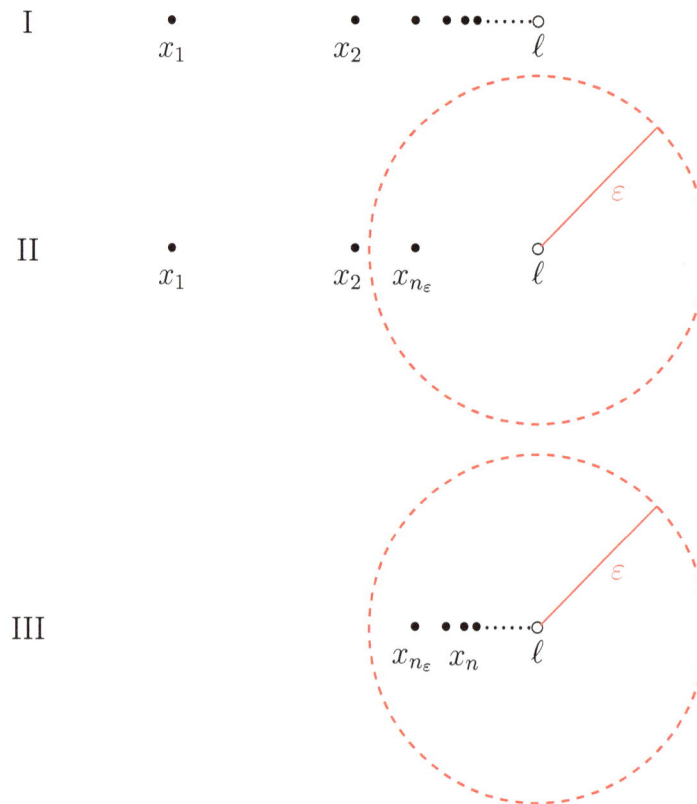

Figure 2.2.1: A visual progression through the definition of sequential limit and convergence (Definition 2.2.1) using the sequence $x_n = 3140 - (1/n)$ from Example 2.2.5 whose limit is $\ell = 3140$. **I:** Consider the range of the sequence and the candidate for the limit ℓ. **II:** The variable $\varepsilon > 0$ represents a distance around ℓ, giving us the ε-neighborhood $V_\varepsilon(\ell)$. The term x_{n_ε} is within ε of ℓ. **III:** Consider the n_ε-tail by ignoring the terms with indices $n = 1, 2, \ldots, n_\varepsilon - 1$. The index n_ε is a threshold for the convergence since each x_n where $n \geq n_\varepsilon$ is within ε of ℓ. Equivalently, $V_\varepsilon(\ell)$ contains the n_ε-tail $(x_{n \geq n_\varepsilon})$.

Time for some examples.

Example 2.2.5: A convergent sequence of real numbers

Consider the sequence (x_n) given by $x_n = 3140 - (1/n)$ for each $n \in \mathbb{N}$. See Figure 2.2.1. We have

$$\lim_{n \to \infty} x_n = \ell = 3140. \tag{2.2.6}$$

Scratch Work 2.2.6: Convergence of a sequence

To prove $\lim_{n \to \infty} x_n = \ell = 3140$ via the definition of sequential limits and convergence (Definition 2.2.1), we must show every neighborhood of 3140 contains a tail of x_n. The idea of the scratch work is to find a candidate for the threshold n_ε much in the way we found suitable indices throughout Chapter 1 and Section 2.1 when working to show points arbitrarily close to sets and sequences. However, for convergence we must further verify our index n_ε defines a whole *tail* that's contained in the neighborhood $V_\varepsilon(3140)$ by showing all terms with indices large enough to have $n \geq n_\varepsilon$ are within ε of $\ell = 3140$. See Figure 2.2.1.

To consider *every* neighborhood of 3140, we can work with an unspecified $\varepsilon > 0$ (which establishes how close to $\ell = 3140$ we'd like to get). The value of n_ε will depend on ε, hence the subscript.

As mentioned in and around Definition 1.5.8, neighborhoods in the real line are defined by an inequality involving the absolute value of a difference. We have

$$x_{n_\varepsilon} \in V_\varepsilon(3140) \qquad \Longleftrightarrow \qquad d_\mathbb{R}(x_{n_\varepsilon}, \ell) = |x_{n_\varepsilon} - 3140| < \varepsilon. \tag{2.2.7}$$

Since $x_n = 3140 - (1/n)$ in our case, we have

$$d_\mathbb{R}(x_{n_\varepsilon}, \ell) = |x_{n_\varepsilon} - \ell| = \left|3140 - \frac{1}{n_\varepsilon} - 3140\right| = \frac{1}{n_\varepsilon} < \varepsilon. \tag{2.2.8}$$

The inequality on the right satisfies the equivalence

$$\frac{1}{n_\varepsilon} < \varepsilon \qquad \Longleftrightarrow \qquad n_\varepsilon > \frac{1}{\varepsilon}. \tag{2.2.9}$$

Note that a precise value of n_ε is not specified by the inequalities in (2.2.9) since any positive integer greater than $1/\varepsilon$ would suffice. But that's okay! We only need the existence of a suitable threshold, so we can choose n_ε to be any positive integer that's large enough.

Once we choose $n_\varepsilon \in \mathbb{N}$ in the proof, we need to follow through and verify the whole n_ε-tail $(x_{n \geq n_\varepsilon})$ is contained in the neighborhood $V_\varepsilon(3140)$. That is, we need to verify

$$n \geq n_\varepsilon \qquad \Longrightarrow \qquad d_\mathbb{R}(x_n, \ell) = |x_n - \ell| < \varepsilon. \tag{2.2.10}$$

We now have enough to attempt a proof. Once again, by not specifying a particular value of ε, we are accounting for *all* positive distances—therefore all neighborhoods—at the same time.

Proof for $\lim_{n \to \infty} x_n = \ell = 3140$ *in Example 2.2.5.* Let $\varepsilon > 0$. Choose an index $n_\varepsilon \in \mathbb{N}$ where $n_\varepsilon > 1/\varepsilon$. Then for every index $n \in \mathbb{N}$ we have

$$n \geq n_\varepsilon \qquad \Longrightarrow \qquad \frac{1}{n} \leq \frac{1}{n_\varepsilon} < \varepsilon. \tag{2.2.11}$$

To verify n_ε is a threshold for the convergence of (x_n) to $\ell = 3140$, note that $n \geq n_\varepsilon$ implies

$$d_\mathbb{R}(x_n, \ell) = |x_n - \ell| = \left|\left(3140 - \frac{1}{n}\right) - 3140\right| = \frac{1}{n} \leq \frac{1}{n_\varepsilon} < \varepsilon. \tag{2.2.12}$$

Therefore, $\lim_{n\to\infty} x_n = \ell = 3140$. Additionally, we have shown that for every $\varepsilon > 0$, the neighborhood $V_\varepsilon(3140)$ contains the n_ε-tail of (x_n). □

Example 2.2.7: Another convergent sequence of real numbers

Recall the sequence (a_n) from Examples 1.1.17, 2.1.4, 2.1.8, and 2.1.10 whose terms are defined for each index $n \in \mathbb{N}$ by

$$a_n = 2 - \frac{1}{\sqrt{n}}. \tag{2.2.13}$$

We have $\lim_{n\to\infty} a_n = 2$. See Figures 2.1.1, 2.1.2, and 2.1.4.

Scratch Work 2.2.8: Arbitrarily close has already been shown

In Example 2.1.10, we proved every ε-neighborhood of 2 contains a tail of (a_n). So, as noted in Remark 2.2.2, this is equivalent to proving (a_n) converges to 2. Actually, we found a suitable threshold n_ε earlier on with Scratch Work 1.1.20 following Example 1.1.17, but we had not yet developed the notation and terminology.

The choice of an index $n_\varepsilon \in \mathbb{N}$ satisfying $n_\varepsilon > 1/\varepsilon^2$ was developed to show $2\,\mathrm{acl}\,A$ as well as $2\,\mathrm{acl}\,(a_n)$ in Scratch Work 2.1.5 for Example 2.1.4. Now we show the same choice of index n_ε is a threshold for the convergence of (a_n) to 2 and therefore $\lim_{n\to\infty} a_n = 2$.

Proof for $\lim_{n\to\infty} a_n = 2$ *in Example 2.2.7.* Let $\varepsilon > 0$. Based on Scratch Work 2.1.5 developed to show $2\,\mathrm{acl}\,(a_n)$ in Example 2.1.4, we can choose an index $n_\varepsilon \in \mathbb{N}$ large enough to give us

$$n_\varepsilon > \frac{1}{\varepsilon^2} \quad\Longleftrightarrow\quad \frac{1}{\sqrt{n_\varepsilon}} < \varepsilon. \tag{2.2.14}$$

Note $0 \le x \le y$ implies $\sqrt{x} \le \sqrt{y}$ (that is, the square root function is increasing on the domain $[0,\infty)$). So, for every index $n \in \mathbb{N}$ we have

$$n \ge n_\varepsilon \quad\Longrightarrow\quad \frac{1}{\sqrt{n}} \le \frac{1}{\sqrt{n_\varepsilon}} < \varepsilon. \tag{2.2.15}$$

Thus, n_ε is a threshold since every index $n \ge n_\varepsilon$ satisfies

$$d_\mathbb{R}(a_n, 2) = |a_n - 2| = \left|2 - \frac{1}{\sqrt{n}} - 2\right| = \frac{1}{\sqrt{n}} \le \frac{1}{\sqrt{n_\varepsilon}} < \varepsilon. \tag{2.2.16}$$

Therefore, $\lim_{n\to\infty} a_n = 2$ and $V_\varepsilon(2)$ contains the n_ε-tail of (a_n). □

The next example explores a convergent sequence in the plane $\mathbb{R}^2$. The algebra behind determining a suitable threshold is more complicated, but we can leverage properties of inequalities to make the process a bit smoother.

$$\mathbf{x}_1 \; \bullet$$
$$(\mathbf{x}_n) \qquad \mathbf{x}_2 \; \bullet_{\!\bullet\cdots\circ} \; \mathbf{y}$$

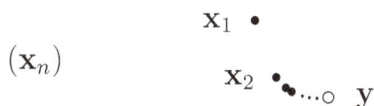

Figure 2.2.2: The sequence $(\mathbf{x}_n)$ and point $\mathbf{y}$ from Example 2.2.9.

Example 2.2.9: A convergent sequence in the plane

Consider the sequence $(\mathbf{x}_n)$ and point $\mathbf{y}$ in the plane $\mathbb{R}^2$ defined by

$$\mathbf{x}_n = \begin{bmatrix} 2 - (1/\sqrt{n}) \\ 1 + (1/n^2) \end{bmatrix} \quad \text{for each } n \in \mathbb{N}, \quad \text{and} \quad \mathbf{y} = \begin{bmatrix} 2 \\ 1 \end{bmatrix}. \qquad (2.2.17)$$

See Figure 2.2.2. We have $\lim_{n \to \infty} \mathbf{x}_n = \mathbf{y}$.

Scratch Work 2.2.10: Start at the end

For each distance $\varepsilon > 0$, we want to end up with an n_ε-tail of the sequence $(\mathbf{x}_n)$ that is within ε of $\mathbf{y}$. We can start by finding an index $n_\varepsilon \in \mathbb{N}$ where

$$d_2(\mathbf{x}_{n_\varepsilon}, \mathbf{y}) = \|\mathbf{x}_{n_\varepsilon} - \mathbf{y}\|_2 \qquad (2.2.18)$$

$$= \sqrt{\left(\frac{-1}{\sqrt{n_\varepsilon}}\right)^2 + \left(\frac{1}{n_\varepsilon^2}\right)^2} \qquad (2.2.19)$$

$$= \sqrt{\frac{1}{n_\varepsilon} + \frac{1}{n_\varepsilon^4}} \qquad (2.2.20)$$

$$< \varepsilon. \qquad (2.2.21)$$

However, solving directly for n_ε in the inequality

$$\sqrt{\frac{1}{n_\varepsilon} + \frac{1}{n_\varepsilon^4}} < \varepsilon \qquad (2.2.22)$$

is not really feasible, but thankfully we don't need to. Any positive integer n_ε that's large enough to ensure $\mathbf{x}_{n_\varepsilon}$ is within ε of $\mathbf{y}$ will do. There's no need for n_ε to be as small as possible or anything like that. With this in mind, we can try to find a simpler inequality to work with that will still get the job done.

Note that for every index $n \in \mathbb{N}$ we have

$$\frac{1}{n^4} \le \frac{1}{n} \qquad \Longrightarrow \qquad \frac{1}{n} + \frac{1}{n^4} \le \frac{2}{n}. \qquad (2.2.23)$$

Also, since $0 \le x \le y$ implies $\sqrt{x} \le \sqrt{y}$ (the square root function is increasing, another consequence of Theorem 1.3.2), we have

$$\sqrt{\frac{1}{n} + \frac{1}{n^4}} \le \sqrt{\frac{2}{n}}. \qquad (2.2.24)$$

So all we really need is a positive integer n_ε large enough to give us

$$\sqrt{\frac{2}{n_\varepsilon}} < \varepsilon. \tag{2.2.25}$$

Solving for n_ε in the more manageable (2.2.25) yields

$$n_\varepsilon > \frac{2}{\varepsilon^2}. \tag{2.2.26}$$

In the proof, we want to verify this choice for n_ε is truly a threshold for the convergence by showing $n \geq n_\varepsilon$ implies $\mathbf{x}_n$ is within ε of $\mathbf{y}$, thus $V_\varepsilon(\mathbf{y})$ contains the n_ε-tail of $(\mathbf{x}_n)$.

Proof for $\lim_{n\to\infty} \mathbf{x}_n = \mathbf{y}$ *in Example 2.2.9.* Let $\varepsilon > 0$. (Once again, we let $\varepsilon > 0$ be arbitrary so we are considering every distance or neighborhood around $\mathbf{y}$.) Based on our Scratch Work 2.2.10 and thanks to the Archimedean Property (Theorem 1.4.6), there is an index n_ε large enough so that

$$n_\varepsilon > \frac{2}{\varepsilon^2} \qquad \Longleftrightarrow \qquad \sqrt{\frac{2}{n_\varepsilon}} < \varepsilon. \tag{2.2.27}$$

For each index $n \geq n_\varepsilon$ we also have $n \geq 1$, hence

$$\frac{1}{n^4} \leq \frac{1}{n} \implies \frac{1}{n} + \frac{1}{n^4} \leq \frac{2}{n}. \tag{2.2.28}$$

Furthermore, since $0 \leq x \leq y$ implies $\sqrt{x} \leq \sqrt{y}$ (the square root function is increasing), we have

$$\sqrt{\frac{1}{n} + \frac{1}{n^4}} \leq \sqrt{\frac{2}{n}} \leq \sqrt{\frac{2}{n_\varepsilon}} < \varepsilon. \tag{2.2.29}$$

Therefore, for every index $n \geq n_\varepsilon$ we have

$$d_2(\mathbf{x}_n, \mathbf{y}) = \|\mathbf{x}_n - \mathbf{y}\|_2 \tag{2.2.30}$$

$$= \sqrt{\left(\frac{-1}{\sqrt{n}}\right)^2 + \left(\frac{1}{n^2}\right)^2} \quad \text{(by (1.2.27))} \tag{2.2.31}$$

$$= \sqrt{\frac{1}{n} + \frac{1}{n^4}} \tag{2.2.32}$$

$$\leq \sqrt{\frac{2}{n}} \quad \text{(by (2.2.29))} \tag{2.2.33}$$

$$\leq \sqrt{\frac{2}{n_\varepsilon}} \tag{2.2.34}$$

$$< \varepsilon. \tag{2.2.35}$$

Since $\varepsilon > 0$ is arbitrary, we have $\lim_{n\to\infty} \mathbf{x}_n = \mathbf{y}$ and the n_ε-tail of $(\mathbf{x}_n)$ is contained in $V_\varepsilon(\mathbf{y})$. $\square$

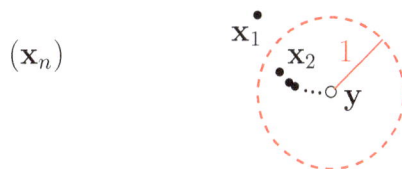

$(\mathbf{x}_n)$

Figure 2.2.3: In Example 2.2.9, all of the terms $\mathbf{x}_n$ with index $n \geq 2$ are within a distance of 1 away from the point $\mathbf{y}$. Thus, the 2-tail of $(\mathbf{x}_n)$ is contained in $V_1(\mathbf{y})$.

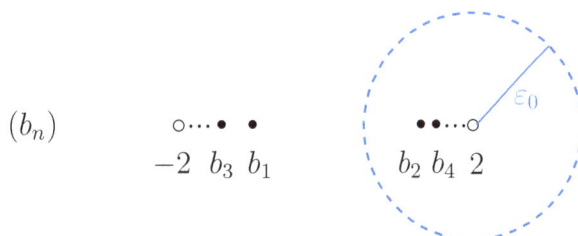

(b_n)

$-2 \quad b_3 \ b_1 \qquad b_2 \ b_4 \ 2$

Figure 2.2.4: The sequence (b_n) from Examples 2.1.4, 2.1.8, and 2.2.4 does not converge to 2 (or anything else for that matter). Here, an infinite collection of the terms b_n are $\varepsilon_0 = 3/2$ or more away from 2.

Remark 2.2.11: Thresholds need not be optimal

The inequality (2.2.23) we used to determine a threshold n_ε is not optimal, but it is good enough. For instance, in Figure 2.2.3, $\varepsilon_1 = 1$ provides a radius around $\mathbf{y}$. Based on the inequality (2.2.23), we can choose the index

$$n_1 = 3 > \frac{2}{\varepsilon_1^2} = \frac{2}{1^2} = 2, \tag{2.2.36}$$

which ensures $\mathbf{x}_{n_1} = \mathbf{x}_3$ is within $\varepsilon_1 = 1$ of $\mathbf{y}$. However, as Figure 2.2.3 indicates[a], the point $\mathbf{x}_2$ is also within $\varepsilon_1 = 1$ of $\mathbf{y}$, so we could've chosen $n_\varepsilon = n_1 = 2$ to be our threshold. But this doesn't matter. Convergence requires the existence of a sufficient threshold, not necessarily the smallest one.

[a]In fact, $d_2(\mathbf{x}_2, \mathbf{y}) = \sqrt{9/16} = 3/4 < 1 = \varepsilon_1$.

Example 2.2.12: The limit is not 2

Once again, consider the following sequence of real numbers (b_n) from Example 2.1.4 defined for each $n \in \mathbb{N}$ by

$$b_n = \left(2 - \frac{1}{\sqrt{n}}\right)(-1)^n. \tag{2.2.37}$$

See Figure 2.2.4. It turns out 2 is not the limit of (b_n).

Proof for $\lim_{n\to\infty} b_n \neq 2$. An infinite number of the terms of (b_n), specifically those whose index is odd, are more than $\varepsilon_0 = 3/2$ away from the point 2. For any given index n, there is an odd integer $j_n \geq n$ that yields

$$d_{\mathbb{R}}(b_{j_n}, 2) = |b_{j_n} - 2| = 4 - \frac{1}{\sqrt{j_n}} > \frac{3}{2} = \varepsilon_0. \tag{2.2.38}$$

Therefore, the distance $\varepsilon_0 = 3/2$ has no corresponding threshold since $V_{3/2}(2)$ does not contain any tails of (b_n). Hence, (b_n) does not converge to 2 and $\lim_{n\to\infty} b_n \neq 2$. $\qquad\square$

The above proof only shows that (b_n) does not converge to 2. A similar proof shows that (b_n) also does not converge to -2. The arguments rely on an idea mentioned in Remark 2.2.3 due to my former student Dylan Alvarenga: When a sequence converges and we have any $\varepsilon > 0$, only a finite number of the terms can be more than ε away from the limit.

But such an argument is not enough to say (b_n) *diverges*. For that, we could show (b_n) does not converge to any real number at all. Instead, let's wait until we have more tools at our disposal.

To wrap up the section, there are more ways to think of the role the threshold n_ε plays for us in the definition of limits and convergence for sequences (Definition 2.2.1). For instance, the existence of thresholds allows us to say a property *eventually* holds, and the thresholds' relationship with the distance ε establishes a sequence's *rate of convergence*.

Definition 2.2.13: Eventually, for large enough

A statement or property $P(\cdot)$ is said to hold *eventually* or *for large enough* $n \in \mathbb{N}$ if there is a threshold n_0 such that $n \geq n_0$ implies $P(n)$ is true.

So when $\lim_{n\to\infty} \mathbf{x}_n = \mathbf{y}$, we can say for any $\varepsilon > 0$ that for large enough n, the terms $\mathbf{x}_n$ are eventually within ε of $\mathbf{y}$.

Remark 2.2.14: Rate of convergence

Instead of formally defining the phrase *rate of convergence*, think of it as the relationship between the distance $\varepsilon > 0$ and the indices $n \in \mathbb{N}$ codified by the key implication

$$n \geq n_\varepsilon \quad \implies \quad d_m(\mathbf{x}_n, \mathbf{y}) = \|\mathbf{x}_n - \mathbf{y}\|_m < \varepsilon. \tag{2.2.39}$$

See Definition 2.2.1 and Remark 2.2.4. Essentially, the rate of convergence tells us through the threshold n_ε how large the indices $n \in \mathbb{N}$ should be to ensure a term $\mathbf{x}_n$ is within ε of the limit $\mathbf{y}$. Typically, the smaller we take ε to be, the larger n needs to be.

Example 2.2.15: Converging to zero at different rates

Consider the sequences (a_n) and (b_n) given by

$$a_n = \frac{1}{\sqrt{n}} \quad \text{and} \quad b_n = \frac{1}{n^2}. \tag{2.2.40}$$

We have $\lim_{n\to\infty} a_n = 0 = \lim_{n\to\infty} b_n$, but their rates of convergence are quite different.

Following the guide in Remark 2.2.4 and given $\varepsilon > 0$, we have

$$|a_n - 0| = \frac{1}{\sqrt{n}} < \varepsilon \quad \Longleftrightarrow \quad \sqrt{n} > \frac{1}{\varepsilon} \quad \Longleftrightarrow \quad n > \frac{1}{\varepsilon^2} \qquad (2.2.41)$$

while

$$|b_n - 0| = \frac{1}{n^2} < \varepsilon \quad \Longleftrightarrow \quad n^2 > \frac{1}{\varepsilon} \quad \Longleftrightarrow \quad n > \frac{1}{\sqrt{\varepsilon}}. \qquad (2.2.42)$$

So, when $\varepsilon = 1/100$ for example, we have

$$|a_n - 0| = \frac{1}{\sqrt{n}} < \frac{1}{100} \quad \Longleftrightarrow \quad n > \frac{1}{(1/100)^2} = 10,000 \qquad (2.2.43)$$

while

$$|b_n - 0| = \frac{1}{n^2} < \varepsilon \quad \Longleftrightarrow \quad n > \frac{1}{\sqrt{1/100}} = 10. \qquad (2.2.44)$$

That is, every index $n \geq 11$ ensures b_n is within $1/100$ of 0, but an index needs to be at least $10,001$ to ensure a_n is within ε of 0. Hence, the rates of convergence of (a_n) and (b_n) are significantly different with (b_n) converging to 0 *faster than* (a_n).

The next section explores properties of convergent sequences and their limits, especially the linearity of limits.

Exercises

2.2.1. Consider the sequence of real numbers (x_n) defined for each index $n \in \mathbb{N}$ by

$$x_n = \frac{4(-1)^n}{\sqrt{n+1}}. \qquad (2.2.45)$$

Exercise 2.1.1 shows us $0 \operatorname{acl}(x_n)$ and more. Here, prove $\lim_{n\to\infty} x_n = 0$.

2.2.2. Consider the sequence of real numbers (z_n) defined for each index $n \in \mathbb{N}$ by

$$z_n = \frac{4(-1)^n}{\sqrt{n+1}} + 2(-1)^n. \qquad (2.2.46)$$

Exercise 2.1.2 shows us $2 \operatorname{acl}(z_n)$ and more. Here, prove $\lim_{n\to\infty} z_n \neq 2$.

2.2.3. Use the definition of limit for sequences (Definition 2.2.1) to prove each of the following sequences of real numbers converges to their given limit.

(i) $\displaystyle\lim_{n\to\infty}\frac{3n+4}{4n+5}=\frac{3}{4}$.

(ii) $\displaystyle\lim_{n\to\infty}\frac{n^2-1}{n^4+2n^2}=0$.

(iii) $\displaystyle\lim_{n\to\infty}\frac{\sqrt[3]{n-1}}{\sqrt{n+1}}=0$.

2.2.4. Given a threshold for the convergence of a sequence, any larger positive integer is also a threshold (see Definition 2.2.1). To prove this, suppose $(\mathbf{x}_n)$ is a sequence of points in $\mathbb{R}^m$ that converges $\mathbf{y}$ with threshold $n_\varepsilon \in \mathbb{N}$ responding to the distance $\varepsilon > 0$. Prove that if $k_\varepsilon \in \mathbb{N}$ satisfies $k_\varepsilon \geq n_\varepsilon$, then k_ε is also a threshold for the convergence of $(\mathbf{x}_n)$ to $\mathbf{y}$ in response to ε.

2.2.5. Consider the sequences $(x_n), (y_n)$, and (z_n) defined for each $n \in \mathbb{N}$ by

$$x_n = \frac{50}{n}, \quad y_n = \frac{50}{n^2}, \quad \text{and} \quad z_n = \frac{50}{\sqrt{n}}, \quad \text{respectively.} \tag{2.2.47}$$

(i) Use the definition of sequential limit (Definition 2.2.1) to prove

$$\lim_{n\to\infty} x_n = \lim_{n\to\infty} y_n = \lim_{n\to\infty} z_n = 0, \tag{2.2.48}$$

and keep track of the threshold you use for each sequence.

(ii) Compare the thresholds you found for part (i) to determine which sequence converges to 0 the fastest and which is slowest.

(iii) Use the definition of sequential limit (Definition 2.2.1) to prove

$$\lim_{n\to\infty} (x_n + y_n + z_n) = 0. \tag{2.2.49}$$

What did you use for a threshold?

2.2.6. Determine the closure of each of the following sets. Don't prove anything, but draw stuff.

(a) $A = \mathbb{N}$

(b) $B = \{8 - (-1)^n/n : n \in \mathbb{N}\}$

(c) $C = \{3(-1)^n + 1/n : n \in \mathbb{N}\}$

(d) $D = \{.9, .99, .999, \ldots\}$

(e) $E = (-1, 1]$

(f) $F = [-1, 1]$

(g) $G = (-1, 1] \cup \{-3 + (1/n) : n \in \mathbb{N}\}$

(h) $H = \{0, 1, 0, 1/2, 0, 1/3, \ldots\}$

Next, consider only the above sets with closures that add at least one new point. For each new point, find a sequence from the corresponding set that converges to the point. (See Theorem 2.3.1.)

2.2.7. A sequence of real numbers (x_n) is a said to be *strictly increasing* if $x_n < x_{n+1}$ for each $n \in \mathbb{N}$.

(i) Prove that if $U \subseteq \mathbb{R}$, $\sup U$ exists, and $\sup U \notin U$, then there is a strictly increasing sequence (x_n) of points in U such that

$$\lim_{n \to \infty} x_n = \sup U. \tag{2.2.50}$$

(ii) Find an example of a set $V \subseteq \mathbb{R}$ where $\sup V$ exists but there is no strictly increasing sequence of points in V whose limit is $\sup V$.

2.2.8. Suppose $(x_n) \subseteq \mathbb{R}$ is a sequence where $\lim_{n \to \infty} x_n = 0$ and let

$$z_n = x_n + 3140 + (-1)^n \quad \text{for each} \quad n \in \mathbb{N}. \tag{2.2.51}$$

(Compare with Example 2.2.12.)

(i) Prove $3139 \operatorname{acl}(z_n)$ and $3141 \operatorname{acl}(z_n)$.

(ii) Despite the results of part (i), prove

$$\lim_{n \to \infty} z_n \neq 3139 \quad \text{and} \quad \lim_{n \to \infty} z_n \neq 3141. \tag{2.2.52}$$

(iii) Prove (z_n) diverges by showing for every real number c we have

$$\lim_{n \to \infty} z_n \neq c. \tag{2.2.53}$$

The upcoming Divergence Criteria for Sequences 2.6.9 would make short work of this proof, but the goal here is to get some practice working with the negation of Definition 2.2.1.

2.3 Properties of sequential limits

Now that we have a formal definition for the limit and convergence of a sequence provided by Definition 2.2.1, we can prove a bunch of results from calculus and multivariable calculus. Lots of scratch work and details are provided with the proofs. Much of the effort comes from trying to determine appropriate thresholds.

First, the following theorem establishes a fundamental connection between the definitions of limit and arbitrarily close. See Figure 2.3.1.

> **Theorem 2.3.1: Fundamental connection between arbitrarily close and convergence**
>
> Let $\mathbf{y} \in \mathbb{R}^m$ and $S \subseteq \mathbb{R}^m$. Then $\mathbf{y}$ is arbitrarily close to S if and only if there is a sequence $(\mathbf{x}_n)$ of points in S whose limit is $\mathbf{y}$.

The proof of Theorem 2.3.1 is an important exercise for mathematicians who are dealing with the definition of convergence for the first time. I strongly encourage you to try it yourself before reading the proof provided here. Your understanding of the definitions for arbitrarily close and convergence (Definitions 1.5.1 and 2.2.1) will be strengthened by giving this proof a shot.

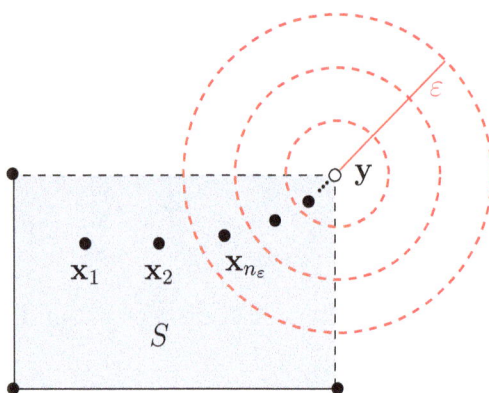

Figure 2.3.1: A set S and a point $\mathbf{y}$ in the plane $\mathbb{R}^2$ where $\mathbf{y}\,\mathrm{acl}\,S$. By the fundamental connection between arbitrarily close and convergence (Theorem 2.3.1), there is a sequence $(\mathbf{x}_n)$ of points in S where $\lim_{n\to\infty}\mathbf{x}_n = \mathbf{y}$.

Scratch Work 2.3.2: A subtle use of definitions

The proof requires nothing more than the definitions for arbitrarily close and convergence (Definitions 1.5.1 and 2.2.1), but there are some subtleties to deal with. When we assume $\mathbf{y}\,\mathrm{acl}\,S$, we need to somehow build a sequence $(\mathbf{x}_n)$ that converges to $\mathbf{y}$ with terms that belong to S. However, $\mathbf{y}\,\mathrm{acl}\,S$ only provides a single point in S for each distance $\varepsilon > 0$.

So, how can we find a whole sequence of points to work with? The idea is to consider a sequence of distances provided by treating the reciprocal of $1/n$ as a distance for each $n \in \mathbb{N}$. Each positive distance $1/n$ provides a term $\mathbf{x}_n$ from the set S thanks to the assumption that $\mathbf{y}\,\mathrm{acl}\,S$. Essentially, I am treating the ε as $1/n$, one index n at a time, in the definition for arbitrarily close (Definition 1.5.1). Actually, any sequence of positive real numbers that converges to 0 could be made to work, but $(1/n)$ is good enough.

The converse is shorter. Convergence gives us a whole n_ε-tail of the given sequence within a given ε of the limit, but we only need one term for each $\varepsilon > 0$ to show the limit is arbitrarily close to the set.

Proof of Theorem 2.3.1. Assume $\mathbf{y}$ is arbitrarily close to S and note that $1/n > 0$ for each $n \in \mathbb{N}$. So by the definition of arbitrarily close (Definition 1.5.1), for each $n \in \mathbb{N}$ there is some $\mathbf{x}_n$ in S where

$$d_m(\mathbf{x}_n, \mathbf{y}) = \|\mathbf{x}_n - \mathbf{y}\|_m < \frac{1}{n}. \tag{2.3.1}$$

Now, let $\varepsilon > 0$. (By not specifying a particular value of ε, we are accounting for all neighborhoods of $\mathbf{y}$ at the same time.) By the Corollary of the Archimedean Property 1.4.8, there is an index $n_\varepsilon \in \mathbb{N}$ large enough to give us $1/n_\varepsilon < \varepsilon$. Then for $n \in \mathbb{N}$ we have

$$n \geq n_\varepsilon \quad \Longrightarrow \quad \frac{1}{n} \leq \frac{1}{n_\varepsilon} < \varepsilon. \tag{2.3.2}$$

Hence, n_ε is a threshold for the convergence of $(\mathbf{x}_n)$ to its limit $\mathbf{y}$ since

$$n \geq n_\varepsilon \quad \Longrightarrow \quad d_m(\mathbf{x}_n, \mathbf{y}) = \|\mathbf{x}_n - \mathbf{y}\|_m < \frac{1}{n} \leq \frac{1}{n_\varepsilon} < \varepsilon. \tag{2.3.3}$$

Therefore, $(\mathbf{x}_n)$ is a sequence of points in S where $\lim_{n\to\infty} \mathbf{x}_n = \mathbf{y}$.

For the converse, assume there is a sequence $(\mathbf{x}_n)$ of points in S whose limit is $\mathbf{y}$ and let $\varepsilon > 0$. By the definition of sequential limit and convergence (Definition 2.2.1), there is a threshold n_ε such that $\mathbf{x}_{n_\varepsilon}$ is in S and

$$d_m(\mathbf{x}_{n_\varepsilon}, \mathbf{y}) = \|\mathbf{x}_{n_\varepsilon} - \mathbf{y}\|_m < \varepsilon. \tag{2.3.4}$$

Therefore, $\mathbf{y}$ acl S. $\qquad\square$

With such a deep connection between arbitrarily close and limits in the context of sequences established in Theorem 2.3.1, it should hopefully come as no surprise that many of the concepts explored in calculus, analysis, and even topology can be discussed in terms of points arbitrarily close to sets.

Remark 2.3.3: Arbitrarily close does not imply convergence

A word of caution: Theorem 2.3.1 does not say that if a sequence is arbitrarily close to a given point, then the limit exists and is equal to the given point. For instance, the sequence (b_n) from Examples 2.1.4, 2.1.10, and 2.2.12 is arbitrarily close to 2 but does not converge to 2.

Theorem 2.3.1 allows us to prove many facts about limits of sequences, including the following corollary. More results follow later in this chapter and throughout the book.

Corollary 2.3.4: Suprema and infima are limits

Suppose $S, T \subseteq \mathbb{R}$ where $u = \sup S$ and $v = \inf T$. Then there is a sequence (x_n) of real numbers in S whose limit is u and there is a sequence (w_n) of real numbers in T whose limit is v.

Proof of Corollary 2.3.4. The statement follows from the definitions of supremum and infimum (Definition 1.1.14) and the fundamental connection between arbitrarily close and convergence connection (Theorem 2.3.1). Since a supremum and an infimum are arbitrarily close to their respective sets, each is the limit of a sequence of points in their respective sets. $\qquad\square$

Next, the limit of a convergent sequence in a Euclidean space is unique.

Theorem 2.3.5: Sequential limits in Euclidean spaces are unique

Every convergent sequence in $\mathbb{R}^m$ has a unique limit.

Scratch Work 2.3.6: Thresholds without explicit formulas

My idea is to show any two points $\mathbf{y}$ and $\mathbf{z}$ satisfying the definition of sequential limit for the same sequence are arbitrarily close to each other, so they must be the same point (see Lemma 1.5.5). The definition of limit and convergence for sequences (Definition 2.2.1) ensures that we can respond to *any positive distance* we like.

However, unlike the examples explored in Section 2.2, we do not have explicit formulas for the thresholds to work with. Instead, we are in a more abstract setting where we will both *use* assumptions that certain limits exist and *show* other related limits exist. Also, whenever $\varepsilon > 0$, we have $\varepsilon/2 > 0$ as well. So given any $\varepsilon > 0$, we can use the definition of sequential limit to take both $\mathbf{y}$ and $\mathbf{z}$ to be within $\varepsilon/2$ of the sequence, each coming with their own threshold. The larger of these two thresholds (corresponding to what would be the slower rate of convergence) is a threshold for the convergence of the sequence to both $\mathbf{y}$ and $\mathbf{z}$. The points $\mathbf{y}$ and $\mathbf{z}$ would then be within any ε of each other thanks to the triangle inequality (1.2.32) combined with one particular term in the sequence.

Proof of Theorem 2.3.5. Suppose $(\mathbf{x}_n)$ is a convergent sequence in $\mathbb{R}^m$ where

$$\mathbf{y} = \lim_{n\to\infty} \mathbf{x}_n \qquad \text{and} \qquad \mathbf{z} = \lim_{n\to\infty} \mathbf{x}_n. \tag{2.3.5}$$

Let $\varepsilon > 0$. Then $\varepsilon/2 > 0$ and by the definition of sequential limit (Definition 2.2.1), there are two positive integer thresholds $j_{\varepsilon/2}$ and $k_{\varepsilon/2}$ where

$$n \geq j_{\varepsilon/2} \implies \|\mathbf{x}_n - \mathbf{y}\|_m < \varepsilon/2 \qquad \text{and} \tag{2.3.6}$$
$$n \geq k_{\varepsilon/2} \implies \|\mathbf{x}_n - \mathbf{z}\|_m < \varepsilon/2. \tag{2.3.7}$$

Now consider the index $n_\varepsilon = \max\{j_{\varepsilon/2}, k_{\varepsilon/2}\}$ (so n_ε is the larger of the two). We have both $n_\varepsilon \geq j_{\varepsilon/2}$ and $n_\varepsilon \geq k_{\varepsilon/2}$, therefore

$$\|\mathbf{y} - \mathbf{z}\|_m \leq \|\mathbf{y} - \mathbf{x}_{n_\varepsilon}\|_m + \|\mathbf{x}_{n_\varepsilon} - \mathbf{z}\|_m \qquad \text{(tri. ineq. (1.2.32))} \tag{2.3.8}$$
$$< \frac{\varepsilon}{2} + \frac{\varepsilon}{2} \qquad \text{((2.3.6) and (2.3.7))} \tag{2.3.9}$$
$$= \varepsilon. \tag{2.3.10}$$

Hence, $\mathbf{y}\,\mathrm{acl}\{\mathbf{z}\}$ and by Lemma 1.5.5, $\mathbf{y} = \mathbf{z}$. $\qquad\square$

Remark 2.3.7: Subscripts can help us track details

Does the above proof make sense? Without more detailed scratch work, it may be hard to tell why each step is there. However, you should be able to tell how each line of the proof follows from assertions and conclusions that come before. This is easier said than done and takes some time. So, please take your time. If you don't feel comfortable with this proof, try writing up a walkthrough. How would you write the proof?

Can you see a reason to use the subscripts for n_ε, $j_{\varepsilon/2}$, and $k_{\varepsilon/2}$ as they are? Each of these positive integers is an index for the sequence $(\mathbf{x}_n)$ given in response to the distance ε or $\varepsilon/2$, accordingly. These distances help us make careful use of the definition of arbitrarily close (Definition 1.5.1) and the definition of limit and convergence for sequences (Definition 2.2.1).

Remark 2.3.8: Linearity of convergent sequences

As mentioned in Section 1.6 and Remark 1.6.18, linearity pervades analysis in a number of interesting ways. In the context of convergent sequences from calculus, linearity manifests as "the limit of a sum is the sum of the limits" (additivity) and "constants factor out" (homogeneity). We now have the mathematical tools to formally prove the linearity of sequential limits.

Theorem 2.3.9: Linearity of sequential limits

Suppose $c \in \mathbb{R}$ and suppose $(\mathbf{a}_n)$ and $(\mathbf{b}_n)$ are convergent sequences in $\mathbb{R}^m$ where $\lim_{n\to\infty} \mathbf{a}_n = \mathbf{a}$ and $\lim_{n\to\infty} \mathbf{b}_n = \mathbf{b}$. Then

(i) $\displaystyle\lim_{n\to\infty} (\mathbf{a}_n + \mathbf{b}_n) = \lim_{n\to\infty} \mathbf{a}_n + \lim_{n\to\infty} \mathbf{b}_n = \mathbf{a} + \mathbf{b}$ (*additivity*); and

(ii) $\displaystyle\lim_{n\to\infty} (c\,\mathbf{a}_n) = c \lim_{n\to\infty} \mathbf{a}_n = c\,\mathbf{a}$ (*homogeneity*).

Remark 2.3.10: No explicit formula for thresholds

To prove each statement in Theorem 2.3.9, we can verify the definition of sequential limit (Definition 2.2.1) holds by considering an arbitrary $\varepsilon > 0$ and finding a suitable threshold n_ε. This threshold can be shown to ensure the terms with indices $n \geq n_\varepsilon$ are within ε of the proposed limit.

As in the proof of Theorem 2.3.5 but unlike the examples explored in Section 2.2, we do not have explicit formulas to work with. We are in a more general setting where we will both *use* assumptions that certain limits exist along with their corresponding thresholds, and *show* other related limits exist by defining new thresholds as needed.

Scratch Work 2.3.11: Additivity of sequential limits

To derive some scratch work for directly proving

$$\lim_{n\to\infty} (\mathbf{a}_n + \mathbf{b}_n) = \mathbf{a} + \mathbf{b}, \tag{2.3.11}$$

let's start at the end. Given $\varepsilon > 0$, we want to end up with

$$\|(\mathbf{a}_n + \mathbf{b}_n) - (\mathbf{a} + \mathbf{b})\|_m < \varepsilon. \tag{2.3.12}$$

We can assume

$$\lim_{n\to\infty} \mathbf{a}_n = \mathbf{a} \qquad \text{and} \qquad \lim_{n\to\infty} \mathbf{b}_n = \mathbf{b}, \tag{2.3.13}$$

so by Definition 2.2.1, given any $\varepsilon > 0$ there are thresholds j_ε and k_ε where $n \geq j_\varepsilon$ implies

$$\|\mathbf{a}_n - \mathbf{a}\|_m < \varepsilon \tag{2.3.14}$$

and $n \geq k_\varepsilon$ implies

$$\|\mathbf{b}_n - \mathbf{b}\|_m < \varepsilon. \tag{2.3.15}$$

Combining (2.3.14) and (2.3.15) by taking n large enough so that both $n \geq j_\varepsilon$ and $n \geq k_\varepsilon$, and applying triangle inequality (1.2.35), gives us

$$\|(\mathbf{a}_n + \mathbf{b}_n) - (\mathbf{a} + \mathbf{b})\|_m = \|(\mathbf{a}_n - \mathbf{a}) + (\mathbf{b}_n - \mathbf{b})\|_m \tag{2.3.16}$$
$$\leq \|\mathbf{a}_n - \mathbf{a}\|_m + \|\mathbf{b}_n - \mathbf{b}\|_m \tag{2.3.17}$$
$$< \varepsilon + \varepsilon \tag{2.3.18}$$
$$= 2\varepsilon. \tag{2.3.19}$$

This isn't quite our goal (2.3.12), but we can adapt: The definition of limit and convergence for sequences (Definition 2.2.1) ensures that we can respond to *any positive distance* we like with suitable thresholds. So, as in the proof of Theorem 2.3.5, $\varepsilon/2 > 0$ whenever $\varepsilon > 0$. We can take advantage of the assumptions

$$\lim_{n\to\infty} \mathbf{a}_n = \mathbf{a} \qquad \text{and} \qquad \lim_{n\to\infty} \mathbf{b}_n = \mathbf{b} \tag{2.3.20}$$

by considering indices n and thresholds $j_{\varepsilon/2}$ and $k_{\varepsilon/2}$ where both

$$n \geq j_{\varepsilon/2} \implies \|\mathbf{a}_n - \mathbf{a}\|_m < \frac{\varepsilon}{2} \qquad \text{and} \tag{2.3.21}$$
$$n \geq k_{\varepsilon/2} \implies \|\mathbf{b}_n - \mathbf{b}\|_m < \frac{\varepsilon}{2}. \tag{2.3.22}$$

See Figure 2.3.2. Choosing $n_\varepsilon = \max\{j_{\varepsilon/2}, k_{\varepsilon/2}\}$ (so n_ε is the larger of the two) yields a sufficient threshold for the sums.

Proof of additivity in Theorem 2.3.9. Assume

$$\lim_{n\to\infty} \mathbf{a}_n = \mathbf{a} \qquad \text{and} \qquad \lim_{n\to\infty} \mathbf{b}_n = \mathbf{b}, \tag{2.3.23}$$

and let $\varepsilon > 0$. Then $\varepsilon/2 > 0$ and there are thresholds $j_{\varepsilon/2}$ and $k_{\varepsilon/2}$ where

$$n \geq j_{\varepsilon/2} \implies \|\mathbf{a}_n - \mathbf{a}\|_m < \frac{\varepsilon}{2} \qquad \text{and} \tag{2.3.24}$$
$$n \geq k_{\varepsilon/2} \implies \|\mathbf{b}_n - \mathbf{b}\|_m < \frac{\varepsilon}{2}. \tag{2.3.25}$$

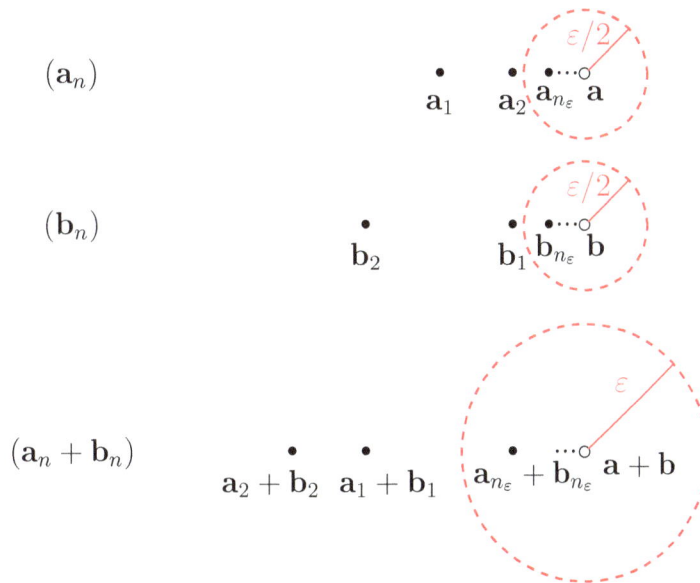

Figure 2.3.2: A figure to accompany the proof of Theorem 2.3.9 showing $\lim_{n\to\infty} \mathbf{a}_n = \mathbf{a}$ and $\lim_{n\to\infty} \mathbf{b}_n = \mathbf{b}$ implies $\lim_{n\to\infty}(\mathbf{a}_n + \mathbf{b}_n) = \mathbf{a} + \mathbf{b}$.

Define $n_\varepsilon = \max\{j_{\varepsilon/2}, k_{\varepsilon/2}\}$ (so n_ε is the larger of the two). Then every index n where $n \geq n_\varepsilon$ is large enough to give us both $n \geq j_{\varepsilon/2}$ and $n \geq k_{\varepsilon/2}$. So by the triangle inequality (1.2.35), (2.3.24), and (2.3.25), we have $n \geq n_\varepsilon$ implies

$$\|(\mathbf{a}_n + \mathbf{b}_n) - (\mathbf{a} + \mathbf{b})\|_m = \|(\mathbf{a}_n - \mathbf{a}) + (\mathbf{b}_n - \mathbf{b})\|_m \tag{2.3.26}$$

$$\leq \|\mathbf{a}_n - \mathbf{a}\|_m + \|\mathbf{b}_n - \mathbf{b}\|_m \tag{2.3.27}$$

$$< \frac{\varepsilon}{2} + \frac{\varepsilon}{2} \tag{2.3.28}$$

$$= \varepsilon. \tag{2.3.29}$$

Therefore, $\lim_{n\to\infty}(\mathbf{a}_n + \mathbf{b}_n) = \mathbf{a} + \mathbf{b}$. $\qquad\square$

Next, let's prove $\lim_{n\to\infty} c\,\mathbf{a}_n = c\,\mathbf{a}$, the homogeneity half of Theorem 2.3.9.

Scratch Work 2.3.12: Homogeneity of sequential limits

Once again, let's start at the end. Given $\varepsilon > 0$, we want to end up with

$$\|c\,\mathbf{a}_n - c\,\mathbf{a}\|_m < \varepsilon. \tag{2.3.30}$$

By (1.2.33) we can consider

$$\|c\,\mathbf{a}_n - c\,\mathbf{a}\|_m = |c|\|\mathbf{a}_n - \mathbf{a}\|_m < \varepsilon. \tag{2.3.31}$$

So if $c \neq 0$, then $|c| \neq 0$ and we can divide both sides of the rightmost inequality in (2.3.31) to get

$$\|\mathbf{a}_n - \mathbf{a}\|_m < \frac{\varepsilon}{|c|}. \tag{2.3.32}$$

So, by assuming $\lim_{n\to\infty} \mathbf{a}_n = \mathbf{a}$, Definition 2.2.1 ensures a threshold $n_{\varepsilon/|c|}$ can be found in response to the positive distance $\varepsilon/|c|$ that will suffice, as long as $c \neq 0$.

Proof of homogeneity in Theorem 2.3.9. This proof has two cases: (i) $c = 0$ and (ii) $c \neq 0$.

$\underline{\text{Case (i)}}$: Suppose $c = 0$ and $\varepsilon > 0$. Define $n_\varepsilon = 7$. Then for every index $n \geq n_\varepsilon = 7$ we have

$$\|c\,\mathbf{a}_n - c\,\mathbf{a}\|_m = \|\mathbf{0} - \mathbf{0}\|_m = 0 < \varepsilon. \tag{2.3.33}$$

Therefore, $\lim_{n\to\infty} c\,\mathbf{a}_n = c\,\mathbf{a} = \mathbf{0}$.

$\underline{\text{Case (ii)}}$: Suppose $c \neq 0$, $\lim_{n\to\infty} \mathbf{a}_n = \mathbf{a}$, and $\varepsilon > 0$. Then $\varepsilon/|c| > 0$ and by the definition of sequential limit (Definition 2.2.1), there is a threshold n_ε such that

$$n \geq n_\varepsilon \quad \Longrightarrow \quad \|\mathbf{a}_n - \mathbf{a}\|_m < \frac{\varepsilon}{|c|}. \tag{2.3.34}$$

By (1.2.33) and (2.3.34), for all indices $n \geq n_\varepsilon$ we have

$$\|c\,\mathbf{a}_n - c\,\mathbf{a}\|_m = |c|\|\mathbf{a}_n - \mathbf{a}\|_m \tag{2.3.35}$$

$$< |c|\frac{\varepsilon}{|c|} \tag{2.3.36}$$

$$= \varepsilon. \tag{2.3.37}$$

Therefore, $\lim_{n\to\infty} c\,\mathbf{a}_n = c\,\mathbf{a}$. $\square$

As mentioned in Remark 1.6.18, a corollary of the linearity of sequential limits holds for linear combinations. As with the proof Corollary 1.6.16 on arbitrarily close and linear combinations of sets, the proof of Corollary 2.3.13 follows from induction. So, it is left as an exercise. Here, the notation $\mathbf{a}_{j,n}$ indicates the nth term of the jth sequence.

Corollary 2.3.13: Linear combinations of sequential limits

Suppose $k \in \mathbb{N}$ and for each $j = 1, \ldots, k$ we have $c_j \in \mathbb{R}$ and the sequence $(\mathbf{a}_{j,n}) \subseteq \mathbb{R}^m$ converges. Then

$$\lim_{n\to\infty}\left(\sum_{j=1}^{k} c_j \mathbf{a}_{j,n}\right) = \sum_{j=1}^{k}\left(c_j \lim_{n\to\infty} \mathbf{a}_{j,n}\right). \tag{2.3.38}$$

The definition of bounded sets (Definition 1.5.20) adapts to sequences.

Definition 2.3.14: Bounded sequence in a Euclidean space

A sequence $(\mathbf{x}_n)$ of points in $\mathbb{R}^m$ is *bounded* if its range is a bounded set. That is, $(\mathbf{x}_n)$ is bounded if there is a real number $b \geq 0$ such that for every index $n \in \mathbb{N}$ we have

$$\|\mathbf{x}_n\|_m \leq b. \tag{2.3.39}$$

In this case, we say b is a *bound* for the sequence $(\mathbf{x}_n)$. If a sequence is not bounded, we say it is *unbounded*.

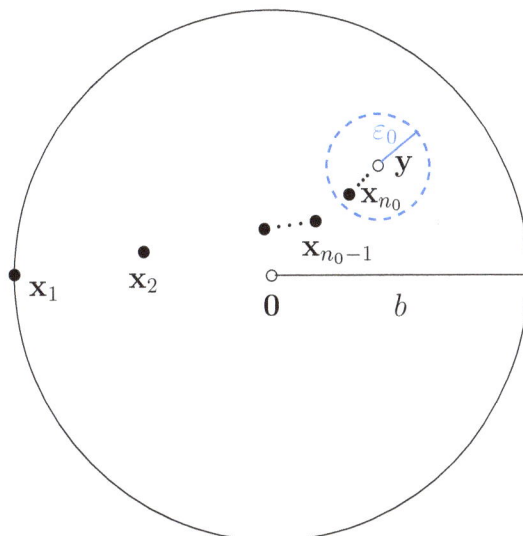

Figure 2.3.3: A convergent sequence $(\mathbf{x}_n)$ with limit $\mathbf{y}$ in $\mathbb{R}^m$ where the first term $\mathbf{x}_1$ has the largest norm. See Scratch Work 2.3.16 and the proof of Theorem 2.3.15.

Convergent sequences must be bounded. Since every neighborhood of the limit contains a tail, we can pick any neighborhood to work with and show that the tail is bounded near the limit while only a finite number of the terms are potentially larger than the limit.

Theorem 2.3.15: Convergent sequences are bounded

Every convergent sequence in $\mathbb{R}^m$ is bounded.

Scratch Work 2.3.16: Find a bound by choosing a distance

Consider the following pair of figures of convergent sequences. In Figure 2.3.3, the term $\mathbf{x}_1$ happens to have the largest norm out of all of the terms in the sequence. In Figure 2.3.4, no particular term in the sequence has the largest norm, but for some $\varepsilon_0 > 0$ the real number $\varepsilon_0 + \|\mathbf{y}\|$ is a bound for the sequence.

In any case, when a sequence $(\mathbf{x}_n)$ converges to $\mathbf{y}$ in $\mathbb{R}^m$, we can choose any $\varepsilon_0 > 0$ and get a threshold n_0 to ensure all of the terms $\mathbf{x}_n$ where $n \geq n_0$ are within ε_0 of $\mathbf{y}$. From there, the real number b defined by

$$b = \max\{\|\mathbf{x}_1\|_m, \|\mathbf{x}_2\|_m, \ldots, \|\mathbf{x}_{n_0-1}\|_m, \varepsilon_0 + \|\mathbf{y}\|_m\} \qquad (2.3.40)$$

is a bound for the sequence.

Proof of Theorem 2.3.15. Suppose $(\mathbf{x}_n)$ is a sequence in $\mathbb{R}^m$ whose limit is $\mathbf{y}$. Consider the distance $\varepsilon_0 = 7$. (There's nothing special about 7, except that it's positive and defines a suitable neighborhood around the limit.) Since $\mathbf{y} = \lim_{n \to \infty} \mathbf{x}_n$, there is a threshold n_0 where $n \geq n_0$

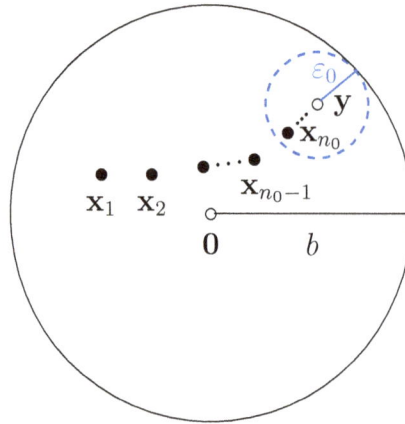

Figure 2.3.4: A convergent sequence $(\mathbf{x}_n)$ with limit $\mathbf{y}$ in $\mathbb{R}^m$ where no particular term has the largest norm but $\|\mathbf{y}\| + \varepsilon_0$ for some positive ε_0 serves as a bound for $(\mathbf{x}_n)$. See Scratch Work 2.3.16 and the proof of Theorem 2.3.15.

implies

$$d_m(\mathbf{x}_n, \mathbf{y}) = \|\mathbf{x}_n - \mathbf{y}\|_m < \varepsilon_0 = 7. \tag{2.3.41}$$

Therefore, for $n \geq n_0$,

$$\|\mathbf{x}_n\|_m = \|\mathbf{x}_n \underbrace{-\mathbf{y} + \mathbf{y}}_{\text{add } \mathbf{0}}\|_m \tag{2.3.42}$$

$$\leq \|\mathbf{x}_n - \mathbf{y}\|_m + \|\mathbf{y}\|_m \qquad \text{(tri. ineq. (1.2.34))} \tag{2.3.43}$$

$$< 7 + \|\mathbf{y}\|_m. \qquad ((2.3.41)) \tag{2.3.44}$$

Now define b by

$$b = \max\{\|\mathbf{x}_1\|_m, \|\mathbf{x}_2\|_m, \ldots, \|\mathbf{x}_{n_0-1}\|_m, 7 + \|\mathbf{y}\|_m\}. \tag{2.3.45}$$

Then $b \geq 0$ and for every index $n \in \mathbb{N}$ we have

$$\|\mathbf{x}_n\|_m \leq b. \tag{2.3.46}$$

Therefore, (x_n) is bounded. $\square$

There are more algebraic properties for limits beyond linearity when we restrict our attention to the real line $\mathbb{R}$. In this context, we can play with multiplication and division. Do you remember hearing a statement like "the limit of a product is the product of the limits"?

Theorem 2.3.17: Products of sequential limits in $\mathbb{R}$

Suppose (a_n) and (b_n) are convergent sequences of real numbers where $\lim_{n\to\infty} a_n = a$ and $\lim_{n\to\infty} b_n = b$. Then

$$\lim_{n\to\infty} (a_n b_n) = \left(\lim_{n\to\infty} a_n\right)\left(\lim_{n\to\infty} b_n\right) = ab. \tag{2.3.47}$$

Scratch Work 2.3.18: Add zero, bound a factor

To derive some scratch work for showing $\lim_{n\to\infty} a_n b_n = ab$ in Theorem 2.3.17, let's start at the end. Given $\varepsilon > 0$, we want to end up with

$$|a_n b_n - ab| < \varepsilon. \tag{2.3.48}$$

In order to get (2.3.48), we can try to play around with expressions involving both $|a_n - a|$ and $|b_n - b|$ since they appear when we assume $\lim_{n\to\infty} a_n = a$ and $\lim_{n\to\infty} b_n = b$.

There are plenty of tools at our disposal to help us here, including the addition of the particular version of zero given by

$$0 = -ab_n + ab_n. \tag{2.3.49}$$

Pairing this especially nice version of zero with the triangle inequality (1.2.34) yields the following string of inequalities:

$$|a_n b_n - ab| = |a_n b_n \underbrace{-ab_n + ab_n}_{\text{add } 0} - ab| \tag{2.3.50}$$

$$\leq |a_n b_n - ab_n| + |ab_n - ab| \qquad (\text{tri. ineq. } (1.2.34)) \tag{2.3.51}$$

$$= |b_n||a_n - a| + |a||b_n - b|. \tag{2.3.52}$$

From there, assuming $\lim_{n\to\infty} a_n = a$ and $\lim_{n\to\infty} b_n = b$ allows us to respond to a given $\varepsilon > 0$ with thresholds j_ε and k_ε where

$$n \geq j_\varepsilon \quad \Longrightarrow \quad |a_n - a| < \varepsilon \qquad \text{and} \tag{2.3.53}$$

$$n \geq k_\varepsilon \quad \Longrightarrow \quad |b_n - b| < \varepsilon. \tag{2.3.54}$$

Combining (2.3.50) through (2.3.54) yields

$$|a_n b_n - ab| \leq |b_n||a_n - a| + |a||b_n - b| \tag{2.3.55}$$

$$< |b_n|\varepsilon + |a|\varepsilon \tag{2.3.56}$$

$$= \varepsilon(|b_n| + |a|). \tag{2.3.57}$$

From here, we can use a common bound on a and the sequence (b_n) to help find a suitable threshold for $(a_n b_n)$. By Theorem 2.3.15, the convergence of (a_n) and (b_n) ensures they are bounded by some real numbers $u \geq 0$ and $v \geq 0$, respectively. We can consider the sum $q = u + v$, which is a bound for both (a_n) and (b_n). The limit a is arbitrarily close to the sequence (a_n) by Theorem 2.3.1, so by Lemma 1.5.24 a respects that same bound and we have $|a| \leq u \leq q$. (A similar statement holds for b, but we won't need it.) Next, by choosing $u > 0$ or $v > 0$ to ensure $q > 0$, we can then consider the positive distance $\varepsilon/2q$. The thresholds $j_{\varepsilon/2q}$ for (a_n) and $k_{\varepsilon/2q}$ for (b_n) combine to create a threshold for $(a_n b_n)$, for instance we can use $n_\varepsilon = j_{\varepsilon/2q} + k_{\varepsilon/2q}$.

We have the pieces. Time for a proof.

Proof of Theorem 2.3.17. Assume (a_n) and (b_n) are convergent sequences of real numbers with $\lim_{n\to\infty} a_n = a$ and $\lim_{n\to\infty} b_n = b$. By Theorem 2.3.15, (a_n) and (b_n) are both bounded by a some positive real number q. The limit a is arbitrarily close to the sequence (a_n) by Theorem 2.3.1, so we have $|a| \leq q$ by Lemma 1.5.24. Thus,

$$|a| \leq q \quad \text{and} \quad |b_n| \leq q \quad \text{for every index } n \in \mathbb{N}. \tag{2.3.58}$$

Now, let $\varepsilon > 0$. Since $2q > 0$, we also have $\varepsilon/2q > 0$. By the definition of limit and convergence for sequences (Definition 2.2.1), there are thresholds $j_{\varepsilon/2q}$ and $k_{\varepsilon/2q}$ where

$$n \geq j_{\varepsilon/2q} \quad \Longrightarrow \quad |a_n - a| < \frac{\varepsilon}{2q} \quad \text{and} \tag{2.3.59}$$

$$n \geq k_{\varepsilon/2q} \quad \Longrightarrow \quad |b_n - b| < \frac{\varepsilon}{2q}. \tag{2.3.60}$$

Define $n_\varepsilon = j_{\varepsilon/2q} + k_{\varepsilon/2q}$. Then for any index $n \geq n_\varepsilon$ we have both $n \geq j_{\varepsilon/2q}$ and $n \geq k_{\varepsilon/2q}$. Hence,

$$|a_n b_n - ab| = |a_n b_n \underbrace{-ab_n + ab_n}_{\text{add } 0} - ab| \tag{2.3.61}$$

$$\leq |a_n b_n - ab_n| + |ab_n - ab| \quad \text{(tri. ineq. (1.2.34))} \tag{2.3.62}$$

$$= |b_n||a_n - a| + |a||b_n - b| \tag{2.3.63}$$

$$\leq q|a_n - a| + q|b_n - b| \quad \text{(by (2.3.58))} \tag{2.3.64}$$

$$< q\left(\frac{\varepsilon}{2q}\right) + q\left(\frac{\varepsilon}{2q}\right) \quad \text{(by (2.3.59) and (2.3.60))} \tag{2.3.65}$$

$$= \varepsilon. \tag{2.3.66}$$

Therefore, $\lim_{n\to\infty} a_n b_n = ab$. $\qquad\square$

Next up, "the limit of a reciprocal is the reciprocal of the limit".

Lemma 2.3.19: Reciprocals of sequential limits in $\mathbb{R}$

Suppose (b_n) is a convergent sequence of real numbers where $\lim_{n\to\infty} b_n = b$, $b \neq 0$, and $b_n \neq 0$ for every index n, then

$$\lim_{n\to\infty} \frac{1}{b_n} = \frac{1}{\lim_{n\to\infty} b_n} = \frac{1}{b}. \tag{2.3.67}$$

Scratch Work 2.3.20: Common denominator, bound a reciprocal

Given $\varepsilon > 0$, we want to end up with

$$\left|\frac{1}{b_n} - \frac{1}{b}\right| < \varepsilon \tag{2.3.68}$$

for large enough $n \in \mathbb{N}$. Finding the common denominator of left-hand side yields

$$\left|\frac{1}{b_n} - \frac{1}{b}\right| = \frac{1}{|b_n b|}|b - b_n| < \varepsilon. \tag{2.3.69}$$

Note that since $b_n \neq 0$ and $b \neq 0$, the reciprocal $1/|b_n b|$ is defined. The convergence of (b_n) to b helps out twice here: We can get a bound for $1/|b_n b|$ then make $|b - b_n|$ as small as we like to compensate.

Proof of Lemma 2.3.19. Suppose (b_n) is a convergent sequence of real numbers where we have $\lim_{n \to \infty} b_n = b$, $b \neq 0$, and $b_n \neq 0$ for every index $n \in \mathbb{N}$. Since $b \neq 0$, we have $|b|/2 > 0$ and can treat this as a value of ε in the definition of sequential limits (Definition 2.2.1). Pairing this with the reverse triangle inequality (1.2.37) produces an index $n_0 \in \mathbb{N}$ where $n \geq n_0$ implies

$$|b| - |b_n| \leq |b_n - b| < \frac{|b|}{2}. \tag{2.3.70}$$

Solving for the reciprocal and multiplying by $1/|b| > 0$ gives us the useful bound

$$\frac{1}{|b_n b|} < \frac{2}{|b|^2}. \tag{2.3.71}$$

Now let $\varepsilon > 0$. Since $|b|^2 \varepsilon / 2 > 0$, the definition of sequential limit (Definition 2.2.1) applied to the convergence of (b_n) to b gives a threshold $n_1 \in \mathbb{N}$ such that $n \geq n_1$ implies

$$|b - b_n| < \frac{|b|^2 \varepsilon}{2}. \tag{2.3.72}$$

Define $n_\varepsilon = \max\{n_0, n_1\}$. Then for every index $n \in \mathbb{N}$ where $n \geq n_\varepsilon$, both (2.3.71) and (2.3.72) hold. Hence, we also have

$$\left| \frac{1}{b_n} - \frac{1}{b} \right| = \frac{1}{|b_n b|} |b - b_n| < \frac{2}{|b|^2} |b - b_n| < \frac{2}{|b|^2} \cdot \frac{|b|^2 \varepsilon}{2} = \varepsilon \tag{2.3.73}$$

for all $n \geq n_\varepsilon$. Therefore, $(1/b_n)$ converges to $1/b$. $\qquad \square$

The idea that "the limit of a quotient is the quotient of the limits" follows from combining Theorem 2.3.17 and Lemma 2.3.19.

Theorem 2.3.21: Quotients of sequential limits in $\mathbb{R}$

Suppose (a_n) and (b_n) are convergent sequences of real numbers where $\lim_{n \to \infty} a_n = a$, $\lim_{n \to \infty} b_n = b$, $b \neq 0$, and $b_n \neq 0$ for every index $n \in \mathbb{N}$. Then

$$\lim_{n \to \infty} \frac{a_n}{b_n} = \frac{\lim_{n \to \infty} a_n}{\lim_{n \to \infty} b_n} = \frac{a}{b}. \tag{2.3.74}$$

Proof of Theorem 2.3.21. Suppose the hypotheses hold. By Theorem 2.3.17 and Lemma 2.3.19 we have

$$\lim_{n \to \infty} \frac{a_n}{b_n} = \lim_{n \to \infty} \left(a_n \cdot \frac{1}{b_n} \right) = \left(\lim_{n \to \infty} a_n \right) \left(\lim_{n \to \infty} \frac{1}{b_n} \right) = a \cdot \frac{1}{b} = \frac{a}{b}. \tag{2.3.75}$$

$\qquad \square$

The final result of the section takes advantage of the deep connection between sequential limits and arbitrarily close in Theorem 2.3.1 and what we have proven regarding upper and lower bounds in the real line in Lemma 1.5.23.

$$(x_n) \qquad \circ \qquad \bullet \qquad\quad \bullet \ \ \bullet\cdots\circ \ \circ$$
$$ a \qquad x_1 \qquad\quad x_2 \quad \ell \ \ b$$

Figure 2.3.5: A plot of a convergent sequence (x_n) with limit ℓ, lower bound a, and upper bound b as in Corollary 2.3.22.

Corollary 2.3.22: Order properties for sequential limits in $\mathbb{R}$

Suppose a and b are real numbers and (x_n) is a convergent sequence of real numbers.

   (i) If $x_n \leq b$ for every index $n \in \mathbb{N}$, then $\lim_{n\to\infty} x_n = \ell \leq b$.

   (ii) If $x_n \geq a$ for every index $n \in \mathbb{N}$, then $\lim_{n\to\infty} x_n = \ell \geq a$.

Proof of Corollary 2.3.22. Suppose $\lim_{n\to\infty} x_n = \ell$. Then by Theorem 2.3.1, $\ell \operatorname{acl} (x_n)$.

Suppose $x_n \leq b$ for every index $n \in \mathbb{N}$. Then b is an upper bound for (x_n) and by part (i) of Lemma 1.5.23, we have $\ell \leq b$.

Now suppose $x_n \geq a$ for every index $n \in \mathbb{N}$. Then a is a lower bound for (x_n) and by part (ii) of Lemma 1.5.23, we have $\ell \geq a$. $\qquad\qquad\square$

The next section explores properties of sequences that guarantee convergence.

Exercises

2.3.1. Write up a walkthrough for the proof of Theorem 2.3.1. (This fundamental exercise connects the definition of arbitrarily close to the definition of sequential limits and convergence.)

2.3.2. Prove that if $(\mathbf{x}_n) \subseteq \mathbb{R}^m$ converges, then the sequence of norms $(\|\mathbf{x}_n\|_m) \subseteq \mathbb{R}$ converges as well.

2.3.3. Find an example of a sequence of real numbers (z_n) where $(|z_n|)$ converges but (z_n) diverges. (Hence, the converse of the previous exercise is false.)

2.3.4. Prove

$$\lim_{n\to\infty} \sqrt{1 + \frac{1}{n}} = 1. \tag{2.3.76}$$

2.3.5. Let (s_n) be the sequence of real numbers defined by

$$s_n = \sqrt{n^2 + n} - n \quad \text{for each} \quad n \in \mathbb{N}. \tag{2.3.77}$$

Prove $\lim_{n\to\infty} s_n = 1/2$. Hint: Use the previous exercise.

2.3.6. Suppose (x_n) is a sequence of real numbers where $x_n \geq 0$ for every $n \in \mathbb{N}$.

(i) Prove $\lim_{n\to\infty} x_n = 0$ implies $\lim_{n\to\infty} \sqrt{x_n} = 0$.

(ii) Prove $\lim_{n\to\infty} x_n = \ell$ implies $\lim_{n\to\infty} \sqrt{x_n} = \sqrt{\ell}$.

2.3.7. Suppose (y_n) is a sequence of real numbers where $y_n \geq 0$ for every $n \in \mathbb{N}$ and $\lim_{n\to\infty} y_n = \ell$. Prove that for each fixed $k \in \mathbb{N}$ we have

$$\lim_{n\to\infty} \sqrt[k]{y_n} = \sqrt[k]{\ell}. \tag{2.3.78}$$

2.3.8. Suppose $p : \mathbb{R} \to \mathbb{R}$ is a polynomial (Definition 1.6.8) and (x_n) is a sequence of real numbers where $\lim_{n\to\infty} x_n = \ell$. Prove

$$\lim_{n\to\infty} p(x_n) = p(\ell). \tag{2.3.79}$$

(This shows polynomials are *sequentially continuous* as in Definition 4.4.5.)

2.3.9. Let S denote the set of convergent sequences in $\mathbb{R}^m$. Use Lemma 1.6.7 to prove S is a vector space.

2.3.10. Suppose $\mathbf{y} \in \mathbb{R}^m$ and let $S_{\mathbf{y}}$ denote the set of sequences in $\mathbb{R}^m$ that converge to $\mathbf{y}$. Use Lemma 1.6.7 to prove $S_{\mathbf{y}}$ is a vector space if and only if $\mathbf{y} = \mathbf{0}$.

2.4 Ensuring convergence

Section 2.3 features results where the convergence of some given sequences is assumed and leads to the convergence of other related sequences. This section explores other ways to ensure the convergence of a sequence by considering various properties the sequence or a related set might have.

To motivate a guarantee property of convergence from calculus, consider the following example along with Figures 2.4.1 and 2.4.2.

Example 2.4.1: A squeezed sequence

Consider the sequence of real numbers (y_n) defined for each $n \in \mathbb{N}$ by

$$y_n = \frac{\sin \sqrt{n^2 + 1}}{n}. \tag{2.4.1}$$

Then $\lim_{n\to\infty} y_n = 0$. See Figures 2.4.1 and 2.4.2.

Remark 2.4.2: Motivating the Squeeze Theorem

A direct proof for Example 2.4.1 using the definition of sequential limit (Definition 2.2.1) would follow from the fact from trigonometry that for every real number x we have

$$-1 \leq \sin x \leq 1 \tag{2.4.2}$$

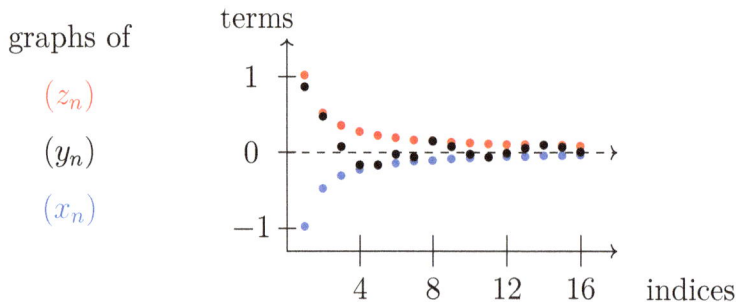

Figure 2.4.1: Graphs of the sequences $(x_n), (y_n)$, and (z_n) where $x_n = -1/n$, $y_n = \sin\left(\sqrt{n^2 + 1}\right)/n$, and $z_n = 1/n$ for each $n \in \mathbb{N}$. Note the term (i.e., height) y_n is between x_n and z_n for each index n. See the Squeeze Theorem 2.4.3, Example 2.4.1, and Figure 2.4.2.

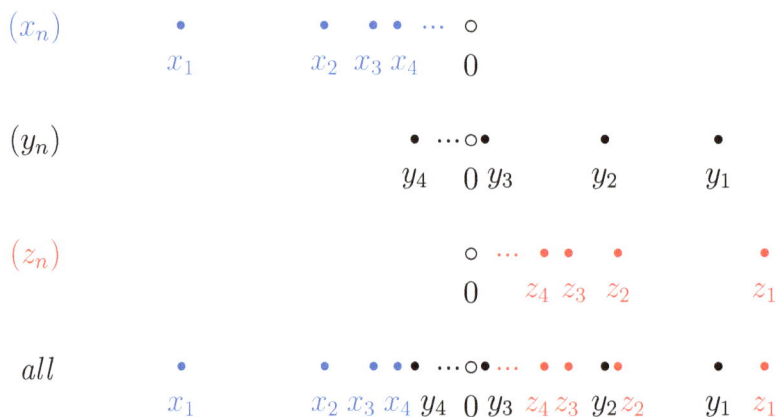

Figure 2.4.2: Ranges of the sequences $(x_n), (y_n)$, and (z_n) where $x_n = -1/n$, $y_n = \sin\left(\sqrt{n^2 + 1}\right)/n$, and $z_n = 1/n$ for each $n \in \mathbb{N}$. See the Squeeze Theorem 2.4.3 and Example 2.4.1. Between Figures 2.4.1 and 2.4.2, which best showcases the "squeezing" that gives the Squeeze Theorem 2.4.3 its name?

combined with choosing a positive integer $n_\varepsilon > 1/\varepsilon$ as a threshold in response to the distance $\varepsilon > 0$. However, a proof using the Squeeze Theorem 2.4.3 is in order so we can see how to put it to use.

Theorem 2.4.3: Squeeze Theorem for sequences

Suppose $(x_n), (y_n)$, and (z_n) are sequences of real numbers where

(i) for each $n \in \mathbb{N}$ we have $x_n \leq y_n \leq z_n$, and

(ii) $\lim\limits_{n \to \infty} x_n = \lim\limits_{n \to \infty} z_n = \ell$.

Then (y_n) converges with $\lim\limits_{n \to \infty} y_n = \ell$.

Scratch Work 2.4.4: Splitting an absolute value

Let's try to prove Theorem 2.4.3 directly from the assumptions and the definition of limit and convergence for sequences (Definition 2.2.1). Also, see Figures 2.4.1 and 2.4.2. The goal is to find a threshold $n_\varepsilon \in \mathbb{N}$ where $n \geq n_\varepsilon$ implies

$$|y_n - \ell| < \varepsilon. \tag{2.4.3}$$

How does (2.4.3) relate to the assumption $x_n \leq y_n \leq z_n$? We can subtract ℓ and get

$$x_n - \ell \leq y_n - \ell \leq z_n - \ell, \tag{2.4.4}$$

but this expression does not involve absolute values like the definition of limit and convergence for sequences in the real line (Definition 2.2.1). To connect with the assumption that (x_n) and (z_n) converge, note we also have

$$-|x_n - \ell| \leq x_n - \ell \leq y_n - \ell \leq z_n - \ell \leq |z_n - \ell|. \tag{2.4.5}$$

From here, each convergent sequence (x_n) and (z_n) has a corresponding threshold we can use. These two thresholds allow us to define a threshold for (y_n) and squeeze the terms from both sides via properties of inequalities. Also, Lemma 1.5.10 tells us how inequalities with and without absolute values are related to one another.

Proof of the Squeeze Theorem for sequences 2.4.3. Assume $x_n \leq y_n \leq z_n$ for each $n \in \mathbb{N}$ and

$$\lim_{n\to\infty} x_n = \lim_{n\to\infty} z_n = \ell. \tag{2.4.6}$$

Let $\varepsilon > 0$. Since (x_n) and (z_n) converge to ℓ, there are thresholds j_ε and k_ε where $n \geq j_\varepsilon$ and $n \geq k_\varepsilon$ imply

$$|x_n - \ell| < \varepsilon \qquad \text{and} \qquad |z_n - \ell| < \varepsilon, \tag{2.4.7}$$

respectively. Now define $n_\varepsilon = \max\{j_\varepsilon, k_\varepsilon\}$. Then for every $n \geq n_\varepsilon$ we have both $n \geq j_\varepsilon$ and $n \geq k_\varepsilon$. Therefore, by Lemma 1.5.10 and other properties of inequalities we have

$$-\varepsilon < -|x_n - \ell| \leq x_n - \ell \leq y_n - \ell \leq z_n - \ell \leq |z_n - \ell| < \varepsilon. \tag{2.4.8}$$

In particular, we have

$$-\varepsilon < y_n - \ell < \varepsilon. \tag{2.4.9}$$

So, n_ε is a threshold for (y_n) since by Lemma 1.5.10 we have

$$|y_n - \ell| < \varepsilon. \tag{2.4.10}$$

Therefore, (y_n) converges and $\lim_{n\to\infty} y_n = \ell$. $\qquad\square$

We're now prepared to prove the result claimed in Example 2.4.1.

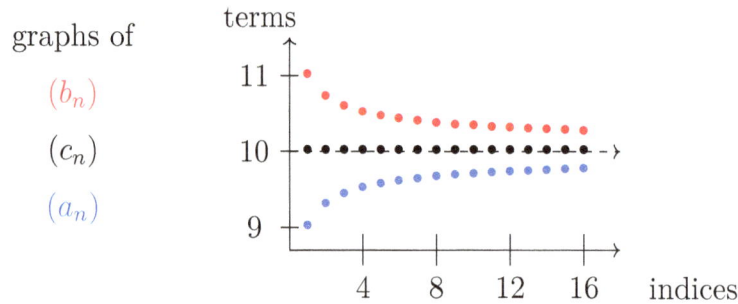

Figure 2.4.3: Graphs of the sequences (a_n), (b_n), and (c_n) from Example 2.4.6. All three sequences are monotone and converge to 10.

Proof for Example 2.4.1. A classic result from trigonometry helps us here. For every real number θ we have

$$-1 \leq \sin \theta \leq 1. \tag{2.4.11}$$

Therefore, for each $n \in \mathbb{N}$ we have

$$x_n = -\frac{1}{n} \leq y_n = \frac{\sin \sqrt{n^2 + 1}}{n} \leq \frac{1}{n} = y_n. \tag{2.4.12}$$

We also have

$$\lim_{n \to \infty} \left(-\frac{1}{n} \right) = \lim_{n \to \infty} \left(\frac{1}{n} \right) = 0. \tag{2.4.13}$$

So by the Squeeze Theorem 2.4.3, we have $\lim_{n \to \infty} y_n = 0$. $\qquad \square$

Monotone sequences play a special role in the development of our results.

Definition 2.4.5: Increasing, decreasing, and monotone sequences

A sequence of real numbers (x_n) is *increasing* if $x_n \leq x_{n+1}$ for every $n \in \mathbb{N}$. Similarly, (x_n) is *strictly increasing* if $x_n < x_{n+1}$ for every $n \in \mathbb{N}$.

A sequence of real numbers (y_n) is *decreasing* if $y_n \geq y_{n+1}$ for every $n \in \mathbb{N}$. Similarly, (y_n) is *strictly decreasing* if $y_n > y_{n+1}$ for every $n \in \mathbb{N}$.

A sequence of real numbers is *monotone* if it is increasing or decreasing.

Example 2.4.6: Some monotone sequences

Consider the sequences of real numbers (a_n), (b_n), and (c_n) defined for each $n \in \mathbb{N}$ by

$$a_n = 10 - \frac{1}{\sqrt{n}}, \quad b_n = 10 + \frac{1}{\sqrt{n}}, \quad \text{and} \quad c_n = 10. \tag{2.4.14}$$

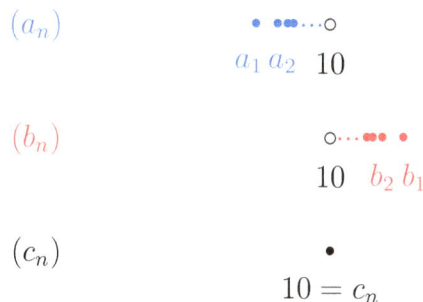

Figure 2.4.4: Ranges of the sequences (a_n), (b_n), and (c_n) from Example 2.4.6. All three sequences are monotone and converge to 10.

See Figures 2.4.3 and 2.4.4. For every $n \in \mathbb{N}$ we have

$$-\frac{1}{\sqrt{n}} \leq -\frac{1}{\sqrt{n+1}} \qquad \text{and} \qquad \frac{1}{\sqrt{n}} \geq \frac{1}{\sqrt{n+1}}. \tag{2.4.15}$$

Hence, (a_n) is increasing while (b_n) is decreasing.

For every $n \in \mathbb{N}$ we have both

$$c_n = 10 \leq 10 = c_{n+1} \qquad \text{and} \qquad c_n = 10 \geq 10 = c_{n+1}. \tag{2.4.16}$$

So, despite how strange it may sound, (c_n) is both increasing and decreasing. The same is true for all constant sequences of real numbers.

The following lemma gives us an alternative way to think about monotonicity: Instead of considering consecutive terms like x_n and x_{n+1}, we can compare two terms based on the order of their indices.

Lemma 2.4.7: An equivalent form of monotonicity

A sequence of real numbers (x_n) is increasing if and only if for every pair of positive integers j and k we have

$$j < k \qquad \Longrightarrow \qquad x_j \leq x_k. \tag{2.4.17}$$

Likewise, a sequence of real numbers (y_n) is decreasing if and only if for every pair of positive integers j and k we have

$$j < k \qquad \Longrightarrow \qquad y_j \geq y_k. \tag{2.4.18}$$

Remark 2.4.8: One proof is provided, the other is similar

Proofs for the two cases in Lemma 2.4.7 can be very similar, so the case for increasing sequences is proven here but the case for decreasing sequences is left as an exercise. An induction argument helps with one of the implications.

Proof for the increasing case in Lemma 2.4.7. Suppose for every pair of positive integers j and k we have

$$j < k \quad \Longrightarrow \quad x_j \leq x_k. \tag{2.4.19}$$

Since $n < n + 1$ for every positive integer n, we have

$$x_n \leq x_{n+1}. \tag{2.4.20}$$

Hence, (x_n) is increasing.

Now suppose (x_n) is increasing and fix a positive integer j. To establish a base case for an induction argument, we have

$$x_j \leq x_{j+1} \tag{2.4.21}$$

by the definition of an increasing sequence (Definition 2.4.5).

To establish an inductive case for the same fixed j, suppose k is a positive integer where $j < k$ and we have

$$x_j \leq x_k. \tag{2.4.22}$$

By the definition of an increasing sequence (Definition 2.4.5), we have

$$x_k \leq x_{k+1} \tag{2.4.23}$$

and since $j < k < k + 1$, we also have

$$x_j \leq x_k \leq x_{k+1}. \tag{2.4.24}$$

Furthermore, since j represents an arbitrary positive integer, for every pair of positive integers j and k we have

$$j < k \quad \Longrightarrow \quad x_j \leq x_k. \tag{2.4.25}$$

$\square$

Monotonicity and boundedness combine to ensure convergence. See Figure 2.4.5.

Theorem 2.4.9: Monotone and Bounded Convergence Theorem

If (x_n) is a monotone and bounded sequence of real numbers, then (x_n) converges. Furthermore, if (x_n) is increasing and bounded, then

$$\lim_{n \to \infty} x_n = \sup\{x_n : n \in \mathbb{N}\}. \tag{2.4.26}$$

If (x_n) is decreasing and bounded, then

$$\lim_{n \to \infty} x_n = \inf\{x_n : n \in \mathbb{N}\}. \tag{2.4.27}$$

Scratch Work 2.4.10: First, establish the existence of the limit

Proofs for the two cases in Theorem 2.4.9 can be very similar, so the case for increasing sequences is explored here while the case for decreasing sequences is left as an exercise.

A subtlety worth noting is that the candidate for the limit of an increasing sequence—the supremum—does not exist unless the sequence is bounded above. This concern deserves attention. In general, we need to be careful and ensure the tools and concepts we use are justified.

By assuming the sequence (x_n) is bounded and is therefore bounded above, its supremum u is assured to exist by the Axiom of Completeness 1.3.8. So, we can use u as a candidate for the limit, as follows. Our goal is now to find a threshold n_ε where

$$n \geq n_\varepsilon \quad \Longrightarrow \quad |x_n - u| < \varepsilon. \tag{2.4.28}$$

Since the supremum of a sequence is arbitrarily close to the sequence, for every $\varepsilon > 0$ there is an index n_ε where

$$|x_{n_\varepsilon} - u| < \varepsilon. \tag{2.4.29}$$

From there, the assumption that (x_n) is increasing and bounded above by u combines with properties of absolute value and inequalities to give us our goal (2.4.28).

Proof for the increasing case in Theorem 2.4.9. Suppose (x_n) is an increasing and bounded sequence of real numbers. (See Figure 2.4.5.) Then the Axiom of Completeness 1.3.8 ensures the existence of the supremum

$$u = \sup\{x_n : n \in \mathbb{N}\}. \tag{2.4.30}$$

By the definition of supremum (Definition 1.1.14), u is an upper bound for the range of (x_n). Hence, for every $k \in \mathbb{N}$ we have

$$x_k \leq u. \tag{2.4.31}$$

Also by the definition of supremum (Definition 1.1.14) we have $u \operatorname{acl}(x_n)$. So by the definition of arbitrarily close (Definition 1.5.1), for every $\varepsilon > 0$ there is an index n_ε where the term x_{n_ε} satisfies

$$|x_{n_\varepsilon} - u| < \varepsilon. \tag{2.4.32}$$

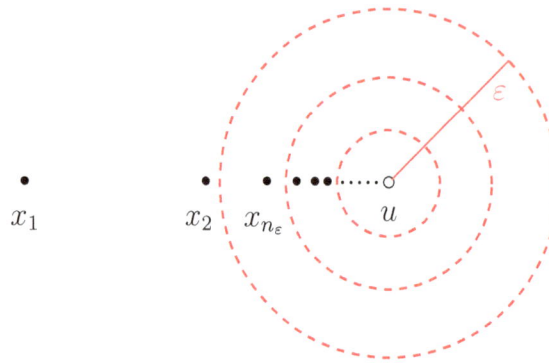

Figure 2.4.5: An increasing sequence of real numbers (x_n) bounded above by its supremum u where the index n_ε is a threshold responding to some distance $\varepsilon > 0$. See Scratch Work 2.4.10 for the proof of the increasing (and bounded) case of the Monotone and Bounded Convergence Theorem 2.4.9.

Inequality (2.4.31) ensures $0 \leq u - x_k$ for every $k \in \mathbb{N}$, so by the definition of absolute value (Definition 1.1.6) we have

$$|x_k - u| = u - x_k. \tag{2.4.33}$$

Since (x_n) is increasing, by Lemma 2.4.7 for every positive integer n where $n \geq n_\varepsilon$ we have

$$x_{n_\varepsilon} \leq x_n. \tag{2.4.34}$$

So, by (2.4.31), (2.4.32), (2.4.33), (2.4.34), and properies of inequalities, for every index $n \in \mathbb{N}$ where $n \geq n_\varepsilon$ we have

$$|x_n - u| = u - x_n \leq u - x_{n_\varepsilon} = |x_{n_\varepsilon} - u| < \varepsilon. \tag{2.4.35}$$

Therefore, n_ε is a threshold for the convergence (x_n) and

$$\lim_{n \to \infty} x_n = u = \sup\{x_n : n \in \mathbb{N}\}. \tag{2.4.36}$$

$\square$

Out of necessity, the context of Theorem 2.4.9 is limited to sequences in the real line $\mathbb{R}$ due to the inequalities (i.e., the order of the real line). The following theorem shows the convergence of a sequence in a Euclidean space $\mathbb{R}^m$ ensures the convergence of its components in $\mathbb{R}$, and vice versa. See Figure 2.4.6.

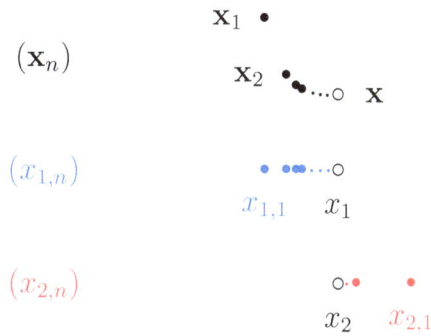

Figure 2.4.6: A plot of the sequence $(\mathbf{x}_n)$ and point $\mathbf{x}$ in the plane $\mathbb{R}^2$ from Theorem 2.4.11 and 2.2.9 where $\lim_{n\to\infty} \mathbf{x}_n = \mathbf{x}$ along with their components in their own horizontal copies of the real line $\mathbb{R}$. The horizontal components $(x_{1,n})$ increase towards x_1 while the vertical components $(x_{2,n})$ decrease towards x_2.

Theorem 2.4.11: Equivalence of convergence and componentwise convergence in Euclidean spaces

Suppose $\mathbf{x}$ is a point and $(\mathbf{x}_n)$ is a sequence of points in $\mathbb{R}^m$ where for each index n we have

$$\mathbf{x} = \begin{bmatrix} x_1 \\ x_2 \\ \vdots \\ x_m \end{bmatrix} \quad \text{and} \quad \mathbf{x}_n = \begin{bmatrix} x_{1,n} \\ x_{2,n} \\ \vdots \\ x_{m,n} \end{bmatrix}. \tag{2.4.37}$$

Then

$$\lim_{n\to\infty} \mathbf{x}_n = \mathbf{x} \tag{2.4.38}$$

if and only if for every $k = 1, 2, \ldots, m$ we have

$$\lim_{n\to\infty} x_{k,n} = x_k. \tag{2.4.39}$$

Remark 2.4.12: Limits and componentwise convergence

When we have the convergence

$$\lim_{n\to\infty} x_{k,n} = x_k \tag{2.4.40}$$

for each $k = 1, 2, \ldots, m$, we say $(\mathbf{x}_n)$ *converges componentwise*. Theorem 2.4.11 yields the

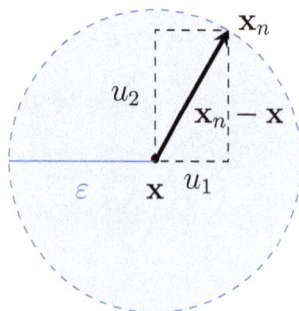

Figure 2.4.7: In the plane $\mathbb{R}^2$, any point $\mathbf{x}_n$ within a positive distance ε of the point $\mathbf{x}$ creates a vector $\mathbf{x}_n - \mathbf{x}$ whose components $u_1 = x_{1,n} - x_1$ and $u_2 = x_{2,n} - x_2$ have lengths (absolute values) strictly less than ε. See the proof of Theorem 2.4.11.

following equations when either of its hypotheses are satisfied:

$$\lim_{n\to\infty} \mathbf{x}_n = \lim_{n\to\infty} \begin{bmatrix} x_{1,n} \\ x_{2,n} \\ \vdots \\ x_{m,n} \end{bmatrix} = \begin{bmatrix} \lim_{n\to\infty} x_{1,n} \\ \lim_{n\to\infty} x_{2,n} \\ \vdots \\ \lim_{n\to\infty} x_{m,n} \end{bmatrix} = \begin{bmatrix} x_1 \\ x_2 \\ \vdots \\ x_m \end{bmatrix} = \mathbf{x}. \tag{2.4.41}$$

In (2.4.41), the limit symbol can be thought of as moving in and out of the brackets.

Scratch Work 2.4.13: Adapting thresholds componentwise

When $\lim_{n\to\infty} \mathbf{x}_n = \mathbf{x}$, there is a threshold n_ε that ensures $\mathbf{x}_n$ is within ε of $\mathbf{x}$ in all directions at the same time, including each component's direction. So, the same threshold n_ε suffices for the convergence in each component. See Figure 2.4.7.

On the other hand, when $\lim_{n\to\infty} x_{k,n} = x_k$ for each $k = 1,\ldots,m$, then given any positive distance there are m thresholds, one for each component. So given $\varepsilon > 0$, each of the m thresholds can be adapted to respond to a suitable proportion of ε. From there, the maximum of the set of m adapted thresholds serves as a threshold to ensure $\lim_{n\to\infty} \mathbf{x}_n = \mathbf{x}$.

Proof of Theorem 2.4.11. Suppose $\lim_{n\to\infty} \mathbf{x}_n = \mathbf{x}$ where the convergence is in $\mathbb{R}^m$. Let $\varepsilon > 0$. By the definition of limit and convergence (Definition 2.2.1), there is a threshold n_ε where $n \geq n_\varepsilon$ ensures

$$d_m(\mathbf{x}_n, \mathbf{x}) = \|\mathbf{x}_n - \mathbf{x}\|_m < \varepsilon. \tag{2.4.42}$$

The threshold n_ε for the sequence $(\mathbf{x}_n)$ also serves as a suitable threshold for each component sequence $(x_{k,n})$. Indeed, since $0 \leq x \leq y$ implies $\sqrt{x} \leq \sqrt{y}$, we have for each $k = 1,\ldots,m$ and every $n \geq n_\varepsilon$ that

$$d_{\mathbb{R}}(x_{k,n}, x_k) = |x_{k,n} - x_k| \tag{2.4.43}$$

$$= \sqrt{(x_{k,n} - x_k)^2} \tag{2.4.44}$$

$$\leq \sqrt{\sum_{j=1}^{m}(x_{j,n} - x_j)^2} \tag{2.4.45}$$

$$= \|\mathbf{x}_n - \mathbf{x}\|_m \tag{2.4.46}$$

$$< \varepsilon. \tag{2.4.47}$$

See Figure 2.4.7. Hence, for every $k = 1, 2, \ldots, m$ we have

$$\lim_{n \to \infty} x_{k,n} = x_k. \tag{2.4.48}$$

Now suppose for every $k = 1, 2, \ldots, m$ we have

$$\lim_{n \to \infty} x_{k,n} = x_k. \tag{2.4.49}$$

Let $\varepsilon > 0$ and note $\varepsilon/\sqrt{m} > 0$. In response, there is a threshold n_k for each $k = 1, 2, \ldots, m$ where for every $n \geq n_k$ we have

$$d_{\mathbb{R}}(x_{k,n}, x_k) = |x_{k,n} - x_k| < \frac{\varepsilon}{\sqrt{m}}. \tag{2.4.50}$$

Define $n_\varepsilon = \max\{n_1, n_2, \ldots, n_m\}$. Then for every $n \geq n_\varepsilon$ we have $n \geq n_k$ for each $k = 1, 2, \ldots, m$, thus inequality (2.4.50) holds for each k. Therefore, since $0 \leq x < y$ implies $\sqrt{x} < \sqrt{y}$, we also have

$$d_m(\mathbf{x}_n, \mathbf{x}) = \|\mathbf{x}_n - \mathbf{x}\|_m \tag{2.4.51}$$

$$= \sqrt{\sum_{j=1}^{m}(x_{k,n} - x_k)^2} \tag{2.4.52}$$

$$< \sqrt{\sum_{j=1}^{m}\left(\frac{\varepsilon}{\sqrt{m}}\right)^2} \tag{2.4.53}$$

$$= \sqrt{m\left(\frac{\varepsilon^2}{m}\right)} \tag{2.4.54}$$

$$= \varepsilon. \tag{2.4.55}$$

Hence, $\lim_{n \to \infty} \mathbf{x}_n = \mathbf{x}$. $\qquad\square$

Subsequences have already made an appearance, but a formal definition is in order so we can start proving results about their convergence.

Definition 2.4.14: Subsequence

Let $(\mathbf{x}_n)$ be a sequence of points in $\mathbb{R}^m$ and let (n_k) be a strictly increasing sequence of positive integers. That is,

$$n_1 < n_2 < n_3 < \cdots \tag{2.4.56}$$

The sequence $(\mathbf{x}_{n_k})$ is called a *subsequence* of $(\mathbf{x}_n)$.

Remark 2.4.15: Subsequence notation

The notation used to define subsequences in Definition 2.4.14 can be confusing, especially with the double subscripts. However, it ensures subsequences comprise only terms from the original sequence and the terms stay in order.

Definition 2.4.16: Subsequential limits

A point $\mathbf{y}$ is a *subsequential limit* of a sequence $(\mathbf{x}_n)$ if $(\mathbf{x}_n)$ has a subsequence whose limit is $\mathbf{y}$. The set of subsequential limits of $(\mathbf{x}_n)$ is denoted by $\mathrm{Slim}(\mathbf{x}_n)$.

Example 2.4.17: A divergent sequence in the plane

Consider the sequence $(\mathbf{z}_n)$ from Example 2.1.14 in $\mathbb{R}^2$ given by

$$\mathbf{z}_n = \begin{cases} \begin{bmatrix} 2 + (2/n) \\ 1 \end{bmatrix}, & \text{if } n \text{ is odd,} \\[2em] \begin{bmatrix} -1 \\ 3 - (2/n) \end{bmatrix}, & \text{if } n \text{ is even.} \end{cases} \tag{2.4.57}$$

See Figure 2.4.8. Also, consider the points $\mathbf{u}$ and $\mathbf{v}$ given by

$$\mathbf{u} = \begin{bmatrix} 2 \\ 1 \end{bmatrix} \qquad \text{and} \qquad \mathbf{v} = \begin{bmatrix} -1 \\ 3 \end{bmatrix}, \tag{2.4.58}$$

as well as the subsequences $(\mathbf{z}_{2k-1}), (\mathbf{z}_{2k})$, and $(\mathbf{z}_{3k})$. We have

$$\mathbf{z}_{2k-1} = \begin{bmatrix} 2 + (2/(2k-1)) \\ 1 \end{bmatrix}, \quad \mathbf{z}_{2k} = \begin{bmatrix} -1 \\ 3 - (1/k) \end{bmatrix} \quad \text{and,} \tag{2.4.59}$$

$$\mathbf{z}_{3k} = \begin{cases} \begin{bmatrix} 2 + (2/3k) \\ 1 \end{bmatrix}, & \text{if } k \text{ is odd,} \\[2em] \begin{bmatrix} -1 \\ 3 - (2/3k) \end{bmatrix}, & \text{if } k \text{ is even.} \end{cases} \tag{2.4.60}$$

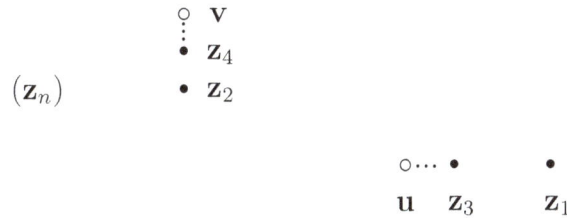

Figure 2.4.8: A plot of the sequence $(\mathbf{z}_n)$ along with $\mathbf{u}$ and $\mathbf{v}$ from Example 2.4.17.

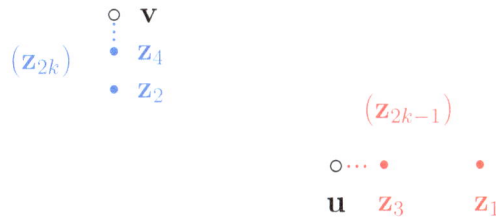

Figure 2.4.9: A plot of the subsequences $(\mathbf{z}_{2k-1})$ in red and $(\mathbf{z}_{2k})$ in blue, along with $\mathbf{u}$ and $\mathbf{v}$ from Example 2.4.17.

See Figure 2.4.9. The subsequence of odd indices $(\mathbf{z}_{2k-1})$ converges to $\mathbf{u}$ with a threshold s_ε satisfying

$$s_\varepsilon > \frac{1}{\varepsilon} + \frac{1}{2}. \tag{2.4.61}$$

The subsequence of even indices $(\mathbf{z}_{2k})$ converges to $\mathbf{v}$ with a threshold t_ε satisfying

$$t_\varepsilon > \frac{1}{\varepsilon}. \tag{2.4.62}$$

The subsequence $(\mathbf{z}_{3k})$ diverges, as will be proven later.

Ultimately, we have $\mathrm{Slim}\,(\mathbf{z}_n) = \{\mathbf{u}, \mathbf{v}\}$. The thresholds s_ε and t_ε can be used to show $\mathbf{u}$ and $\mathbf{v}$ belong to $\mathrm{Slim}\,(\mathbf{z}_n)$ by proving

$$\lim_{k\to\infty} \mathbf{z}_{2k-1} = \mathbf{u} \quad \text{and} \quad \lim_{k\to\infty} \mathbf{z}_{2k} = \mathbf{v}, \quad \text{respectively.} \tag{2.4.63}$$

But why are no other points in $\mathrm{Slim}\,(\mathbf{z}_n)$? Every point in $\mathbb{R}^2$ aside from $\mathbf{u}$ and $\mathbf{v}$ is either away from $(\mathbf{z}_n)$ or is a term of $(\mathbf{z}_n)$ away from all the others. This takes some effort to prove and is left as an exercise.

The following theorem solidifies what I believe to be an intuitive idea.

Theorem 2.4.18: Convergence implies subsequential convergence

Every subsequence of a convergent sequence in $\mathbb{R}^m$ converges to the same limit.

Proof of Theorem 2.4.18. Suppose $(\mathbf{x}_n)$ is a convergent sequence of points in $\mathbb{R}^m$ whose limit is $\mathbf{y}$, and suppose $(\mathbf{x}_{n_k})$ is a subsequence of $(\mathbf{x}_n)$. Let $\varepsilon > 0$. By the definition of sequential limit (Definition 2.2.1), there is a threshold j_ε where $n \geq j_\varepsilon$ implies

$$d_m(\mathbf{x}_n, \mathbf{y}) = \|\mathbf{x}_n - \mathbf{y}\|_m < \varepsilon. \tag{2.4.64}$$

Then j_ε is also a suitable threshold for the subsequence $(\mathbf{x}_{n_k})$, as follows. Since the indices of $(\mathbf{x}_{n_k})$ form a strictly increasing sequence by definition of subsequence (Definition 2.4.14), we have $n_k \geq k$ for every positive integer k. Hence, we have $n_k \geq k \geq j_\varepsilon$ implies

$$d_m(\mathbf{x}_{n_k}, \mathbf{y}) = \|\mathbf{x}_{n_k} - \mathbf{y}\|_m < \varepsilon. \tag{2.4.65}$$

Therefore, $(\mathbf{x}_{n_k})$ converges to $\mathbf{y}$.   □

The following corollary of Theorem 2.4.18 formalizes an idea from calculus.

Corollary 2.4.19: Powers of a small constant

Suppose $|c| < 1$. Then $\lim_{n \to \infty} c^n = 0$.

Scratch Work 2.4.20: Invoking the Squeeze Theorem

Corollary 2.4.19 does not follow from Corollary 2.3.22 which provides bounds on where limits could be (between -1 and 1) but not enough information to determine the limit precisely. Theorem 2.4.18 provides a way to do so by taking advantage of a particular subsequence as well as properties unique to zero, at least when $c \geq 0$. From there, the result for $|c| < 1$ follows from an application of the Squeeze Theorem 2.4.3. As a result, the proof does not address the rates of convergence of sequences of the form (c^n) since we avoid arguments involving thresholds. Still, the thresholds for such sequences are interesting.

Proof of Corollary 2.4.19. Suppose $0 \leq c < 1$. Then for every positive integer n we have

$$c^n \geq c^n \cdot c = c^{n+1}. \tag{2.4.66}$$

So by Corollary 2.3.22, (c^n) is a decreasing sequence bounded below by 0. By the Monotone and Bounded Convergence Theorem 2.4.9, (c^n) converges to some real number ℓ. Then the subsequence (c^{2n}) also converges to ℓ by Theorem 2.4.18. By Theorem 2.3.17 we have

$$\ell = \lim_{n \to \infty} c^{2n} = \lim_{n \to \infty} (c^n \cdot c^n) = \left(\lim_{n \to \infty} c^n \right) \left(\lim_{n \to \infty} c^n \right) = \ell^2. \tag{2.4.67}$$

Hence, either $\ell = 0$ or $\ell = 1$. However, by Lemma 2.4.7 and Corollary 2.3.22 we have

$$0 \leq \ell = \lim_{n \to \infty} c^n \leq c < 1. \tag{2.4.68}$$

So, $0 \leq \ell < 1$ and we must have $\lim_{n \to \infty} c^n = \ell = 0$.

Now suppose $0 \leq |c| < 1$. Then for every index $n \in \mathbb{N}$ we have

$$-|c|^n \leq c^n \leq |c|^n. \tag{2.4.69}$$

Since $0 \leq |c| < 1$ and by the linearity of sequential limits (Theorem 2.3.9) we have

$$\lim_{n \to \infty} -|c|^n = 0 = \lim_{n \to \infty} |c|^n. \tag{2.4.70}$$

Therefore, by the Squeeze Theorem 2.4.3 we have

$$\lim_{n \to \infty} c^n = 0. \tag{2.4.71}$$

$\square$

The following section further develops and proves some significant results about ensuring convergence of sequences.

Exercises

2.4.1. Prove the following statement is false: There is a convergent sequence of real numbers with an infinite number of zeroes whose limit is not zero.

2.4.2. Suppose $|c| \geq 1$. Prove (c^n) diverges.

2.4.3. Suppose $c > 0$. Prove $\lim_{n \to \infty} \sqrt[n]{c} = 1$. Hint: First consider $c > 1$ and let $a_n = \sqrt[n]{c} - 1$. Use the Binomial Theorem 1.2.24 to show that for each $n \in \mathbb{N}$ we have

$$c = (1 + a_n)^n \geq 1 + n a_n. \tag{2.4.72}$$

From here, use the Squeeze Theorem for sequences 2.4.3.

2.4.4. Prove $\lim_{n \to \infty} \sqrt[n]{n} = 1$ by completing the following steps.

  (i) For each $n \in \mathbb{N}$, set $\sqrt[n]{n} = 1 + \delta_n$ (note $\delta_n > 0$) and use the Binomial Theorem 1.2.24 to prove

$$0 < \delta_n < \sqrt{\frac{2}{n-1}} \tag{2.4.73}$$

  for all $n \in \mathbb{N}$ where $n > 1$.

  (ii) Use (i) to prove

$$\lim_{n \to \infty} \sqrt[n]{n} = \lim_{n \to \infty} (1 + \delta_n) = 1. \tag{2.4.74}$$

2.4.5. Suppose $c > 1$ and $p \in \mathbb{R}$. Prove

$$\lim_{n \to \infty} \frac{n^p}{c^n} = 0. \tag{2.4.75}$$

Hint: Let $k \in \mathbb{N}$ such that $k > c$ and $k > 0$. First use the equation $c = 1 + (c - 1)$ with the Binomial Theorem 1.2.24 to show that for $n > 2k$ we have

$$c^n > \binom{n}{k}(c-1)^k = \frac{n(n-1)\cdots(n-k+1)}{k!}(c-1)^k > \frac{n^k(c-1)^k}{2^k k!}. \tag{2.4.76}$$

2.4.6. For each $n \in \mathbb{N}$, recall $n! = 1(2)(3) \cdots (n)$. Prove that for every $c \in \mathbb{R}$ we have

$$\lim_{n \to \infty} \frac{c^n}{n!} = 0. \tag{2.4.77}$$

2.4.7. Let (x_n) be the sequence of real numbers defined recursively by

$$x_1 = \sqrt{2} \quad \text{and} \quad x_{n+1} = \sqrt{2 + x_n} \quad \text{for each } n \in \mathbb{N} \text{ with } n \geq 2. \tag{2.4.78}$$

(i) Prove $x_n \leq 2$ for all $n \in \mathbb{N}$.

(ii) Prove (x_n) is increasing.

(iii) Prove (x_n) converges and $\lim_{n \to \infty} x_n = 2$.

2.4.8. Prove the following statement is false by finding a counterexample: For any two sequences $(\mathbf{a}_n)$ and $(\mathbf{b}_n)$ in $\mathbb{R}^m$ we have

$$\lim_{n \to \infty} (\mathbf{a}_n - \mathbf{b}_n) = \mathbf{0} \quad \iff \quad \lim_{n \to \infty} \mathbf{a}_n = \lim_{n \to \infty} \mathbf{b}_n. \tag{2.4.79}$$

2.4.9. Prove the following statements are false by finding a counterexample for each one.

(i) If (a_n) and (b_n) diverge, then the sequence of sums $(a_n + b_n)$ diverges as well.

(ii) If (a_n) and (b_n) diverge, then the sequence of products $(a_n b_n)$ diverges as well.

(iii) If (a_n) and (b_n) diverge, then the sequence of quotients (a_n/b_n) diverges as well.

(iv) If (a_n) and (b_n) diverge, then at least one of $(a_n + b_n)$ or $(a_n b_n)$ diverges.

2.4.10. Suppose $(\mathbf{x}_n) \subseteq \mathbb{R}^m$ is bounded and $(c_n) \subseteq \mathbb{R}$ where $\lim_{n \to \infty} c_n = 0$. Prove

$$\lim_{n \to \infty} c_n \mathbf{x}_n = \mathbf{0}. \tag{2.4.80}$$

2.4.11. Suppose $(b_n) \subseteq \mathbb{R}$, $(\mathbf{x}_n) \subseteq \mathbb{R}^m$, and $\mathbf{y} \in \mathbb{R}^m$ satisfy

$$b_n \geq 0 \quad \text{and} \quad \|\mathbf{x}_n - \mathbf{y}\|_m \leq b_n \quad \text{for each } n \in \mathbb{N}. \tag{2.4.81}$$

Prove that if $\lim_{n \to \infty} b_n = 0$, then $\lim_{n \to \infty} \mathbf{x}_n = \mathbf{y}$.

2.4.12. Given a sequence $(\mathbf{x}_n) \subseteq \mathbb{R}^m$, for each $n \in \mathbb{N}$ the average $\mathbf{y}_n$ given by

$$\mathbf{y}_n = \frac{\sum_{k=1}^{n} \mathbf{x}_k}{n} = \frac{\mathbf{x}_1 + \mathbf{x}_2 + \cdots + \mathbf{x}_n}{n} \tag{2.4.82}$$

is called the nth *Cesaro mean*.

(i) Prove that if $\lim_{n \to \infty} \mathbf{x}_n = \mathbf{x}$, then $\lim_{n \to \infty} \mathbf{y}_n = \mathbf{x}$ as well.

(ii) Find an example of a divergent sequence of real numbers (x_n) whose sequence of Cesaro means (y_n) converges.

2.4.13. Prove that if all subsequences of $(\mathbf{x}_n) \subseteq \mathbb{R}^m$ converge, then $(\mathbf{x}_n)$ converges, too.

2.5 The Bolzano-Weierstrass Theorem

A fundamental consequence of the completeness of the real line and Euclidean spaces is the existence of suitable candidates for limits of sequences. In particular, we have the Bolzano-Weierstrass Theorem:

Every bounded sequence has a convergent subsequence.

The aim of this section is to prove two versions of the Bolzano-Weierstrass Theorem. The first (Theorem 2.5.6) establishes the result in the real line, while the second (Theorem 2.5.13) establishes a generalized result for Euclidean spaces.

Both proofs are included based on feedback from my students. They are quite involved and make use of other theorems which are interesting in their own right. There are certainly similarities between the two cases, but the proof for Euclidean spaces has a lot more notation and nuance to deal with.

Ideas common to both proofs stem from *nested* closed and bounded sets with nonempty intersections which generate candidates for limits, and the use of repeated bisection to create suitable collections of sets.

There's a lot to do, so let's start with Example 2.5.1 which shows us that intersections of nonempty and overlapping (nested) sets can be empty.

Example 2.5.1: Nested intervals with empty intersection

Consider the intervals defined for each $n \in \mathbb{N}$ by

$$(0, 1/n] = \{x \in \mathbb{R} : 0 < x \leq 1/n\}. \tag{2.5.1}$$

We have $\bigcap_{n=1}^{\infty}(0, 1/n] = \varnothing$ (the intersection of all these intervals is empty).

Scratch Work 2.5.2: No real number is in every interval

To prove the intersection in Example 2.5.1 is empty, we should show that no real number is in the intersection. Since a given point needs to be in *every* set in order to be in the intersection, we only need to find *one* set where a given point doesn't belong.

If $x \leq 0$, then x is not in any of the intervals, so it's not in the intersection. But what if $x > 0$? Things are trickier, but we can handle it. For instance, $x = 1/100$ is in the first 100 intervals, but not the 101st since $1/101 < 1/100$ and we have $1/100 \notin (0, 1/101]$. It might help to note

$$(0, 1/101] = \{y \in \mathbb{R} : 0 < y \leq 1/101\}. \tag{2.5.2}$$

The Corollary of the Archimedean Property 1.4.8 allows us to apply this type of argument for any positive number.

Time for a proof, which can be done in two cases.

Proof for Example 2.5.1. Case (i): Suppose $x \leq 0$. Then $x \notin (0, 1]$. So,

$$x \notin \bigcap_{n=1}^{\infty} (0, 1/n]. \tag{2.5.3}$$

Case (ii): Suppose $x > 0$. By Corollary 1.4.8, there is an index $n_x \in \mathbb{N}$ where

$$0 < 1/n_x < x. \tag{2.5.4}$$

So, x is too large to be in the interval $(0, 1/n_x]$. Hence,

$$x \notin \bigcap_{n=1}^{\infty} (0, 1/n]. \tag{2.5.5}$$

Whether $x \leq 0$ or $x > 0$, we have x is not in the intersection. Therefore,

$$\bigcap_{n=1}^{\infty} (0, 1/n] = \varnothing. \tag{2.5.6}$$

$\square$

Example 2.5.1 holds despite the fact that the intervals are *nested*: Each interval contains the next.

Definition 2.5.3: Nested sets

A sequence (S_n) of sets is *nested* if for every index $n \in \mathbb{N}$ we have $S_n \supseteq S_{n+1}$.

If we add a couple of conditions to the nested property, namely closed and bounded, we can ensure the intersection of intervals is nonempty.

Theorem 2.5.4: NCBI Property

Every nested sequence of closed and bounded intervals has a nonempty intersection.[a]

---

   [a]NCBI stands for "nested, closed, bounded intervals".

Scratch Work 2.5.5: Finding a point in the intersection

To show the intersection is nonempty, we need to find just one point which is in all of the intervals. See Figure 2.5.1. By using nested intervals that are both closed and bounded, the intersection of any finite number of them is always nonempty. The Axiom of Completeness 1.3.8 ensures of the existence of a point in the intersection of all of the intervals, namely the supremum of the set of left endpoints.

Proof of Theorem 2.5.4. Suppose for each index $n \in \mathbb{N}$ we have

$$[a_n, b_n] = \{x \in \mathbb{R} : a_n \leq x \leq b_n\} \quad \text{and} \tag{2.5.7}$$
$$[a_1, b_1] \supseteq [a_2, b_2] \supseteq [a_3, b_3] \supseteq \cdots. \tag{2.5.8}$$

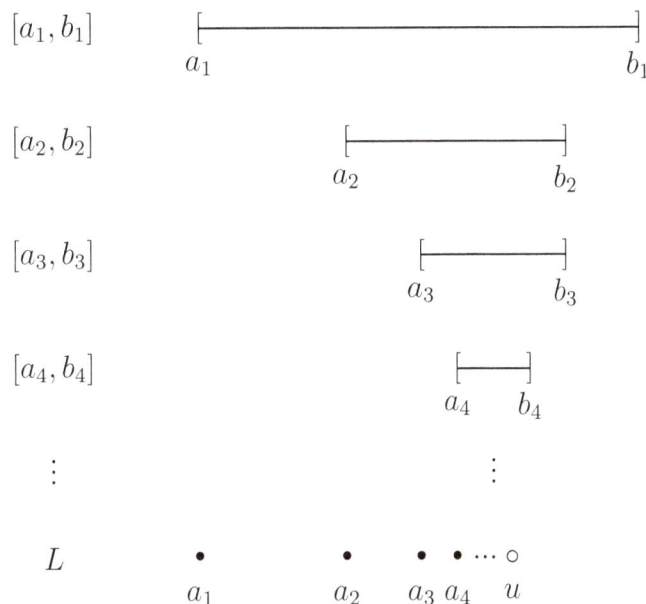

Figure 2.5.1: A sequence of nested, closed, bounded intervals to accompany the NCBI Property (Theorem 2.5.4) along with the set L comprising the left endpoints of the intervals and its supremum u. This supremum u is in the intersection of the intervals but not necessarily in L.

Line (2.5.7) ensures the intervals are closed and bounded while line (2.5.8) ensures they form a nested sequence.

Let L denote the set of left-endpoints of intervals $[a_n, b_n]$. So,

$$L = \{a_n : n \in \mathbb{N}\}. \tag{2.5.9}$$

Since the first interval $[a_1, b_1]$ contains all the others, we have $a_n \leq b_1$ for every $n \in \mathbb{N}$. Thus, L is bounded above by b_1. Since $a_1 \in L$, L is nonempty. By the Axiom of Completeness 1.3.8, $u = \sup L$ exists. See Figure 2.5.1.

It turns out every b_n is an upper bound for L. If not, there would be some $a_k \in L$ where $b_n < a_k$, but this would contradict the nested property in line (2.5.8). Now, since $u = \sup L$ is the *least* upper bound of L by Theorem 1.3.10, we have $u \leq b_n$ for every $n \in \mathbb{N}$. Since $u = \sup L$ is an upper bound for L by Definition 1.1.14, we also have $a_n \leq u$ for every $n \in \mathbb{N}$. Hence, for every $n \in \mathbb{N}$ we have

$$a_n \leq u \leq b_n. \tag{2.5.10}$$

Therefore, $u \in [a_n, b_n]$ for every $n \in \mathbb{N}$ and so

$$\bigcap_{n=1}^{\infty} [a_n, b_n] \neq \varnothing. \tag{2.5.11}$$

$\square$

The following theorem is a significant result in the analysis of the real line. The proof below relies a careful application of the NCBI Property (Theorem 2.5.4).

Theorem 2.5.6: Bolzano-Weierstrass Theorem in $\mathbb{R}$

Every bounded sequence in the real line $\mathbb{R}$ has a convergent subsequence.

Scratch Work 2.5.7: Finding a suitable candidate for the limit

The Bolzano-Weierstrass Theorem in the real line requires a bit of work to prove. All we have to start with is a bounded sequence in $\mathbb{R}$, so we need to ensure the existence of a suitable subsequence as well as a suitable candidate for the limit. From there, we need to prove the candidate really is the limit. The proof below takes advantage of the NCBI Property (Theorem 2.5.4) to ensure the existence of a point which serves as a suitable candidate for the limit. But this follows from a carefully chosen sequence of intervals constructed in a recursive manner: After an initial set of steps, the process is repeated ad infinitum.

The idea is to first take a closed interval that's big enough to contain the bounded sequence, then recursively bisect to produce smaller and smaller subintervals as in Figure 2.5.2. Each bisection produces two intervals with half the original length, allowing us to identify a convergent subsequence thanks to a key fact: With each bisection, at least one of the subintervals must contain an infinite number of terms. From there, we can choose one such smaller subinterval and one term to add to a subsequence, then repeat. By bisecting with each new step, the terms we select are forced closer and closer together, ensuring our chosen subsequence converges. See Figure 2.5.3.

Proof of the Bolzano-Weierstrass Theorem in the real line 2.5.6. Suppose (x_n) is a bounded sequence in the real line $\mathbb{R}$. Consider a bound $b > 0$ that defines a closed interval $I_1 = [-b, b]$ large enough to contain the range of the sequence (x_n). See Figure 2.5.2.

Choose a term x_{n_1} to serve as the first term of our desired subsequence. Next, bisect I_1 by considering the two closed intervals of the form $[-b, 0]$ and $[0, b]$ whose union is I_1. At least one of these subintervals contains an infinite number of the terms of (x_n), so choose one such interval and name it I_2. Note that the length of I_2 is b, half the length of each side of I_1. From there, choose a second term x_{n_2} from the sequence where $n_1 < n_2$ and x_{n_2} is in I_2. See Figure 2.5.3.

To proceed recursively, suppose $k > 2$ is a positive integer for which a closed and bounded interval I_{k-1} and term $x_{n_{k-1}}$ have been chosen where $n_{k-2} < n_{k-1}$, $x_{n_{k-1}}$ is in I_{k-1}, $I_{k-2} \supseteq I_{k-1}$, and the length of I_{k-1} is half the length of I_{k-2}. Note that for each $k > 2$, the length of I_{k-1} is $b/2^{k-3}$, corresponding to having bisected the original I_1 interval—whose length is $2b$—a total of $k - 2$ times.

Now, bisect I_{k-1} to produce two closed subintervals whose union is I_{k-1}. At least one of these closed subintervals contains an infinite number of the terms of the sequence (x_n), so choose one such interval and rename it I_k. Note that the length of I_k is $b/2^{k-2}$, half the length of I_{k-1}. From there, choose a kth term x_{n_k} from the given sequence where $n_{k-1} < n_k$ and x_{n_k} is in I_k.

Our recursive process yields a subsequence (x_{n_k}) and a sequence of closed and bounded intervals (I_k) where, for each positive integer k, we have

$$I_{k-1} \supseteq I_k \tag{2.5.12}$$

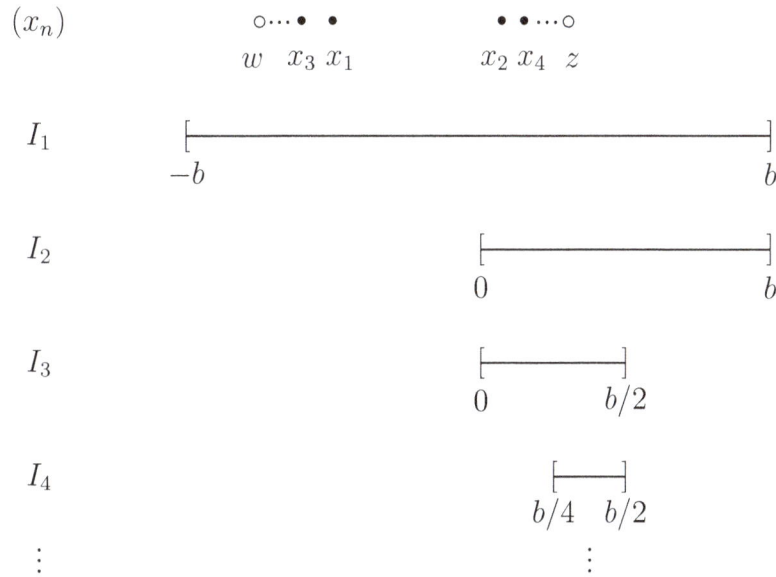

Figure 2.5.2: According to the Bolzano-Weierstrass Theorem in the real line (Theorem 2.5.6), a bounded sequence of real numbers (x_n) must have a convergent subsequence. Here, the closed and bounded interval $I_1 = [-b, b]$ contains all of the terms of (x_n) while the subintervals I_2, I_3, I_4, etc.—obtained via repeated bisection—are chosen to ensure each contains an infinite number of the terms.

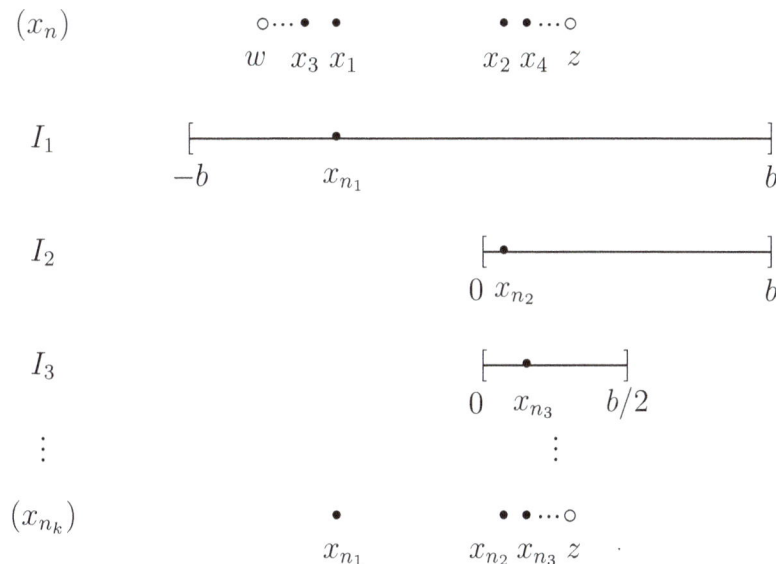

Figure 2.5.3: A bounded sequence (x_n) in the real line $\mathbb{R}$ along with a nested sequence of closed and bounded intervals (I_k) constructed via bisection. For each $k \in \mathbb{N}$, the subinterval I_k contains a term x_{n_k}. The terms are recursively chosen to ensure $n_1 < n_2 < n_3 < \ldots$ and to produce a convergent subsequence (x_{n_k}). Here, the subsequence converges to the real number z.

and the length of I_k is $b/2^{k-2}$. Thus, (I_k) is a nested sequence of closed and bounded intervals. By the NCBI Property (Theorem 2.5.4), there is point z in $\cap_{k=1}^{\infty} I_k$.

It remains to show z is the limit of the subsequence (x_{n_k}). By construction, for each positive integer k we have both z and x_{n_k} are in I_k whose length is $b/2^{k-2}$. Hence, for every positive integer k we have

$$|x_{n_k} - z| \leq \frac{b}{2^{k-2}}. \tag{2.5.13}$$

By Corollary 2.4.19 and the homogeneity of sequential limits (part (ii) of Theorem 2.3.9), we have

$$\lim_{k \to \infty} \frac{b}{2^{k-2}} = 0. \tag{2.5.14}$$

Now let $\varepsilon > 0$. By the definition of limit and convergence for sequences (Definition 2.2.1), there is a threshold k_ε where $k \geq k_\varepsilon$ implies

$$|x_{n_k} - z| \leq \frac{b}{2^{k-2}} \leq \frac{b}{2^{k_\varepsilon - 2}} < \varepsilon. \tag{2.5.15}$$

Therefore, $\lim_{k \to \infty} x_{n_k} = z$. □

Both the NCBI Property (Theorem 2.5.4) and the Bolzano-Weierstrass Theorem in the real line (Theorem 2.5.6) generalize to higher dimensions by considering *boxes* in place of intervals: In the real line $\mathbb{R}$, a box is an interval; in the plane $\mathbb{R}^2$, a box is a rectangle; in $\mathbb{R}^3$, a box is a rectangular parallelepiped. The remainder of this sections is dedicated to proving these generalized results which appear below as the NCBB[2] Property (Theorem 2.5.10) and the Bolzano-Weierstrass Theorem in Euclidean spaces (Theorem 2.5.13).

First, a formal definition for *boxes*.

Definition 2.5.8: Boxes

A set $B \subseteq \mathbb{R}^m$ is a *box* if for each $j = 1, \ldots, m$ there is an interval I_j where

$$B = I_1 \times I_2 \times \cdots \times I_m \tag{2.5.16}$$

$$= \left\{ \mathbf{x} = \begin{bmatrix} x_1 \\ x_2 \\ \vdots \\ x_m \end{bmatrix} : x_j \in I_j \text{ for each } j = 1, \ldots, m \right\}. \tag{2.5.17}$$

Furthermore, B is a *closed box* if each I_j is a closed interval. Similarly, B is an *open box* if each I_j is an open interval.

---

[2]NCBB stands for "nested, closed, bounded boxes".

Notation 2.5.9: Sequences of boxes

Since the NCBB Property (Theorem 2.5.10) deals with a sequence of boxes, some carefully chosen notation will help us be precise. Given a sequence of boxes (B_n), for each positive integer n and each $j = 1, \ldots, m$, let $I_{j,n}$ be an interval where

$$B_n = I_{1,n} \times I_{2,n} \times \cdots \times I_{m,n} \tag{2.5.18}$$

$$= \left\{ \mathbf{x} = \begin{bmatrix} x_1 \\ x_2 \\ \vdots \\ x_m \end{bmatrix} : x_j \in I_{j,n} \text{ for each } j = 1, \ldots, m \right\}. \tag{2.5.19}$$

The intersection of the boxes amounts to the cross product of the intersections of their component intervals:

$$\bigcap_{n=1}^{\infty} B_n = \bigcap_{n=1}^{\infty} I_{1,n} \times \bigcap_{n=1}^{\infty} I_{2,n} \times \cdots \times \bigcap_{n=1}^{\infty} I_{m,n} \tag{2.5.20}$$

$$= \left\{ \mathbf{x} = \begin{bmatrix} x_1 \\ x_2 \\ \vdots \\ x_m \end{bmatrix} : x_j \in \bigcap_{n=1}^{\infty} I_{j,n} \text{ for each } j = 1, \ldots, m \right\}. \tag{2.5.21}$$

Similarly, containment in $\mathbb{R}^m$ amounts to containment in $\mathbb{R}$ of each component. That is,

$$B_n \supseteq B_{n+1} \iff I_{j,n} \supseteq I_{j,n+1} \text{ for each } j = 1, \ldots, m, \tag{2.5.22}$$

so (B_n) is nested if and only if $(I_{j,n})$ is nested for each $j = 1, \ldots, m$.

Theorem 2.5.10: NCBB Property

Every nested sequence of closed and bounded boxes in $\mathbb{R}^m$ has a nonempty intersection.

Scratch Work 2.5.11: Nested boxes and componentwise convergence

My approach is motivated by Theorem 2.4.11 where convergence of a sequence in a Euclidean space $\mathbb{R}^m$ is ensured by the convergence in the real line $\mathbb{R}$ of each of its components (and vice versa). Here, the NCBI Property (Theorem 2.5.4) applies to the closed intervals that define the closed boxes in question, allowing us to find a point in the intersection of these boxes by controlling their component intervals.

Proof of Theorem 2.5.10. Suppose (B_n) is a nested sequence of closed and bounded boxes in $\mathbb{R}^m$ whose component intervals are given by $I_{j,n}$ for each positive integer n and each $j = 1, \ldots, m$. Then for each $j = 1, \ldots, m$, the sequence $(I_{j,n})$ is a nested sequence of closed and bounded intervals. So by the NCBI Property (Theorem 2.5.4), we have $\cap_{n=1}^{\infty} I_{j,n}$ is nonempty for each

$j = 1, \ldots, m$. Thus, for each $j = 1, \ldots, m$ there is a real number u_j in $\cap_{n=1}^{\infty} I_{j,n}$. Now define

$$\mathbf{u} = \begin{bmatrix} u_1 \\ u_2 \\ \vdots \\ u_m \end{bmatrix}. \tag{2.5.23}$$

Then $\mathbf{u} \in \cap_{n=1}^{\infty} B_n$, so $\cap_{n=1}^{\infty} B_n$ is nonempty. □

Problem 2.5.12: Disks in squares

Draw a big disk along with a smaller square contained completely inside the big disk. Then, draw a disk small enough to be contained completely within the square. Can you see how this process can be repeated indefinitely, no matter how small the disks and squares are?

This problem is designed to help you connect the definition of limit and convergence of sequences (Definition 2.2.1), which involves spheres and disks in the form of ε-neighborhoods, to boxes in various Euclidean spaces. Boxes and convergence combine in the proof of the Bolzano-Weierstrass Theorem 2.5.13. Compare and contrast Figures 1.5.2, 2.5.4, and 2.5.5.

The following theorem is a major result in analysis which guarantees the existence of a convergent subsequence for a given bounded sequence in a Euclidean space. The proof is very similar to that of the corresponding result in the real line (Theorem 2.5.6), but here we consider boxes instead of intervals as well as componentwise convergence of the sequences.

Theorem 2.5.13: Bolzano-Weierstrass Theorem in $\mathbb{R}^m$

Every bounded sequence in a Euclidean space $\mathbb{R}^m$ has a convergent subsequence.

Scratch Work 2.5.14: Leveraging Bolzano-Weierstrass in $\mathbb{R}$

The Bolzano-Weierstrass Theorem for Euclidean spaces considers a bounded sequence in $\mathbb{R}^m$, and as with its counterpart for the real line (Theorem 2.5.6), we need to ensure the existence of a suitable subsequence as well as a suitable candidate for the limit. From there, we need to prove the candidate really is the limit of the subsequence. The proof below takes advantage of the NCBB Property (Theorem 2.5.10) to ensure the existence of a suitable candidate for the limit, following the recursive construction via bisection of a nested sequence of boxes.

The idea is to start by considering a box that's big enough to contain the given bounded sequence, then recursively bisect boxes along each of their sides to produce smaller and smaller boxes. In the real line $\mathbb{R}$, bisecting an interval produces two intervals with half the original length. In the plane $\mathbb{R}^2$, bisecting a square along each of its sides produces $2^2 = 4$ squares with half the original side length. (See Figures 2.5.4 and 2.5.5.) In a Euclidean space $\mathbb{R}^m$, bisecting a box along each of its sides produces 2^m boxes whose side lengths are half the original side lengths, respectively.

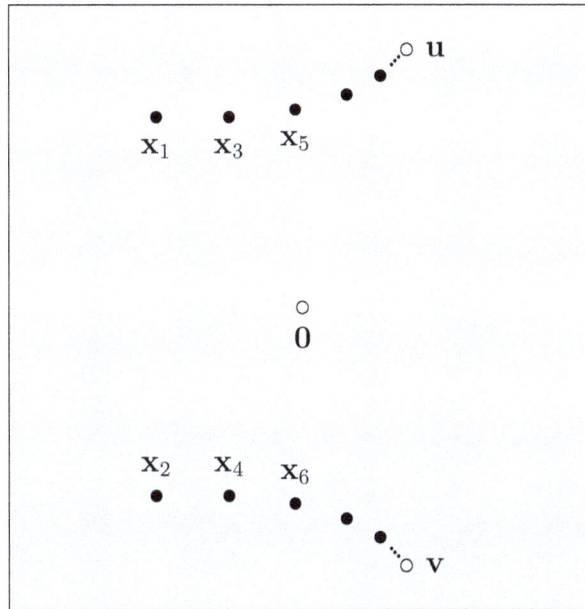

Figure 2.5.4: A bounded sequence $(\mathbf{x}_n)$ in the plane $\mathbb{R}^2$ contained in a closed square with the origin $\mathbf{0}$ in the center. The points $\mathbf{u}$ and $\mathbf{v}$ are arbitrarily close to the sequence $(\mathbf{x}_n)$.

At each step, at least one of the smaller boxes must contain an infinite number of the terms of the sequence. From there, we can choose one such smaller box and one term in the box to add to a subsequence, then repeat. By bisecting with each new step, we ensure our chosen subsequence converges.

Proof of the Bolzano-Weierstrass Theorem in Euclidean spaces 2.5.13. Suppose $(\mathbf{x}_n)$ is a bounded sequence of points in $\mathbb{R}^m$. Let B_1 be a closed box in $\mathbb{R}^m$ defined by

$$B_1 = \underbrace{[-b,b] \times \cdots \times [-b,b]}_{m \text{ copies of } [-b,b]} = [-b,b]^m \tag{2.5.24}$$

where $b > 0$ is large enough to contain the range of the sequence $(\mathbf{x}_n)$, similar to Figure 2.5.4.

Choose a term $\mathbf{x}_{n_1}$ to serve as the first term of our desired subsequence. Next, consider the 2^m distinct closed boxes whose component intervals are of the form $[0,b]$ or $[-b,0]$ and whose union is B_1. At least one of these smaller boxes contains an infinite number of the terms of $(\mathbf{x}_n)$, so choose one such box and name it B_2. Note that the length of each side of B_2 is b, half the length of each side of B_1. From there, choose a second term $\mathbf{x}_{n_2}$ from the sequence where $n_1 < n_2$ and $\mathbf{x}_{n_2}$ is in B_2. See Figure 2.5.5.

Proceeding recursively, suppose $k > 2$ is a positive integer for which a closed box B_{k-1} and term $\mathbf{x}_{n_{k-1}}$ have been chosen where $n_{k-2} < n_{k-1}$, $\mathbf{x}_{n_{k-1}}$ is in B_{k-1}, $B_{k-2} \subseteq B_{k-1}$, and the length of each side of B_{k-1} is $b/2^k$ (half the length of each side of B_{k-2}). Bisect each side of B_{k-1} to produce 2^m distinct closed boxes whose union is B_{k-1}. At least one of these closed boxes contains an infinite number of the terms of the sequence $(\mathbf{x}_n)$, so choose one such box and rename it B_k.

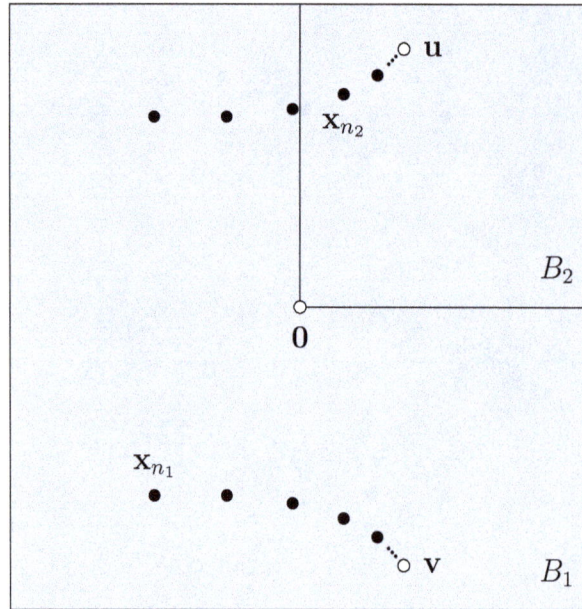

Figure 2.5.5: A bounded sequence $(\mathbf{x}_n)$ in the plane $\mathbb{R}^2$ contained in a closed square B_1 with the origin $\mathbf{0}$ in the center. The terms $\mathbf{x}_{n_1}$ and $\mathbf{x}_{n_2}$ are chosen so that $n_1 < n_2$, $\mathbf{x}_{n_1}$ is in B_1, and $\mathbf{x}_{n_2}$ is in B_2. The points $\mathbf{u}$ and $\mathbf{v}$ are arbitrarily close to the sequence $(\mathbf{x}_n)$.

Note that the length of each side of B_k is $b/2^{k-2}$, half the length of each side of B_{k-1}. From there, choose a kth term $\mathbf{x}_{n_k}$ from the sequence where $n_{k-1} < n_k$ and $\mathbf{x}_{n_k}$ is in B_k.

Our recursive process yields a subsequence $(\mathbf{x}_{n_k})$ and a sequence of closed boxes (B_k) where

$$B_{k-1} \supseteq B_k \tag{2.5.25}$$

and the length of the sides of B_k is $b/2^{k-2}$ for each positive integer k. Thus, (B_k) is a nested sequence of closed and bounded boxes. By the NCBB Property (Theorem 2.5.10), there is point $\mathbf{u}$ in $\cap_{k=1}^{\infty} B_k$.

It remains to show $\mathbf{u}$ is the limit of the subsequence $(\mathbf{x}_{n_k})$. Since our construction makes use of boxes instead of the disks or spheres defined by neighborhoods, consider the componentwise breakdown of $\mathbf{u}$ and $(\mathbf{x}_{n_k})$ given for each positive integer k by

$$\mathbf{u} = \begin{bmatrix} u_1 \\ u_2 \\ \vdots \\ u_m \end{bmatrix} \quad \text{and} \quad \mathbf{x}_{n_k} = \begin{bmatrix} x_{1,n_k} \\ x_{2,n_k} \\ \vdots \\ x_{m,n_k} \end{bmatrix}. \tag{2.5.26}$$

Both $\mathbf{u}$ and $\mathbf{x}_{n_k}$ are in B_k and the lengths of the sides of B_k are $b/2^{k-2}$ for each positive integer k, so we have

$$|x_{j,n_k} - u_j| \leq \frac{b}{2^{k-2}} \tag{2.5.27}$$

for each $j = 1, \ldots, m$ and every positive integer k. By Corollary 2.4.19 and part (ii) of Theorem 2.3.9 we have

$$\lim_{k \to \infty} \frac{b}{2^{k-2}} = 0. \tag{2.5.28}$$

Let $\varepsilon > 0$. By the definition of limit and convergence for sequences (Definition 2.2.1), there is a threshold k_ε where $k \geq k_\varepsilon$ implies

$$|x_{j,n_k} - u_j| \leq \frac{b}{2^{k-2}} \leq \frac{b}{2^{k_\varepsilon - 2}} < \varepsilon \tag{2.5.29}$$

for each $j = 1, \ldots, m$. Therefore, the index k_ε is a threshold for every component sequence (x_{j,n_k}) at the same time and we have

$$\lim_{k \to \infty} x_{j,n_k} = u_j \tag{2.5.30}$$

for each $j = 1, \ldots, m$. Since the components of the subsequence $(\mathbf{x}_{n_k})$ converge, by Theorem 2.4.11 we finally have

$$\lim_{k \to \infty} \mathbf{x}_{n_k} = \mathbf{u}. \tag{2.5.31}$$

$\square$

The next section looks into an equivalent form of convergence and criteria for divergence.

Exercises

2.5.1. Give an example of a sequence in $\mathbb{R}^m$ with no convergent subsequences.

2.5.2. Give an example of a sequence of real numbers which contains neither 1 nor -1 yet has subsequences converging to each of these numbers.

2.5.3. Give an example of a divergent sequence of real numbers with a strictly decreasing subsequence and a strictly increasing subsequence.

2.5.4. Prove there is no bounded sequence with an unbounded subsequence. On the other hand, find an example of an unbounded sequence with a bounded subsequence.

2.5.5. Consider a finite set of points $F \subseteq \mathbb{R}^m$ given by

$$F = \{\mathbf{a}_1, \mathbf{a}_2, \cdots, \mathbf{a}_k\} \tag{2.5.32}$$

for some $k \in \mathbb{N}$.

(i) Describe a sequence $(\mathbf{x}_n) \subseteq \mathbb{R}^m$ where $\text{Slim}\,(\mathbf{x}_n) = F$. (See Definition 2.4.16.)

(ii) Describe a sequence $(\mathbf{y}_n) \subseteq \mathbb{R}^m$ where $\mathbf{y}_n \neq \mathbf{a}_j$ for all $n \in \mathbb{N}$ and $j = 1, \ldots, k$ and yet $\mathrm{Slim}\,(\mathbf{y}_n) = F$.

2.5.6. Consider the set of reciprocals $1/\mathbb{N}$ given by

$$\frac{1}{\mathbb{N}} = \left\{ \frac{1}{n} : n \in \mathbb{N} \right\} = \left\{ 1, \frac{1}{2}, \frac{1}{3}, \ldots \right\}. \tag{2.5.33}$$

(i) Describe a sequence $(x_k) \subseteq \mathbb{R}$ where $1/\mathbb{N} \subseteq \mathrm{Slim}\,(x_k)$. (See Definition 2.4.16.)

(ii) Prove that your example for part (i) satisfies $0\,\mathrm{acl}(x_k)$ and $0 \in \mathrm{Slim}(x_k)$.

(iii) Is there a sequence $(y_k) \subseteq \mathbb{R}$ where $1/\mathbb{N} = \mathrm{Slim}\,(y_k)$? Why or why not?

2.6   Cauchy and Divergence Criteria

The Bolzano-Weierstrass Theorem 2.5.13 ensures the existence of a limit even without a particular candidate in mind or readily available. This is quite different from many of the results pertaining to limits and convergence discussed earlier in this chapter. In fact, the definition of limit and convergence for sequences (Definition 2.2.1) explicitly relies on a candidate for the limit. The Cauchy criterion for sequences (Theorem 2.6.5) is a powerful result which ensures the existence of a limit even when we do not have a candidate.

Definition 2.6.1: Cauchy sequence

A sequence $(\mathbf{x}_n)$ of points in $\mathbb{R}^m$ is *Cauchy* if for every $\varepsilon > 0$ there is a positive integer n_ε such that for positive integers n and k we have

$$n, k \geq n_\varepsilon \quad \Longrightarrow \quad d_m(\mathbf{x}_n, \mathbf{x}_k) = \|\mathbf{x}_n - \mathbf{x}_k\|_m < \varepsilon. \tag{2.6.1}$$

In this case, n_ε is called a *threshold* and its value depends on ε.

Remark 2.6.2: Convergence versus Cauchy

In Euclidean spaces, the concepts of convergence and Cauchy are equivalent. Consider quantified versions of the statements where "$\forall$" means "for all" and "$\exists$" means "there exists":

$(\mathbf{x}_n)$ is Cauchy	$\mathbf{y} = \lim_{n \to \infty}(\mathbf{x}_n)$
$\forall\,\varepsilon > 0,$ $\exists\,n_\varepsilon \in \mathbb{N}$ such that $n, k \geq n_\varepsilon \implies \|\mathbf{x}_n - \mathbf{x}_k\|_m < \varepsilon.$	$\forall\,\varepsilon > 0,$ $\exists\,n_\varepsilon \in \mathbb{N}$ such that $n \geq n_\varepsilon \implies \|\mathbf{x}_n - \mathbf{y}\|_m < \varepsilon.$

In both settings, $\varepsilon > 0$ tells us how close we would like pairs of objects to be while the positive integer n_ε is a threshold that ensures the objects are within ε of each other. The

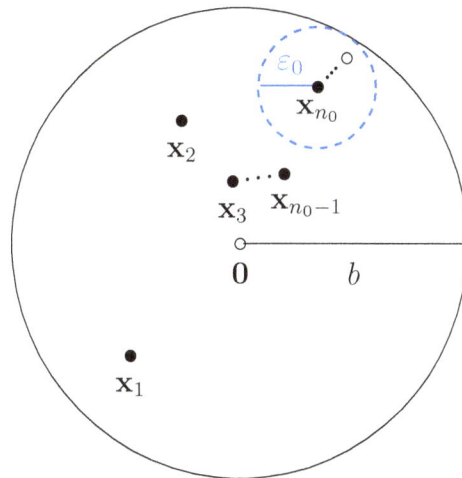

Figure 2.6.1: A Cauchy sequence $(\mathbf{x}_n)$ in the plane $\mathbb{R}^2$. By Lemma 2.6.3, $(\mathbf{x}_n)$ must be bounded by some nonnegative real number b. Note how the $\circ$ near the term $\mathbf{x}_{n_0}$ looks like it could represent the limit of the sequence $(\mathbf{x}_n)$, but the definition of a Cauchy sequence makes no mention of a limit at all. See Definition 2.6.1, Scratch Work 2.6.4, and the proof of Theorem 2.3.15.

key difference is the objects taken into consideration: Convergence compares terms of a sequence with the limit or a candidate for the limit; Cauchy compares pairs of terms of a sequence and does not consider a candidate for the limit at all.

To me, when a sequence is both convergent and Cauchy, the terms of the sequence get close and stay close to the limit and to each other. To prove they are equivalent, an additonal property of Cauchy sequences will help.

Lemma 2.6.3: Cauchy sequences are bounded

Every Cauchy sequence in $\mathbb{R}^m$ is bounded.

Scratch Work 2.6.4: All but a finite number of terms are close

The proof of Lemma 2.6.3 is very similar to the proof of Theorem 2.3.15 which says convergent sequences are bounded. Both amount to solidifying the idea that, eventually, the terms of the sequence are as close together as we like. In Figure 2.6.1, no particular term in the sequence has the largest norm, but for some $\varepsilon_0 > 0$ there is a threshold n_0 that produces the real number $\varepsilon_0 + \|\mathbf{x}_{n_0}\|_m$ which is a bound for the sequence. However, this may not be true in general since a term whose index is less than n_0 may have a larger norm.

Proof of Lemma 2.6.3. Suppose $(\mathbf{x}_n)$ is a Cauchy sequence in $\mathbb{R}^m$. Let $\varepsilon_0 = 11$. By the definition of a Cauchy sequence 2.6.1, there is a threshold n_0 where $n \geq n_0$ implies

$$d_m(\mathbf{x}_n, \mathbf{x}_{n_0}) = \|\mathbf{x}_n - \mathbf{x}_{n_0}\|_m < \varepsilon_0 = 11. \tag{2.6.2}$$

Hence, for $n \geq n_0$,

$$\|\mathbf{x}_n\|_m = \|\mathbf{x}_n \underbrace{-\mathbf{x}_{n_0} + \mathbf{x}_{n_0}}_{\text{add zero}}\|_m \tag{2.6.3}$$

$$\leq \|\mathbf{x}_n - \mathbf{x}_{n_0}\|_m + \|\mathbf{x}_{n_0}\|_m \qquad \text{(tri. ineq. (1.2.34))} \tag{2.6.4}$$

$$< 11 + \|\mathbf{x}_{n_0}\|_m. \qquad ((2.6.2)) \tag{2.6.5}$$

Now define b by

$$b = \max\{\|\mathbf{x}_1\|_m, \|\mathbf{x}_2\|_m, \ldots, \|\mathbf{x}_{n_0-1}\|_m, 11 + \|\mathbf{x}_{n_0}\|_m\}. \tag{2.6.6}$$

Then $b \geq 0$ and for every index $n \in \mathbb{N}$ we have

$$\|\mathbf{x}_n\|_m < b. \tag{2.6.7}$$

Therefore, (x_n) is bounded. $\square$

The following Cauchy criterion for sequences (Theorem 2.6.5) is another major result in analysis that depends heavily on the completeness of the real line and Euclidean spaces. It provides a characterization of the existence of limits and the convergence of sequences, all without having an explicit candidates for the limits in mind.

Theorem 2.6.5: Cauchy criterion for sequences

Suppose $(\mathbf{x}_n)$ is a sequence of points in $\mathbb{R}^m$. Then $(\mathbf{x}_n)$ converges if and only if $(\mathbf{x}_n)$ is Cauchy.

Scratch Work 2.6.6: Find a candidate for the limit

The proof is lopsided: One direction follows from the definitions plus the triangle inequality (1.2.34), the other sets up and takes advantage of the Bolzano-Weierstrass Theorem 2.5.13 to ensure the existence of a candidate for the limit before the definitions and the triangle inequality (1.2.34) are used.

Proof of the Cauchy criterion 2.6.5. First, suppose $\varepsilon > 0$ and $(\mathbf{x}_n)$ converges to $\mathbf{y}$ in $\mathbb{R}^m$. By the definition of sequential limit (Definition 2.2.1), there is a threshold $n_{\varepsilon/2}$ where $n \geq n_{\varepsilon/2}$ ensures

$$d_m(\mathbf{x}_n, \mathbf{y}) = \|\mathbf{x}_n - \mathbf{y}\|_m < \frac{\varepsilon}{2}. \tag{2.6.8}$$

Now suppose n and k are positive integers where $n, k \geq n_{\varepsilon/2}$. Then we have

$$d_m(\mathbf{x}_n, \mathbf{x}_k) = \|\mathbf{x}_n - \mathbf{x}_k\|_m \tag{2.6.9}$$

$$= \|\mathbf{x}_n \underbrace{-\mathbf{y} + \mathbf{y}}_{\text{add zero}} -\mathbf{x}_k\|_m \tag{2.6.10}$$

$$\leq \|\mathbf{x}_n - \mathbf{y}\|_m + \|\mathbf{y} - \mathbf{x}_k\|_m \tag{2.6.11}$$

$$< \frac{\varepsilon}{2} + \frac{\varepsilon}{2} \tag{2.6.12}$$

$$= \varepsilon. \tag{2.6.13}$$

Therefore, $n_{\varepsilon/2}$ is a suitable threshold and $(\mathbf{x}_n)$ is Cauchy.

To prove the converse, suppose $(\mathbf{x}_n)$ is Cauchy. In order to find a suitable candidate for the limit, note that by Lemma 2.6.3, $(\mathbf{x}_n)$ is bounded. So by the Bolzano-Weierstrass Theorem 2.5.13, there is a point $\mathbf{y}$ in $\mathbb{R}^m$ and subsequence $(\mathbf{x}_{n_k})$ whose limit is $\mathbf{y}$.

To show $\mathbf{y}$ is the limit of the original sequence $(\mathbf{x}_n)$, let $\varepsilon > 0$. By the definition of a Cauchy sequence (Definition 2.6.1), there is a threshold $j_{\varepsilon/2}$ such that $n, j \geq j_{\varepsilon/2}$ ensures

$$d_m(\mathbf{x}_n, \mathbf{x}_j) = \|\mathbf{x}_n - \mathbf{x}_j\|_m < \frac{\varepsilon}{2}. \tag{2.6.14}$$

Since the subsequence $(\mathbf{x}_{n_k})$ converges to $\mathbf{y}$, by the definition of limit and convergence for sequences (Definition 2.2.1), there is a threshold $k_{\varepsilon/2}$ such that $n_k \geq k \geq k_{\varepsilon/2}$ ensures

$$d_m(\mathbf{x}_{n_k}, \mathbf{y}) = \|\mathbf{x}_{n_k} - \mathbf{y}\|_m < \frac{\varepsilon}{2}. \tag{2.6.15}$$

Now, there is a positive integer k_0 large enough so that both $n_{k_0} \geq k_0 \geq k_{\varepsilon/2}$ and n_{k_0} is an index for the subsequence where $n_{k_0} \geq j_{\varepsilon/2}$. Then for every $n \geq n_{k_0}$ we have

$$d_m(\mathbf{x}_n, \mathbf{y}) = \|\mathbf{x}_n - \mathbf{y}\|_m \tag{2.6.16}$$

$$= \|\mathbf{x}_n \underbrace{- \mathbf{x}_{n_{k_0}} + \mathbf{x}_{n_{k_0}}}_{\text{add zero}} - \mathbf{y}\|_m \tag{2.6.17}$$

$$\leq \|\mathbf{x}_n - \mathbf{x}_{n_{k_0}}\|_m + \|\mathbf{x}_{n_{k_0}} - \mathbf{y}\|_m \tag{2.6.18}$$

$$< \frac{\varepsilon}{2} + \frac{\varepsilon}{2} \tag{2.6.19}$$

$$= \varepsilon. \tag{2.6.20}$$

Therefore, n_{k_0} is a suitable threshold and $(\mathbf{x}_n)$ converges to $\mathbf{y}$.   □

The next result solidifies the notion that if two sequences approach each other and one is Cauchy, then they both converge to the same limit.

Corollary 2.6.7: Two sequences with the same limit

Suppose $(\mathbf{a}_n)$ and $(\mathbf{b}_n)$ are sequences of points in $\mathbb{R}^m$ where $(\mathbf{a}_n)$ is Cauchy and

$$\lim_{n \to \infty} d_m(\mathbf{a}_n, \mathbf{b}_n) = \lim_{n \to \infty} \|\mathbf{a}_n - \mathbf{b}_n\|_m = 0. \tag{2.6.21}$$

Then $(\mathbf{a}_n)$ and $(\mathbf{b}_n)$ converge to the same limit.

Scratch Work 2.6.8: Apply the Cauchy criterion

By the Cauchy criterion 2.6.5, if $(\mathbf{a}_n)$ is Cauchy then $(\mathbf{a}_n)$ converges to some limit $\mathbf{a}$. From there, the condition of having the sequences approach each other, namely $\lim_{n \to \infty} d_m(\mathbf{a}_n, \mathbf{b}_n) = 0$, can be used to show $(\mathbf{b}_n)$ converges to $\mathbf{a}$ as well. Some of the techniques used in the proof should look familiar.

Proof of Corollary 2.6.7. Suppose $(\mathbf{a}_n)$ is Cauchy. Then by the Cauchy criterion 2.6.5, $(\mathbf{a}_n)$ converges to some limit $\mathbf{a}$.

Furthermore, suppose $\lim_{n \to \infty} d_m(\mathbf{a}_n, \mathbf{b}_n) = 0$ and let $\varepsilon > 0$. Then we also have $\varepsilon/2 > 0$ and by two applications of the definition of limit and convergence for sequences (Definition 2.2.1), there are thresholds $j_{\varepsilon/2}$ and $k_{\varepsilon/2}$ where

$$n \geq j_{\varepsilon/2} \quad \Longrightarrow \quad d_m(\mathbf{a}_n, \mathbf{a}) = \|\mathbf{a}_n - \mathbf{a}\|_m < \frac{\varepsilon}{2} \quad \text{and} \tag{2.6.22}$$

$$n \geq k_{\varepsilon/2} \quad \Longrightarrow \quad |d_m(\mathbf{a}_n, \mathbf{b}_n) - 0| = \|\mathbf{a}_n - \mathbf{b}_n\|_m < \frac{\varepsilon}{2}. \tag{2.6.23}$$

Define $n_\varepsilon = \max\{j_{\varepsilon/2}, k_{\varepsilon/2}\}$. Then for all $n \geq n_\varepsilon$ we have both $n \geq j_{\varepsilon/2}$ and $n \geq k_{\varepsilon/2}$. Therefore, $n \geq n_\varepsilon$ implies

$$d_m(\mathbf{b}_n, \mathbf{a}) = \|\mathbf{b}_n - \mathbf{a}\|_m \tag{2.6.24}$$

$$= \|\mathbf{b}_n \underbrace{-\mathbf{a}_n + \mathbf{a}_n}_{\text{add zero}} -\mathbf{a}\|_m \tag{2.6.25}$$

$$\leq \|\mathbf{b}_n - \mathbf{a}_n\|_m + \|\mathbf{a}_n - \mathbf{a}\|_m \tag{2.6.26}$$

$$< \frac{\varepsilon}{2} + \frac{\varepsilon}{2} \tag{2.6.27}$$

$$= \varepsilon. \tag{2.6.28}$$

Hence, n_ε is a threshold for the convergence of $(\mathbf{b}_n)$ to $\mathbf{a}$, and

$$\lim_{n \to \infty} \mathbf{a}_n = \mathbf{a} = \lim_{n \to \infty} \mathbf{b}_n. \tag{2.6.29}$$

$\square$

The contrapositions of Theorems 2.3.15 and 2.4.18 as well as the Cauchy criterion for sequences (Theorem 2.6.5) provide conditions to ensure divergence. As such, the proof of the following corollary is omitted.

Corollary 2.6.9: Divergence Criteria for Sequences

Suppose $(\mathbf{x}_n) \subseteq \mathbb{R}^m$.

(i) If $(\mathbf{x}_n)$ is unbounded, then $(\mathbf{x}_n)$ diverges.

(ii) If $(\mathbf{x}_n)$ has subsequences with different limits, then $(\mathbf{x}_n)$ diverges.

(iii) If $(\mathbf{x}_n)$ has a divergent subsequence or tail, then $(\mathbf{x}_n)$ diverges.

(iv) If $(\mathbf{x}_n)$ has a subsequence or tail which is not Cauchy, then $(\mathbf{x}_n)$ diverges.

Example 2.6.10: Examples of divergent sequences

The sequences (b_n) and $(\mathbf{z}_n)$ from Examples 2.1.4 and 2.1.14, respectively, diverge. The sequence (c_n) defined by $c_n = n(-1)^n$ diverges as well.

Proofs for Example 2.6.10. Since $c_n = n(-1)^n$ for each positive integer n, we have $|c_n| = n$. So by the Archimedean Property 1.4.6, (c_n) is unbounded. Hence, by part (i) of the Divergence Criteria for Sequences 2.6.9, (c_n) diverges.

The subsequences (b_{2k-1}) and (b_{2k}) satisfy

$$b_{2k-1} = -\left(2 - \frac{1}{\sqrt{2k-1}}\right) \qquad \text{and} \qquad b_{2k} = 2 - \frac{1}{\sqrt{2k}} \tag{2.6.30}$$

for every positive integer k. Hence,

$$\lim_{k\to\infty} b_{2k-1} = -2 \qquad \text{and} \qquad \lim_{k\to\infty} b_{2k} = 2. \tag{2.6.31}$$

Therefore, (b_n) diverges by part (ii) of the Divergence Criteria for Sequences 2.6.9.

A shown in Example 2.4.17, the subsequences $(\mathbf{z}_{2k-1})$ and $(\mathbf{z}_{2k})$ satisfy

$$\lim_{k\to\infty} \mathbf{z}_{2k-1} = \mathbf{u} = \begin{bmatrix} 2 \\ 1 \end{bmatrix} \qquad \text{and} \qquad \lim_{k\to\infty} \mathbf{z}_{2k} = \mathbf{v} = \begin{bmatrix} -1 \\ 3 \end{bmatrix}. \tag{2.6.32}$$

Therefore, $(\mathbf{z}_n)$ diverges by part (ii) of the Divergence Criteria for Sequences 2.6.9. $\qquad \square$

Divergence of unbounded sequences has a variety of flavors illuminated by different notions of *diverging to infinity.*

Definition 2.6.11: Diverge to ∞ in $\mathbb{R}^m$ for $m \geq 2$

A sequence $(\mathbf{x}_n) \subseteq \mathbb{R}^m$ *diverges to infinity* if for every $b > 0$ there is a *threshold* $n_\varepsilon \in \mathbb{N}$ such that

$$n \in \mathbb{N} \text{ with } n \geq n_\varepsilon \qquad \Longrightarrow \qquad \|\mathbf{x}_n\|_m > b. \tag{2.6.33}$$

In this case, we write $\lim_{n\to\infty} \mathbf{x}_n = \infty$.

The case is somewhat different and specialized for the real line $\mathbb{R}$. Note the lack of absolute values follwing the key implications. This means Definition 2.6.11 is *not* a generalization of Definition 2.6.12.

Definition 2.6.12: Diverge to $\pm\infty$ in $\mathbb{R}$

A sequence $(x_n) \subseteq \mathbb{R}$ *diverges to positive infinity* if for every $b > 0$ there is a *threshold* $n_\varepsilon \in \mathbb{N}$ such that

$$n \in \mathbb{N} \text{ with } n \geq n_\varepsilon \qquad \Longrightarrow \qquad x_n \geq b. \tag{2.6.34}$$

In this case, we write $\lim_{n\to\infty} x_n = \infty$.

Similarly, a sequence $(y_n) \subseteq \mathbb{R}$ *diverges to negative infinity* if for every $c < 0$ there is a *threshold* $n_\varepsilon \in \mathbb{N}$ such that

$$n \in \mathbb{N} \text{ with } n \geq n_\varepsilon \qquad \Longrightarrow \qquad y_n \leq c. \tag{2.6.35}$$

In this case, we write $\lim_{n\to\infty} y_n = -\infty$.

Example 2.6.13: Approaching $-\infty$

Consider the sequence $(z_n) \subseteq \mathbb{R}$ given by $z_n = -n$ for each $n \in \mathbb{N}$. We have

$$\lim_{n\to\infty} z_n = \lim_{n\to\infty} -n = -\infty \qquad (2.6.36)$$

Proof for Example 2.6.13. Suppose $c < 0$. By the Archimedean Property 1.4.6, there is some $n_{|c|} \in \mathbb{N}$ such that

$$|c| < n_{|c|}. \qquad (2.6.37)$$

Since $c < 0$, we have $c = -|c|$. So for all $n \in \mathbb{N}$ where $n \geq n_{|c|}$ it follows that

$$z_n = -n \leq -n_{|c|} < -|c| = c. \qquad (2.6.38)$$

Therefore, $\lim_{n\to\infty} z_n = \lim_{n\to\infty} -n = -\infty$. $\qquad\square$

The next section explores *decimals* and their relationship with *geometric sums*.

Exercises

2.6.1. Suppose $(\mathbf{x}_n)$ and $(\mathbf{y}_n)$ are Cauchy sequences in $\mathbb{R}^m$. Prove the sequence $(z_n) \subseteq \mathbb{R}$ given by

$$z_n = \|\mathbf{x}_n - \mathbf{y}_n\|_m \quad \text{for all} \quad n \in \mathbb{N} \qquad (2.6.39)$$

is Cauchy.

2.6.2. Use the definition of a Cauchy sequence (Definition 2.6.1) to prove directly that Cauchy sequences are linear in the following sense: Suppose $c \in \mathbb{R}$ and suppose $(\mathbf{x}_n)$ and $(\mathbf{y}_n)$ are Cauchy sequences in $\mathbb{R}^m$. Then

  (i) $(\mathbf{x}_n + \mathbf{y}_n)$ is Cauchy (*additivity*); and

  (ii) $(c\mathbf{x}_n)$ is Cauchy (*homogeneity*).

2.6.3. Find a decreasing sequence of real numbers which is not Cauchy.

2.6.4. Find an example of a sequence $(\mathbf{x}_n) \subseteq \mathbb{R}^m$ where $((-1)^n\mathbf{x}_n)$ is Cauchy but $(\mathbf{x}_n)$ is not.

2.6.5. Suppose $(\mathbf{y}_n) \subseteq \mathbb{R}^m$ satisfies the condition that for every $\varepsilon > 0$, there is an index $j_\varepsilon \in \mathbb{N}$ such that

$$\|\mathbf{y}_{j_\varepsilon} - \mathbf{y}_{j_\varepsilon+1}\|_m < \varepsilon. \qquad (2.6.40)$$

Prove the sequence of square roots $(\sqrt{n})$ satisfies this condition even though $(\sqrt{n})$ is *not* Cauchy as in Definition 2.6.1.

2.6.6. This exercise showcases a well-known recursive process for approximating square roots. Fix $c > 0$ and choose $x_1 \in \mathbb{R}$ where $x_1 > 0$ and $x_1^2 > c$. From there, define (x_n) by

$$x_{n+1} = \frac{1}{2}\left(x_n + \frac{c}{x_n}\right). \tag{2.6.41}$$

Prove (x_n) is decreasing and $\lim\limits_{n \to \infty} x_n = \sqrt{c}$.

2.6.7. Give an example of an unbounded sequence of real numbers which diverges to neither ∞ nor $-\infty$.

2.6.8. Prove that for any $r \in \mathbb{R}$, there is a divergent sequence (x_n) where $\mathrm{Slim}(x_n) = \{r\}$.

2.6.9. Give an example of a sequence of real numbers with a subsequence that converges to 0, a subsequence that diverges to ∞, and a subsequence that diverges to $-\infty$. (See Definition 2.6.12.)

2.6.10. Fix $\mathbf{a} \in \mathbb{R}^m$. Give an example of a sequence $(\mathbf{a}_n) \subseteq \mathbb{R}^m$ with a subsequence that converges to $\mathbf{a}$ and a subsequence that diverges to infinity as in Definition 2.6.11.

2.7 Geometric sums and decimals

Decimals provide a way to represent real numbers, something rational numbers cannot do. Although they are probably familiar, there is a lot to say in terms of what they are in a detailed and mathematically concrete manner.

To motivate the definitions found in this sections, specifically finite and infinite *decimal expansions*, let's start with a classic result known as the *Geometric Sum Formula* (Theorem 2.7.2).

Definition 2.7.1: Geometric sum

A *geometric sum* is the linear combination of powers of a given real number r with a constant weight a resulting in the form

$$\sum_{k=0}^{n} ar^k = a + ar + ar^2 + \cdots + ar^n \tag{2.7.1}$$

for some $n \in \mathbb{N} \cup \{0\}$. The real number a is called the *initial term* and r is called the *common ratio*. Also, we use the convention $0^0 = 1$ and the indices begin with $k = 0$, so there are actually $n + 1$ summands (terms added together).

Geometric sums have a nice closed formula.

Theorem 2.7.2: Geometric Sum Formula

If $a, r \in \mathbb{R}$ with $r \neq 1$ and n is a nonnegative integer, then

$$\sum_{k=0}^{n} ar^k = a + ar + ar^2 + \cdots + ar^n = \frac{a(1 - r^{n+1})}{1 - r}. \tag{2.7.2}$$

Scratch Work 2.7.3: A classic trick

The key step in this proof is a classic trick everyone should see. Multiplying both versions of the sum in (2.7.2) by $1 - r$ creates two copies with one increasing the powers of r and making summands negative. This results in a lot of cancellation via "telescoping" and a sleek form for the sum.

Proof of the Geometric Sum Formula 2.7.2. Suppose $a, r \in \mathbb{R}$ with $r \neq 1$ and n is a nonnegative integer. We have

$$(1 - r) \sum_{k=0}^{n} ar^k = (1 - r)(a + ar + ar^2 + \cdots + ar^n) \tag{2.7.3}$$

$$= a + ar + ar^2 + \cdots + ar^n \tag{2.7.4}$$

$$- ar - ar^2 - \cdots - ar^n - ar^{n+1} \tag{2.7.5}$$

$$= a - ar^{n+1}. \tag{2.7.6}$$

Since $r \neq 1$, dividing by $1 - r$ yields

$$\sum_{k=0}^{n} ar^k = a + ar + ar^2 + \cdots + ar^n = \frac{a(1 - r^{n+1})}{1 - r}. \tag{2.7.7}$$

$\square$

Remark 2.7.4: Closed form of a geometric sum

The rightmost expression

$$\frac{a(1 - r^{n+1})}{1 - r} \tag{2.7.8}$$

in the conclusion of the Geometric Sum Formula 2.7.2 is called the *closed form* of the corresponding geometric sum, but the conclusion only holds for $r \neq 1$. For the common ratio $r = 1$ have a different closed form:

$$\sum_{k=0}^{n} a(1)^k = \underbrace{a + a + \cdots + a}_{n+1 \text{ copies of } a} = a(n + 1). \tag{2.7.9}$$

> **Example 2.7.5: Finite decimal expansion of nines**
>
> For a positive integer n, consider the sum of n terms given by
>
> $$\frac{9}{10} + \frac{9}{10^2} + \frac{9}{10^3} + \cdots + \frac{9}{10^n}. \tag{2.7.10}$$
>
> This is actually a geometric sum defined by $n-1$ (yielding n summands), $a = 9/10$, and common ratio $r = 1/10$. We have
>
> $$\frac{9}{10} + \frac{9}{10^2} + \frac{9}{10^3} + \cdots + \frac{9}{10^n} \tag{2.7.11}$$
>
> $$= \frac{9}{10} + \left(\frac{9}{10}\right)\left(\frac{1}{10}\right) + \left(\frac{9}{10}\right)\left(\frac{1}{10}\right)^2 + \cdots + \left(\frac{9}{10}\right)\left(\frac{1}{10}\right)^{n-1} \tag{2.7.12}$$
>
> $$= \sum_{k=0}^{n-1} \left(\frac{9}{10}\right)\left(\frac{1}{10}\right)^k \tag{2.7.13}$$
>
> $$= \frac{(9/10)(1 - (1/10)^n)}{1 - (1/10)} \tag{2.7.14}$$
>
> $$= \frac{(9/10)(1 - (1/10)^n)}{(9/10)} \tag{2.7.15}$$
>
> $$= 1 - \frac{1}{10^n}. \tag{2.7.16}$$

Do you recognize the sum (2.7.10)? It defines the decimal with n digits that at are all 9:

$$0.\underbrace{999\ldots9}_{n \text{ digits}} = \frac{9}{10} + \frac{9}{10^2} + \frac{9}{10^3} + \cdots + \frac{9}{10^n}. \tag{2.7.17}$$

More to the point, sums like this one define decimals.

> **Definition 2.7.6: Finite decimal expansion**
>
> A *finite decimal expansion* is a linear combination of powers of $1/10$ of the form
>
> $$0.x_1x_2\ldots x_n = \frac{x_1}{10} + \frac{x_2}{10^2} + \cdots + \frac{x_n}{10^n} \tag{2.7.18}$$
>
> where $n \in \mathbb{N}$ and
>
> $$x_j \in \{0,1,2,\ldots,9\} \quad \text{for each} \quad j \in \{1,2,\ldots,n\}. \tag{2.7.19}$$
>
> In this case, the weights x_j are called *digits*.

Limits of finite decimal expansions built one summand at a time always converge. This allows us to define infinite decimal expansions.

Definition 2.7.7: Infinite decimal expansion

An *infinite decimal expansion* is the limit of a sequence of finite decimal expansions of the form

$$0.x_1 x_2 \ldots = \lim_{n \to \infty} 0.x_1 x_2 \ldots x_n = \lim_{n \to \infty} \left(\frac{x_1}{10} + \frac{x_2}{10^2} + \cdots + \frac{x_n}{10^n} \right) \qquad (2.7.20)$$

where (x_n) is a sequence of digits satisfying

$$x_j \in \{0, 1, 2, \ldots, 9\} \quad \text{for each} \quad j \in \mathbb{N}. \qquad (2.7.21)$$

Due diligence suggests we should make sure these limits exist.

Theorem 2.7.8: Infinite decimal expansions converge

The limits defining infinite decimal expansions exist.

Scratch Work 2.7.9: Consider monotoncity and boundedness

The proof follows from a careful application of the Monotone and Bounded Convergence Theorem 2.4.9.

Proof of Theorem 2.7.8. Suppose (x_n) is a sequence of digits satisfying

$$x_j \in \{0, 1, 2, \ldots, 9\} \quad \text{for each} \quad j \in \mathbb{N} \qquad (2.7.22)$$

and consider the corresponding sequence of finite decimal expansions given for each index $n \in \mathbb{N}$ by

$$0.x_1 x_2 \ldots x_n = \frac{x_1}{10} + \frac{x_2}{10^2} + \cdots + \frac{x_n}{10^n}. \qquad (2.7.23)$$

Since every digit x_{n+1} is nonnegative ($x_{n+1} \geq 0$ for each index $n \in \mathbb{N}$), we have

$$
\begin{aligned}
0.x_1 x_2 \ldots x_n &= \frac{x_1}{10} + \frac{x_2}{10^2} + \cdots + \frac{x_n}{10^n} & (2.7.24) \\
&\leq \frac{x_1}{10} + \frac{x_2}{10^2} + \cdots + \frac{x_n}{10^n} + \frac{x_{n+1}}{10^{n+1}} & (2.7.25) \\
&= 0.x_1 x_2 \ldots x_n x_{n+1}. & (2.7.26)
\end{aligned}
$$

Hence, our sequence of finite decimal expansions is increasing.

Also, since each digit x_j is bounded above by 9 ($x_j \leq 9$ for each $j \in \mathbb{N}$), Example 2.7.5 yields

$$0.x_1 x_2 \ldots x_n = \frac{x_1}{10} + \frac{x_2}{10^2} + \cdots + \frac{x_n}{10^n} \qquad (2.7.27)$$

$$\leq \frac{9}{10} + \frac{9}{10^2} + \cdots + \frac{9}{10^n} \qquad (2.7.28)$$

$$= 0.\underbrace{999\ldots 9}_{n\text{ digits}} \qquad (2.7.29)$$

$$= 1 - \frac{1}{10^n} \qquad (2.7.30)$$

$$< 1. \qquad (2.7.31)$$

Hence, our sequence of finite decimal expansions is bounded above by 1.

Therefore, by the Monotone and Bounded Convergence Theorem 2.4.9, the limit of our sequence of finite decimal expansions exists and we have

$$0.x_1 x_2 \ldots = \lim_{n \to \infty} 0.x_1 x_2 \ldots x_n = \sup\{0.x_1 x_2 \ldots x_n : n \in \mathbb{N}\}. \qquad (2.7.32)$$

$\square$

Decimal expansions lead to some interesting representations of real numbers. Here's a notorious example.

Example 2.7.10: A controversial fact

The infinite decimal expansion whose digits are all 9 is equal to one. That is,

$$0.999\ldots = 1. \qquad (2.7.33)$$

The definition of the expansion $0.999\ldots$, the comment immediately following Example 2.7.5, and properties of convergent sequences give us

$$0.999\ldots = \lim_{n \to \infty} 0.\underbrace{99\ldots 9}_{n\text{ digits}} \qquad (2.7.34)$$

$$= \lim_{n \to \infty} \left(\frac{9}{10} + \frac{9}{10^2} + \cdots + \frac{9}{10^n} \right) \qquad (2.7.35)$$

$$= \lim_{n \to \infty} \left(1 - \frac{1}{10^n} \right) \qquad (2.7.36)$$

$$= 1. \qquad (2.7.37)$$

By the way, the last equation above is justified by the choice of a threshold $n_\varepsilon \in \mathbb{N}$ where, given $\varepsilon > 0$, we make sure

$$n_\varepsilon > \log_{10}\left(\frac{1}{\varepsilon} \right). \qquad (2.7.38)$$

Example 2.7.10 serves as a warning before proceeding further: Some decimals with different digits represent the same real number, like this:

$$0.999\ldots = 1 = 1.000\ldots. \qquad (2.7.39)$$

However, if we limit ourselves to the digits 0 and 1, then infinite decimal expansions with different digits are distinct from one another. This special class of decimals, along with Examples 2.7.10 and 2.7.11, allow us to prove the real line $\mathbb{R}$ is uncountable. The next example helps set the stage.

Example 2.7.11: Ones and one ninth

The infinite decimal expansion whose digits are all 1 is equal to 1/9. That is,

$$0.111\ldots = \frac{1}{9}. \tag{2.7.40}$$

Thanks to the Geometric Sum Formula 2.7.2 where $a = r = 1/10$, and similar to Example 2.7.5, for each $n \in \mathbb{N}$ we have

$$0.\underbrace{11\ldots1}_{n \text{ digits}} = \frac{1}{10} + \frac{1}{10^2} + \cdots + \frac{1}{10^n} \tag{2.7.41}$$

$$= \sum_{k=0}^{n-1} \left(\frac{1}{10}\right)\left(\frac{1}{10^k}\right) \tag{2.7.42}$$

$$= \frac{(1/10)(1 - (1/10)^n)}{1 - (1/10)} \tag{2.7.43}$$

$$= \frac{(1/10)(1 - (1/10)^n)}{9/10} \tag{2.7.44}$$

$$= \frac{1}{9}\left(1 - \frac{1}{10^n}\right). \tag{2.7.45}$$

So, by the definition of infinite decimal expansions and properties of limits we have

$$0.111\ldots = \lim_{n \to \infty} 0.\underbrace{11\ldots1}_{n \text{ digits}} \tag{2.7.46}$$

$$= \lim_{n \to \infty} \left(\frac{1}{9}\left(1 - \frac{1}{10^n}\right)\right) \tag{2.7.47}$$

$$= \frac{1}{9}. \tag{2.7.48}$$

We can use Example 2.7.11 to show that if we restrict the digits to 0 and 1 only, then infinite decimal expansions with different digits are distinct from one another.

Lemma 2.7.12: Binary decimals are distinct

Suppose $0.x_1x_2\ldots$ and $0.y_1y_2\ldots$ are infinite decimal expansions such that for every index $n \in \mathbb{N}$ we have $x_n, y_n \in \{0, 1\}$ and for at least one index $j_0 \in \mathbb{N}$ we have $x_{j_0} \neq y_{j_0}$. Then

$$0.x_1x_2\ldots \neq 0.y_1y_2\ldots. \tag{2.7.49}$$

> **Scratch Work 2.7.13: Find a little distance**
>
> My goal is to use the first digit where the two infinite decimal expansions differ to force a positive distance between their finite decimal expansions, then apply properties of convergent sequences along with results from this section to complete the proof. This turned out to be harder than I though! Ultimately, the proof shows us that for some index $j_0 \in \mathbb{N}$ we have
>
> $$|0.x_1x_2\ldots - 0.y_1y_2\ldots| \geq \frac{8}{9 \cdot 10^{j_0}} > 0, \tag{2.7.50}$$
>
> but it takes a while to get there.

Proof of Lemma 2.7.12. Suppose $0.x_1x_2\ldots$ and $0.y_1y_2\ldots$ are infinite decimal expansions such that for each index $n \in \mathbb{N}$ we have $x_n, y_n \in \{0, 1\}$ and, without loss of generality, $j_0 \in \mathbb{N}$ is the smallest index where the digits differ and we have $x_{j_0} < y_{j_0}$. Then we must also have

$$0 = x_{j_0} < y_{j_0} = 1. \tag{2.7.51}$$

Hence, for any index $n \in \mathbb{N}$ where $n > j_0$ we have

$$0.x_1x_2\ldots x_n \tag{2.7.52}$$

$$= 0.x_1x_2\ldots x_{j_0-1}x_{j_0}x_{j_0+1}\ldots x_n \tag{2.7.53}$$

$$\leq 0.x_1x_2\ldots x_{j_0-1}0\underbrace{11\ldots 1}_{n-j_0 \text{ digits}} \tag{2.7.54}$$

$$= 0.x_1x_2\ldots x_{j_0-1} + \sum_{k=0}^{n-j_0-1}\left(\frac{1}{10^{j_0+1}}\right)\left(\frac{1}{10^k}\right) \tag{2.7.55}$$

where the digits with indices $j_0 + 1$ to n are all 1 in line (2.7.54). The summation in line (2.7.55) simplifies thanks to the Geometric Sum Formula 2.7.2 with $a = 1/10^{j_0+1}$ and $r = 1/10$. For each $n > j_0$ we have

$$\sum_{k=0}^{n-j_0-1}\left(\frac{1}{10^{j_0+1}}\right)\left(\frac{1}{10^k}\right) = \left(\frac{1}{10^{j_0+1}}\right)\left(\frac{1 - (1/10)^{n-j_0}}{1 - (1/10)}\right) \tag{2.7.56}$$

$$\leq \left(\frac{1}{10^{j_0+1}}\right)\left(\frac{1}{1 - (1/10)}\right) \tag{2.7.57}$$

$$= \left(\frac{1}{10^{j_0+1}}\right)\left(\frac{10}{9}\right) \tag{2.7.58}$$

$$= \frac{1}{9 \cdot 10^{j_0}}. \tag{2.7.59}$$

Hence, for every index $n \geq j_0$ we have

$$0.x_1x_2\ldots x_n \leq 0.x_1x_2\ldots x_{j_0-1} + \frac{1}{9 \cdot 10^{j_0}}. \tag{2.7.60}$$

Therefore, by the order properties for limits of sequences in $\mathbb{R}$ (Corollary 2.3.22) we have

$$0.x_1x_2\ldots = \lim_{n\to\infty} 0.x_1x_2\ldots x_n \leq 0.x_1x_2\ldots x_{j_0-1} + \frac{1}{9 \cdot 10^{j_0}}. \tag{2.7.61}$$

Now, since $y_{j_0} = 1$ and the first $j_0 - 1$ digits are the same for both $0.x_1x_2\ldots$ and $0.y_1y_2\ldots$, for every index $n > j_0$ we have we have

$$0.x_1x_2\ldots x_{j_0-1} + \frac{1}{10^{j_0}} = 0.x_1x_2\ldots x_{j_0-1}1 \tag{2.7.62}$$

$$\leq 0.x_1x_2\ldots x_{j_0-1}y_{j_0}y_{j_0+1}\ldots y_n \tag{2.7.63}$$

$$= 0.y_1y_2\ldots y_n. \tag{2.7.64}$$

Therefore, by the order properties for limits of sequences in $\mathbb{R}$ (Corollary 2.3.22) we have

$$0.y_1y_2\ldots = \lim_{n\to\infty} 0.y_1y_2\ldots y_n \tag{2.7.65}$$

$$\geq 0.x_1x_2\ldots x_{j_0-1} + \frac{1}{10^{j_0}}. \tag{2.7.66}$$

Bringing these results together, and especially keeping in mind the first $j_0 - 1$ digits are the same, we have

$$|0.x_1x_2\ldots - 0.y_1y_2\ldots| = 0.y_1y_2\ldots - 0.x_1x_2\ldots \tag{2.7.67}$$

$$\geq \frac{1}{10^{j_0}} - \frac{1}{9\cdot 10^{j_0}} \tag{2.7.68}$$

$$= \frac{8}{9\cdot 10^{j_0}} \tag{2.7.69}$$

$$> 0. \tag{2.7.70}$$

Therefore (and finally), $0.x_1x_2\ldots$ and $0.y_1y_2\ldots$ are distinct. $\square$

The next section uses properties of sequences and infinite decimal expansions to explore how large sets can be, including identifying different types of infinity.

Exercises

2.7.1. Explain why every real number between 0 and 1 has an infinite decimal expansion. That is, for every $x \in [0, 1]$, show there is a sequence of digits (x_n) with $x_j \in \{0, 1, \ldots, 9\}$ for each $j \in \mathbb{N}$ such that

$$x = 0.x_1x_2x_3\ldots. \tag{2.7.71}$$

2.7.2. Prove every rational number between 0 and 1 has a *repeating* infinite decimal expansion. That is, for every $r \in [0, 1] \cap \mathbb{Q}$, there is a finite set of digits $r_1, r_2, \ldots, r_k \subseteq \{0, 1, \ldots, 9\}$ for each $j \in \mathbb{N}$ such that

$$r = 0.r_1r_2\ldots r_j \underbrace{r_1r_2\ldots r_j}_{\text{repeat}} \underbrace{r_1r_2\ldots r_j}_{\text{repeat}} \ldots. \tag{2.7.72}$$

2.7.3. An *infinite binary expansion* is the limit of the form

$$0.b_1 b_2 \ldots_2 = \lim_{n \to \infty} \left(\frac{b_1}{2} + \frac{b_2}{2^2} + \cdots + \frac{b_n}{2^n} \right) \tag{2.7.73}$$

where $b_k \in \{0, 1\}$ for each $k \in \mathbb{N}$. In contrast to Lemma 2.7.12, prove infinite binary expansions are not distinct in the following sense:

(i) Prove the limits defining infinite binary expansions exist.

(ii) Prove there are sequences (a_n) and (c_n) where

- $a_k, c_k \in \{0, 1\}$ for each $k \in \mathbb{N}$,
- $a_j \neq c_j$ for at least one $j \in \mathbb{N}$, and yet
- $0.a_1 a_2 \ldots_2 = 0.c_1 c_2 \ldots_2$.

2.8   Countable and uncountable sets

The following mathematical fact blew my mind when I first found my own way to understand it:

> There are different types of infinity.

Some sets can have all of their elements counted by associating each with its own positive integer. Others are so large and have so many elements that they are *uncountable*: There are not enough positive integers to count them all.

To help you find your own meaning for the different types of infinity and set the stage for distinguishing between different types of infinite sets, let's first see what we can say about finite sets beyond Definition 1.2.22 by making use of the definitions for bijection (Definition 1.2.19) and onto (Definition 1.2.16).

Remark 2.8.1: Finite versus infinite

A nonempty set B is finite if there is some $n_0 \in \mathbb{N}$ and a bijection f where

$$f : \{1, 2, \ldots, n_0\} \to B. \tag{2.8.1}$$

Basically, f counts the elements of B from a first to an n_0th. We have

$$B = \{b_1, b_2, \ldots, b_{n_0}\} \quad \text{where} \quad f(1) = b_1, \ f(2) = b_2, \ \ldots, \ f(n_0) = b_{n_0}. \tag{2.8.2}$$

Furthermore, f is onto. On the other hand, a set J is infinite if for every $n \in \mathbb{N}$ and any function $g : \{1, 2, \ldots, n\} \to J$, there is some point $y \in J$ which is not an output of g since there are not enough numbers in $\{1, 2, \ldots, n\}$ for g to count every element of J. In other words, g is not onto.

Infinite sets come in two flavors: *countable* and *uncountable*. Countable sets are the focus of this section while uncountable sets are explored in the next.

The idea for countable sets extends finite sets (which have a bijection with some set $\{1, \ldots, n\}$) to include sets which can be indexed by the set of positive integers $\mathbb{N}$. In other words, countable sets are the range of a sequence. It may help to revisit the definitions for sequence and onto (Definitions 2.1.1 and 1.2.16).

Definition 2.8.2: Countable and uncountable

A set is *countable* if it is the empty set or the range of a sequence. So, a nonempty set is countable when there is an onto function from $\mathbb{N}$ to the set. A set is *uncountable* if it is not countable.

The notion of countable sets give us a way to view the set of positive integers $\mathbb{N}$ and the set of integers $\mathbb{Z}$ as the same type of infinite set. Even though $\mathbb{N}$ is a proper subset of $\mathbb{Z}$, there are enough positive integers to count for all of integers.

Example 2.8.3: $\mathbb{Z}$ is countable

The set of integers $\mathbb{Z}$ is countable.

Scratch Work 2.8.4: Exploit the parity of integers

My idea takes advantage of the parity of even and odd positive integers, and that there infinitely many of each. I'd like to define a function $f : \mathbb{N} \to \mathbb{Z}$ which uses the odd positive integers to account for the positive integers and zero (nonnegative integers) and use the even positive integers to account for the negative integers, like this:

$$f(1) = 0 \qquad f(3) = 1 \qquad f(5) = 2 \qquad \cdots$$
$$f(2) = -1 \qquad f(4) = -2 \qquad f(6) = -3 \qquad \cdots$$

So, the outputs of f should alternate between the nonnegative and negative integers. With the goal of finding a suitable function f in mind and after playing around with formulas for a bit, I think the sequence (z_n) defined by the function $f : \mathbb{N} \to \mathbb{Z}$ where

$$z_n = f(n) = \begin{cases} \dfrac{n-1}{2}, & \text{if } n \text{ is odd,} \\[2ex] -\dfrac{n}{2}, & \text{if } n \text{ is even} \end{cases} \tag{2.8.3}$$

is onto. This formula for f yields nonnegative outputs when n is odd and negative outputs when n is even. The proof takes this a step further by showing every integer in $\mathbb{Z}$ is one of the terms of (z_n), that is, every integer is an output of f. Since f splits $\mathbb{Z}$ into nonnegative and negative cases, the proof follows suit.

Proof for Example 2.8.3. Consider the sequence (z_n) defined by the function $f : \mathbb{N} \to \mathbb{Z}$ given by

$$z_n = f(n) = \begin{cases} \dfrac{n-1}{2}, & \text{if } n \text{ is odd,} \\[3mm] -\dfrac{n}{2}, & \text{if } n \text{ is even.} \end{cases} \tag{2.8.4}$$

The goal is to show f is onto with two cases.

$\underline{\text{Case (i) } w \geq 0}$: Suppose $w \in \mathbb{Z}$ where $w \geq 0$. Define

$$n_w = 2w + 1. \tag{2.8.5}$$

Note that n_w is positive and odd, so $n_w \in \mathbb{N}$. Then

$$z_{n_w} = f(n_w) = \frac{n_w - 1}{2} = \frac{2w + 1 - 1}{2} = w. \tag{2.8.6}$$

$\underline{\text{Case (ii) } w < 0}$: Suppose $w \in \mathbb{Z}$ where $w < 0$. Define

$$n_w = -2w. \tag{2.8.7}$$

Note that n_w is positive and even, so $n_w \in \mathbb{N}$. Then

$$z_{n_w} = f(n_w) = -\frac{n_w}{2} = -\frac{2w}{2} = w. \tag{2.8.8}$$

Hence, for every $w \in \mathbb{Z}$ there is an index $n_w \in \mathbb{N}$ where

$$z_{n_w} = f(n_w) = w. \tag{2.8.9}$$

Therefore, f is onto and $\mathbb{Z}$ is the range of the sequence (z_n), so $\mathbb{Z}$ is countable. $\square$

What about the set of rational numbers $\mathbb{Q}$? Or the real line $\mathbb{R}$? We'll get to them in a bit. First, consider a subset of the plane whose components are integers and positive integers, respectively.

Example 2.8.5: $\mathbb{Z} \times \mathbb{N}$

Consider the set $\mathbb{Z} \times \mathbb{N}$ defined by

$$\mathbb{Z} \times \mathbb{N} = \left\{ \mathbf{x} = \begin{bmatrix} m \\ n \end{bmatrix} : m \in \mathbb{Z} \text{ and } n \in \mathbb{N} \right\} \subseteq \mathbb{R}^2. \tag{2.8.10}$$

I claim $\mathbb{Z} \times \mathbb{N}$ is countable. To partially justify my claim, I can provide evidence that there is a sequence $(\mathbf{x}_k)$ whose range is $\mathbb{Z} \times \mathbb{N}$. However, I don't have an explicit formula to define $(\mathbf{x}_k)$. All I offer is the path tracing through $\mathbb{Z} \times \mathbb{N}$ one point at a time in Figure 2.8.1, and the assurance that every point in $\mathbb{Z} \times \mathbb{N}$ is eventually identified as a term of $(\mathbf{x}_k)$. By that

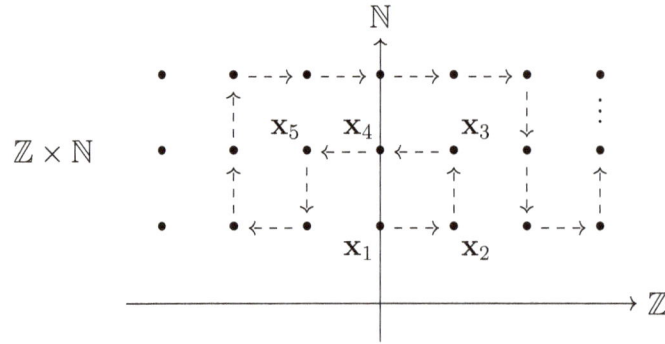

Figure 2.8.1: The range of a sequence $(\mathbf{x}_k)$ is the set $\mathbb{Z} \times \mathbb{N}$, which is therefore countable. See Example 2.8.5. Every point in $\mathbb{Z} \times \mathbb{N}$ is identified as a term of $(\mathbf{x}_k)$.

I mean there is an onto function $h : \mathbb{N} \to \mathbb{Z} \times \mathbb{N}$ such that for every

$$\begin{bmatrix} m \\ n \end{bmatrix} \in \mathbb{Z} \times \mathbb{N}, \tag{2.8.11}$$

there is an index $k \in \mathbb{N}$ where

$$\mathbf{x}_k = h(k) = \begin{bmatrix} m \\ n \end{bmatrix}. \tag{2.8.12}$$

Example 2.8.6: $\mathbb{Q}$ is countable

Now consider the set of rational numbers $\mathbb{Q}$ defined by

$$\mathbb{Q} = \left\{ \frac{m}{n} : m \in \mathbb{Z} \text{ and } n \in \mathbb{N} \right\}. \tag{2.8.13}$$

To show $\mathbb{Q}$ is countable, we can show $\mathbb{Q}$ is the range of a sequence.

Let $h : \mathbb{N} \to \mathbb{Z} \times \mathbb{N}$ be an onto function defining a sequence $(\mathbf{x}_k)$ whose range is $\mathbb{Z} \times \mathbb{N}$, like the one described in Example 2.8.5 and Figure 2.8.1. Since h is onto, every point $\begin{bmatrix} m \\ n \end{bmatrix} \in \mathbb{Z} \times \mathbb{N}$ is the image of some index $k \in \mathbb{N}$ under h where

$$h(k) = \mathbf{x}_k = \begin{bmatrix} m \\ n \end{bmatrix}. \tag{2.8.14}$$

From here, we can use the function $g : \mathbb{Z} \times \mathbb{N} \to \mathbb{Q}$ defined by

$$g\left(\begin{bmatrix} m \\ n \end{bmatrix} \right) = \frac{m}{n}. \tag{2.8.15}$$

Note that g is onto since every rational number m/n in the codomain $\mathbb{Q}$ is the image of its input $\begin{bmatrix} m \\ n \end{bmatrix}$ from the domain $\mathbb{Z} \times \mathbb{N}$.

Since the range of h is the domain of g, their composition $g \circ h : \mathbb{N} \to \mathbb{Q}$ is well-defined. Furthermore, $g \circ h$ defines a sequence whose range is $\mathbb{Q}$ since both h and g are onto: Every rational number m/n is the image of a point $\begin{bmatrix} m \\ n \end{bmatrix} \in \mathbb{Z} \times \mathbb{N}$ which is the image of an index $k \in \mathbb{N}$. We have

$$\frac{m}{n} = g\left(\begin{bmatrix} m \\ n \end{bmatrix} \right) = g(\mathbf{x}_k) = g(h(k)) = (g \circ h)(k). \tag{2.8.16}$$

Therefore, $(g(\mathbf{x}_k))$ is a sequence whose range is $\mathbb{Q}$, and $\mathbb{Q}$ is countable.

So, the set of integers $\mathbb{Z}$ and the set of rational numbers $\mathbb{Q}$ are both countable since each is the range of a sequence.

However, the real line $\mathbb{R}$ is uncountable[3]. It's too large to be the range of a sequence. My approach to prove this is to consider *any* function $f : \mathbb{N} \to \mathbb{R}$ and show f is not onto: No matter how f is defined, some real number y is not an output of f.

There is a fair amount of material to build up before diving into the full argument. For now, there is more to say about countable sets.

Lemma 2.8.7: Finite sets are countable

Every finite set is countable.

Scratch Work 2.8.8: Repeat the last element

For a nonempty finite set, my idea is to construct a sequence that runs through the elements in the set, then when those elements are exhausted, continue defining terms by repeating the last element indefinitely, like this:

$$(x_n) = (s_1, s_2, \ldots, s_{n_0}, s_{n_0}, s_{n_0}, \ldots) \tag{2.8.17}$$

Proof of Lemma 2.8.7. The empty set $\varnothing$ is countable by Definition 2.8.2. Now, suppose S is a nonempty finite set. Then there is a positive integer $n_0 \in \mathbb{N}$ where

$$S = \{s_1, s_2, \ldots, s_{n_0}\}. \tag{2.8.18}$$

For each index $n = 1, 2, \ldots, n_0 - 1$, define

$$x_n = s_n. \tag{2.8.19}$$

For every other index $n \geq n_0$, define

$$x_n = s_{n_0}. \tag{2.8.20}$$

---

[3]I think this is really cool.

Then the range of the sequence (x_n) is S, so S is countable.   □

The sets $\mathbb{N}$, $\mathbb{Z}$, $\mathbb{Z} \times \mathbb{N}$, and $\mathbb{Q}$ are examples of infinite countable sets. That is, they are *countably infinite*.

> ### Definition 2.8.9: Countably infinite
>
> A set is *countably infinite* if it is infinite and countable.

The distinction between finite and countably infinite sets allows for a distinction between their underlying functions.

> ### Definition 2.8.10: Enumeration
>
> An *enumeration* of a nonempty finite set S with n_0 elements is a bijection from $\{1, 2, \ldots, n_0\}$ to S. An *enumeration* of a countably infinite set T is a bijection from $\mathbb{N}$ to T.

> ### Example 2.8.11: An enumeration of the integers
>
> The function $f : \mathbb{N} \to \mathbb{Z}$ in Example 2.8.3 assures us $\mathbb{Z}$ is countable since $\mathbb{Z}$ the range of the sequence defined by f. This means f is onto, but it is also a bijection. That is, f is an enumeration of $\mathbb{Z}$.

Proof for Example 2.8.11. Once again, define $f : \mathbb{N} \to \mathbb{Z}$ for each $n \in \mathbb{N}$ by

$$f(n) = \begin{cases} \dfrac{n-1}{2}, & \text{if } n \text{ is odd,} \\[2ex] -\dfrac{n}{2}, & \text{if } n \text{ is even.} \end{cases} \tag{2.8.21}$$

To show f is a bijection from $\mathbb{N}$ to $\mathbb{Z}$, it suffices to show f is both one-to-one and onto. The fact that f is onto is established in the proof for Example 2.8.3. So, it remains to show f is one-to-one (Definition 1.2.18).

   Case (i): Suppose $j, k \in \mathbb{N}$ where j is odd and k is even. Then

$$f(j) = \frac{j-1}{2} \geq 0 \quad \text{and} f(k) \qquad\qquad = -\frac{k}{2} < 0. \tag{2.8.22}$$

Hence, $f(j)$ is nonnegative while $f(k)$ is negative, so $f(j) \neq f(k)$.

   Case (ii): Suppose $j, k \in \mathbb{N}$ where j and k are both odd and, without loss of generality, $j < k$. Then

$$f(j) = \frac{j-1}{2} < \frac{k-1}{2} = f(k). \tag{2.8.23}$$

Hence, $f(j) \neq f(k)$.

   Case (iii): Suppose $j, k \in \mathbb{N}$ where j and k are both even and, without loss of generality, $j < k$. Then

$$f(j) = -\frac{j}{2} > -\frac{k}{2} = f(k). \tag{2.8.24}$$

Hence, $f(j) \neq f(k)$.

Therefore, $j \neq k$ implies $f(j) \neq f(k)$, so f is one-to-one. Furthermore, f is a bijection and an enumeration of $\mathbb{Z}$. $\qquad \square$

Example 2.8.12: Not an enumeration of $\mathbb{Q}$

The function $h : \mathbb{N} \to \mathbb{Z} \times \mathbb{N}$ described in Example 2.8.6 is an enumeration of $\mathbb{Z} \times \mathbb{N}$. However, the composition $g \circ h : \mathbb{N} \to \mathbb{Q}$ also described in Example 2.8.6 is not an enumeration of $\mathbb{Q}$. Since h is onto (or at least seems to be), there are indices $j, k \in \mathbb{N}$ where $j \neq k$ and

$$h(j) = \begin{bmatrix} 1 \\ 2 \end{bmatrix} \in \mathbb{Z} \times \mathbb{N} \quad \text{and} \quad h(k) = \begin{bmatrix} 2 \\ 4 \end{bmatrix} \in \mathbb{Z} \times \mathbb{N}. \tag{2.8.25}$$

Hence, $h(j) \neq h(k)$ but we have

$$(g \circ h)(j) = g(h(j)) = \frac{1}{2} = \frac{2}{4} = g(h(k)) = (g \circ h)(k), \tag{2.8.26}$$

so $g \circ h$ is not one-to-one. Therefore, $g \circ h$ is not an enumeration of $\mathbb{Q}$.

To wrap the section up, let's prove the real line $\mathbb{R}$ is uncountable. The approach we take has a name: *Cantor's Diagonalization*. Note that the approach is named, not the result. The following example showcases this classic idea.

Example 2.8.13: Cantor's Diagonalization

Consider a sequence (t_n) of infinite decimal expansions whose digits are 2 or 3 only and where

$$t_1 = f(1) = 0.t_{11}t_{12}t_{13}t_{14}\ldots = 0.2323\ldots \tag{2.8.27}$$
$$t_2 = f(2) = 0.t_{21}t_{22}t_{23}t_{24}\ldots = 0.2222\ldots \tag{2.8.28}$$
$$t_3 = f(3) = 0.t_{31}t_{32}t_{33}t_{34}\ldots = 0.3333\ldots \tag{2.8.29}$$

$$\vdots$$

Note that f is some function from $\mathbb{N}$ to the set of infinite decimal expansions whose digits are only 2 or 3.

We can use Cantor's Diagonalization to find another infinite decimal expansion using digits 2 and 3 only which is not the same as any of the t_n, and so not an output of f. Let $s = 0.s_1s_2s_3\ldots$ where we define

$$s_1 = 3 \neq 2 = t_{11}, \tag{2.8.30}$$
$$s_2 = 3 \neq 2 = t_{22}, \tag{2.8.31}$$
$$s_3 = 2 \neq 3 = t_{33}, \tag{2.8.32}$$

$$\vdots \tag{2.8.33}$$

Can you see the pattern? We basically define each digit s_n as the opposite of the digit t_{nn} in the nth row and nth column—the "diagonal"—of the list (t_n).

More specifically, for each index $n \in \mathbb{N}$, define the digit s_n by

$$s_n = \begin{cases} 2, & \text{if } t_{nn} = 3, \\ 3, & \text{if } t_{nn} = 2. \end{cases} \tag{2.8.34}$$

This scheme guarantees $s = 0.s_1 s_2 s_3 n \ldots$ is a different infinite decimal expansion from all of the $t_n = 0.t_{n1} t_{n2} t_{n3} \ldots$, even though every digit of s is either 2 or 3. That is, for each index $n \in \mathbb{N}$, the digits s_n and t_{nn} are different. So s is not in the range of the function f that defines (t_n).

It's taken a while to get this point, but we can now show that the real line $\mathbb{R}$ is a different kind of infinite set.

Theorem 2.8.14: $\mathbb{R}$ is uncountable

The real line $\mathbb{R}$ is uncountable.

Scratch Work 2.8.15: Cantor's Diagonalization to show uncountable

Based on the definition of uncountable sets (Definition 2.8.2), the goal is to show $\mathbb{R}$ cannot be the range of a sequence. This means *every* function that defines a sequence of real numbers, so where $f : \mathbb{N} \to \mathbb{R}$, is not onto. We can use Cantor's Diagonalization to help. However, unlike Example 2.8.13, we cannot consider just one function and find a real number which is not in its range.

Thankfully, this is not a big issue. We can modify Cantor's Diagonalization to work for any function f from $\mathbb{N}$ to a suitable subset of $\mathbb{R}$, specifically the set B of infinite decimal expansions with digits limited to 0 and 1. The goal is to show that every function from $\mathbb{N}$ to B fails to be onto and, therefore, no function from $\mathbb{N}$ to R can be onto.

Proof of Theorem 2.8.14. Suppose B denotes the set of infinite decimal expansions with digits 0 and 1 only, and suppose a function $f : \mathbb{N} \to B$ defines a sequence such that for each index $n \in \mathbb{N}$ we have

$$f(1) = 0.x_{11} x_{12} x_{13} \ldots x_{1n} \ldots \tag{2.8.35}$$
$$f(2) = 0.x_{21} x_{22} x_{23} \ldots x_{2n} \ldots \tag{2.8.36}$$
$$f(3) = 0.x_{31} x_{32} x_{33} \ldots x_{3n} \ldots \tag{2.8.37}$$
$$\vdots \tag{2.8.38}$$
$$f(n) = 0.x_{n1} x_{n2} x_{n3} \ldots x_{nn} \ldots \tag{2.8.39}$$
$$\vdots \tag{2.8.40}$$

where, for every pair of indices $j, k \in \mathbb{N}$, we have $x_{jk} \in \{0, 1\}$. (This ensures every output of f has digits that are 0 or 1 only.) Since infinite decimal expansions are real numbers (Theorem 2.7.8), every output of f is a real number.

Now, to show f is not onto, use Cantor's Diagonalization to define the digit y_n for each index $n \in \mathbb{N}$ by

$$y_n = \begin{cases} 0, & \text{if } x_{nn} = 1, \\ 1, & \text{if } x_{nn} = 0. \end{cases} \tag{2.8.41}$$

Let y denote the resulting infinite decimal expansion

$$y = 0.y_1 y_2 y_3 \ldots \tag{2.8.42}$$

Then y is also real number by Theorem 2.7.8.

Furthermore, y is different from every output of f. To see this, note that for every index $n \in \mathbb{N}$ we have

$$y_n \neq x_{nn}. \tag{2.8.43}$$

That is, the nth digit of y is distinct from from nth digit of $f(n)$. Since both y and $f(n)$ are real numbers given by infinite decimal expansions which use on the digits 0 and 1, Lemma 2.7.12 tells us

$$y \neq f(n). \tag{2.8.44}$$

Hence, f is not onto since $y \in B$ but y is not an output of f. Therefore, B is uncountable. Furthermore, since $B \subseteq \mathbb{R}$ and no function with domain $\mathbb{N}$ maps to every element of B, so no function with domain $\mathbb{N}$ maps onto the real line $\mathbb{R}$. Therefore, $\mathbb{R}$ is uncountable. $\square$

The next section generalizes the notion of convergence to help us get a handle on some divergent sequences.

Exercises

2.8.1. Prove that a countable union of countable sets is countable.

2.8.2. A real number r is an *algebraic number* if r is a the *root* of a polynomial p, meaning $p(r) = 0$, where the coefficients of p are integers. (See Definition 1.6.8.)

(i) Prove the *golden ratio* $\varphi = \dfrac{1 + \sqrt{5}}{2}$ is an algebraic number.

(ii) Prove r is an algebraic number if and only if r is the root of a polynomial q with rational coefficients.

(iii) Let A denote the set of all algebraic numbers. Prove A is countable.

2.8.3. Find a sequence of real numbers $(x_n) \subseteq \mathbb{R}$ satisfying the following conditions:

(i) $\mathrm{Slim}(x_n) = \{-1, 0, 1\}$,

(ii) there is a subsequence diverging to ∞, and

(iii) there is a subsequence diverging to $-\infty$.

2.8.4. Show that for any finite set of real numbers $F \subseteq \mathbb{R}$, there is a sequence (y_n) where $\mathrm{Slim}(x_n) = F$.

2.8.5. Find a sequence of real numbers (y_n) where $\mathrm{Slim}\,(y_n)$ is bounded and countably infinite.

2.8.6. Prove there is a sequence of real numbers whose set of subsequential limits is the whole real line $\mathbb{R}$. That is, find $(a_n) \subseteq \mathbb{R}$ where $\mathrm{Slim}(a_n) = \mathbb{R}$.

2.8.7. Suppose $a, b \in \mathbb{R}$ with $a < b$.

(i) Give an example of a sequence $(z_n) \subseteq \mathbb{R}$ where $\mathrm{Slim}(z_n) = [a, b]$.

(ii) Prove there is no sequence $(y_n) \subseteq \mathbb{R}$ where $\mathrm{Slim}(y_n) = (a, b]$.

2.8.8. Prove nontrivial intervals are uncountable. That is, if $a, b \in \mathbb{R}$ with $a < b$, then any interval with endpoints a and b is uncountable.

2.8.9. The *Cantor set C* is perhaps the simplest example of a fractal. One way to define it is to first define *infinite ternary expansions* given by

$$0.x_1x_2\ldots_3 = \lim_{n\to\infty} \left(\frac{x_1}{3} + \frac{x_2}{3^2} + \cdots + \frac{x_n}{3^n} \right) \tag{2.8.45}$$

where $x_k \in \{0, 1, 2\}$ for each $k \in \mathbb{N}$ are the digits. From there, the Cantor set is the set of points whose ternary expansions have digits 0 or 2 only. That is,

$$C = \{y \in [0, 1] : y = 0.y_1y_2\ldots_3 \text{ where } y_k \in \{0, 2\} \text{ for each } k \in \mathbb{N}\}. \tag{2.8.46}$$

(i) Prove the infinite ternary expansion of a point in the Cantor set is unique.

(ii) Prove the Cantor set is uncountable.

(iii) Prove the Cantor set is a *perfect* set: Every point in the Cantor set is the limit of a sequence of other points from the Cantor set.

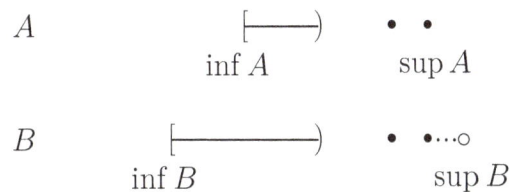

Figure 2.9.1: Bounded sets of real numbers A and B where $A \subseteq B$ and their suprema and infima. By Corollary 1.4.4 or Lemma 2.9.1, we have $\inf B \leq \inf A$ and $\sup A \leq \sup B$.

2.9 lim sup and lim inf

What does it mean for a sequence to "approach" a given point, exactly? One interpretation is provided by the definition for limit and convergence in Definition 2.2.1. But are there more flexible interpretations that define what it means for a sequence to "approach" more than one point? Or even a set? Or nothing in particular?

This section defines and explores additional features of sequences in the real line largely through the use of tails, subsequences, and points arbitrarily close to the original sequence. First, the following lemma helps parse the ideas explored in this section. It is a slight variation of Corollary 1.4.4.

Lemma 2.9.1: Suprema and infima of subsets

Suppose A and B are nonempty sets of real numbers.

(i) If $A \subseteq B$ and B is bounded above, then $\sup A \leq \sup B$.

(ii) If $A \subseteq B$ and B is bounded below, then $\inf B \leq \inf A$.

See Figure 2.9.1 for a pair of bounded sets real numbers A and B where $A \subseteq B$ that exhibit both parts of Lemma 2.9.1.

Proof of Lemma 2.9.1. To show (i), suppose $A \subseteq B$ and B is bounded above. Then any upper bound for B is also an upper bound for A. By the Axiom of Completeness (Axiom 1.3.8), both $\sup A$ and $\sup B$ exist. Since $A \subseteq B$, $\sup B$ is an upper bound for B as well as A. Since $\sup A$ is the *least* upper bound for A, we have

$$\sup A \leq \sup B. \tag{2.9.1}$$

Now, to show (ii), suppose $A \subseteq B$ and B is bounded below. Then any lower bound for B is also a lower bound for A. By Theorem 1.4.1, both $\inf A$ and $\inf B$ exist. Since $A \subseteq B$, $\inf B$ is a lower bound for B as well as A. Since $\inf A$ is the *greatest* lower bound for A, we have

$$\inf B \leq \inf A. \tag{2.9.2}$$

$\square$

Given a sequence of real numbers, the suprema and infima of its tails are interesting and allow us to explore the behavior of a sequence beyond convergence and divergence.

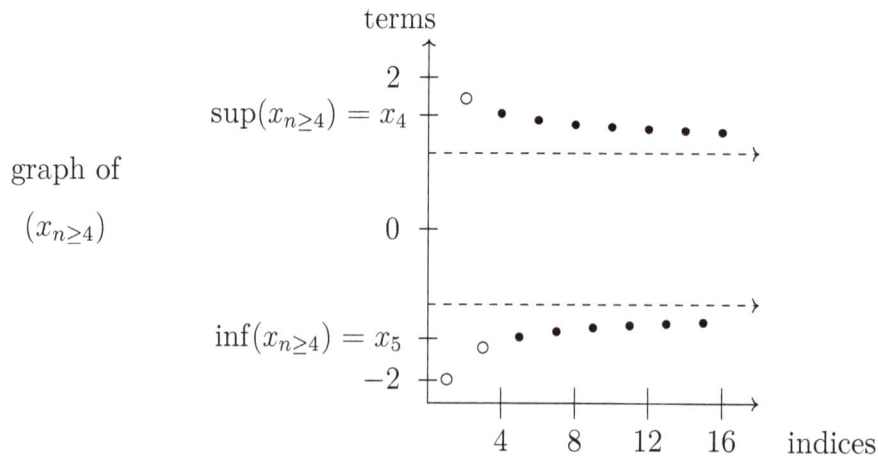

Figure 2.9.2: A graph of $(x_{n\geq4})$, the 4-tail of the sequence (x_n) from Example 2.9.3, along with its supremum x_4 and infimum x_5.

Notation 2.9.2: Suprema and infima of tails

For a sequence of real numbers (x_n) and an index $k \in \mathbb{N}$, denote the supremum of the k-tail by $\sup(x_{n\geq k})$ and denote the infimum of the k-tail by $\inf(x_{n\geq k})$. That is,

$$\sup(x_{n\geq k}) = \sup\{x_n : n \geq k\} = \sup\{x_k, x_{k+1}, \ldots\}, \quad \text{and} \qquad (2.9.3)$$
$$\inf(x_{n\geq k}) = \inf\{x_n : n \geq k\} = \inf\{x_k, x_{k+1}, \ldots\}. \qquad (2.9.4)$$

Example 2.9.3: Supremum and infimum can be terms

Consider the sequence (x_n) of real numbers defined for each $n \in \mathbb{N}$ by

$$x_n = (-1)^n \left(1 + \frac{1}{\sqrt{n}}\right). \qquad (2.9.5)$$

See Figure 2.9.2 for a plot of $(x_{n\geq4})$, the 4-tail of (x_n), along with it supremum and infimum. We have

$$\sup(x_{n\geq4}) = x_4 = 1 + \frac{1}{\sqrt{4}} = \frac{3}{2}, \quad \text{and} \qquad (2.9.6)$$

$$\inf(x_{n\geq4}) = x_5 = -1 - \frac{1}{\sqrt{5}}. \qquad (2.9.7)$$

Lemma 2.9.4: Monotonicity of the suprema and infima of tails

Let (x_n) be a sequence of real numbers.

(i) If (x_n) is bounded above, then the sequence of the suprema of its tails $(\sup(x_{n\geq k}))$ is decreasing.

(ii) If (x_n) is bounded below, then the sequence of the infima of its tails $(\inf(x_{n \geq k}))$ is increasing.

Both parts of Lemma 2.9.4 follow from Lemma 2.9.1 since the range of tail of a sequence is a subset of the range of the whole sequence.

Proof of Lemma 2.9.4. In both cases of this proof, suppose (x_n) is a sequence of real numbers. For every index $k \in \mathbb{N}$ we have

$$(x_{n \geq k+1}) \subseteq (x_{n \geq k}). \tag{2.9.8}$$

(Equivalently, $\{x_n : n \geq k + 1\} \subseteq \{x_n : n \geq k\}$.)

Case (i): Suppose (x_n) is bounded above. Then for every index $k \in \mathbb{N}$, the k-tail $(x_{n \geq k})$ is bounded above by $\sup(x_n)$, so $\sup(x_{n \geq k})$ exists by the Axiom of Completeness (Axiom 1.3.8). Therefore, by Lemma 2.9.1 we have

$$\sup(x_{n \geq k+1}) \leq \sup(x_{n \geq k}). \tag{2.9.9}$$

That is, $(\sup(x_{n \geq k}))$ is decreasing.

Case (ii): Suppose (x_n) is bounded below. Then for every index $k \in \mathbb{N}$, the k-tail $(x_{n \geq k})$ is bounded below by $\inf(x_n)$, so $\inf(x_{n \geq k})$ exists by Theorem 1.4.1. Therefore, by Lemma 2.9.1 we have

$$\inf(x_{n \geq k}) \leq \inf(x_{n \geq k+1}). \tag{2.9.10}$$

That is, $(\inf(x_{n \geq k}))$ is increasing. $\qquad\square$

When a sequence of real numbers is bounded (above and below), the sequences respectively defined by the of the suprema and infima of its tails both converge.

Theorem 2.9.5: Existence of $\limsup$ and $\liminf$ for bounded sequences of real numbers

Suppose (x_n) is a bounded sequence of real numbers. Then the sequences $(\sup(x_{n \geq k}))$ and $(\inf(x_{n \geq k}))$ converge.

Proof of Theorem 2.9.5. Suppose (x_n) is a bounded sequence of real numbers. Since the range of every tail of (x_n) is a subset of the range of (x_n), every tail of (x_n) is bounded as well. By Lemma 2.9.4, the sequence of suprema of tails $(\sup(x_{n \geq k}))$ and the sequence of infima of tails $(\inf(x_{n \geq k}))$ are monotone. So, by the Monotone and Bounded Convergence Theorem (Theorem 2.4.9), both sequences converge. $\qquad\square$

The notions of *limit superior* and *limit inferior* codify the limits of the sequences of suprema and infima of a bounded sequence of real numbers.

Definition 2.9.6: Limit superior and limit inferior

Let (x_n) be a sequence of real numbers.

(i) If (x_n) is bounded above, the *limit superior*, denoted by $\limsup_{n\to\infty} x_n$, is the limit of the sequence of suprema of its tails (assuming this limit exists). That is,

$$\limsup_{n\to\infty} x_n = \lim_{k\to\infty} \left(\sup(x_{n\geq k})\right). \tag{2.9.11}$$

If (x_n) is not bounded above, we write

$$\limsup_{n\to\infty} x_n = \infty. \tag{2.9.12}$$

(ii) If (x_n) is bounded below, the *limit inferior*, denoted by $\liminf_{n\to\infty} x_n$, is the limit of the sequence of infima of its tails (assuming this limit exists). That is,

$$\liminf_{n\to\infty} x_n = \lim_{k\to\infty} \left(\inf(x_{n\geq k})\right). \tag{2.9.13}$$

If (x_n) is not bounded below, we write

$$\liminf_{n\to\infty} x_n = -\infty. \tag{2.9.14}$$

Remark 2.9.7: Strangeness of $\limsup$ and $\liminf$

We can be more specific about the values the limit superior and limit inferior actually attain when they exist, but they're kind of strange: A $\limsup$ is an infimum while a $\liminf$ is a supremum.

In the proof Theorem 2.9.5, we make use of the Monotone and Bounded Convergence Theorem (Theorem 2.4.9) but not fully. When (x_n) is bounded above, the sequence of suprema of its tails $(\sup(x_{n\geq k}))$ is increasing by Lemma 2.9.4. So when (x_n) and therefore $(\sup(x_{n\geq k}))$ are bounded, the Monotone and Bounded Convergence Theorem (Theorem 2.4.9) tells us

$$\limsup_{n\to\infty} x_n = \lim_{k\to\infty} \left(\sup(x_{n\geq k})\right) = \inf\{\sup(x_{n\geq k}) : k \in \mathbb{N}\}. \tag{2.9.15}$$

Similarly, when (x_n) is bounded we have

$$\liminf_{n\to\infty} x_n = \lim_{k\to\infty} \left(\inf(x_{n\geq k})\right) = \sup\{\inf(x_{n\geq k}) : k \in \mathbb{N}\}. \tag{2.9.16}$$

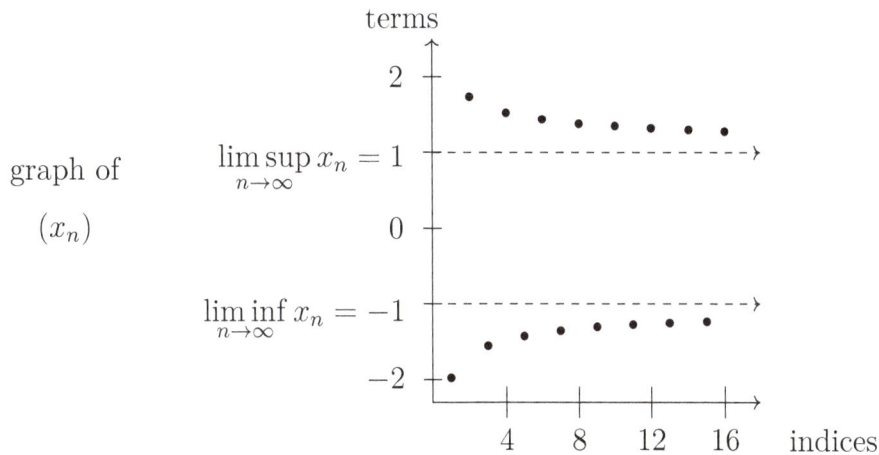

graph of

(x_n)

$$\limsup_{n\to\infty} x_n = 1$$

$$\liminf_{n\to\infty} x_n = -1$$

Figure 2.9.3: A graph of the sequence (x_n) from Example 2.9.8. Note how the terms alternate between being close to -1 and 1 as the indices increase with the suprema of the tails decreasing to 1 and the infima of the tails increasing to -1. We have $\liminf x_n = -1$ and $\limsup x_n = 1$.

Example 2.9.8: Distinct $\limsup$ **and** $\liminf$

Once again, consider the sequence (x_n) of real numbers from Example 2.9.3 defined for each $n \in \mathbb{N}$ by

$$x_n = (-1)^n \left(1 + \frac{1}{\sqrt{n}} \right). \tag{2.9.17}$$

See Figure 2.9.3. We have

$$\liminf_{n\to\infty} x_n = -1 < 1 = \limsup_{n\to\infty} x_n. \tag{2.9.18}$$

Proof for Example 2.9.8. For every index $k \in \mathbb{N}$, the supremum and infimum of the k-tail $(x_{n\geq k})$ depends on the parity of the index k as an even or odd positive integer. We have

$$\sup(x_{n\geq k}) = \begin{cases} x_{k+1}, & \text{if } k \text{ is odd,} \\ x_k, & \text{if } k \text{ is even.} \end{cases} \tag{2.9.19}$$

Note that the value of $\sup(x_{n\geq k})$ is a term whose index is even, whether the index k is even or not. Since (x_n) is bounded, Remark 2.9.7 tells us $\limsup x_n$ is the infimum of the range of the sequence $\sup(x_{n\geq k})$. Since the set of terms of (x_n) with even indices is both bounded below by 1 and arbitrarily close to 1, by the definition of infimum (Definition 1.1.14) we have that 1 is the infimum of the sequence of suprema of tails of (x_n). Hence,

$$\limsup_{n\to\infty} x_n = \inf\{\sup(x_{n\geq k}) : k \in \mathbb{N}\} = 1. \tag{2.9.20}$$

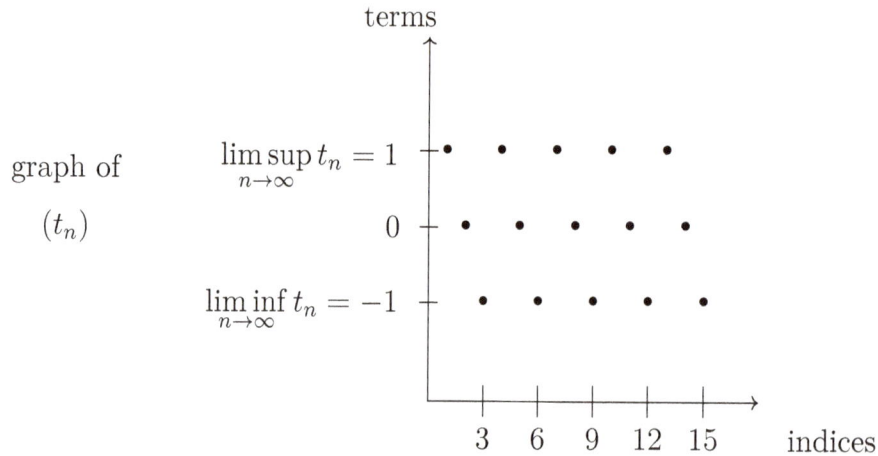

Figure 2.9.4: A graph of the sequence (t_n) from Example 2.9.9. Note how the terms alternate between $1, 0$, and -1. We have $\liminf t_n = -1$ and $\limsup t_n = 1$.

Similarly, for every index $k \in \mathbb{N}$ we have

$$\inf(x_{n \geq k}) = \begin{cases} x_k, & \text{if } k \text{ is odd,} \\ x_{k+1}, & \text{if } k \text{ is even.} \end{cases} \tag{2.9.21}$$

Here, the value of $\inf(x_{n \geq k})$ is a term whose index is odd, whether the index k is odd or not. Since the set of terms of (x_n) with odd indices is both bounded above by -1 and arbitrarily close to -1, by the definition of supremum (Definition 1.1.14) we have

$$\liminf_{n \to \infty} x_n = \sup\{\inf(x_{n \geq k}) : k \in \mathbb{N}\} = -1. \tag{2.9.22}$$

$\square$

Example 2.9.9: A sequence with neither $\limsup$ nor $\liminf$

Consider the sequence (t_n) of real numbers defined for each $n \in \mathbb{N}$ by

$$t_n = \begin{cases} 1, & \text{if } n = 3j - 2 \text{ for some } j \in \mathbb{N}, \\ 0, & \text{if } n = 3j - 1 \text{ for some } j \in \mathbb{N}, \\ -1, & \text{if } n = 3j \text{ for some } j \in \mathbb{N}. \end{cases} \tag{2.9.23}$$

See Figure 2.9.4. We have

$$\liminf_{n \to \infty} t_n = -1 \qquad \text{and} \qquad \limsup_{n \to \infty} t_n = 1. \tag{2.9.24}$$

But what about 0? We have (t_{3j-1}), the subsequence of (t_n) whose indices have remainder 2 when divided by 3, converges to 0:

$$\lim_{j \to \infty} t_{3j-1} = \lim_{j \to \infty} 0 = 0. \tag{2.9.25}$$

terms

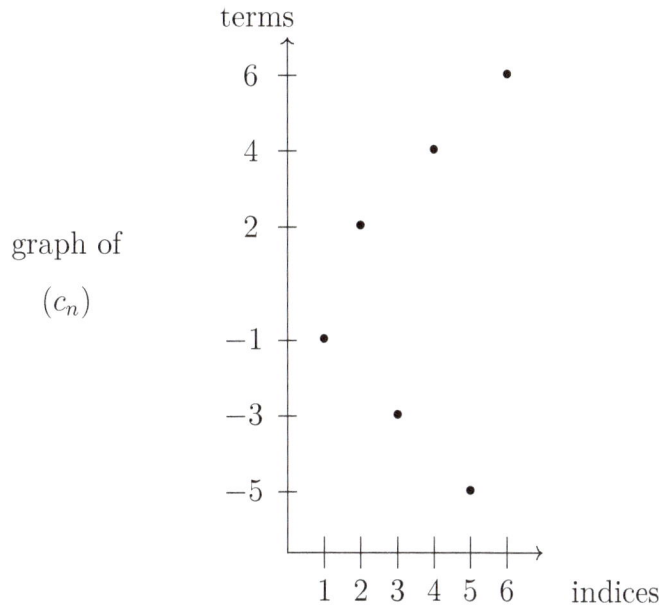

Figure 2.9.5: A graph of the sequence (c_n) from Example 2.9.10. Note how the terms alternate between n and $-n$, and neither $\liminf c_n$ nor $\limsup c_n$ exist.

Hence, 0 is a subsequential limit of (t_n) and so $0 \in \text{Slim}\,(t_n)$. See Definition 2.4.16. In fact, $-1, 0$, and 1 are subsequential limits of (t_n). We have

$$\lim_{j \to \infty} t_{3j-2} = \lim_{j \to \infty} 1 = 1 \qquad \text{and} \tag{2.9.26}$$

$$\lim_{j \to \infty} t_{3j} = \lim_{j \to \infty} -1 = -1. \tag{2.9.27}$$

Example 2.9.10: Empty Slim

Consider the sequence (c_n) of real numbers defined for each $n \in \mathbb{N}$ by

$$c_n = n(-1)^n = \begin{cases} -n, & \text{if } n \text{ is odd,} \\ n, & \text{if } n \text{ is even.} \end{cases} \tag{2.9.28}$$

See Figure 2.9.5. It turns out no real number is a subsequential limit of (c_n) since every real number is away from a tail of (c_n). Thus, $\text{Slim}(c_n) = \varnothing$. Also,

$$\limsup_{n \to \infty} c_n = \infty \qquad \text{and} \qquad \liminf_{n \to \infty} c_n = -\infty. \tag{2.9.29}$$

Proof for Example 2.9.10. Consider any $x \in \mathbb{R}$. By the Archimedean Property (Theorem 1.4.6), there is an index $k_x \in \mathbb{N}$ such that $|x| < k_x$. Then for every index $n \in \mathbb{N}$ where $n \geq k_x$ we have

$$|x| < k_x \leq n = |n(-1)^n| = |c_n|. \tag{2.9.30}$$

By the reverse triangle inequality (1.2.37) and properties of absolute value, we have

$$|x - c_n| \geq ||c_n| - |x|| = |c_n| - |x| = n - |x| > k_x - |x| > 0. \tag{2.9.31}$$

That is, x is away from the k_x-tail of (c_n), which we can write as $x \operatorname{awf}(c_{n \geq k_x})$. Hence, x cannot be the limit of a subsequence of (c_n). Therefore,

$$\operatorname{Slim}(c_n) = \varnothing. \tag{2.9.32}$$

By considering even and odd indices separately, the Archimedean Property 1.4.6 shows (c_n) is neither bounded above nor bounded below. Therefore,

$$\limsup_{n \to \infty} c_n = \infty \qquad \text{and} \qquad \liminf_{n \to \infty} c_n = -\infty. \tag{2.9.33}$$

$\square$

For a bounded sequence of real numbers, its limit superior and limit inferior are examples of points arbitrarily to *every* tail of the sequence. This idea is generalized by and formalized as the *coda* of a sequence. For a sequence in a Euclidean space, the coda serves as a technical notion for the set of all points the sequence "approaches" in some way, in fact exactly the same way as the set of subsequential limits. The relationship between and the coda and the set of subsequential limits is explored in the next section.

Exercises

2.9.1. Let (x_n) be a sequence of real numbers. Prove (x_n) converges if and only if

$$\limsup_{n \to \infty} x_n = \liminf_{n \to \infty} x_n. \tag{2.9.34}$$

2.9.2. Suppose $(x_n) \subseteq \mathbb{R}$ is bounded above.

(i) Prove $\limsup_{n \to \infty} x_n = \max \operatorname{Slim}(x_n)$.

(ii) State and prove a similar result for $\liminf_{n \to \infty} x_n$.

2.9.3. Suppose $a, b \in \mathbb{R}$ with $a < b$. Prove there is a sequence $(y_n) \subseteq \mathbb{R}$ where $\liminf_{n \to \infty} y_n = a$, $\limsup_{n \to \infty} y_n = b$, and $y_n \notin [a, b]$ for all $n \in \mathbb{N}$. What is $\operatorname{Slim}(y_n)$ in this case?

2.9.4. Consider the sequence (z_n) given by

$$(z_n) = \left(1, \frac{1}{2}, \frac{1}{3}, \frac{2}{3}, \frac{1}{4}, \frac{2}{4}, \frac{3}{4}, \ldots\right). \tag{2.9.35}$$

Determine $\liminf_{n \to \infty} z_n$, $\limsup_{n \to \infty} z_n$, and $\operatorname{Slim}(z_n)$.

(a_n)

$(a_{n \geq 2})$

$(a_{n \geq 3})$

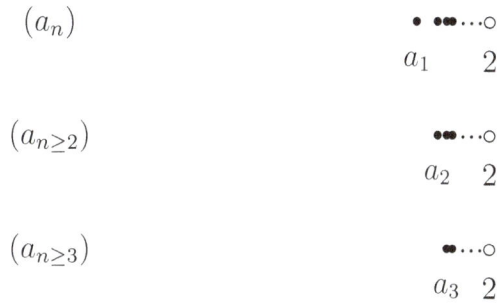

Figure 2.10.1: All of the tails of the sequence (a_n) from Examples 2.1.4 and 2.10.2 are arbitrarily close to 2. Also, 2 is the only real number arbitrarily close to every tail of (a_n), so $\text{Coda}(a_n) = \{2\}$.

2.10   Coda and Slim

Sequences exhibit a wide variety of interesting behaviors with convergence and limits as in Definition 2.2.1 being the most prominent. This section explores other behaviors by making use of tails and subsequences in Euclidean spaces along with sets of points—as opposed to individual points—that are arbitrarily close.

The section kicks off with the *coda* of a sequence, roughly the set of all points the sequence "approaches" in a way that is not as constrained as convergence. Recall that the closure $\overline{S}$ of a set S is the set of points arbitrarily close to S (see Definition 1.5.15).

Definition 2.10.1: Coda of a sequence

Let $(\mathbf{x}_n)$ be a sequence of points in $\mathbb{R}^m$. The *coda* of $(\mathbf{x}_n)$, denoted by $\text{Coda}(\mathbf{x}_n)$, is the set of points arbitrarily close to every tail of $(\mathbf{x}_n)$. Equivalently,

$$\text{Coda}(\mathbf{x}_n) = \{\mathbf{y} \in \mathbb{R}^m : \forall\, k \in \mathbb{N}, \mathbf{y}\,\text{acl}\,(\mathbf{x}_{n \geq k})\,\} \tag{2.10.1}$$

$$= \bigcap_{k \in \mathbb{N}} \overline{\{\mathbf{x}_n : n \geq k\}}. \tag{2.10.2}$$

Example 2.10.2: Codas of two sequences

Recall the sequences of real numbers (a_n) and (b_n) from Example 2.1.4 defined for each $n \in \mathbb{N}$ by

$$a_n = 2 - \left(\frac{1}{\sqrt{n}}\right) \quad \text{and} \quad b_n = \left[2 - \left(\frac{1}{\sqrt{n}}\right)\right](-1)^n. \tag{2.10.3}$$

We have

$$\text{Coda}(a_n) = \{2\} \quad \text{and} \quad \text{Coda}(b_n) = \{-2, 2\}. \tag{2.10.4}$$

See Figures 2.10.1 and Figure 2.10.2.

(b_n) ○···● ●   ●●···○
                -2 b_3 b_1 b_2 b_4 2

$(b_{n\geq 2})$ ○···●   ●●···○
                -2 b_3 b_2 b_4 2

$(b_{n\geq 3})$ ○···●   ●···○
                -2 b_3 b_4 2

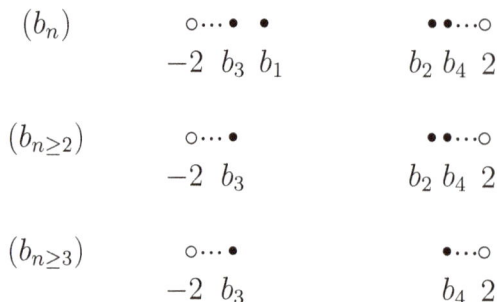

Figure 2.10.2: All of the tails of the sequence (b_n) from Examples 2.1.4 and 2.10.2 are arbitrarily close to both 2 and -2. Also, 2 and -2 are the only real numbers arbitrarily close to every tail of (b_n), so Coda $(b_n) = \{-2, 2\}$.

Scratch Work 2.10.3: Similar to showing arbitrarily close

Finding the coda of a sequence amounts to gathering all points which are not only arbitrarily close to the sequence, but arbitrarily close to *all of the tails*. For Example 2.10.2, only Coda$(a_n) = \{2\}$ is proven here. The proof showing Coda $(b_n) = \{-2, 2\}$ is similar but benefits from additional consideration of the parity of the indices as even or odd.

The proof for Coda$(a_n) = \{2\}$ is similar to showing $2\,\mathrm{acl}\,(a_n)$ as with Example 2.1.4 and partially follows from Scratch Work 2.1.5, but there are two key differences. We need to show 2 is arbitrarily close to every tail of (a_n) and every other real number is away from at least one tail of (a_n).

To capture every tail of (a_n), the first part of the proof begins with an arbitrary distance $\varepsilon > 0$ *and* an arbitrary index $k \in \mathbb{N}$. For the second part, we consider an arbitrary real number that is not equal to 2, then find a tail of (a_n) which is away from it. Here, we take advantage of the fact that (a_n) converges to 2, as shown in Example 2.2.7, which guarantees a tail of (a_n) is closer to 2 than the other number.

Proof for Coda$(a_n) = \{2\}$ *in Example 2.10.2.* Let $\varepsilon > 0$ and $k \in \mathbb{N}$. Following Scratch Work 2.1.5, choose an index $n_\varepsilon \in \mathbb{N}$ large enough to satisfy both $n_\varepsilon > 1/\varepsilon^2$ and $n_\varepsilon \geq k$. Then

$$n_\varepsilon > \frac{1}{\varepsilon^2} \quad \Longleftrightarrow \quad \frac{1}{\sqrt{n_\varepsilon}} < \varepsilon. \tag{2.10.5}$$

Hence, the term $a_{n_\varepsilon} = 2 - (1/\sqrt{n_\varepsilon})$ is in the k-tail $(a_{n\geq k})$ and within ε of 2:

$$d_{\mathbb{R}}(a_{n_\varepsilon}, 2) = |a_{n_\varepsilon} - 2| = \left| \left(2 - \frac{1}{\sqrt{n_\varepsilon}} \right) - 2 \right| = \frac{1}{\sqrt{n_\varepsilon}} < \varepsilon. \tag{2.10.6}$$

Therefore, $2\,\mathrm{acl}\,(a_{n\geq k})$. That is, 2 is arbitrarily close to every tail of (a_n), so $2 \in$ Coda (a_n).

Now suppose $x \in \mathbb{R}$ where $x \neq 2$. Then $|x - 2| > 0$. By Example 2.2.7, we have

$$\lim_{n\to\infty} a_n = 2. \tag{2.10.7}$$

So, by the definition of sequential limit (Definition 2.2.1), there is a threshold $n_x \in \mathbb{N}$ such that for all indices $n \geq n_x$ we have

$$|a_n - 2| < \frac{|x - 2|}{2} \iff -\frac{|x - 2|}{2} < -|a_n - 2|. \qquad (2.10.8)$$

So, the n_x-tail $(a_{n \geq n_x})$ is within $|x - 2|/2$ of 2. Furthermore, by the reverse triangle inequality (1.2.37) and properties of absolute value, for all indices $n \geq n_x$ we have

$$\frac{|x - 2|}{2} = |x - 2| - \frac{|x - 2|}{2} \qquad (2.10.9)$$
$$< |x - 2| - |2 - a_n| \qquad (2.10.10)$$
$$\leq ||x - 2| - |2 - a_n|| \qquad (2.10.11)$$
$$\leq |(x - 2) + (2 - a_n)| \qquad (2.10.12)$$
$$= |x - a_n|. \qquad (2.10.13)$$

Hence, every term in the n_x-tail $(a_{n \geq n_x})$ is $|x - 2|/2$ or more away from x. Therefore, x awf $(a_{n \geq n_x})$ and $\mathrm{Coda}\,(a_n) = \{2\}$. $\qquad\square$

Example 2.10.4: Coda of a convergent sequence

The coda of the sequence of real numbers in Example 1.1.15 given by $x_n = 3140 - (1/n)$ is a singleton and we have $\mathrm{Coda}(x_n) = \{3140\}$. The coda of the sequence $(\mathbf{z}_n)$ in $\mathbb{R}^2$ from Example 2.1.14 comprises two points and we have

$$\mathrm{Coda}(\mathbf{z}_n) = \{\mathbf{u}, \mathbf{v}\} \quad \text{where} \quad \mathbf{u} = \begin{bmatrix} 2 \\ 1 \end{bmatrix} \quad \text{and} \quad \mathbf{v} = \begin{bmatrix} -1 \\ 3 \end{bmatrix}. \qquad (2.10.14)$$

The proofs are left as an exercise, but they are similar to showing $\mathrm{Coda}\,(a_n) = \{2\}$ in Example 2.10.2.

Examples 2.10.2 and 2.10.4 provide just a glimpse into the considerable variety of structures exhibited by codas of sequences.

Example 2.10.5: Empty Coda

It is possible for the coda of a sequence to be empty. As in Examples 2.6.10 and 2.9.10, let $c_n = n(-1)^n$ for each $n \in \mathbb{N}$. In Figure 2.10.3, it looks like the terms c_n do not stay close to any particular real number as the index n increases (that is, as we consider the tails). As a result, $\mathrm{Coda}(c_n)$ is empty. The proof is nearly identical to showing $\mathrm{Slim}(c_n)$ is empty as in Example 2.9.10.

Proof for Example 2.10.5. Consider any $x \in \mathbb{R}$ and let $c_n = n(-1)^n$. By the Archimedean Property (Theorem 1.4.6), there is an index $n_x \in \mathbb{N}$ such that $|x| < n_x$. Then for every index $n \in \mathbb{N}$ where $n \geq n_x$ we have

$$|x| < n_x \leq n = |n(-1)^n| = |c_n|. \qquad (2.10.15)$$

(c_n) ··· •   •   •   •   •   • ···

           -5 -3 -1 2 4 6

$(c_{n \geq 2})$ ··· •   •   •   •   • ···

           -5 -3 2 4 6

$(c_{n \geq 3})$ ··· •   •   •   • ···

           -5 -3 4 6

$(c_{n \geq 4})$ ··· •   •   • ···

           -5 4 6

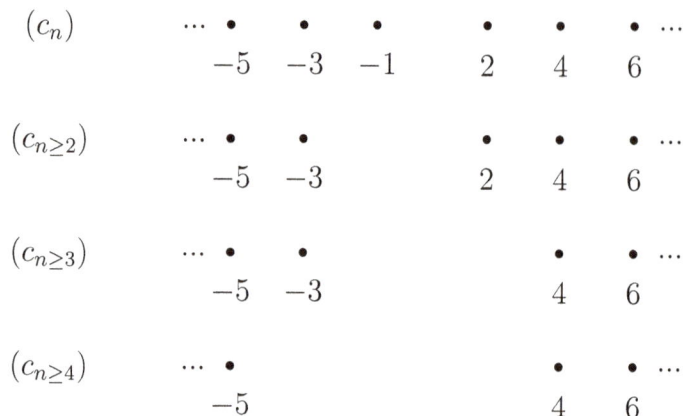

Figure 2.10.3: The tails of the sequence (c_n) from Example 2.10.5 do not seem to approach anything. Ultimately, we have $\mathrm{Coda}(c_n) = \varnothing$.

By the reverse triangle inequality (1.2.37) and properties of absolute value, we have

$$|x - c_n| \geq ||c_n| - |x|| = |c_n| - |x| = n - |x| > n_x - |x| > 0. \tag{2.10.16}$$

That is, x is away from the n_x-tail of (c_n), which we can write as $x \, \mathrm{awf} \, (c_{n \geq n_x})$. Hence, x cannot be the limit of a subsequence of (c_n). Therefore, x is not in $\mathrm{Coda}(c_n)$ and

$$\mathrm{Coda}(c_n) = \varnothing. \tag{2.10.17}$$

$\square$

Codas highlight the uniqueness of limits.

Theorem 2.10.6: Coda of a convergent sequence

If a sequence of points in $\mathbb{R}^m$ converges, then its coda is a singleton comprising the limit.

The first part of the proof shows the limit is in the coda and the second shows no other point is in the coda.

Proof of Theorem 2.10.6. Assume $\lim_{n \to \infty} \mathbf{x}_n = \mathbf{x}$, let $\varepsilon > 0$, and let $k \in \mathbb{N}$. Since $(\mathbf{x}_n)$ converges to $\mathbf{x}$, there is a threshold $n_\varepsilon \in \mathbb{N}$ large enough so that both $n_\varepsilon \geq k$ and

$$d_m(\mathbf{x}_{n_\varepsilon}, \mathbf{x}) < \varepsilon. \tag{2.10.18}$$

Hence, $\mathbf{x}_{n_\varepsilon}$ is in the k-tail $(\mathbf{x}_{n \geq k})$ and so $\mathbf{x} \, \mathrm{acl}(\mathbf{x}_{n \geq k})$. Hence, $\mathbf{x} \in \mathrm{Coda}(\mathbf{x}_n)$.

Now suppose $\mathbf{y} \neq \lim_{n \to \infty} \mathbf{x}_n = \mathbf{x}$. Considering the distance

$$\varepsilon_0 = \frac{1}{2} d_m(\mathbf{x}, \mathbf{y}) > 0, \tag{2.10.19}$$

it follows from the triangle inequality and the definition of sequential limit (Definition 2.2.1) that there is a threshold $n_0 \in \mathbb{N}$ such that for all indices $n \geq n_0$ we have

$$d_m(\mathbf{x}, \mathbf{y}) \leq d_m(\mathbf{x}, \mathbf{x}_n) + d_m(\mathbf{x}_n, \mathbf{y}) < \frac{1}{2} d_m(\mathbf{x}, \mathbf{y}) + d_m(\mathbf{x}_n, \mathbf{y}). \tag{2.10.20}$$

Therefore,

$$0 < \varepsilon_0 = \frac{1}{2} d_m(\mathbf{x}, \mathbf{y}) \leq d_m(\mathbf{x}_n, \mathbf{y}). \tag{2.10.21}$$

Thus, $\mathbf{y}$ is away from the n_0-tail $(\mathbf{x}_{n \geq n_0})$, which we can write as $\mathbf{y} \, \mathrm{awf} \, (\mathbf{x}_{n \geq n_0})$. Therefore, $\mathbf{y} \notin \mathrm{Coda}(\mathbf{x}_n)$ and

$$\mathrm{Coda}(\mathbf{x}_n) = \{\mathbf{x}\}. \tag{2.10.22}$$

$\square$

Remark 2.10.7: Coda does not determine convergence

Care must be taken when connecting the notions of coda and limit of a sequence: They're related but not equivalent. The limit of a sequence does not necessarily exist when the coda is a singleton. That is, the converse of Theorem 2.10.6 is false.

Example 2.10.8: Singleton coda with divergence

Let (w_n) be the sequence of real numbers with terms defined by $w_n = 0$ when n is odd and $w_n = n$ when n is even. That is, for each $n \in \mathbb{N}$ we have

$$w_n = \frac{n + n(-1)^n}{2} = \begin{cases} 0, & \text{if } n \text{ is odd}, \\ n, & \text{if } n \text{ is even}. \end{cases} \tag{2.10.23}$$

See Figure 2.10.4. For each index $k \in \mathbb{N}$ and the range of the corresponding k-tail $(w_{n \geq k})$ we have

$$\{w_n : n \geq k\} \subseteq \{0\} \cup \{k, k+1, \ldots\}. \tag{2.10.24}$$

Furthermore, every real number not in the range $\{w_n : n \geq k\}$ is away from it. So

$$\overline{\{w_n : n \geq k\}} \subseteq \{0\} \cup \{k, k+1, \ldots\}. \tag{2.10.25}$$

Hence, by the definition of coda (Definition 2.10.1) and its equilavent form in terms of closure we have

$$\{0\} \subseteq \mathrm{Coda}(w_n) \subseteq \bigcap_{k \in \mathbb{N}} (\{0\} \cup \{k, k+1, \ldots\}) = \{0\}. \tag{2.10.26}$$

Therefore, $\mathrm{Coda}(w_n) = \{0\}$. However, as an unbounded sequence (w_n) does not converge.

Remark 2.10.9: Coda and Slim

It turns out the coda of a sequence in a Euclidean space is the set of *subsequential limits* of the sequence. The definition of subsequential limit (Definition 2.4.16) uses both limits and subsequences while the definition of coda for sequences uses only arbitrarily close and tails.

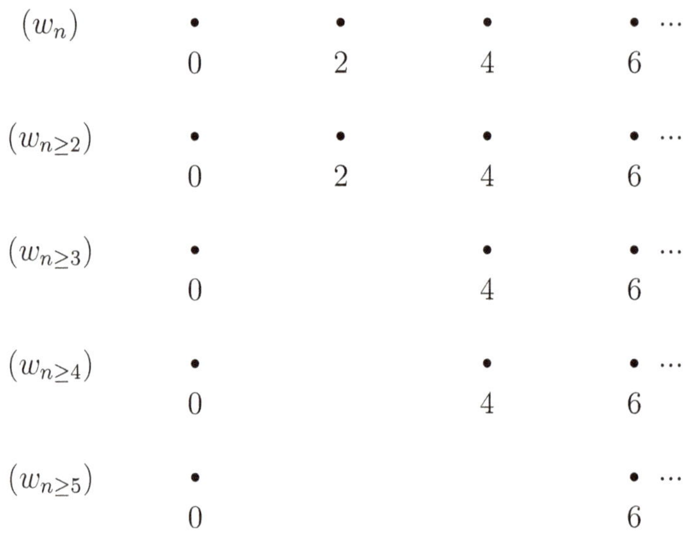

(w_n)		$\bullet$		$\bullet$		$\bullet$		$\bullet$	$\cdots$
		0		2		4		6	

$(w_{n \geq 2})$		$\bullet$		$\bullet$		$\bullet$		$\bullet$	$\cdots$
		0		2		4		6	

$(w_{n \geq 3})$		$\bullet$				$\bullet$		$\bullet$	$\cdots$
		0				4		6	

$(w_{n \geq 4})$		$\bullet$				$\bullet$		$\bullet$	$\cdots$
		0				4		6	

$(w_{n \geq 5})$		$\bullet$						$\bullet$	$\cdots$
		0						6	

Figure 2.10.4: The tails of the sequence (w_n) from Example 2.10.8 all contain 0 but the subsequence with even indices does not not seem to approach anything. Ultimately, we have $\mathrm{Coda}(w_n) = \{0\}$ but (w_n) diverges.

Thus, the definition of coda uses weaker conditions to generate the same set, as summarized in the following result.

Theorem 2.10.10: Equivalence of Coda and Slim

For every sequence of points in $\mathbb{R}^m$, the coda and the set of subsequential limits are the same set. That is, for every $(\mathbf{x}_n) \subseteq \mathbb{R}^m$,

$$\mathrm{Coda}(\mathbf{x}_n) = \mathrm{Slim}(\mathbf{x}_n). \tag{2.10.27}$$

Scratch Work 2.10.11: Tails versus subsequences

The proof of Theorem 2.10.10 amounts to carefully reorganizing terms of a given sequence into tails and subsequences, accordingly. I had a hard time writing this one up! Ultimately, I decided against a notation-heavy proof because it felt like I was making it hard to see the forest for the trees.

Proof of Theorem 2.10.10. First, suppose $\mathbf{z} \in \mathrm{Slim}(\mathbf{x}_n)$. Then there is a subsequence $(\mathbf{x}_{n_k})$ of $(\mathbf{x}_n)$ whose limit is $\mathbf{z}$. Since there are infinitely many terms in any subsequence, we have for any tail of $(\mathbf{x}_n)$ and any distance $\varepsilon > 0$ there is a term from the subsequence within ε of $\mathbf{z}$ whose index is large enough to be in the tail. Hence, $\mathbf{z}$ is arbitrarily close to every tail of $(\mathbf{x}_n)$. Therefore, $\mathbf{z} \in \mathrm{Coda}(\mathbf{x}_n)$.

Next, suppose $\mathbf{y} \in \mathrm{Coda}(\mathbf{x}_n)$. Then $\mathbf{y}$ is arbitrarily close to every tail of $(\mathbf{x}_n)$. Also, for every index $k \in \mathbb{N}$, we have $1/k > 0$. Since every k-tail of $(\mathbf{x}_n)$ contains all of the terms with indices greater than or equal to k, we can recursively construct a sequence $(\mathbf{x}_{n_k})$ where for every k we have

$$d_m(\mathbf{x}_{n_k}, \mathbf{y}) < \frac{1}{k} \qquad \text{and} \qquad n_k < n_{k+1} \tag{2.10.28}$$

by choosing terms from a succession of tails as needed. From there, an application of the corollary to the Archimedean Property (Corollary 1.4.8) allows us to conclude

$$\lim_{k\to\infty} \mathbf{x}_{n_k} = \mathbf{y}. \tag{2.10.29}$$

Hence, $\mathbf{y} \in \mathrm{Slim}(\mathbf{x}_n)$. $\qquad\qquad\square$

The coda of a sequence, and therefore the set of subsequential limits, also connects directly to the classical notions of *limit superior* and *limit inferior* for bounded sequences of real numbers. An exploration of this fact is left as an exercise, but consider the following example.

> ### Example 2.10.12: Coda, Slim, $\limsup$, and $\liminf$
>
> Consider the bounded sequence (x_n) of real numbers defined for each $n \in \mathbb{N}$ by
>
> $$x_n = (-1)^n + \frac{1}{\sqrt{n}}. \tag{2.10.30}$$
>
> See Figure 2.10.5. We have $\min(x_n) = \min\{x_n : n \in \mathbb{N}\}$ does not exist while
>
> $$\max(x_n) = \max\{x_n : n \in \mathbb{N}\} = x_2 = 1 + \frac{1}{\sqrt{2}}. \tag{2.10.31}$$
>
> Furthermore, $\mathrm{Coda}(x_n) = \mathrm{Slim}(x_n) = \{-1, 1\}$ with
>
> $$\liminf x_n = -1 < \limsup x_n = 1 < \max(x_n) = 1 + \frac{1}{\sqrt{2}}. \tag{2.10.32}$$
>
> Additionally,
>
> $$-1 = \liminf x_n = \min \mathrm{Coda}(x_n) = \min \mathrm{Slim}(x_n) \quad \text{and} \tag{2.10.33}$$
> $$1 = \limsup x_n = \max \mathrm{Coda}(x_n) = \max \mathrm{Slim}(x_n). \tag{2.10.34}$$

The deep relationships between codas, subsequential limits, limit superior, and limit inferior are explored in the exercises. Give them a shot!

The next chapter explores the *topology* of Euclidean spaces through a lens provided by points arbitrarily close to or away from sets and their complements.

Exercises

2.10.1. Find the coda of each of the following sequences, each defined by the corresponding formula for every positive integer n. Don't prove anything, but draw stuff!

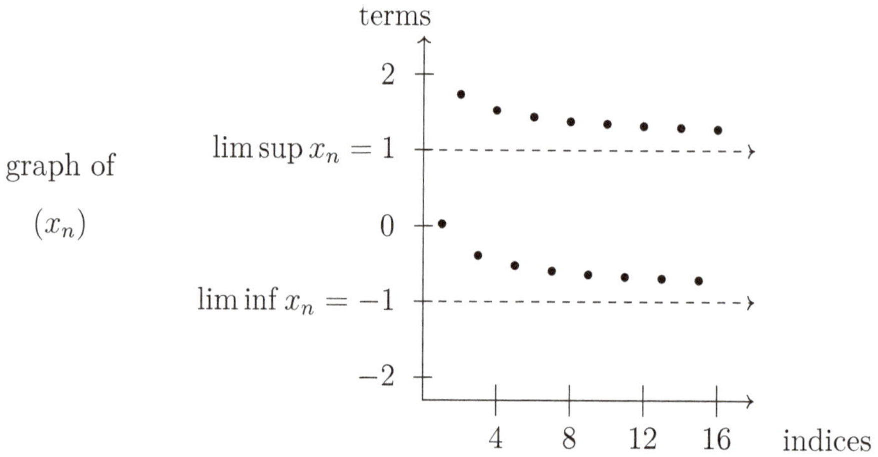

Figure 2.10.5: A graph of the sequence (x_n) from Example 2.10.12. Note how the terms alternate between being close to -1 and 1 as the indices increase. We have $\liminf x_n = -1$ and $\limsup x_n = 1$.

(i) $a_n = n$ (iii) $c_n = 3(-1)^n + 1/n$

(ii) $b_n = 8 - (-1)^n/n$ (iv) $d_n = 1 - 10^{-n}$

2.10.2. Prove the codas of the sequences (x_n) and (z_n) in Example 2.10.4 are as stated. In particular, why aren't other points in these codas?

2.10.3. Prove the following reformulation of the Bolzano-Weierstrass Theorem 2.5.13: Every bounded sequence of points in $\mathbb{R}^m$ has nonempty coda.

2.10.4. Suppose $(\mathbf{x}_n)$ is a sequence of points in $\mathbb{R}^m$. Prove

$$\lim_{n \to \infty} \mathbf{x}_n = \mathbf{y} \tag{2.10.35}$$

if and only if $\mathbf{y}$ is in the coda of every subsequence of $(\mathbf{x}_n)$. (Note that the boundedness of the sequence is not assumed.)

2.10.5. Prove that if (x_n) is a bounded sequence of real numbers, then

$$\limsup x_n = \max \mathrm{Coda}(x_n) = \max \mathrm{Slim}(x_n) \quad \text{and} \tag{2.10.36}$$
$$\liminf x_n = \min \mathrm{Coda}(x_n) = \min \mathrm{Slim}(x_n). \tag{2.10.37}$$

2.10.6. Find a sequence of real numbers whose coda is the whole real line.

Chapter 3

Topology of Euclidean Spaces

The definition of arbitrarily close in Definition 1.5.1 allows us to explore how points can be arbitrarily close to sets, whatever form the sets take. Expanding the setting to include *complements* of sets leads to fundamental aspects of the usual topology on a Euclidean space $\mathbb{R}^m$. The definition of arbitrarily close is essentially topological in nature: We have $\mathbf{y} \, \mathrm{acl} \, B$ if and only if every neighborhood of $\mathbf{y}$ intersects B. See Figure 3.1.1.

3.1   A closed-minded approach to topology

The set of points arbitrarily close to a given set gives rise to classic topological concepts: *closure* and *closed sets*. The definition of closure was already provided in Definition 1.5.15; it's repeated here for convenience. For a reminder about notation and terminology regarding ε-neighborhoods such as $V_\varepsilon(\mathbf{x})$, see Section 1.5, especially Definition 1.5.8, Figure 1.5.2, and Remark 1.5.9.

Definition 3.1.1: Closure and closed

The *closure* of a set $B \subseteq \mathbb{R}^m$ is the set of points arbitrarily close to B and is denoted by $\overline{B}$. That is,

$$\overline{B} = \{\mathbf{y} \in \mathbb{R}^m : \mathbf{y} \, \mathrm{acl} \, B\} = \{\mathbf{y} \in \mathbb{R}^m : \forall \, \varepsilon > 0, V_\varepsilon(\mathbf{y}) \cap B \neq \varnothing\}. \tag{3.1.1}$$

A set $F \subseteq \mathbb{R}^m$ is *closed* if all points arbitrarily close to F are in F. That is, F is closed if

$$\mathbf{y} \, \mathrm{acl} \, F \implies \mathbf{y} \in F \qquad \text{or, equivalently, if} \qquad \overline{F} \subseteq F. \tag{3.1.2}$$

Note that the empty set $\varnothing$ is vacuously closed. Also, by the definitions, closed sets contain their closures. But more is true.

Lemma 3.1.2: Closed sets are their closures

A set $F \subseteq \mathbb{R}^m$ is closed if and only if $F = \overline{F}$.

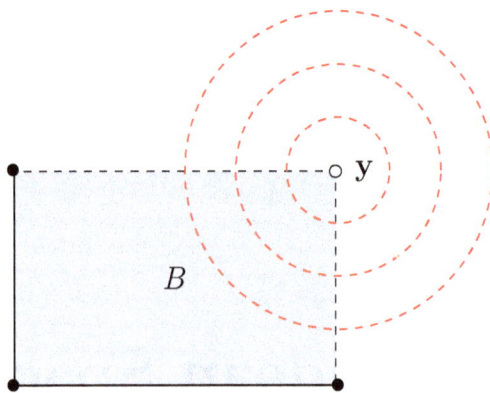

Figure 3.1.1: A point $\mathbf{y}$ and a set B in the plane $\mathbb{R}^2$ where $\mathbf{y}\,\mathrm{acl}\,B$. As such, every neighborhood of $\mathbf{y}$ intersects B. Also, B is not closed since $\mathbf{y}$ is not in B. See Definition 3.1.1.

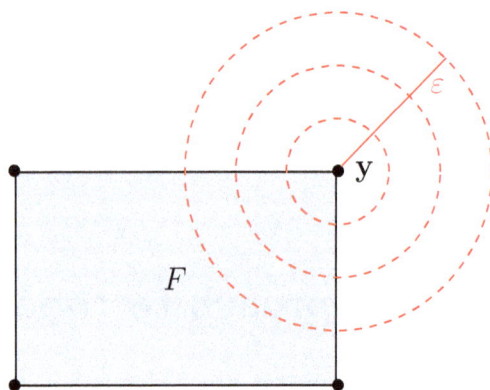

Figure 3.1.2: The set F contains all points in and arbitrarily close to F, including the corner $\mathbf{y}$ and the sides of the rectangle. As such, F is closed. See Definition 3.1.1.

Proof of Lemma 3.1.2. Given $F \subseteq \mathbb{R}^m$, we always have $F \subseteq \overline{F}$ thanks to Lemma 1.5.4 which tells us points in a set are arbitrarily close to the set. On the other hand, if F is closed, we have $\overline{F} \subseteq F$ by Definition 3.1.1. Therefore, F is closed if and only if $F = \overline{F}$. $\qquad\square$

Remark 3.1.3: The classic notion of closure

The definitions of closure and closed sets in Definition 3.1.1 are equivalent to "standard" definitions found in other texts, at least in the context of Euclidean spaces. One benefit of Definition 3.1.1 is how it follows directly from the definition of arbitrarily close (Definition 1.5.1). Classic approaches such as [1, Definition 3.2.7, p.90] require the definitions of more complicated ideas such as limits of sequences or similar concepts like *accumulation points* before defining closure and closed (see Definition 3.6.7). All this provides another reason I consider arbitrarily close to be the kernel of analysis.

In order to prove a set is closed, it often helps to consider the contraposition of Definition

3.1.1: A set F is closed if all points not in F are away from F.

> **Example 3.1.4: $\mathbb{Z}$ is closed**
>
> The set of integers $\mathbb{Z}$ is a closed set.

> **Scratch Work 3.1.5: Contraposition often helps**
>
> To ensure $\mathbb{Z}$ satisfies the definition of a closed set (Definition 3.1.1), we can work with contraposition. More generally, instead of showing $F \subseteq \mathbb{R}$ is closed by directly proving
>
> $$y \operatorname{acl} F \implies y \in F, \tag{3.1.3}$$
>
> we can show
>
> $$x \notin F \implies x \operatorname{awf} F. \tag{3.1.4}$$
>
> For this example, the proof below shows every noninteger is away from $\mathbb{Z}$.

Proof for Example 3.1.4. By way of contraposition, suppose $x \notin \mathbb{Z}$. Then by Corollary 1.4.9, there is an integer m_x such that

$$m_x < x < m_x + 1. \tag{3.1.5}$$

Define $\varepsilon_x = \min\{|m_x - x|, |m_x + 1 - x|\}$ (so ε_x is the shorter of the two distances between x and m_x or between x and $m_x + 1$). Since x is not an integer, we have $\varepsilon_x > 0$. Now suppose $z \in \mathbb{Z}$. Then either

$$z \leq m_x < x \qquad \text{or} \qquad x < m_x + 1 \leq z. \tag{3.1.6}$$

So by part (vi) of Theorem 1.3.2 we have

$$|z - x| \geq |m_x - x| \geq \varepsilon_x > 0 \qquad \text{or} \qquad |z - x| \geq |m_x + 1 - x| \geq \varepsilon_x > 0. \tag{3.1.7}$$

Either way, $x \operatorname{awf} \mathbb{Z}$. Therefore, $\mathbb{Z}$ contains all points arbitrarily close to $\mathbb{Z}$, so $\mathbb{Z}$ is closed. $\square$

The next result is a fundamental property of closed sets.

> **Lemma 3.1.6: Intersections of closed sets are closed**
>
> The intersection of any collection of closed sets is closed.

> **Scratch Work 3.1.7: Any means any**
>
> We need to be careful here. Lemma 3.1.6 refers to *any* collection of closed sets, no matter how large. So, I let A stand for any nonempty index set which could be finite, countably infinite, or uncountable. Aside from this sublety, the result follows from the definitions of a closed set (Definition 3.1.1), arbitrarily close (Definition 1.5.1), and neighborhood (Definition 1.5.8), as well as properties of intersections. It may help to revisit Remark 1.5.9

as well.

Proof of Lemma 3.1.6. Let $\{F_\alpha : \alpha \in A\}$ denote a collection of closed sets in $\mathbb{R}^m$ with a nonempty index set A. If any F_α is empty, then the intersection $\cap_{\alpha \in A} F_\alpha$ is empty as well and is therefore closed.

Next, suppose F_α is nonempty for every $\alpha \in A$ and $\mathbf{y}$ is arbitrarily close to $\cap_{\alpha \in A} F_\alpha$. Let $\varepsilon > 0$ and consider the ε-neighborhood $V_\varepsilon(\mathbf{y})$. By the definition of a closed set (Definition 3.1.1) and the version of arbitrarily close in terms of neighborhoods as in Remark 1.5.9, there is some $\mathbf{x} \in (\cap_{\alpha \in A} F_\alpha) \cap V_\varepsilon(\mathbf{y})$ which means $\mathbf{x} \in F_\alpha \cap V_\varepsilon(\mathbf{y})$ for each α. Since the distance ε was chosen arbitrarily and each F_α is closed, it follows that $\mathbf{y} \in F_\alpha$ for each α; thus $\mathbf{y} \in \cap_{\alpha \in A} F_\alpha$. Therefore, $\cap_{\alpha \in A} F_\alpha$ is closed.   □

Another lemma regarding closed sets indicates closures themselves are closed. Its proof is left as an exercise.

Lemma 3.1.8: Closures are closed

For any set $S \subseteq \mathbb{R}^m$, the closure $\overline{S}$ is a closed set.

There is another immediate result stemming from Lemma 3.1.6, Lemma 3.1.8, and the fact that the coda of a sequence of points in $\mathbb{R}^m$ is defined to be an intersection of closed sets (see Definition 2.10.1).

Corollary 3.1.9: Codas of sequences are closed

The coda—and therefore the set of subsequential limits—of a sequence of points in $\mathbb{R}^m$ is closed.

Proof of Corollary 3.1.9. Suppose is $(\mathbf{x}_n)$ a sequence of points in $\mathbb{R}^m$. By Definition 2.10.1, $\mathrm{Coda}((\mathbf{x}_n))$ is the set of points arbitrarily close to every tail of $(\mathbf{x}_n)$ and we have

$$\mathrm{Coda}((\mathbf{x}_n)) = \bigcap_{k \in \mathbb{N}} \overline{\{\mathbf{x}_n : n \geq k\}}. \tag{3.1.8}$$

By Lemma 3.1.8, we have for each index $k \in \mathbb{N}$ the closure of k-tail $(\mathbf{x}_{n \geq k})$ given by $\overline{\{\mathbf{x}_n : n \geq k\}}$, is closed for each index $k \in \mathbb{N}$. So, line (3.1.8) tells us $\mathrm{Coda}((\mathbf{x}_n))$ is an intersection of closed sets. Therefore, $\mathrm{Coda}((\mathbf{x}_n))$ is closed by Lemma 3.1.6.   □

A classic way to define closed sets in analysis stems from considering the limits of sequences whose terms are in the set. The following theorem provides a first characterization along these lines, but a more classic characterization which is very similar comes from considering what are called *accumulation points* (see Definition 3.6.7).

Theorem 3.1.10: Closed sets contain their limits

A set $F \subseteq \mathbb{R}^m$ is closed if and only if F contains the limits of all convergent sequences of points in F.

Scratch Work 3.1.11: Definitions and a fundamental connection

The proof follows from the definition of a closed set (Definition 3.1.1) along with the fundamental connection between the definition of arbitrarily close (Definition 1.5.1) and the definition of limit and convergence for sequences (Definition 2.2.1) provided by Theorem 2.3.1.

Proof of Theorem 3.1.10. Suppose F is a closed subset of $\mathbb{R}^m$ and let $(\mathbf{x}_n)$ be a convergent sequence of points in F with limit $\mathbf{x}$. By Theorem 2.3.1, we have $\mathbf{x} \operatorname{acl} (\mathbf{x}_n)$. Since every $\mathbf{x}_n$ is in F, we also have $\mathbf{x} \operatorname{acl} F$. So by the definition of a closed set (Definition 3.1.1), the limit $\mathbf{x}$ is in F.

Now suppose F contains the limits of all convergent sequences of points in F and suppose $\mathbf{y} \operatorname{acl} F$. By Theorem 2.3.1, there is a sequence $(\mathbf{y}_n)$ of points in F whose limit is $\mathbf{y}$. So, F contains $\mathbf{y}$ and F is closed. □

Example 3.1.12: A set E and its closure $\overline{E}$

To get a better idea of what's going on with Definition 3.1.1 and the results in this section, consider the set of real numbers $E \subseteq \mathbb{R}$ given by

$$E = [0,1) \cup \{2\} \cup \{3 + (1/n) : n \in \mathbb{N}\} \tag{3.1.9}$$

See Figure 3.1.3. Neither 1 nor 3 is in E, but both are arbitrarily close to E so E is not closed. Since points in a set are arbitrarily close to the set (Lemma 1.5.4), the closure of E contains all the real numbers in E as well as 1 and 3. Ultimately, we have

$$\overline{E} = [0,1] \cup \{2\} \cup \{3\} \cup \{3 + (1/n) : n \in \mathbb{N}\}. \tag{3.1.10}$$

By Lemma 3.1.8, $\overline{E}$ is closed. Theorem 3.1.10 tells us closed sets contain the limits of all convergent sequences, so here are some sequences of real numbers in E whose limits are in the closure $\overline{E}$:

(i) If $x_n = 1 - (1/2n)$ for each index $n \in \mathbb{N}$, then

$$(x_n) \subseteq E \quad \text{and} \quad \lim_{n \to \infty} x_n = 1 \in \overline{E}. \tag{3.1.11}$$

(ii) If $y_n = 3 + (1/n)$ for each index $n \in \mathbb{N}$, then

$$(y_n) \subseteq E \quad \text{and} \quad \lim_{n \to \infty} y_n = 3 \in \overline{E}. \tag{3.1.12}$$

(iii) If $z_n = 2$ for each index $n \in \mathbb{N}$, then

$$(z_n) \subseteq E \quad \text{and} \quad \lim_{n \to \infty} z_n = 2 \in \overline{E}. \tag{3.1.13}$$

The following lemma considers closed boxes of the form

$$B = [a_1, b_1] \times \cdots \times [a_m, b_m] \subseteq \mathbb{R}^m \tag{3.1.14}$$

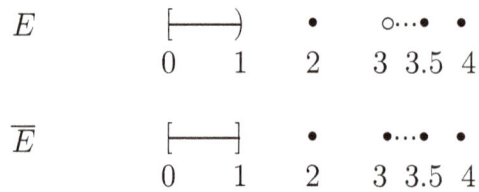

Figure 3.1.3: A set of real numbers E and its closure $\overline{E}$. See Definition 3.1.1 and Example 3.1.12.

where $a_j < b_j$ for each $j = 1, \ldots, m$. See Definition 2.5.8. Even though these are called *closed boxes*, the fact that they are also closed sets according to Definition 3.1.1 still needs to be justified. It turns out boxes of this form are closed and bounded.

> **Lemma 3.1.13: Closed and bounded intervals make closed and bounded boxes**
>
> If $B \subseteq \mathbb{R}^m$ is a box of the form (3.1.14), then B is closed and bounded.

> **Scratch Work 3.1.14: Directly from definitions**
>
> Both parts of the proof are direct in the sense that it shows B directly satisfies the definitions for bounded (Definition 1.5.20) and closed (Definition 3.1.1). For the latter, the proof shows every point not in B is away from B, so B must contain all points arbitrarily close.

Proof of Lemma 3.1.13. Suppose $B \subseteq \mathbb{R}^m$ is a box of the form (3.1.14). For every point $\mathbf{x} \in B$ we have

$$\mathbf{x} = \begin{bmatrix} x_1 \\ x_2 \\ \vdots \\ x_m \end{bmatrix} \qquad \text{and} \qquad a_j \leq x_j \leq b_j \tag{3.1.15}$$

for every component (or coordinate) x_j with $j = 1, \ldots, m$.

To show B is bounded, let

$$u = \max\{|a_1|, \ldots, |a_m|, |b_1|, \ldots, |b_m|\}. \tag{3.1.16}$$

Then for every $j = 1, \ldots, m$ we have $|x_j| \leq u$. Since $0 \leq x \leq y$ implies both $\sqrt{x} \leq \sqrt{y}$ and $x^2 \leq y^2$, for every $\mathbf{x} \in B$ we have

$$\|\mathbf{x}\|_m = \sqrt{x_1^2 + \ldots + x_m^2} \tag{3.1.17}$$

$$\leq \underbrace{\sqrt{u^2 + \ldots + u^2}}_{m \text{ copies of } u^2} \tag{3.1.18}$$

$$= u\sqrt{m}. \tag{3.1.19}$$

Therefore, B is bounded.

To show B is closed, suppose $\mathbf{w} \in \mathbb{R}^m \backslash B$ where

$$\mathbf{w} = \begin{bmatrix} w_1 \\ w_2 \\ \vdots \\ w_m \end{bmatrix}. \tag{3.1.20}$$

Since $\mathbf{w}$ is not in B, at least one of its components (or coordinates) w_{j_0} is less than a_{j_0} or greater than b_{j_0}. Without loss of generality, suppose $w_{j_0} > b_{j_0}$. In this case we have $w_{j_0} - b_{j_0} > 0$. This creates a positive distance between $\mathbf{w}$ and every point in B, as follows.

Since every point $\mathbf{x} \in B$ is of the form (3.1.15), we have every component x_j with $j = 1, \ldots, m$ satisfies $x_j \leq b_j$. For the particular index j_0 we have

$$x_{j_0} \leq b_{j_0} < w_{j_0} \quad \implies \quad w_{j_0} - x_{j_0} \geq w_{j_0} - b_{j_0} > 0, \tag{3.1.21}$$

which follows from multiplying the left side by -1 then adding w_{j_0}. Furthermore, by the reverse triangle inequality (1.2.37) and various properties of absolute value we have

$$\|\mathbf{w} - \mathbf{x}\|_m = \sqrt{(w_1 - x_1)^2 + \ldots + (w_m - x_m)^2} \tag{3.1.22}$$

$$\geq \sqrt{(w_{j_0} - x_{j_0})^2} \tag{3.1.23}$$

$$= |w_{j_0} - x_{j_0}| \tag{3.1.24}$$

$$= w_{j_0} - x_{j_0} \tag{3.1.25}$$

$$\geq w_{j_0} - b_{j_0} \tag{3.1.26}$$

$$> 0. \tag{3.1.27}$$

Hence, $\mathbf{w}$ is away from B and therefore B is closed. □

In the case where $m = 1$, Lemma 3.1.13 tells us closed and bounded intervals are in fact closed sets since they satisfy Definition 3.1.1. Even so, a direct proof of this fact appears as an exercise.

The *complements* of the closed sets in a Euclidean space $\mathbb{R}^m$ form the fundamental objects in the mathematical subject area known as *topology*. These complements are called *open* sets and are the focus of the next section. See Definitions 3.2.1, 3.2.2, and 3.2.9.

Exercises

3.1.1. Consider a closed interval $[a, b] = \{x \in \mathbb{R} : a \leq x \leq b\}$ where a and b are real numbers satisfying $a < b$. Prove $[a, b]$ is closed according to Definition 3.1.1 and draw figures to help. Why isn't this trivial?

3.1.2. Consider an interval $(a, b] = \{x \in \mathbb{R} : a < x \leq b\}$ where a and b are real numbers satisfying $a < b$. Prove $(a, b]$ is not closed.

3.1.3. Prove the set of positive integers $\mathbb{N}$ is closed.

3.1.4. Prove the set of rational numbers $\mathbb{Q}$ is not closed.

3.1.5. Prove the real line $\mathbb{R} = (-\infty, \infty)$ is closed.

3.1.6. Prove Lemma 3.1.8: The closure of a set is a closed set.

3.1.7. Prove the closure of a given set in $\mathbb{R}^m$ is the smallest closed set containing the given set in the following sense: Given a set $S \subseteq \mathbb{R}^m$, every closed set which contains S also contains the closure $\overline{S}$.

3.1.8. Consider a rectangle B in the plane $\mathbb{R}^2$ such as in Figure 3.1.1 that contains some of its sides and corners, but not all of them. Prove B is not closed.

3.1.9. A set $E \subseteq \mathbb{R}^m$ is said to be *coda-closed* if $\mathrm{Coda}((\mathbf{x}_n)) \subseteq E$ for every sequence $(\mathbf{x}_n)$ of points in E. Prove E is coda-closed if and only if E is closed.

3.2   Open sets and topology

This section explores the *topology* of Euclidean spaces. Our approach is somewhat backwards compared to approaches to topology found in other texts in that we covered closed sets first. To get to the heart of topology, *open sets*, a definition for *complement* is in order.

Definition 3.2.1: Complement

Let B be a subset of some set X. The *complement* of B (with respect to X) is the set of points in X but not B. Thus, the complement of B is given by

$$X \backslash B = \{x \in X : x \notin B\}. \tag{3.2.1}$$

To reinforce a key concept in Euclidean spaces, any point $\mathbf{z}$ not in a closed set F is away from F (see Definitions 3.1.1 and 1.5.11). Hence, there is an $\varepsilon_{\mathbf{z}}$-neighborhood $V_{\varepsilon_{\mathbf{z}}}(\mathbf{z})$ which does not intersect F. This is precisely a characterizing property—therefore a defining property—of points in an *open* set (cf. [1, Definition 3.2.1, p.88]). See Figure 3.2.1.

Definition 3.2.2: Open

A set $U \subseteq \mathbb{R}^m$ is *open* if every point in U has a neighborhood contained in U.

Remark 3.2.3: The classic notion of open

Equivalently, U is open if for every $\mathbf{a} \in U$ there is an $\varepsilon_{\mathbf{a}} > 0$ such that $V_{\varepsilon_{\mathbf{a}}}(\mathbf{a}) \subseteq U$ (see Figure 3.2.1). In the language of the negation of arbitrarily close, a set U is open if every point of U is away from the complement $\mathbb{R}^m \backslash U$ (see Definition 1.5.11). In other words, all points in an open set are *interior points* (see Definition 3.6.3 below).

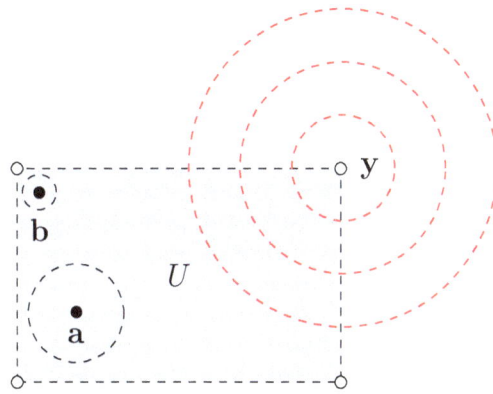

Figure 3.2.1: The set U contains a neighborhood around each of its points, such as **a** and **b**. So, U is an open set (see Definition 3.2.2). Note the neighborhood centered at **b** is necessarily smaller than the neighborhood centered at **a**. The point **y** is arbitrarily close to U, but no neighborhood of **y** is contained in U and **y** is not in U.

The fundamental connection between open and closed sets readily follows (cf. [1, Theorem 3.2.13, p.92]).

Theorem 3.2.4: Open versus closed

A set $U \subseteq \mathbb{R}^m$ is open if and only if its complement $\mathbb{R}^m \backslash U$ is closed.

A proof very similar to the one presented here was created by Rasha Issa as she prepared for a final exam in the summer of 2019. In particular, she used the language of arbitrarily close and preferred this approach over the one used in [1, Theorem 3.2.13, p.92].

Rasha Issa's proof of Theorem 3.2.4. Assume U is open and suppose **y** is arbitrarily close to $\mathbb{R}^m \backslash U$. By way of contradiction, assume $\mathbf{y} \in U$. Since U is open, there is an $\varepsilon_{\mathbf{y}}$-neighborhood of **y** contained in U. Hence, every **x** in $\mathbb{R}^m \backslash U$ lies outside of this $\varepsilon_{\mathbf{y}}$-neighborhood of **y**. Thus, **x** is at least a positive distance $\varepsilon_{\mathbf{y}}$ away from **y**. Hence, **y** is away from $\mathbb{R}^m \backslash U$, a contradiction. Therefore, $\mathbb{R}^m \backslash U$ is closed.

For the converse, assume $\mathbb{R}^m \backslash U$ is closed and let $\mathbf{z} \in U$. Since $\mathbb{R}^m \backslash U$ contains all points arbitrarily close to $\mathbb{R}^m \backslash U$, **z** is away from $\mathbb{R}^m \backslash U$. So there must be some $\varepsilon_{\mathbf{z}} > 0$ where $V_{\varepsilon_{\mathbf{z}}}(\mathbf{z}) \subseteq U$. Therefore, U is open. $\square$

The following pair of examples illustrate the definition of an open set (Definition 3.2.2) and Theorem 3.2.4 in the real line $\mathbb{R}$ by revisiting the closed interval F studied throughout Chapter 1, its complement $\mathbb{R} \backslash F$, and the set E from Example 3.1.12.

Example 3.2.5: Away from a closed interval

Consider the closed interval $F = [0, 3140]$ and its complement

$$\mathbb{R} \backslash F = (-\infty, 0) \cup (3140, \infty) = \{x \in \mathbb{R} : x < 0 \text{ or } x > 3140\}. \tag{3.2.2}$$

$$F \qquad \underset{0}{\vdash\!\!\!\!\!\frac{\qquad\qquad\qquad}{}\!\!\!\!\!\dashv} \underset{3140}{} \qquad (\!\!-\!\!\underset{4710}{+}\!\!-\!\!) \qquad V_{\varepsilon_z}(4710)$$

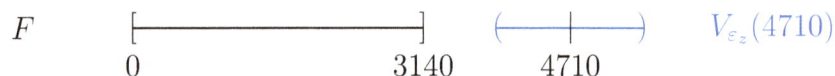

Figure 3.2.2: Every point in the closed interval $F = [0, 3140]$ is more than the distance $\varepsilon_z = 785$ away from the real number $z = 4710$. Hence, the ε_z-neighborhood $V_{\varepsilon_z}(4710)$ is contained in the complement $\mathbb{R}\backslash F$.

The complement $\mathbb{R}\backslash F$ is open. We could make use of Exercise 3.1.1 and Theorem 3.2.4 to prove this, but I think it'd be helpful to work with the definition of an open set directly (Definition 3.2.2). The idea is that every element in $\mathbb{R}\backslash F$ comes with its own ε-neighborhood contained in $\mathbb{R}\backslash F$. See Figure 3.2.2.

Proof for Example 3.2.5. For each $z \in \mathbb{R}\backslash F$, define the positive distance ε_z to be half of the shorter of the two distances between z and the endpoints of $F = [0, 3140]$. That is, define

$$\varepsilon_z = \frac{1}{2}\min\{|z-0|, |z-3140|\} = \begin{cases} \dfrac{|z-0|}{2}, & \text{if } z < 0, \\[2mm] \dfrac{|z-3140|}{2}, & \text{if } z > 3140. \end{cases} \tag{3.2.3}$$

(See Figure 3.2.2 where $z = 4710 > 3140$ and thus $\varepsilon_z = 785$.) Then for each $z \in \mathbb{R}\backslash F$, every point in F is more than ε_z away from z. Hence,

$$V_{\varepsilon_z}(z) = (4710 - \varepsilon_z, 4710) \subseteq \mathbb{R}\backslash F. \tag{3.2.4}$$

Therefore, the complement $\mathbb{R}\backslash F$ is open since for every element z in $\mathbb{R}\backslash F$ there is an ε_z-neighborhood contained in $\mathbb{R}\backslash F$. $\qquad\qquad\square$

Example 3.2.6: E is not open

The set of real numbers E from Example 3.1.12 is not open. We have

$$E = [0, 1) \cup \{2\} \cup \{3 + (1/n) : n \in \mathbb{N}\}. \tag{3.2.5}$$

The real number 2 is in E, but no ε-neighborhood of the form

$$V_\varepsilon(2) = (2 - \varepsilon, 2 + \varepsilon) \tag{3.2.6}$$

is contained in E, no matter how small we take the radius ε to be. See Figure 3.2.3 for one such ε-neighborhood.

In Euclidean spaces, ε-neighborhoods are open sets: Neighborhoods contain neighborhoods of their points.

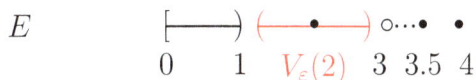

Figure 3.2.3: The set of real numbers E from Examples 3.1.12 and 3.2.6 is not open. Here, the real number 2 is E but no ε-neighborhood of 2, such as the red $V_\varepsilon(2)$ in the figure, is contained in E. See Definition 3.2.2.

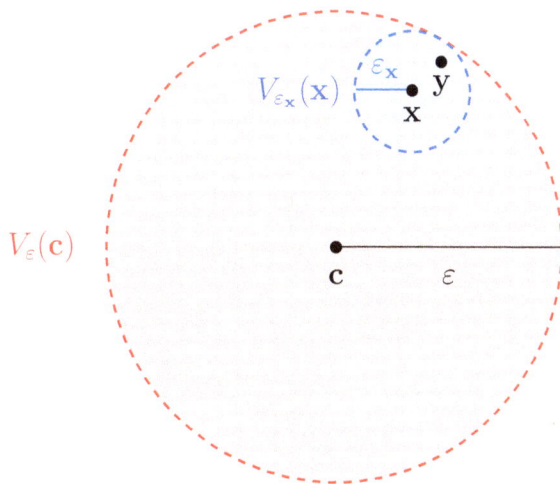

Figure 3.2.4: By Lemma 3.2.7, every ε-neighborhood $V_\varepsilon(\mathbf{c})$ is open since every point $\mathbf{x} \in V_\varepsilon(\mathbf{c})$ comes with an $\varepsilon_\mathbf{x}$-neighborhood $V_{\varepsilon_\mathbf{x}}(\mathbf{x})$ contained in $V_\varepsilon(\mathbf{c})$: We have $V_{\varepsilon_\mathbf{x}}(\mathbf{x}) \subseteq V_\varepsilon(\mathbf{c})$.

Lemma 3.2.7: ε-neighborhoods are open

For every $\mathbf{c} \in \mathbb{R}^m$ and every distance $\varepsilon > 0$, the ε-neighborhood

$$V_\varepsilon(\mathbf{c}) = \{\mathbf{x} \in \mathbb{R}^m : d_m(\mathbf{x}, \mathbf{c}) = \|\mathbf{x} - \mathbf{c}\|_m < \varepsilon\} \tag{3.2.7}$$

is open.

Scratch Work 3.2.8: Distance from difference

The proof follows from defining the distance $\varepsilon_\mathbf{x}$ as the difference between a given $\varepsilon > 0$ and the distance between $\mathbf{x}$ and $\mathbf{c}$ when $\mathbf{x} \in V_\varepsilon(\mathbf{c})$. For a visual idea of what is going on, see Figure 3.2.4.

Proof of Lemma 3.2.7. Suppose $\mathbf{c} \in \mathbb{R}^m$ and $\varepsilon > 0$. Let $\mathbf{x}$ be an arbitrary point in $V_\varepsilon(\mathbf{c})$, which means

$$d_m(\mathbf{x}, \mathbf{c}) = \|\mathbf{x} - \mathbf{c}\|_m < \varepsilon. \tag{3.2.8}$$

(The goal is to show $V_{\varepsilon_\mathbf{x}}(\mathbf{x}) \subseteq V_\varepsilon(\mathbf{c})$ for some distance $\varepsilon_\mathbf{x} > 0$.) Define

$$\varepsilon_\mathbf{x} = \varepsilon - \|\mathbf{x} - \mathbf{c}\|_m > 0 \tag{3.2.9}$$

and let $\mathbf{y}$ be an arbitrary point in $V_{\varepsilon_{\mathbf{x}}}(\mathbf{x})$, which means

$$d_m(\mathbf{y}, \mathbf{x}) = \|\mathbf{y} - \mathbf{x}\|_m < \varepsilon_{\mathbf{x}} = \varepsilon - \|\mathbf{x} - \mathbf{c}\|_m. \tag{3.2.10}$$

By adding zero and applying the triangle inequality ((1.2.34) and (1.2.32)), we have

$$
\begin{align}
d_m(\mathbf{y}, \mathbf{c}) &= \|\mathbf{y} - \mathbf{c}\|_m \tag{3.2.11}\\
&= \|\mathbf{y} \underbrace{-\mathbf{x} + \mathbf{x}}_{\text{add zero}} - \mathbf{c}\|_m \tag{3.2.12}\\
&\leq \|\mathbf{y} - \mathbf{x}\|_m + \|\mathbf{x} - \mathbf{c}\|_m \tag{3.2.13}\\
&< \varepsilon_{\mathbf{x}} + \|\mathbf{x} - \mathbf{c}\|_m \tag{3.2.14}\\
&= \varepsilon - \|\mathbf{x} - \mathbf{c}\|_m + \|\mathbf{x} - \mathbf{c}\|_m \tag{3.2.15}\\
&= \varepsilon. \tag{3.2.16}
\end{align}
$$

Hence, $\mathbf{y} \in V_{\varepsilon}(\mathbf{c})$, so $V_{\varepsilon_{\mathbf{x}}}(\mathbf{x}) \subseteq V_{\varepsilon}(\mathbf{c})$ and therefore $V_{\varepsilon}(\mathbf{c})$ is open.   □

The word *topology* describes both a mathematical topic and a particular mathematical object. The topic of topology is beautiful, massive, and connects to many other branches of mathematics in beautiful and endless ways. The mathematical object called a topology is a particular collection of subsets of a given set.

Definition 3.2.9: Topology

Let X be a set and let $\mathcal{T}$ be a collection of subsets of X. Then $\mathcal{T}$ is a *topology on X* if the following properties hold:

(i) The empty set $\varnothing$ and the set X are in $\mathcal{T}$.

(ii) The intersection of any finite number of sets in $\mathcal{T}$ is a set in $\mathcal{T}$.

(iii) The union of any collection of sets in $\mathcal{T}$ is a set in $\mathcal{T}$.

When a set X is paired with a topology on X, we call the ordered pair $(X, \mathcal{T})$ a *topological space* but often refer only to X.

Theorem 3.2.10: Topologies comprise open sets

The collection of all open subsets of $\mathbb{R}^m$ is a topology on $\mathbb{R}^m$. That is,

(i) The empty set $\varnothing$ and the set $\mathbb{R}^m$ are open.

(ii) The intersection of any finite number of open sets is open.

(iii) The union of any collection of open sets is open.

Remark 3.2.11: Terminology of open sets in topology

For those of you who have seen topology before, Theorem 3.2.10 may sound like a tautology. After all, in a topology class, open sets are defined to be the sets in a topology. However, our definition for open sets (Definition 3.2.2) precedes the definition of topology (Definition 3.2.9), so the proof of Theorem 3.2.10 amounts to verifying the collection of all open subsets of $\mathbb{R}^m$ satisfies the three properties defining a topology.

Scratch Work 3.2.12: Apply the definitions

All of the results follow from a careful application of the definitions.

Proof of Theorem 3.2.10. Let $\mathcal{S}$ denote the collection of all open subsets of $\mathbb{R}^m$.

Proof of (i): Consider the empty set $\varnothing$. Then $\varnothing$ vacuously satisfies the definition of an open set (Definition 3.2.2) since it has no points in need of a neighborhood. Hence, $\varnothing$ is in $\mathcal{S}$.

Now consider the set $\mathbb{R}^m$ itself. Since $\mathbb{R}^m$ contains all ε-neighborhoods of all points in $\mathbb{R}^m$, we have $\mathbb{R}^m$ is open. (For instance, $\mathbb{R}^m$ contains the 17-neighborhood of $\mathbf{x}$ for every $\mathbf{x} \in \mathbb{R}^m$). Hence, $\mathbb{R}^m$ is in $\mathcal{S}$.

Proof of (ii): Suppose $U_1, U_2, \ldots, U_n$ are open sets in $\mathbb{R}^m$ and let

$$\mathbf{x} \in \bigcap_{j=1}^{n} U_j. \tag{3.2.17}$$

So, $\mathbf{x}$ is in the open set U_j for each $j = 1, \ldots, n$. By the definition of open (Definition 3.2.2), for each $j = 1, \ldots, n$ there is an $\varepsilon_j > 0$ such that the ε_j-neighborhood of $\mathbf{x}$, $V_{\varepsilon_j}(\mathbf{x})$, is contained in U_j. Since we are considering a finite number of open sets, the smallest of these neighborhoods has positive radius $\varepsilon_0 = \min\{\varepsilon_1, \ldots, \varepsilon_n\}$ and is contained in each of the ε_j-neighborhoods of $\mathbf{x}$. That is,

$$V_{\varepsilon_0}(\mathbf{x}) \subseteq \bigcap_{j=1}^{n} V_{\varepsilon_j}(\mathbf{x}) \subseteq \bigcap_{j=1}^{n} U_j. \tag{3.2.18}$$

Hence, the intersection of any finite number of open sets in $\mathbb{R}^m$ is an open set, and so $\bigcap_{j=1}^{n} U_j$ is in $\mathcal{S}$.

Proof of (iii): Suppose $\{U_\alpha : \alpha \in A\}$ is a collection of open sets in $\mathbb{R}^m$ with nonempty index set A and let

$$\mathbf{x} \in \bigcup_{\alpha \in A} U_\alpha. \tag{3.2.19}$$

Then there must be some index $\alpha_\mathbf{x}$ in A where $\mathbf{x}$ is in the open set $U_{\alpha_\mathbf{x}}$. By the definition of open (Definition 3.2.2), there is an $\varepsilon_\mathbf{x} > 0$ where

$$V_{\varepsilon_\mathbf{x}}(\mathbf{x}) \subseteq U_{\alpha_\mathbf{x}} \subseteq \bigcup_{\alpha \in A} U_\alpha. \tag{3.2.20}$$

Therefore, the union of any collection of open sets in $\mathbb{R}^m$ is an open set.

Since all three conditions defining a topology in Definition 3.2.9 are satisfied by $\mathcal{S}$, the collection of open subsets of $\mathbb{R}^m$ is a topology on $\mathbb{R}^m$.   □

Theorem 3.2.10 justifies the following definition. It may sound like a tautology if you are familiar with topology.

Definition 3.2.13: Standard topology on $\mathbb{R}^m$

The *standard topology* on a Euclidean space $\mathbb{R}^m$ is the collection of all open subsets of $\mathbb{R}^m$ as defined in Definition 3.2.2.

Example 3.2.14: Standard topology on $\mathbb{R}$

The standard topology on the real line $\mathbb{R}$ is the collection of all open intervals and all unions of open intervals.

However, the idea that open intervals are actually open sets is not to be taken for granted. The proof is left as an important exercise which encourages you understand the definitions involved. The fact that unions of opens intervals are open follows from part (iii) of Theorem 3.2.10.

Example 3.2.14 is just one piece of a more powerful statement regarding the standard topology on the real line. Its proof is left as a challenging exercise and makes use of the following definition.

Definition 3.2.15: Pairwise disjoint

A collection of sets $\mathcal{C}$ is *pairwise disjoint* if for every pair of sets $A, B \in \mathcal{C}$ where $A \neq B$ we have $A \cap B = \varnothing$.

Theorem 3.2.16: Open sets in the real line $\mathbb{R}$

Every open subset of the real line is the union of a pairwise disjoint countable collection of open intervals. That is, for every open set $U \subseteq \mathbb{R}$, there is a sequence of open intervals (I_n) where the collection $\{I_n : n \in \mathbb{N}\}$ is pairwise disjoint and

$$U = \bigcup_{n=1}^{\infty} I_n. \tag{3.2.21}$$

An analogue of Theorem 3.2.10 holds for collections of closed sets, one of which has already been stated in Lemma 3.1.6.

Theorem 3.2.17: A trio of results on closed sets

The following properties regarding closed sets in $\mathbb{R}^m$ hold:

   (i) The empty set $\varnothing$ and the set $\mathbb{R}^m$ are closed.

(ii) The union of any finite number of closed sets is closed.

(iii) The intersection of any collection of closed sets is closed.

Scratch Work 3.2.18: De Morgan's Laws play a key role

Part (iii) of Theorem 3.2.17 is a rephrased version of Lemma 3.1.6. Parts (i) and (ii) follow from Theorem 3.2.4 and results from set theory on the relationships between complements, intersections, and unions known as *De Morgan's Laws*. These results are stated but not proven below, the proofs of Theorems 3.2.17 3.2.19 are left as an exercise.

Theorem 3.2.19: De Morgan's Laws

Suppose A and B are subsets of some set X. Then:

(i) $X \backslash (A \cap B) = (X \backslash A) \cup (X \backslash B)$; and

(ii) $X \backslash (A \cup B) = (X \backslash A) \cap (X \backslash B)$.

Suppose $\mathcal{C}$ is a collection of subsets of some set X. Then:

(i) $X \backslash \left(\bigcap_{S \in \mathcal{C}} S \right) = \bigcup_{S \in \mathcal{C}} (X \backslash S)$; and

(ii) $X \backslash \left(\bigcup_{S \in \mathcal{C}} S \right) = \bigcap_{S \in \mathcal{C}} (X \backslash S)$.

The following section explores the classic topological notion of *connectedness* using an unconventional definition that stems from the concept of arbitrarily close.

Exercises

3.2.1. Consider an open interval (a, b) given by

$$(a, b) = \{x \in \mathbb{R} : a < x < b\} \tag{3.2.22}$$

where a and b are real numbers satisfying $a < b$. Prove (a, b) is open according to Definition 3.2.2 and draw figures to help. Why isn't this trivial?

3.2.2. Consider an interval $(a, b] = \{x \in \mathbb{R} : a < x \leq b\}$ where a and b are real numbers satisfying $a < b$. Prove $(a, b]$ is not open.

3.2.3. Prove the set of noninteger real numbers $\mathbb{R} \backslash \mathbb{Z}$ is open. HINT: See Example 3.1.4.

3.2.4. Prove the set of rational numbers $\mathbb{Q}$ is not open.

3.2.5. Consider the real line $\mathbb{R} = (-\infty, \infty)$. Prove $\mathbb{R}$ is both open and closed.

3.2.6. Consider a rectangle B in the plane $\mathbb{R}^2$ such as in Figure 3.1.1 which contains some of its sides and corners, but not all of them. Prove B is not open.

3.2.7. Prove Theorem 3.2.16.

3.2.8. Prove Theorem 3.2.17.

3.2.9. Prove De Morgan's Laws (Theorem 3.2.19).

3.3   Connected sets

The notion of *connectedness* is yet another classic topic in analysis and topology which lends itself to a description in terms of arbitrarily close (Definition 1.5.1). Intuitively, a set E is *connected* if it comes in one piece with no separate chunks. For the purpose of writing proofs, a more technical definition is in order: A set is *connected* if every partition into two sets features a point in one set arbitrarily close to the other set.

The definition for connected in this section is parsed using the idea of *coupled* sets, similar to the way convergence and limits for sequences are parsed by arbitrarily close and tails in Chapter 2.

Definition 3.3.1: Coupled

Two sets are *coupled* if there is a point in one set arbitrarily close to the other set. That is, if $A, B \subseteq \mathbb{R}^m$, then A and B are *coupled* if there is a point $\mathbf{x} \in A$ where $\mathbf{x} \operatorname{acl} B$ or there is a point $\mathbf{y} \in B$ where $\mathbf{y} \operatorname{acl} A$.

See Example 3.3.2 and Figure 3.3.1 for two coupled sets I and C in the real line. See Example 3.3.3 and Figure 3.3.2 for a coupled square S and disk D in the plane.

Example 3.3.2: An interval coupled with a countable set

In the real line $\mathbb{R}$, consider the interval I and countable set C in Figure 3.3.1 given by

$$I = (0, 2] \quad \text{and} \quad C = \left\{ 2 + \frac{2}{n} : n \in \mathbb{N} \right\}. \qquad (3.3.1)$$

The sets I and C are coupled.

Proof for Example 3.3.2. The real number 2 is in I since I includes its right endpoint. Also, 2 is arbitrarily close to C since, given any $\varepsilon > 0$, an index $n_\varepsilon \in \mathbb{N}$ where $n_\varepsilon > 2/\varepsilon$ produces a real number c_{n_ε} such that

$$c_{n_\varepsilon} = 2 + \frac{2}{n_\varepsilon} \quad \implies \quad |c_{n_\varepsilon} - 2| = \frac{2}{n_\varepsilon} < \varepsilon. \qquad (3.3.2)$$

Hence, c_{n_ε} is in C and within ε of 2. Therefore, I and C are coupled. $\square$

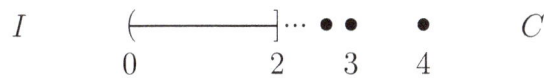

Figure 3.3.1: The interval $I = (0, 2]$ and the countable set $C = \{2 + (2/n) : n \in \mathbb{N}\}$ are coupled since $2 \in I$ and $2 \operatorname{acl} C$. See Definition 3.3.1 and Example 3.3.2.

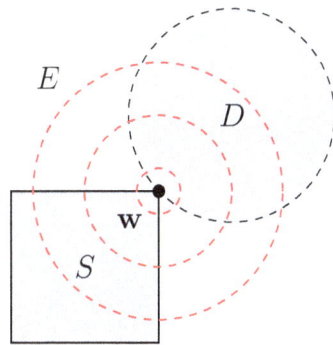

Figure 3.3.2: The closed square S and the open disk D in Example 3.3.3 are coupled since the point $\mathbf{w}$ is in S and arbitrarily close to D, acting like an anchor. Furthermore, their union $E = S \cup D$ is connected, though this is harder to justify. See Definitions 3.3.1 and 3.3.4.

Example 3.3.3: A connected set in the plane $\mathbb{R}^2$

In the plane $\mathbb{R}^2$, consider the closed square S and the open disk D in Figure 3.3.2 given by

$$S = \left\{ \mathbf{x} \in \mathbb{R}^2 : \mathbf{x} = \begin{bmatrix} x \\ y \end{bmatrix}, -1 \leq x \leq 1, \text{ and } -1 \leq y \leq 1 \right\}, \qquad \text{and} \qquad (3.3.3)$$

$$D = V_{\sqrt{2}}(\mathbf{x}_0) = \left\{ \mathbf{x} \in \mathbb{R}^2 : d_2(\mathbf{x}, \mathbf{x}_0) < \sqrt{2} \text{ where } \mathbf{x}_0 = \begin{bmatrix} 2 \\ 2 \end{bmatrix} \right\}. \qquad (3.3.4)$$

S and D are coupled since the point $\mathbf{w} = \begin{bmatrix} 1 \\ 1 \end{bmatrix}$ is both in S and arbitrarily close to D.

We can use the idea of coupled sets to parse the definition for connected sets. Connectedness is a much more subtle idea since it considers *all* possible ways to partition a set into two sets.

Definition 3.3.4: Connected

A set $E \subseteq \mathbb{R}^m$ is *connected* if every pair of nonempty sets A and B where $A \cup B = E$ is coupled.

Remark 3.3.5: The empty set is connected

The empty set $\varnothing$ is vacuously connected since it is not a union of two nonempty sets. When E is nonempty and connected where A and B are nonempty sets with $A \cup B = E$, then there is either some $\mathbf{x} \in A \cap \overline{B}$ or some $\mathbf{y} \in B \cap \overline{A}$.

Remark 3.3.6: From coupled to connected

As mentioned with Figure 3.3.2, the square S and disk D in Example 3.3.3 form a connected union $E = S \cup D$, but why? A proof at this point would be cumbersome and probably not instructive, so we can revisit the idea later after more tools are developed.

Still, what is connectedness? For the set $E = S \cup D$, we should consider not just the point $\mathbf{w}$ which ensures S and D are coupled, but *any* partition of E into two sets should result in a coupled pair. For example, try drawing a figure to accompany this process: What if we cut the square S along a diagonal from its upper left to its lower right? We could keep the points on this diagonal in the resulting closed triangle T and leave everything else to the set $E \backslash T$. Then a point on this diagonal would be in T and arbitrarily close to $E \backslash T$, meaning T and $E \backslash T$ are coupled.

But what about splitting E into one subset comprising points with rational components and another comprising points with at least one irrational component? The density of both the rational and the irrationals in the real line could play a role here, but to ensure E is connected this new pair of sets should also be coupled. In general, it can be very difficult to prove a set is connected by directly verifying Definition 3.3.4 holds since we need to account for *every* partition into two sets.

Example 3.3.7: Disconnected interval with countable set

In the real line $\mathbb{R}$, consider the union W of the interval I and countable set C from Example 3.3.2 and Figure 3.3.1. See Figure 3.3.3. We have

$$W = (0,2] \cup \left\{ 2 + \frac{2}{n} : n \in \mathbb{N} \right\}. \tag{3.3.5}$$

Even though the sets I and C are coupled, their union W is not connected. Consider the pair of sets given by the singleton $\{4\}$ and the set $W \backslash \{4\}$. Since $n = 1$ produces the real number

$$2 + \frac{2}{n} = 2 + \frac{2}{1} = 4, \tag{3.3.6}$$

so 4 is in W and the pair $\{4\}$ and $W \backslash \{4\}$ are two nonempty sets whose union is W. However, as indicated in Figure 3.3.3, the real number 4 is more than a distance of $\varepsilon_4 = 1/2$ away from every other point in W. Hence, $\{4\}$ and $W \backslash \{4\}$ are not coupled, meaning W is not connected.

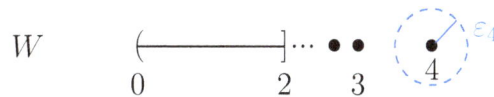

Figure 3.3.3: The set of real numbers W is not connected since the only point in the singleton $\{4\}$ is more than the distance $\varepsilon_4 = 1/2$ away from the rest of W, meaning $\{4\}$ and $W\backslash\{4\}$ are uncoupled. See Example 3.3.7 and Definitions 3.3.1, 3.3.4 and 3.3.8.

The negation of coupled is important enough to merit its own definition.

Definition 3.3.8: Uncoupled

Two sets A and B are *uncoupled* if every point of A is away from B and every point of B is away from A.

There is a straightforward way to show a given pair of sets is uncoupled: Find disjoint open sets that split the pair.

Lemma 3.3.9: Uncoupled when split by disjoint open sets

Two sets A and B are uncoupled if and only if there are disjoint open sets U and V such that $A \subseteq U$ and $B \subseteq V$.

Scratch Work 3.3.10: Uncoupled when split by disjoint open sets

The proof relies on a common consequence of the definitions of away from (Definition 1.5.11) and open (Definition 3.2.2): Points in a set have neighborhoods also contained in the set.

Proof of Lemma 3.3.9. First, suppose A and B are uncoupled. Then for each $\mathbf{a} \in A$ there is some $\varepsilon_{\mathbf{a}} > 0$ such that $V_{\varepsilon_{\mathbf{a}}}(\mathbf{a}) \cap B = \varnothing$. Similarly, for each $\mathbf{b} \in B$ there is some $\varepsilon_{\mathbf{b}} > 0$ such that $V_{\varepsilon_{\mathbf{b}}}(\mathbf{b}) \cap A = \varnothing$. Define

$$U = \bigcup_{\mathbf{a} \in A} V_{\varepsilon_{\mathbf{a}}}(\mathbf{a}) \qquad \text{and} \qquad V = \bigcup_{\mathbf{b} \in B} V_{\varepsilon_{\mathbf{b}}}(\mathbf{b}). \tag{3.3.7}$$

Then $A \subseteq U$, $B \subseteq V$, and $U \cap V = \varnothing$. Since ε-neighborhoods are open (Lemma 3.2.7) and unions of open sets are open (Theorem 3.2.10), U and V are open.

To prove the converse, suppose there are disjoint open sets U and V such that $A \subseteq U$ and $B \subseteq V$. Since U is open and $\mathbf{a} \in U$ for each $\mathbf{a} \in A$, there is some $\varepsilon_{\mathbf{a}} > 0$ such that $V_{\varepsilon_{\mathbf{a}}}(\mathbf{a}) \subseteq U$. Since U and V are disjoint and $B \subseteq V$, we have $V_{\varepsilon_{\mathbf{a}}}(\mathbf{a}) \cap B = \varnothing$. Hence, every element of A is away from B. A similar argument shows every element of B is away from A. Therefore, A and B are uncoupled. $\qquad\square$

A slight change in the definition of the square S (from closed to open) in Example 3.3.3 creates a new pair of sets S^o and D which are uncoupled. See Example 3.3.11 and Figure 3.3.4.

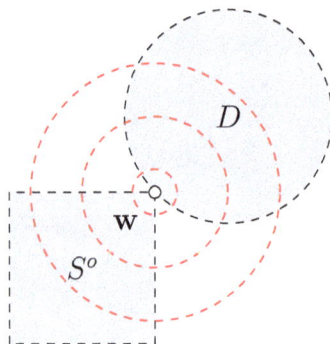

Figure 3.3.4: The open square S^o and the open disk D in Example 3.3.11 are uncoupled, so their union $U = S \cup D$ is disconnected. See Definitions 3.3.8 and 3.3.12. The point $\mathbf{w}$ (the upper right corner of S^o) is *not* in S^o but is arbitrarily close to both S^o and D.

Example 3.3.11: Disconnected set in the plane

Now consider the open square S^o given by

$$S^o = \left\{ \mathbf{x} \in \mathbb{R}^2 : \mathbf{x} = \begin{bmatrix} x \\ y \end{bmatrix}, -1 < x < 1, \text{ and } -1 < y < 1 \right\} \qquad (3.3.8)$$

and once again consider the open disk D from Example 3.3.3. See Figure 3.3.4. In this case, the point

$$\mathbf{w} = \begin{bmatrix} 1 \\ 1 \end{bmatrix} \qquad (3.3.9)$$

is in neither S^o nor D. Moreover, S^o and D are disjoint ($S^o \cap D = \varnothing$) and they are both open (though I won't prove it here). So, by Lemma 3.3.9, every point in one of the sets is away from the other, so S^o and D are uncoupled.

The union of S^o and D is *disconnected*.

Definition 3.3.12: Disconnected and separation

A set E is *disconnected* if there are nonempty uncoupled sets A and B where $A \cup B = E$. In this case, A and B form a *separation* of E.

Remark 3.3.13: Classic notion of connected

Standard approaches to defining connectedness first define disconnected using separation in one way or another, then take the negation of disconnected to define connected. Here are two such examples:

(i) A set $F \subseteq \mathbb{R}^m$ is *disconnected* if there is a pair of nonempty sets A and B where $A \cup B = F, A \cap \overline{B} = \varnothing$, and $B \cap \overline{A} = \varnothing$. Such a pairing of nonempty sets A and B

is called a *separation* of F. From there, a set $E \subseteq \mathbb{R}^m$ is said to be connected if it is not disconnected. (Cf. [1, Definition 3.4.4, p.104].)

(ii) A set $F \subseteq \mathbb{R}$ is *disconnected* if there is a pair of nonempty open sets U and V where $F \subseteq U \cup V$, $U \cap V = \varnothing$, $U \cap F \neq \varnothing$, and $V \cap F \neq \varnothing$. Such a pairing of sets U and V is called a *separation* of F. From there, a set $E \subseteq \mathbb{R}^m$ is said to be connected if it is not disconnected.

The fact that these two approaches to defining connectedness are equivalent to Definition 3.3.4 is left as an exercise.

Thanks to the deep connection between limits of convergent sequences and the notion of arbitrarily close established in Theorem 2.3.1, there is a characterization of connectedness in terms of limits.

Corollary 3.3.14: Connected and arbitrarily close

A nonempty set $E \subseteq \mathbb{R}^m$ is connected if and only if for any pair of nonempty sets A and B where $A \cup B = E$, there is a sequence of points in A whose limit is in B, or vice versa.

Scratch Work 3.3.15: Connected and arbitrarily close

The result follows from the definitions for coupled and connected (Definitions 3.3.1 and 3.3.4) along with both directions of Theorem 2.3.1 (the fundamental connection between arbitrarily close and limits of sequences).

Proof of Corollary 3.3.14. First, suppose E is both nonempty and connected. Then any pair of nonempty sets A and B where $A \cup B = E$ is coupled. So, without loss of generality, there is a point $\mathbf{y}$ in B where $\mathbf{y}\,\mathrm{acl}\,A$. By Theorem 2.3.1, some sequence $(\mathbf{x}_n)$ of points in A converges to $\mathbf{y}$.

Now suppose A and B are nonempty sets where $A \cup B = E$. Also, without loss of generality, suppose there is a sequence $(\mathbf{x}_n)$ of points in A whose limit is $\mathbf{y}$ where $\mathbf{y}$ is in B. By Theorem 2.3.1, $\mathbf{y}\,\mathrm{acl}\,A$ and therefore A and B are coupled. So, E is nonempty and connected. □

Theorem 3.3.17 is yet another special feature of the real line $\mathbb{R}$: Intervals and singletons are the only connected subsets of the real line.

In the case of intervals, their characterization in Lemma 3.3.16 provides a helpful perspective to consider as when trying to prove Theorem 3.3.17. The proof of Lemma 3.3.16 amounts to checking the definition of an interval in Definition 1.2.9, so it is omitted. After all, intervals are defined to be subsets of the real line comprising the points between endpoints and possibly the endpoints themselves. See Figure 1.2.1.

Lemma 3.3.16: A characterization of intervals

A subset I of the real line $\mathbb{R}$ is an interval if and only if whenever $x \in I, y \in I$, and $x < z < y$, then $z \in I$ as well.

Theorem 3.3.17: Intervals are the connected subsets of the real line $\mathbb{R}$

A nonempty subset of the real line $\mathbb{R}$ is connected if and only if the subset is an interval or a singleton.

The proof of the forward direction is handled via contraposition. But first, the backward direction is handled directly with a pair of cases.

Proof of Theorem 3.3.17. Suppose E is a nonempty subset of the real line $\mathbb{R}$.

Case (i): Suppose is a singleton where $E = \{c\}$. Then any nonempty subsets A and B of E must also be the same singleton, that is

$$A = B = \{c\}. \tag{3.3.10}$$

Since $c \in A$ and $c \in B$, by Lemma 1.5.4 we have $c \,\mathrm{acl}\, A$ and $c \,\mathrm{acl}\, B$. Hence, A and B are coupled and therfore E is connected (see Definitions 3.3.4 and 3.3.4).

Case (ii): Suppose E is an interval with nonempty subsets A and B where $E = A \cup B$ and, without loss of generality, there are real numbers $a_0 \in A$ and $b_0 \in B$ where $a_0 < b_0$.

Now consider the closed interval $I_0 = [a_0, b_0]$ which is a subset of E since E is an interval. Following a bisection method much like the proof of the NCBI Property 2.5.4, the midpoint $(a_0 + b_0)/2$ is in E. Therefore, $(a_0 + b_0)/2$ is in A or B. Define $I_1 = [a_1, b_1]$ by either $I_1 = [a_0, (a_0 + b_0)/2]$ or $I_1 = [(a_0 + b_0)/2, b_0]$, chosen so that $a_1 \in A$ and $b_1 \in B$. Proceeding recursively, define a sequence of closed and bounded intervals I_n where: the midpoint of I_n is an endpoint of I_{n+1}; I_{n+1} is chosen so its left endpoint a_{n+1} is in A, and its right endpoint b_{n+1} is in B. Then the sequence of intervals (I_n) is nested, so by the NCBI Property 2.5.4, there is a point x where

$$x \in \bigcap_{n=1}^{\infty} I_n \subseteq E. \tag{3.3.11}$$

Since the I_n are constructed via bisection, their lengths are successively cut in half. So, for every $n \in \mathbb{N}$ we have

$$a_n \leq x \leq b_n \qquad \text{and} \qquad |a_n - b_n| = \frac{|a_0 - b_0|}{2^n}. \tag{3.3.12}$$

By the linearity of limits for sequences and Corollary 2.4.19, we have

$$\lim_{n \to \infty} \frac{|a_0 - b_0|}{2^n} = 0. \tag{3.3.13}$$

Now let $\varepsilon > 0$. Since $a_n \leq x \leq b_n$, properties inequalities and the definition of limit and convergence (Definition 2.2.1), there is a threshold n_ε where for every $n \geq n_\varepsilon$ we have both

$$|a_n - x| \leq |a_n - b_n| = \frac{|a_0 - b_0|}{2^n} < \varepsilon \quad \text{and} \tag{3.3.14}$$

$$|b_n - x| \leq |a_n - b_n| = \frac{|a_0 - b_0|}{2^n} < \varepsilon. \tag{3.3.15}$$

Since $a_n \in A$ and $b_n \in B$ for every $n \in \mathbb{N}$, we have $x \,\mathrm{acl}\, A$ and $x \,\mathrm{acl}\, B$. Since $x \in E$ and $E = A \cup B$, we have $x \in A$ or $x \in B$. Therefore, A and B are coupled and E is connected.

Next, to show a nonempty connected subset of the real line is an interval, let's argue via contraposition. Suppose E contains at least two points and is not an interval (otherwise, E contains just one point and is thus a singleton). By Lemma 3.3.16 and without loss of generality, there are $x \in E$ and $y \in E$ where $x < y$ as well as $z \in (x, y)$ where $z \notin E$. Define

$$L_z = (-\infty, z) \cap E \quad \text{and} \quad R_z = (z, \infty) \cap E. \tag{3.3.16}$$

Then $E = L_z \cup R_z$, $x \in L_z$, and $y \in R_z$. For every $\ell \in L_z$ and every $r \in R_z$ we have

$$\ell < z < r. \tag{3.3.17}$$

So, for every $\ell \in L_z$ and every $r \in R_z$ we have both

$$|\ell - r| > |\ell - z| > 0 \quad \text{and} \tag{3.3.18}$$
$$|\ell - r| > |r - z| > 0. \tag{3.3.19}$$

Hence, every $\ell \in L_z$ is away from R_z and every $r \in R_z$ is away from L_z. Therefore, E is disconnected (see Definition 3.3.12). □

Next up, consider the classic example of the *topologist's sine curve* thought of as a subset of the plane.

Example 3.3.18: Topologist's sine curve

Let G be the graph of the function $g : \mathbb{R}^+ \to \mathbb{R}$ given by

$$g(x) = \sin\left(\frac{1}{x}\right). \tag{3.3.20}$$

That is,

$$G = \left\{ \mathbf{x} \in \mathbb{R}^2 : \mathbf{x} = \begin{bmatrix} x \\ y \end{bmatrix} \text{ where } x > 0 \text{ and } y = \sin\left(\frac{1}{x}\right) \right\}. \tag{3.3.21}$$

Try using free online software such as Desmos, GeoGebra, or WolframAlpha to plot G. Also, let L be the line segment given by

$$L = \left\{ \mathbf{x} \in \mathbb{R}^2 : \mathbf{x} = \begin{bmatrix} x \\ y \end{bmatrix} \text{ where } x = 0 \text{ and } -1 \leq y \leq 1 \right\}. \tag{3.3.22}$$

Even though $G \cap L = \varnothing$, the set $E = G \cup L$ is connected. The proof that $E = G \cup L$ is connected will be handled later when we have more tools at our disposal. For now, we can prove every point in L is the limit of a convergent sequence of points in G.

Partial proof of Example 3.3.18. Let $\mathbf{p} = \begin{bmatrix} 0 \\ y_0 \end{bmatrix}$ be a point in L. Then (3.3.22) ensures we have $-1 \leq y_0 \leq 1$, so there is some $x_0 > 0$ where $\sin(x_0) = y_0$.

Now consider the sequence of positive real numbers defined by

$$a_n = \frac{1}{x_0 + 2\pi n} \tag{3.3.23}$$

for each positive integer n. Then, thanks to the periodicity of the sine function, we have

$$g(a_n) = \sin\left(\frac{1}{a_n}\right) = \sin(x_0 + 2\pi n) = y_0. \tag{3.3.24}$$

From there, consider the sequence $(\mathbf{x}_n)$ in the plane defined by

$$\mathbf{x}_n = \begin{bmatrix} a_n \\ g(a_n) \end{bmatrix} = \begin{bmatrix} a_n \\ y_0 \end{bmatrix} \tag{3.3.25}$$

for each positive integer n and $(\mathbf{x}_n)$ is a sequence of points in G. Since

$$\lim_{n\to\infty} a_n = \lim_{n\to\infty} \left(\frac{1}{x_0 + 2\pi n}\right) = 0 \qquad \text{and} \tag{3.3.26}$$

$$\lim_{n\to\infty} g(a_n) = \lim_{n\to\infty} \sin\left(\frac{1}{a_n}\right) = \lim_{n\to\infty} y_0 = y_0, \tag{3.3.27}$$

Theorem 2.4.11 (regarding componentwise convergence) applies and tells us

$$\lim_{n\to\infty} \mathbf{x}_n = \begin{bmatrix} \lim_{n\to\infty} a_n \\ \lim_{n\to\infty} y_0 \end{bmatrix} = \begin{bmatrix} 0 \\ y_0 \end{bmatrix} = \mathbf{p}. \tag{3.3.28}$$

So, every point in L is the limit of a sequence of points in G. □

To conclude the section, the following definition provides a specific meaning for the concept of having two *sets* arbitrarily close to one anothern, so not comparing a set to a point but rather another set. Former students Jeffrey Robbins, Lekha Patil, and Ryan Aniceto each thought of equivalent versions of this definition.

Definition 3.3.19: Two sets arbitrarily close

Suppose $A, B \subseteq \mathbb{R}^m$. The sets A and B are said to be *arbitrarily close* if there is a point $\mathbf{y} \in \mathbb{R}^m$ where both $\mathbf{y}\,\mathrm{acl}\,A$ and $\mathbf{y}\,\mathrm{acl}\,B$.

In other words, when two sets A and B are arbitrarily close, their closures intersect and we have $\overline{A} \cap \overline{B} \neq \varnothing$.

There is a nice pair of one-way relationships between sets that intersect, are coupled, or are arbitrarily close.

Theorem 3.3.20: Intersecting, coupled, and arbitrarily close sets

Suppose A and B are subsets of $\mathbb{R}^m$.

(i) If A and B intersect (i.e., $A \cap B \neq \varnothing$), then they are coupled.

(ii) If A and B are coupled, then they are arbitrarily close.

Proof of Theorem 3.3.20. For both proofs below, suppose A and B are subsets of $\mathbb{R}^m$.

Proof of (i): Suppose A and B intersect. Then there is a point in both sets. By Lemma 1.5.4, a point in a set is arbitrarily close to the set. So, A and B are coupled.

Proof of (ii): Suppose A and B are coupled. Then there is a point $\mathbf{y}$ in one set that's arbitrarily close to the other. By Lemma 1.5.4, a point in a set is arbitrarily close to the set, so $\mathbf{y}$ is arbitrarily close to both A and B. Hence, A and B are arbitrarily close. $\square$

The converses of statements (i) and (ii) in Theorem 3.3.20 are false in general.

Example 3.3.21: Sets arbitrarily close yet disconnected

The sets S and D from Example 3.3.3 are coupled, but their intersection is empty. On the other hand, the open sets S^o and D in Example 3.3.11 are arbitrarily close in the sense of Definition 3.3.19 since the point $\mathbf{w}$ is arbitrarily close to both sets in the sense of Definition 1.5.1. However, S^o and D are uncoupled since neither contains $\mathbf{w}$ and each of their points is away from the other set.

Remark 3.3.22: Arbitrarily close for pairs of points

A perspective on having two *points* arbitrarily close to one another is provided by Lemma 1.5.5: In $\mathbb{R}^m$, two points are arbitrarily close to one another if and only if they are the same point. In the more general setting of topological spaces, this is not necessarily the case.

Example 3.3.21 shows us two sets can be arbitrarily close while having a disconnected union. The following corollary of Theorem 3.3.20 tells us that when a union of two sets is connected, the two sets are arbitrarily close. The proof is omitted since it follows directly from Theorem 3.3.20 and the corresponding definitions.

Corollary 3.3.23: Connected implies sets arbitrarily close

Suppose $E \subseteq \mathbb{R}^m$ is connected. Then for every pair of nonempty sets A and B where $E = A \cup B$ we have $A \operatorname{acl} B$.

The next section explores *compactness*, another important classic topic in analysis and topology with a difficult definition. I have not been able to determine a useful way to parse the definition of compactness using arbitrarily close directly, but I think it helps to parse compactness by considering *open covers* for a bit beforehand.

Exercises

3.3.1. Prove $\mathbb{Q}$ and $\mathbb{R}\backslash\mathbb{Q}$ are coupled.

3.3.2. Prove $\mathbb{N}, \mathbb{Z}, \mathbb{Q}$, and $\mathbb{R} \backslash \mathbb{Q}$ are disconnected.

3.3.3. Prove the range of a sequence of real numbers is connected if and only if the sequence is constant.

3.3.4. Suppose $A, B \subseteq \mathbb{R}^m$ are connected and $A \cap B \neq \varnothing$. Prove $A \cup B$ is connected.

3.3.5. Prove the claim made in Remark 3.3.13: The definition for connected sets in Definition 3.3.4 and the classic definitions (i) and (ii) in Remark 3.3.13 are equivalent.

3.3.6. Prove every ε-neighborhood is connected. That is, for every $\mathbf{c} \in \mathbb{R}^m$ and every $\varepsilon > 0$, the set $V_\varepsilon(\mathbf{c})$ given by

$$V_\varepsilon(\mathbf{c}) = \{\mathbf{x} \in \mathbb{R}^m : \|\mathbf{x} - \mathbf{c}\|_m < \varepsilon\} \tag{3.3.29}$$

is connected.

3.3.7. Prove the closure of a connected set is connected.

3.3.8. The open subsets of the real line $\mathbb{R}$ have a special characterization: A subset of $\mathbb{R}$ is open if and only if it is a union of countably many disjoint open intervals. More formally, $U \subseteq \mathbb{R}$ is open if and only if there is a sequence of open intervals (I_n) such that

$$U = \bigcup_{n=1}^{\infty} I_n \qquad \text{and} \qquad I_j \cap I_k = \varnothing \text{ when } j \neq k. \tag{3.3.30}$$

Prove this claim.

3.4   Open covers and compact sets

Compactness is a topological property with implications across analysis and topology. It proves to be a powerful concept and plays a key role in many of the nice results throughout the book. But its definition can be difficult to understand and appreciate at first, and I find it difficult to motivate.

As my former student Ryan Aniceto says:

> Compactness is the next best thing to finiteness.

This section aims to interpret Ryan's idea in a mathematically precise way so we can prove stuff. To motivate the formal definition of compactness in Definition 3.4.12), let's first see what we can say about finite sets in the real line $\mathbb{R}$.

Theorem 3.4.1: Facts about finite sets of real numbers

Suppose S is a nonempty and finite set of real numbers. Then:

(i) S is bounded.

(ii) Both $\max S$ and $\min S$ exist.

(iii) S is closed.

(iv) Every sequence of real numbers in S has a constant subsequence.

Thanks to the order structure of the real line, we can always list the elements in a finite set of real numbers from least to greatest.

Proof of Theorem 3.4.1. Let S be a nonempty and finite set of real numbers S with n_0 elements. Without loss of generality, we have

$$S = \{s_1, s_2, \ldots, s_{n_0}\} \quad \text{where} \quad s_1 < s_2 < \cdots < s_{n_0}. \tag{3.4.1}$$

Hence, s_1 is a lower bound for S which is in S and s_{n_0} is an upper bound for S which is in S. So, S is bounded with $\min S = s_1$ and $\max S = s_{n_0}$.

To see why S is closed, note that by Lemma 1.5.4, each of the elements in S is arbitrarily close to S. Also, every other real number is away from S. To that end, suppose x is a real number that is not in S. Then for every $k = 1, \ldots, n_0$ we have

$$x \neq s_k \quad \text{and so} \quad d_{\mathbb{R}}(x, s_k) = |x - s_k| > 0. \tag{3.4.2}$$

Now let $\varepsilon_x = \min\{|x - s_k| : k = 1, \ldots, n_0\}$. Since there are only finite distances to consider, we have $\varepsilon_x > 0$. Also, for every $k = 1, \ldots, n_0$ we have

$$d_{\mathbb{R}}(x, s_k) = |x - s_k| \geq \varepsilon_x > 0. \tag{3.4.3}$$

Therefore, x awf S. Since S contains all points arbitrarily close to S, S is closed.

Finally, suppose (x_n) is a sequence of points in S. Since there are infinitely many terms of the sequence (x_n) but only a finite number of real numbers in S, at least one of the real numbers in S must be repeated an infinite number of times. Hence, for some index $j_0 \in \{1, 2, \ldots, n_0\}$ and its corresponding element $s_{j_0} \in S$, there is a constant subsequence (x_{n_k}) such that for every index n_k we have $x_{n_k} = s_{j_0}$. □

Each of the properties in Theorem 3.4.1 fails to hold in general for infinite subsets of the real line and Euclidean spaces. For an example of a set of real numbers where none of these properties hold, consider the set of rational numbers $\mathbb{Q}$.

Example 3.4.2: Some properties of $\mathbb{Q}$

The set of rational numbers $\mathbb{Q}$ is unbounded. Moreover, neither $\sup \mathbb{Q}$ nor $\inf \mathbb{Q}$ exist, so neither $\max \mathbb{Q}$ and $\min \mathbb{Q}$ exist. $\mathbb{Q}$ is not closed since, for instance, $\sqrt{2}$ is arbitrarily close to $\mathbb{Q}$ but not in $\mathbb{Q}$. Also, some sequences of rational numbers have no constant subsequences. For instance, consider the sequence (c_n) defined by $c_n = n$ for each positive integer n: Each $c_n = n$ is a rational number, but each one appears in any given subsequence of (c_n) at most once and cannot be repeated. Therefore, given any subsequence of (c_n), no term is repeated an infinite number of times. Hence, (c_n) has no constant subsequences.

Sets which are both closed and bounded provide a first glimpse of compactness and represent a collection of infinite sets with properties very similar to those of finite sets in Theorem 3.4.1.

Theorem 3.4.3: Properties of closed and bounded intervals

Suppose $a, b \in \mathbb{R}$ with $a < b$. Then the interval $I = [a, b]$ satisfies the following properties:

(i) I is bounded.

(ii) Both $\max I$ and $\min I$ exist.

(iii) I is closed.

(iv) Every sequence of real numbers in I has a convergent subsequence whose limit is in I.

Remark 3.4.4: Properties of closed and bounded intervals

The only difference between the properties for finite sets of real numbers in Theorem 3.4.1 and closed and bounded intervals in Theorem 3.4.3 lies with the fourth property, respectively.

Proof of Theorem 3.4.3. For the interval $I = [a, b]$, a is a lower bound for I which is in I and b is an upper bound for I which is in I. Hence, I is bounded with $\min I = a$ and $\max I = b$.

The proof that I is closed is left an important exercise. See Exercise 3.1.1.

This leaves property (iv). Suppose (x_n) is a sequence of real numbers in I. Since I is bounded, (x_n) is bounded as well. By the Bolzano-Weierstrass Theorem in the real line (Theorem 2.5.6), (x_n) has a convergent subsequence (x_{n_k}) with limit ℓ. By Theorem 2.3.1, since ℓ is arbitrarily close to (x_{n_k}), ℓ is arbitrarily close to I as well. Since I is closed, ℓ is in I. $\square$

There is another key property of finite sets emulated by compact sets to explore before getting to the definition for compactness (Definition 3.4.12), but it's the hardest property for me to motivate. So let's consider an example.

Example 3.4.5: A notion of finiteness

Consider the closed and bounded interval

$$H = [0, 100] = \{x \in \mathbb{R} : 0 \le x \le 100\}. \tag{3.4.4}$$

This set is uncountable, which follows from applying the same proof that the real line is uncountable in Theorem 2.8.14 to the interval H. Even so, we can use collections of open sets to imbue H with a finite structure.

Consider the following collections of open sets:

$$\mathcal{A} = \{V_5(x) = (x-5, x+5) : x \in H\}, \tag{3.4.5}$$

$$\mathcal{B} = \{(-10, 10), (99, 101)\} \cup \left\{ \left(\frac{100}{n+1}, \frac{100}{n-1} \right) : n \geq 2, n \in \mathbb{N} \right\}, \quad \text{and} \tag{3.4.6}$$

$$\mathcal{C} = \{V_{\delta_x}(x) = (x - \delta_x, x + \delta_x) : x \in H \text{ and } \delta_x > 0\}. \tag{3.4.7}$$

$\mathcal{A}$ comprises open sets of the same size at each point in H, namely their 5-neighborhoods. So, $\mathcal{A}$ is an uncountable collection since H is uncountable. $\mathcal{B}$ is a countable collection of open sets of various sizes. $\mathcal{C}$ is an uncountable collection of open sets of various and unspecified sizes.

I *strongly* suggest you draw stuff! Unlike many of the other examples in this book, I think you will get a better feel for what's going on if you draw figures yourself. What I've come up with ends up looking cluttered in the end, but the process of drawing stuff helped me justify the rest of the work in this example.

Even though they're infinite collections, each comes with a finite subcollection that accounts for all the points in H. We can specify suitable subcollections for $\mathcal{A}$ and $\mathcal{B}$ since we have detailed descriptions of their sets. We'll deal with $\mathcal{C}$ later.

For the collection $\mathcal{A}$, consider the subcollection $\mathcal{A}_0$ defined by integers in H which are multiples of 5:

$$\mathcal{A}_0 = \{V_5(5j) : j = 0, 1, 2, \ldots, 19, 20\} \tag{3.4.8}$$
$$= \{(-5, 5), (0, 10), (5, 15), \ldots, (90, 100), (95, 105)\}. \tag{3.4.9}$$

We have every real number in H is contained in one or two of the open sets in $\mathcal{A}_0$. Therefore, H is a contained in the union of a finite number of the open sets in $\mathcal{A}$ and we have

$$H \subseteq \bigcup_{j=1}^{20} V_5(5j). \tag{3.4.10}$$

For the collection $\mathcal{B}$, consider the subcollection $\mathcal{B}_0$ that contains $(-10, 10)$, $(99, 101)$, and enough of the other open sets to make sure we account for all the real numbers in H. Since the intervals with consecutive indices n and $n+1$ overlap and $(-10, 10)$ already contains all the points in H from 0 to just shy of 10, let's solve for an index $n_0 \in \mathbb{N}$ where

$$10 \in \left(\frac{100}{n_0+1}, \frac{100}{n_0-1} \right) \iff \frac{100}{n_0+1} < 10 < \frac{100}{n_0-1}. \tag{3.4.11}$$

This will allow us to cover all the real numbers in H with one or two of the open sets from $\mathcal{B}$. We have

$$\frac{100}{n_0+1} < 10 \iff n_0 > \frac{100}{10} - 1 = 9. \tag{3.4.12}$$

So, $n_0 = 10$ will work. Define the subcollection $\mathcal{B}_0$ by

$$\mathcal{B}_0 = \{(-10, 10), (99, 101)\} \cup \left\{\left(\frac{100}{j+1}, \frac{100}{j-1}\right) : j = 2, 3, \ldots, 10\right\}. \qquad (3.4.13)$$

Then H is a contained in the union of a finite number of the open sets in $\mathcal{B}$ and we have

$$H \subseteq (-10, 10) \cup (99, 101) \cup \left(\bigcup_{j=1}^{10} \left(\frac{100}{j+1}, \frac{100}{j-1}\right)\right). \qquad (3.4.14)$$

It is also true that H is a contained in the union of a finite number of the open sets in $\mathcal{C}$. But how can we prove this? More definitions and results will get us there.

Remark 3.4.6: The next best thing to finiteness

How far can we push the analogy of finiteness to infinite sets? Compactness is one way to answer this question, and it uses collections of open sets to imbue a modicum of finiteness on infinite sets, just like the closed and bounded interval H and the collections of open sets $\mathcal{A}$ and $\mathcal{B}$ in Example 3.4.5. Loosely speaking, compactness replaces the idea of having a finite number of elements in a set with the notion of approximating the set with a finite number of open sets in a peculiar way. The collections of open sets should be substantial enough to account for all points in the would-be compact sets, leading to the definition of *open covers*.

Definition 3.4.7: Open cover

Let S be a subset of $\mathbb{R}^m$. An *open cover* for S is a collection $\mathcal{W}$ where

  (i) every object in $\mathcal{W}$ is an open set; and

  (ii) $S \subseteq \bigcup_{U \in \mathcal{W}} U$, in which case we say $\mathcal{W}$ *covers* S.

Remark 3.4.8: Examples of open covers

In Example 3.4.5, the collections of open sets $\mathcal{A}, \mathcal{B}$, and $\mathcal{C}$ are open covers for the closed and bounded interval $H = [0, 100]$.

The following example provides a countable set of real numbers and a pair of collections of open sets, but only one suffices to be an open cover.

Example 3.4.9: Open covers or not

Consider the set of real numbers F given by

$$F = \{0\} \cup \frac{1}{\mathbb{N}} = \{0\} \cup \left\{\frac{1}{n} : n \in \mathbb{N}\right\} = \left\{0, 1, \frac{1}{2}, \frac{1}{3}, \ldots\right\}. \qquad (3.4.15)$$

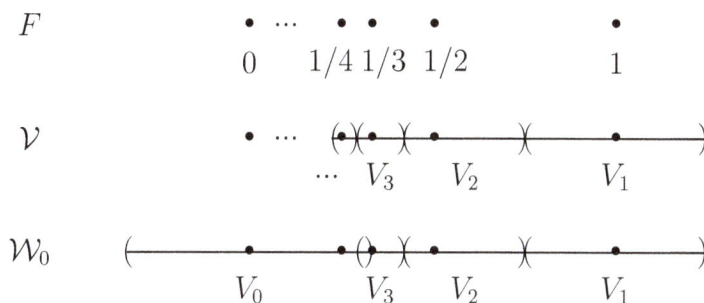

Figure 3.4.1: The set of real numbers F along with collections of open sets $\mathcal{V}$ and $\mathcal{W}_0$ from Example 3.4.9 and Remark 3.4.10. $\mathcal{V}$ is not a cover for F, but $\mathcal{W}_0$ is.

See Figure 3.4.1. Consider the open sets

$$V_1 = \left(\frac{3}{4}, \frac{5}{4} \right) \qquad \text{and} \qquad V_n = (a_n, b_n) \qquad (3.4.16)$$

where V_n is the open interval defined for each index $n \in \mathbb{N}$ where $n \geq 2$ by taking a_n to be the midpoint between $1/n$ and $1/(n+1)$ and b_n to be the midpoint between $1/(n-1)$ and $1/n$. From there, consider the collection of open sets $\mathcal{V}$ given by

$$\mathcal{V} = \{V_n : n \in \mathbb{N}\} = \{V_1, V_2, \ldots\}. \qquad (3.4.17)$$

See Figure 3.4.1.

It turns out $\mathcal{V}$ is *not* an open cover for F since 0 is in F but 0 is not in any of the V_n, thus $\mathcal{V}$ does not cover F since

$$F \not\subseteq \bigcup_{n=1}^{\infty} V_n = \bigcup_{U \in \mathcal{V}} U. \qquad (3.4.18)$$

However, by including one more open set that contains 0 to the collection, we get an open cover $\mathcal{W}$ for the set F. To that end, define

$$V_0 = \left(-\frac{1}{3}, \frac{1}{3} \right) \qquad \text{and} \qquad \mathcal{W} = \mathcal{V} \cup \{V_0\} = \{V_0, V_1, V_2, \ldots\}. \qquad (3.4.19)$$

Then $0 \in V_0$ and $1/n \in V_n$ for each $n \in \mathbb{N}$. Therefore,

$$F \subseteq \bigcup_{n=0}^{\infty} V_n = \bigcup_{U \in \mathcal{W}} U. \qquad (3.4.20)$$

Since $\mathcal{W}$ covers F and V_0 and every V_n is open, $\mathcal{W}$ is an open cover for F.

Remark 3.4.10: Open covers or not

By adding the single open set V_0 to the countably infinite collection $\mathcal{V}$ in Example 3.4.9, we obtain a collection $\mathcal{W}$ which is an open cover for F. Furthermore, after adding V_0 to the collection, we only needed a finite number of open sets to contain F. More specifically, consider the finite collection

$$\mathcal{W}_0 = \{V_0, V_1, V_2, V_3\}. \tag{3.4.21}$$

See Figure 3.4.1 once again. Then $0 \in V_0$, $1/n \in V_n$ for $n = 1, 2, 3$, and $1/n \in V_0$ for every $n \geq 4$. Hence,

$$F \subseteq \bigcup_{n=0}^{3} V_n = \bigcup_{U \in \mathcal{W}_0} U. \tag{3.4.22}$$

Therefore, $\mathcal{W}_0$ is a finite subcollection of $\mathcal{W}$ whose objects are open sets with a union that still contains F. That is, $\mathcal{W}_0$ is itself an open cover for F, so $\mathcal{W}_0$ is a *finite subcover*.

Definition 3.4.11: Finite subcover

Let $S \subseteq \mathbb{R}^m$ and let $\mathcal{W}$ be an open cover for S. A subcollection $\mathcal{W}_0$ (of $\mathcal{W}$) is a *finite subcover* if there are a finite number of open sets $U_1, \ldots, U_{n_0}$ such that

(i) $\mathcal{W}_0 = \{U_1, \ldots, U_{n_0}\} \subseteq \mathcal{W}$, and

(ii) $S \subseteq \displaystyle\bigcup_{n=1}^{n_0} U_n = \bigcup_{U \in \mathcal{W}_0} U$. (That is, $\mathcal{W}_0$ covers S.)

In Example 3.4.5, the open covers $\mathcal{A}$ and $\mathcal{B}$ for H have finite subcovers $\mathcal{A}_0$ and $\mathcal{B}_0$. In Example 3.4.9, the collection $\mathcal{W}_0$ provides a way for us to represent the infinite collection of points in F with a finite number of open sets: Every real number in F can be represented by an open set in the collection $\mathcal{W}_0$ containing the real number. Specifically, V_n represents $1/n$ for each $n = 1, 2, 3$ while V_0 simultaneously represents 0 and all of the $1/n$ where $n \geq 4$. This is an interpretation of the "next best thing to finiteness" idea and leads us to the following definition for compactness.

Definition 3.4.12: Compact

A set $K \subseteq \mathbb{R}^m$ is *compact* if every open cover for K has a finite subcover.

Remark 3.4.13: Compactness beyond the examples

The definition of compactness (Definition 3.4.12) goes beyond the finite sets in Theorem 3.4.1, the closed and bounded intervals in Example 3.4.5 and Theorem 3.4.3, and the closed and countable set F in Example 3.4.9 in a couple of important ways: (i) K is not necessarily a subset of the real line; and (ii) *every* open cover for K has a finite subcover (not just some). This is not an easy definition to process!

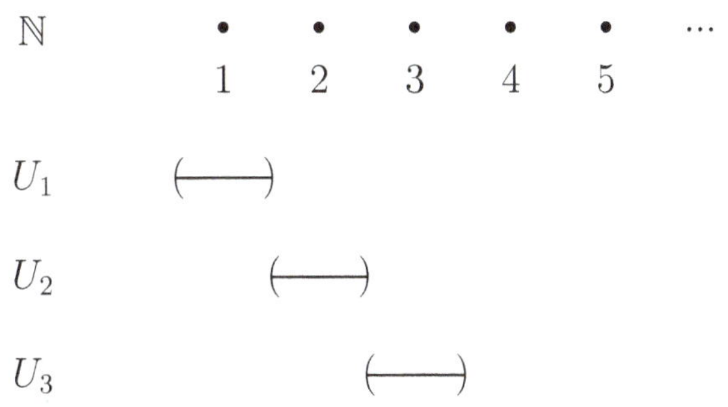

Figure 3.4.2: The set of positive integers $\mathbb{N}$ and the first few intervals in the open cover $\mathcal{U} = \{U_1, U_2, \ldots\}$ from Example 3.4.14.

The next example contains two sets which, for different reasons, are not compact.

Example 3.4.14: Not compact

Consider the set of positive integers $\mathbb{N}$ and the set of reciprocals of positive integers S given by

$$\mathbb{N} = \{n : n \in \mathbb{N}\} = \{1, 2, 3, \ldots\} \qquad \text{and} \tag{3.4.23}$$

$$S = \frac{1}{\mathbb{N}} = \left\{\frac{1}{n} : n \in \mathbb{N}\right\} = \left\{1, \frac{1}{2}, \frac{1}{3}, \ldots\right\}. \tag{3.4.24}$$

Neither $\mathbb{N}$ nor S is compact. To prove this, it suffices to find an open cover with no finite subcover.

Proof for Example 3.4.14. For the set of positive integers $\mathbb{N}$, consider the open cover given by the collection $\mathcal{U} = \{U_1, U_2, \ldots\}$ comprising open intervals defined for each positive integer n by

$$U_n = \left(n - \frac{1}{2}, n + \frac{1}{2}\right). \tag{3.4.25}$$

See Figure 3.4.2. For each positive integer n, the only open interval in the collection $\mathcal{U}$ that contains n is U_n. Now let $\mathcal{V}$ be any finite subcollection of $\mathcal{U}$. Then there is some positive integer k_0 where

$$\bigcup_{U \in \mathcal{V}} U \subseteq \bigcup_{n=1}^{k_0} U_n \subseteq \left(\frac{1}{2}, k_0 + \frac{1}{2}\right). \tag{3.4.26}$$

Since $k_0 + 1$ is a positive integer but $k_0 + \frac{1}{2} < k_0 + 1$, we have

$$k_0 + 1 \notin \bigcup_{U \in \mathcal{V}} U. \tag{3.4.27}$$

Therefore, $\mathcal{V}$ is not an open cover for $\mathbb{N}$. Since $\mathcal{V}$ represents an arbitrary finite subcollection of $\mathcal{U}$, we have that $\mathcal{U}$ has no finite subcover. Therefore, $\mathbb{N}$ is not compact.

For the set S, consider the open cover $\mathcal{W} = \{W_1, W_2, \ldots\}$ given by

$$W_1 = \left(\frac{1}{2}, 2\right) \quad \text{and} \quad W_n = \left(\frac{1}{n+1}, \frac{1}{n-1}\right) \quad \text{for} \quad n \geq 2. \tag{3.4.28}$$

(Try drawing a figure for S and $\mathcal{W}$, it should be similar to Figure 3.4.1.) For every index $n \in \mathbb{N}$, $1/n$ is in W_n and only W_n. Now let $\mathcal{T}$ be any finite subcollection of $\mathcal{W}$. Then there is some positive integer j_0 where

$$\bigcup_{W \in \mathcal{T}} W \subseteq \bigcup_{n=1}^{j_0} W_n \subseteq \left(\frac{1}{j_0+1}, 2\right). \tag{3.4.29}$$

So, $1/(j_0 + 2) \in S$ but $1/(j_0 + 2) \notin \cup_{W \in \mathcal{T}} W$. Therefore, $\mathcal{T}$ is not an open cover for S. Since $\mathcal{T}$ represents an arbitrary finite subcollection of $\mathcal{W}$, we have that $\mathcal{W}$ has no finite subcover. Therefore, S is not compact. $\qquad\square$

Remark 3.4.15: Not compact from either not closed or not bounded

The open covers with no finite subcovers in the proof of Example 3.4.14 exploit features of the underlying sets: $\mathcal{U}$ takes advantage of the fact that $\mathbb{N}$ is unbounded while $\mathcal{W}$ takes advantage of the fact that S is not closed since 0 is arbitrarily close to S but not in S. These ideas will help with the exercises.

The section concludes with an example of a compact set along with a proof.

Example 3.4.16: A compact set

Once again, consider the set of real numbers F from Figure 3.4.1, Example 3.4.9, and Remark 3.4.10 given by

$$F = \{0\} \cup \frac{1}{\mathbb{N}} = \{0\} \cup \left\{\frac{1}{n} : n \in \mathbb{N}\right\} = \left\{0, 1, \frac{1}{2}, \frac{1}{3}, \ldots\right\}. \tag{3.4.30}$$

F is compact.

Scratch Work 3.4.17: Generalize an example

The argument in Example 3.4.9 can be generalized to prove F is compact. Below, we take advantage of the convergence of $(1/n)$ to 0 to generate a finite subcover of a given open cover. The idea is that any open cover for F must cover 0 with a neighborhood, and such a neighborhood contains all but a finite number of the $1/n$ in F. A suitable finite subcover comprises an open set from the original open cover that contains both 0 and a neighborhood of 0 along with open sets for each of finite number of points in F not in that neighborhood.

Proof for Example 3.4.16. Suppose $\mathcal{U}$ is an open cover for F. Then there is an open set $U_0 \in \mathcal{U}$ where $0 \in U_0$ and, for each $n \in \mathbb{N}$, there is an open set $U_n \in \mathcal{U}$ where $1/n \in U_n$. Since U_0 is open, by Definition 3.2.2 there is some $\varepsilon_0 > 0$ where

$$|x - 0| = |x| < \varepsilon_0 \qquad \Longrightarrow \qquad x \in U_0. \tag{3.4.31}$$

Since $\lim_{n \to \infty}(1/n) = 0$, by Definition 2.2.1 there is a threshold $n_0 \in \mathbb{N}$ such that for all $n \in \mathbb{N}$ where $n \geq n_0$ we have

$$\left| \frac{1}{n} - 0 \right| = \frac{1}{n} \leq \frac{1}{n_0} < \varepsilon_0 \qquad \implies \qquad \frac{1}{n} \in U_0. \tag{3.4.32}$$

Now consider the finite subcollection $\mathcal{S}$ of the open cover $\mathcal{U}$ given by

$$\mathcal{S} = \{U_0, U_1, \ldots, U_{n_0 - 1}\}. \tag{3.4.33}$$

Then 0 and each $1/n$ where $n \geq n_0$ is contained in U_0, and if needed, each $1/n$ where $n = 1, \ldots, n_0 - 1$ is contained in its own U_n. Hence,

$$F \subseteq \bigcup_{n=0}^{n_0 - 1} U_n. \tag{3.4.34}$$

Therefore, $\mathcal{S}$ is a finite subcover of $\mathcal{U}$, and so F is compact by Definition 3.4.12. $\qquad \square$

The Heine-Borel Theorem 3.5.1, the focus of the next section, provides a variety of equivalent perspectives on compactness which help us identify and work with compact sets in Euclidean spaces.

Exercises

3.4.1. Prove finite subsets of Euclidean spaces are compact.

3.4.2. Consider the set K given by

$$K = \{5\} \cup \left\{ 5 + \frac{(-1)^n}{\sqrt{n}} : n \in \mathbb{N} \right\}. \tag{3.4.35}$$

Prove K is compact.

3.4.3. Suppose $(\mathbf{x}_n) \subseteq \mathbb{R}^m$ converges to $\mathbf{y} \in \mathbb{R}^m$. Prove

$$F = \{\mathbf{x}_n : n \in \mathbb{N}\} \cup \{\mathbf{y}\} \tag{3.4.36}$$

is compact.

3.4.4. Prove the open unit interval $(0, 1)$ is not compact.

3.4.5. Prove the unbounded interval $[0, \infty)$ is closed but not compact.

3.4.6. Prove $[0, 1] \cap \mathbb{Q}$ is not compact.

3.4.7. Prove that if $(x_n) \subseteq \mathbb{R}$ is bounded and strictly decreasing, then the range of (x_n) is not compact.

3.4.8. Suppose $(\mathbf{x}_n) \subseteq \mathbb{R}^m$ is bounded and its range does not contain $\mathrm{Slim}(\mathbf{x}_n)$, the set of subsequential limits. Prove the range of $(\mathbf{x}_n)$ is not compact.

3.5   The Heine-Borel Theorem

In general, how can we check whether or not a given set is compact? Definition 3.4.12 is so difficult to work and the question is so tough to answer they merit this additional section. The answer itself comes in the form of the Heine-Borel Theorem 3.5.1 which establishes various equivalent forms of compactness for Euclidean spaces.

Throughout this section, please take both a stand-alone "compact" and the phrase "topologically compact" to stand for Definition 3.4.12, as it does here.

Theorem 3.5.1: Heine-Borel

For a subset K of a Euclidean space $\mathbb{R}^m$, the following are equivalent:

(i) K is *topologically compact*: Every open cover for K has a finite subcover.

(ii) K is closed and bounded.

(iii) K is *sequentially compact*: Every sequence in K has a convergent subsequence whose limit is in K.

The word "topologically" supplements "compact" in part (i) of the Heine-Borel Theorem to help me distinguish between the two other statements. In the context of Euclidean spaces, (i), (ii), and (iii) are interchangeable and each one stands for compactness. However, the statement "every open cover has a finite subcover" defines compactness in all topological settings, which is why the phrase "topologically compact" is attributed to it.

Each version of compactness in the Heine-Borel Theorem 3.5.1 has its own pros and cons, and all three will be used throughout the book.

This entire section is dedicated to proving the Heine-Borel Theorem 3.5.1. Instead of proving it all at once, we parse the statements and prove them in a series of lemmas which are interesting enough on their own anyway. A complete proof of the Heine-Borel Theorem 3.5.1 appears at the end of the section.

Lemma 3.5.2: Compact implies bounded

Every compact set in $\mathbb{R}^m$ is bounded.

Scratch Work 3.5.3: Contraposition via unbounded

The idea behind the proof of Lemma 3.5.2 is very similar to the proof that the set of positive integers $\mathbb{N}$ is not compact in Example 3.4.14: Unbounded sets have open covers made up of bounded open sets which have no finite subcovers. The approach argues via contraposition and defines a generic open cover for arbitrary unbounded sets which cannot have a finite subcover.

Proof of Lemma 3.5.2. Suppose S is an unbounded set of points in $\mathbb{R}^m$. Also, consider the collection of open sets $\mathcal{V} = \{V_n(\mathbf{0}) : n \in \mathbb{N}\}$ where for each positive integer n we have

$$V_n(\mathbf{0}) = \{\mathbf{x} \in \mathbb{R}^m : \|\mathbf{x}\|_m < n\}. \tag{3.5.1}$$

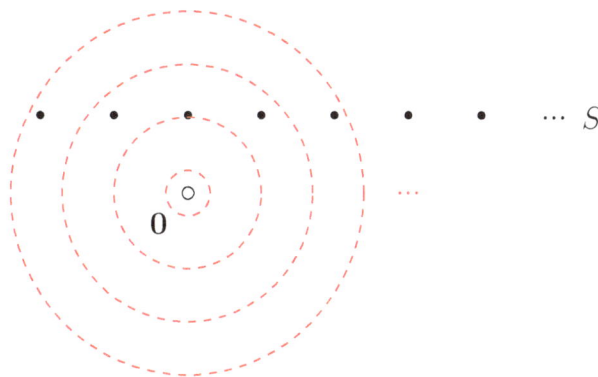

Figure 3.5.1: A unbounded set S in the plane $\mathbb{R}^2$ and open cover $\mathcal{V} = \{V_n(\mathbf{0}) : n \in \mathbb{N}\}$ of open disks centered at the origin (in red) with radii given by positive integers. S is not compact since $\mathcal{V}$ has no finite subcovers. See Lemma 3.5.2 and its proof.

(See Figure 3.5.1 for a version of S and an open cover $\mathcal{V}$ in the plane $\mathbb{R}^2$.) Then for every point $\mathbf{x}$ in S, by the Archimedean Property 1.4.6 there is a positive integer $n_\mathbf{x}$ large enough so that $\|\mathbf{x}\|_m < n_\mathbf{x}$. Therefore, $\mathbf{x}$ is in $V_{n_\mathbf{x}}(\mathbf{0})$ and so $\mathcal{V}$ is an open cover for S.

Now suppose $\mathcal{W}$ is a finite subcollection of the open sets in $\mathcal{V}$. Then there is a positive integer k_0 where we have

$$\bigcup_{V \in \mathcal{W}} V \subseteq \bigcup_{n=1}^{k_0} V_n(\mathbf{0}) = V_{k_0}(\mathbf{0}) \tag{3.5.2}$$

Since S is unbounded, there is a point $\mathbf{x}_0$ in S where $\|\mathbf{x}_0\|_m > k_0$. As such,

$$\mathbf{x}_0 \notin V_{k_0}(\mathbf{0}) \quad \Longrightarrow \quad \mathbf{x}_0 \notin \bigcup_{V \in \mathcal{W}} V. \tag{3.5.3}$$

So, $\mathcal{W}$ is not an open cover for S. Since $\mathcal{W}$ is an arbitrary finite subcollection of $\mathcal{V}$, we have S is not compact. $\qquad\square$

Lemma 3.5.4: Compact implies closed

Every compact set in $\mathbb{R}^m$ is closed.

Scratch Work 3.5.5: Contraposition via not closed

The idea behind the proof of Lemma 3.5.4 is very similar to the proof that the set S is not compact in Example 3.4.14: Sets that are not closed have a point whose neighborhoods have complements (well, almost complements) which can be used to create an open cover for S with no finite subcover.

Proof of Lemma 3.5.4. To argue via contraposition, suppose T is a subset of $\mathbb{R}^m$ that is not closed. Then there is a point $\mathbf{y}$ in $\mathbb{R}^m$ such that $\mathbf{y} \operatorname{acl} T$ but $\mathbf{y}$ is not in T. Now, for each positive integer

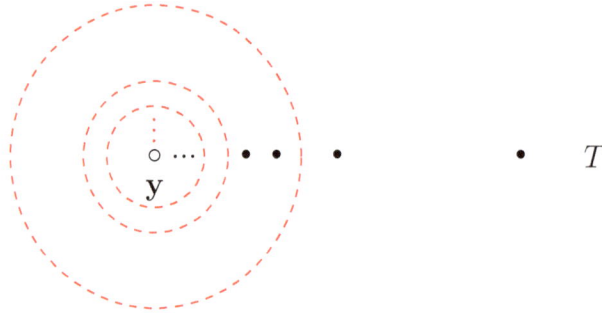

Figure 3.5.2: A set T and a point $\mathbf{y}$ in the plane $\mathbb{R}^2$ where $\mathbf{y}$ is arbitrarily close to T but $\mathbf{y}$ is not in T, so T is not closed. The circles in red represent the neighborhoods centered at $\mathbf{y}$ whose radii are the reciprocals of positive integers. However, the complements of the closures of these neighborhoods $\mathbb{R}^2\backslash\overline{V_{1/n}(\mathbf{y})}$ form the open cover used to show T is not compact in the proof of Lemma 3.5.4.

n, consider the open set $\mathbb{R}^m\backslash\overline{V_{1/n}(\mathbf{y})}$ where

$$\mathbb{R}^m\backslash\overline{V_{1/n}(\mathbf{y})} = \{\mathbf{x} \in \mathbb{R}^m : d_m(\mathbf{x},\mathbf{y}) = \|\mathbf{x}-\mathbf{y}\|_m > 1/n\}. \tag{3.5.4}$$

See Figure 3.5.2. Note that $\mathbb{R}^m\backslash\overline{V_{1/n}(\mathbf{y})}$ is the complement of the closed $1/n$-neighborhood of $\mathbf{y}$, so $\mathbb{R}^m\backslash\overline{V_{1/n}(\mathbf{y})}$ is open by Theorem 3.2.4. Since $\mathbf{y}$ is not in T, for every point $\mathbf{t}$ in T we have

$$d_m(\mathbf{t},\mathbf{y}) = \|\mathbf{t}-\mathbf{y}\|_m > 0. \tag{3.5.5}$$

So, by the Corollary of the Archimedean Property (Corollary 1.4.8), there is some positive integer $n_{\mathbf{t}}$ large enough so that

$$\frac{1}{n_{\mathbf{t}}} < d_m(\mathbf{t},\mathbf{y}) = \|\mathbf{t}-\mathbf{y}\|_m. \tag{3.5.6}$$

Thus, $\mathbf{t}$ is in $\mathbb{R}^m\backslash\overline{V_{1/n_{\mathbf{t}}}(\mathbf{y})}$ and therefore

$$\mathcal{V} = \left\{\mathbb{R}^m\backslash\overline{V_{1/n}(\mathbf{y})} : n \in \mathbb{N}\right\} \tag{3.5.7}$$

is an open cover for T.

Now suppose $\mathcal{W}$ is a finite subcollection of the open sets in $\mathcal{V}$. Then there is a positive integer k_0 where we have

$$\bigcup_{V\in\mathcal{W}} V \subseteq \bigcup_{n=1}^{k_0}(\mathbb{R}^m\backslash\overline{V_{1/n}(\mathbf{y})}) = \mathbb{R}^m\backslash\overline{V_{1/k_0}(\mathbf{y})}. \tag{3.5.8}$$

Since $\mathbf{y}\operatorname{acl} T$, there is a point $\mathbf{t}_0$ in T such that

$$d_m(\mathbf{t}_0,\mathbf{y}) = \|\mathbf{t}_0-\mathbf{y}\|_m < \frac{1}{k_0}. \tag{3.5.9}$$

Therefore,

$$\mathbf{t}_0 \in V_{1/k_0}(\mathbf{y}) \implies \mathbf{t}_0 \notin \mathbb{R}^m\backslash\overline{V_{1/k_0}(\mathbf{y})} \implies \mathbf{t}_0 \notin \cup_{V\in\mathcal{W}}V. \tag{3.5.10}$$

Hence, $\mathcal{W}$ is not an open cover for T. Since $\mathcal{W}$ is an arbitrary finite subcollection of $\mathcal{V}$, we have T is not compact. $\square$

At this point, Lemmas 3.5.2 and 3.5.4 tell us topologically compact sets are closed and bounded. That is, we have proven (i) implies (ii) in the Heine-Borel Theorem 3.5.1. The next two lemmas allow us to prove (ii) implies (i) with an interesting combination of special cases at the end of the section.

Lemma 3.5.6: Closed and bounded boxes are compact

If $B_0 \subseteq \mathbb{R}^m$ is a closed box of the form

$$B_0 = \underbrace{[-a, a] \times \cdots \times [-a, a]}_{m \text{ copies of } [-a,a]} = [-a, a]^m \tag{3.5.11}$$

for some $a > 0$, then B_0 is compact.

Scratch Work 3.5.7: An open cover with no finite subcover

The proof argues via contraposition and uses the negation of the definition for compactness, Definition 3.4.12, to provide the existence of an open cover with no finite subcover. A careful argument based on repeated bisection of boxes follows from there, similar to the proof of the Bolzano-Weierstrass Theorem in Euclidean spaces (Theorem 2.5.13). However, the details regarding the choice of a suitable threshold are omitted.

Proof Lemma 3.5.6. Suppose $B_0 \subseteq \mathbb{R}^m$ is a closed box of the form (3.5.11), which is closed by Lemma 3.1.13. See Figure 3.5.3.

By way of contradiction, suppose B_0 is not compact. Then there is an open cover $\mathcal{W}$ for the set B_0 which has no finite subcovers. Bisecting each of the sides of B_0 creates 2^m smaller boxes of the form (3.1.14) whose sides are half the length of the sides of B_0. Then at least one of the 2^m smaller boxes, call it B_1, can only be covered by an infinite subcollection of the open sets in $\mathcal{W}$, otherwise $\mathcal{W}$ would have a finite subcover. (Such a finite subcover could be obtained by gathering the open sets in the finite subcollections for all of the 2^m smaller boxes.) Another application of Lemma 3.1.13 tells us B_1 is closed and bounded.

Repeating this bisection and smaller-box-selection process yields a nested sequence of closed and bounded boxes where

$$B_0 \supseteq B_1 \supseteq B_2 \supseteq \ldots \tag{3.5.12}$$

and each of these boxes can only be covered by an infinite subcollection of the open sets in $\mathcal{W}$. By the NCBB Property (Theorem 2.5.10), there is a point $\mathbf{y}$ where

$$\mathbf{y} \in \bigcap_{n=0}^{\infty} B_n. \tag{3.5.13}$$

See Figure 3.5.3.

Since $\mathcal{W}$ covers B_0 and $\mathbf{y} \in B_0$, there is an open set U in the open cover $\mathcal{W}$ which contains $\mathbf{y}$. Since U is open, there is a positive distance $\varepsilon_{\mathbf{y}} > 0$ such that $V_{\varepsilon_{\mathbf{y}}}(\mathbf{y})$, the $\varepsilon_{\mathbf{y}}$-neighborhood of $\mathbf{y}$ is contained in U.

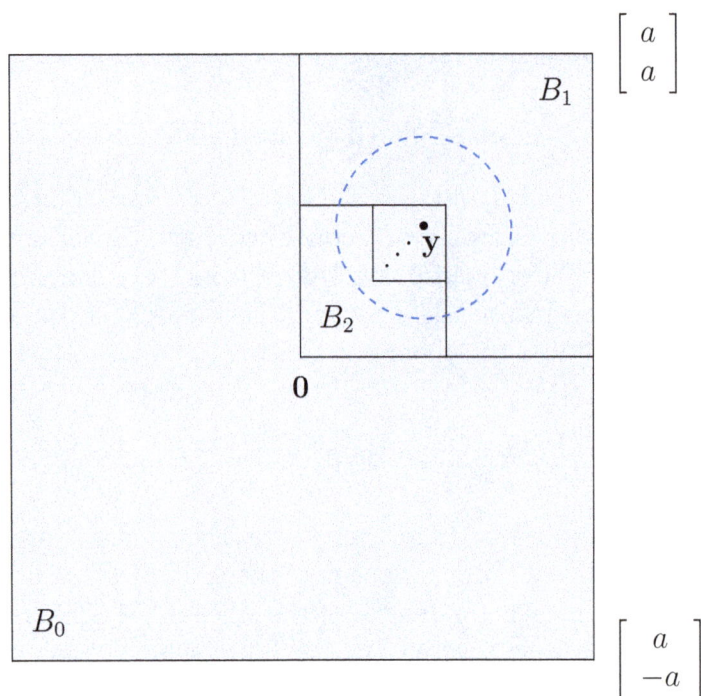

Figure 3.5.3: A box of the form $B_0 = [-a, a]^2$ in the plane $\mathbb{R}^2$ to go along with the proof of Lemma 3.5.6. The boxes B_1, B_2, and the small one containing **y** are obtained by repeatedly bisecting sides and choosing a smaller box which requires an infinite subcover. The point **y** has a neighborhood containing one of the smaller boxes, facilitating a contradiction.

However, for each index $n \in \mathbb{N}$, the length of each of the sides of B_n is half that of the previous box B_{n-1}. Since B_0 is bounded, this implies the sequence of lengths of the sides converges to 0, as in the proof of the Bolzano-Weierstrass Theorem in Euclidean spaces (Theorem 2.5.13). Hence, there is a threshold $n_{\mathbf{y}} \in \mathbb{N}$ for which the box $B_{n_{\mathbf{y}}}$ is small enough to be contained in U. That is, for large enough $n_{\mathbf{y}}$ we have

$$B_{n_{\mathbf{y}}} \subseteq U. \tag{3.5.14}$$

(See the smallest box containing $\mathbf{y}$ in Figure 3.5.3.) However, this means $\{U\}$ is a finite subcollection of $\mathcal{W}$ which covers $B_{n_{\mathbf{y}}}$, contradicting the assertion that $B_{n_{\mathbf{y}}}$ can only be covered by an infinite subcollection of $\mathcal{W}$.

Therefore, B_0 is compact. $\qquad\square$

The previous lemma directly pairs with the next in the proof of the Heine-Borel Theorem 3.5.18 at the end of the section.

Lemma 3.5.8: Closed subsets of compact sets are compact

Every closed subset of a compact set in $\mathbb{R}^m$ is compact.

Proof of Lemma 3.5.8. Suppose K is a compact set in $\mathbb{R}^m$ and F is a closed set in $\mathbb{R}^m$ where $F \subseteq K$.

Let $\mathcal{W}$ be an open cover for F. To take advantage of the compactness of K, we can create an open cover for K by expanding $\mathcal{W}$ with one more open set to pick up any points that might be in $K \backslash F$. By Theorem 3.2.4, $\mathbb{R}^m \backslash F$ is an open set. Furthermore, the collection of open sets $\mathcal{W} \cup \{\mathbb{R}^m \backslash F\}$ is an open cover for K since

$$F \subseteq \bigcup_{U \in \mathcal{W}} U \tag{3.5.15}$$

implies

$$K \subseteq \mathbb{R}^m = F \cup (\mathbb{R}^m \backslash F) \subseteq \left(\bigcup_{U \in \mathcal{W}} U \right) \cup (\mathbb{R}^m \backslash F). \tag{3.5.16}$$

By the definition of compact (Definition 3.4.12), there is a finite subcover $\mathcal{W}_0 = \{U_1, \ldots, U_{n_0}\}$ of $\mathcal{W} \cup \{\mathbb{R}^m \backslash F\}$ where

$$K \subseteq \bigcup_{n=1}^{n_0} U_n. \tag{3.5.17}$$

If $\mathbb{R}^m \backslash F = U_j$ for some $j = 1, \ldots, n_0$, then $\mathcal{W}_0 \backslash \{\mathbb{R}^m \backslash F\}$ is a finite subcover of $\mathcal{W}$ which covers F. Otherwise, $\mathcal{W}_0$ is a finite subcover of $\mathcal{W}$ which covers F. Either way, since $\mathcal{W}$ is an arbitrary open cover for F, we have F is compact. $\qquad\square$

Lemmas 3.5.2 through 3.5.8 can be used to prove a set is topologically compact if and only if the set is closed and bounded (showing statements (i) and (ii) in the Heine-Borel Theorem 3.5.1 are equivalent). However, there is still a little work is left to do.

Lemma 3.5.9: Closed and bounded implies compact

Every set in $\mathbb{R}^m$ which is both closed and bounded is compact.

Scratch Work 3.5.10: Apply the previous lemmas

The goal is to find a closed box large enough to contain K, then apply the two previous lemmas to conclude K is topologically compact.

Proof of Lemma 3.5.9. Suppose K is a closed and bounded subset of $\mathbb{R}^m$. Since K is bounded, there is a bound $u > 0$ such that for every point $\mathbf{x} \in K$ we have $\|\mathbf{x}\|_m \leq u$. Define the closed box B by

$$B = \underbrace{[-u, u] \times \cdots \times [-u, u]}_{m \text{ copies of } [-u,u]} = [-u, u]^m. \tag{3.5.18}$$

To see that $K \subseteq B$, for every index $j = 1, \ldots, m$ let x_j denote the j-th component (or coordinate) of $\mathbf{x}$. For every $\mathbf{x} \in B$ and index $j = 1, \ldots, m$ we have

$$|x_j| = \sqrt{x_j^2} \leq \sqrt{x_1^2 + \ldots + x_m^2} = \|\mathbf{x}\|_m \leq u. \tag{3.5.19}$$

Hence, $x_j \in [-u, u]$ for each j and so $\mathbf{x} \in B$. Therefore, $K \subseteq B$.

Now, since B is a closed box, B is topologically compact by Lemma 3.5.6. Since K is a closed subset of a topologically compact set, K is also topologically compact by Lemma 3.5.8.   □

Next up is a definition for and a brief discussion of *sequential compactness*, statement (iii) of the Heine-Borel Theorem 3.5.1.

Definition 3.5.11: Sequentially compact

A set $K \subseteq \mathbb{R}^m$ is *sequentially compact* if every sequence in K has a convergent subsequence whose limit is in K.

Example 3.5.12: Sequential compactness

Once again, consider the of real numbers E and its closure $\overline{E}$ from Example 3.1.12. We have

$$E = [0, 1) \cup \{2\} \cup \{3 + (1/n) : n \in \mathbb{N}\} \quad \text{and} \tag{3.5.20}$$
$$\overline{E} = [0, 1] \cup \{2\} \cup \{3\} \cup \{3 + (1/n) : n \in \mathbb{N}\}. \tag{3.5.21}$$

See Figure 3.5.4. To get an idea for sequential compactness, consider the sequences (x_n) and (y_n) defined for each index $n \in \mathbb{N}$ by

$$x_n = 3 + \frac{1}{n} \quad \text{and} \quad y_n = \frac{1}{2} + \frac{(n+1)(-1)^n}{4n}. \tag{3.5.22}$$

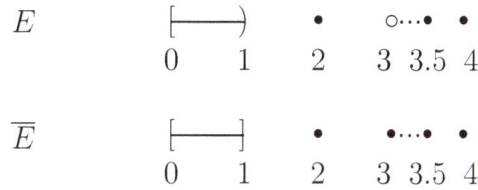

Figure 3.5.4: The set of real numbers E and its closure $\overline{E}$ from Examples 3.1.12 and 3.5.12. The closure $\overline{E}$ is sequentially compact, topologically compact, closed, and bounded.

Then (x_n) is a sequence of points in both E and $\overline{E}$ where

$$\lim_{n \to \infty} x_n = \lim_{n \to \infty} \left(3 + \frac{1}{n} \right) = 3. \tag{3.5.23}$$

So, (x_n) converges to 3 and, thanks to Theorem 2.4.18, all of the subsequences of (x_n) converge to 3 as well. However, 3 is not in E, so (x_n) is a sequence in E whose subsequences all converge to a limit which is not in E. Therefore, E is not sequentially compact. On the other hand, 3 is in the closure $\overline{E}$.

Also, (y_n) is also a sequence of points in both E and $\overline{E}$, as follows. For every index $n \in \mathbb{N}$, we add at least $-1/2$ but no more than $1/4$ to $1/2$, so we have

$$y_n = \frac{1}{2} + \frac{(n+1)(-1)^n}{4n} \in [0,1) \subseteq E \subseteq \overline{E}. \tag{3.5.24}$$

The sequence (y_n) diverges by Divergence Criteria 2.6.9 since its subsequences (y_{2k}) and (y_{2k-1}) have different limits:

$$\lim_{k \to \infty} y_{2k} = \lim_{k \to \infty} \left(\frac{1}{2} + \frac{2k+1}{4(2k)} \right) = \frac{3}{4} \qquad \text{while} \tag{3.5.25}$$

$$\lim_{k \to \infty} y_{2k-1} = \lim_{k \to \infty} \left(\frac{1}{2} + \frac{-2k}{4(2k-1)} \right) = \frac{1}{4}. \tag{3.5.26}$$

Both of these subsequential limits are in the closure $\overline{E}$.

Moreover, $\overline{E}$ is sequentially compact, but proving this directly is difficult since we'd need to consider *any* sequence of real numbers in $\overline{E}$, find a convergent subsequence and its limit, then show this limit is in the set.

At this point, it is much easier to show $\overline{E}$ is closed and bounded: As the closure of a set, $\overline{E}$ is closed by Lemma 3.1.8; and $\overline{E}$ is bounded by 4.

Example 3.5.12 leads us to a string of lemmas showing statements (ii) and (iii) in the Heine-Borel Theorem 3.5.1 are equivalent: A subset of a Euclidean space is sequentially compact if and only if it is closed and bounded.

Lemma 3.5.13: Sequentially compact implies closed

Every sequentially compact set in $\mathbb{R}^m$ is closed.

Scratch Work 3.5.14: Contraposition via convergence outside of the set

The proof argues via contraposition to construct a nice sequence, but it also deals with the subtlety of considering *every* subsequence of the sequence we construct. The trick is to show all of these subsequences converge to a point that is not in the set, which we get from assuming our set in not closed.

Proof of Lemma 3.5.13. Suppose $A \subseteq \mathbb{R}^m$ is not closed. Then there is a point $\mathbf{y} \in \mathbb{R}^m$ where $\mathbf{y}$ is not in A but $\mathbf{y}$ acl A. By the fundamental connection between arbitrarily close and convergence (Theorem 2.3.1), there is a sequence of points in A whose limit is $\mathbf{y}$. By Theorem 2.4.18, all subsequences of a convergent sequence in $\mathbb{R}^m$ converge to the same limit. In this case, they all converge to $\mathbf{y}$. So, none of the subsequences of have a limit in A. Therefore, A is not sequentially compact. $\qquad \square$

Lemma 3.5.15: Sequentially compact implies bounded

Every sequentially compact set in $\mathbb{R}^m$ is bounded.

Scratch Work 3.5.16: Contraposition via an unbounded sequence

The proof argues via contraposition to construct a nice sequence, but like the previous proof it also deals with the subtlety of considering *every* subsequence of the sequence we construct. The idea here is to show all of these subsequences are unbounded, in which case they diverge by Divergence Criteria 2.6.9.

Proof of Lemma 3.5.15. Suppose $U \subseteq \mathbb{R}^m$ is unbounded. Then for every positive integer $n \in \mathbb{N}$, there is a point $\mathbf{u}_n$ in U where

$$n < \|\mathbf{u}_n\|_m. \tag{3.5.27}$$

The goal from here is to show every subsequence of $(\mathbf{u}_n)$ is unbounded. So, consider an arbitrary subsequence $(\mathbf{u}_{n_k})$ and any real number $x \in \mathbb{R}$. By the Archimedean Property (Theorem 1.4.6) along with the definition of subsequences (Definition 2.4.14) and inequality (3.5.27), there is an index $k \in \mathbb{N}$ large enough to give us

$$x < k \le n_k < \|\mathbf{u}_{n_k}\|_m. \tag{3.5.28}$$

Hence, the subsequence $(\mathbf{u}_{n_k})$ is unbounded and by the Divergence Criteria 2.6.9, $(\mathbf{u}_{n_k})$ diverges.

Moreover, $(\mathbf{u}_n)$ is a sequence of points in U whose subsequences all diverge, so none of them have a limit in U. Therefore, U is not sequentially compact. $\qquad \square$

Up next is the final lemma of the section before stringing the results together to prove the Heine-Borel Theorem 3.5.1.

Lemma 3.5.17: Closed and bounded implies sequentially compact

Every set in $\mathbb{R}^m$ which is both closed and bounded is sequentially compact.

Proof of Lemma 3.5.17. Suppose K is a closed and bounded subset of $\mathbb{R}^m$ and $(\mathbf{x}_n)$ is a sequence of points in K. Since K is bounded, $(\mathbf{x}_n)$ is bounded as well. By the Bolzano-Weierstrass Theorem in Euclidean spaces (Theorem 2.5.13), $(\mathbf{x}_n)$ has a convergent subsequence with a limit $\mathbf{y}$. Since K is closed, the subsequential limit $\mathbf{y}$ must be in K by Theorem 3.1.10 (closed sets contain their limits). Therefore, K is sequentially compact since every sequence in K has a convergent subsequence whose limit is also in K. □

Time to prove the Heine-Borel Theorem 3.5.1, copied here for convenience.

Theorem 3.5.18: A copy of the Heine-Borel Theorem

For a subset K of a Euclidean space $\mathbb{R}^m$, the following are equivalent:

(i) K is *topologically compact*: Every open cover for K has a finite subcover.

(ii) K is closed and bounded.

(iii) K is *sequentially compact*: Every sequence in K has a convergent subsequence whose limit is in K.

Scratch Work 3.5.19: A string of equivalences

The proof below gets the job done by showing (i) is equivalent to (ii) and (ii) is equivalent to (iii), so (i) and (iii) are also equivalent.

Proof of the Heine-Borel Theorem 3.5.1. Throughout this proof, suppose K is a subset of a Euclidean space $\mathbb{R}^m$.

(i) $\implies$ (ii): Suppose K is topologically compact. By Lemma 3.5.4, K is closed. By Lemma 3.5.2 K is bounded.

(ii) $\implies$ (i): Suppose K is closed and bounded. By Lemma 3.5.9, K is topologically compact. Hence, (i) $\iff$ (ii) and topological compactness is equivalent to closed and bounded.

(ii) $\implies$ (iii): Suppose K is closed and bounded. By Lemma 3.5.17, K is sequentially compact.

(iii) $\implies$ (ii): Suppose K is sequentially compact. By Lemma 3.5.13, K is closed. By Lemma 3.5.15, K is bounded.

Hence, (iii) $\iff$ (ii) and sequential compactness is equivalent to closed and bounded.

(i) $\iff$ (iii): Since topological and sequential compactness are both equivalent to closed and bounded, they're equivalent to each other. That is, since (i) $\iff$ (ii) and (iii) $\iff$ (ii), we have (i) $\iff$ (iii). □

Compactness is a hard idea to process! So much so that this section and the previous one were designed to share the burden. Moving forward, the Heine-Borel Theorem 3.5.1 gives us the flexibility to think of compact subsets of Euclidean spaces in three equivalent ways.

The next section provides a change of pace by introducing a variety of ways to use arbitrarily close and away from to explore other classic properties of sets in Euclidean spaces.

Exercises

3.5.1. An analogue of Theorem 3.2.17 holds for collections of compact sets. Prove the following properties regarding compact sets in $\mathbb{R}^m$ hold:

(i) The empty set $\varnothing$ is compact, but $\mathbb{R}^m$ is not.

(ii) The union of any finite number of compact sets is compact.

(iii) The intersection of any nonempty collection of compact sets is compact.

3.5.2. Suppose $a < b$ and consider the set $[a, b] \setminus \mathbb{Q}$, the set of irrational numbers between a and b. Prove $[a, b] \setminus \mathbb{Q}$ is not compact.

3.5.3. Prove that if $(\mathbf{x}_n) \subseteq \mathbb{R}^m$ is bounded, then the set of subsequential limits $\mathrm{Slim}(\mathbf{x}_n)$ is compact.

3.5.4. Recall the Cantor set C from Exercise 2.8.9.

(i) Prove the Cantor set C is compact.

(ii) Prove the Cantor set C is arbitrarily small in the following sense: For every $\varepsilon > 0$, there is a finite collection of compact subintervals $[a_1, b_1], \dots, [a_n, b_n] \subseteq [0, 1]$ where

$$C \subseteq \bigcup_{j=1}^n [a_j, b_j] \qquad \text{and} \qquad \sum_{j=1}^n (b_j - a_j) < \varepsilon. \tag{3.5.29}$$

Hint: Do an online search for the "middle third" construction of the Cantor set and find a geometric sum for $\sum_{j=1}^n (b_j - a_j)$.

3.5.5. There is another equivalent form of compactness in Euclidean spaces: A set is *coda-compact* if every sequence in the set has nonempty coda contained in the set. Prove $K \subseteq \mathbb{R}^m$ is compact if and only K is coda-compact.

3.6 Other topological properties

The lenses provided by *arbitrarily close* and *away from* in Definitions 1.5.1 and 1.5.11 allow us to explore classic concepts in an unconventional way. But please know I did not make the decision to buck convention lightly.

I believe arbitrarily close is the kernel of analysis. This perspective drives the way everything in this textbook is designed. It gives me hope that the intuition and technical skills gained from studying *arbitrarily close* and *away from* will make the process of finding your own understanding of analysis a little smoother.

This section continues the trend of bucking convention by defining some classic concepts from topology and analysis using arbitrarily close and away from. The theorems, lemmas, exercises, etc., connect the concepts with their classic versions.

Let's start with the set of points arbitrarily close to both a given set and its complement.

Definition 3.6.1: Boundary and boundary point

The *boundary* of a set $B \subseteq \mathbb{R}^m$, denoted by ∂B, is the set of points arbitrarily close to both B and its complement $\mathbb{R}^m \backslash B$. That is,

$$\partial B = \{\mathbf{y} \in \mathbb{R}^m : \mathbf{y} \operatorname{acl} B \text{ and } \mathbf{y} \operatorname{acl} (\mathbb{R}^m \backslash B)\}. \qquad (3.6.1)$$

A point $\mathbf{y} \in \partial B$ is called a *boundary point* of B.

Remark 3.6.2: Multiple figures

Figures 3.6.1, 3.6.2, and 3.6.3 along with Example 3.6.9 feature a set of real numbers E and a set B in the plane along with their closures, boundaries, and other related sets defined in this section. Please revisit them as you work through the definitions.

For a comparison of our definition for the boundary of a set with a classic approach, see [2, Definition 2.13, p.65] but also the discussions on page 9 of that text where the phrase "arbitrarily close" is used but not formally defined, and page 65 regarding "points that lie close to both the inside and the outside of the set".

The points in a set that come with some distance away from the complement form the *interior* of the set.

Definition 3.6.3: Interior and interior point

The *interior* of a set $B \subseteq \mathbb{R}^m$, denoted by B^o, is the set of points away from the complement $\mathbb{R}^m \backslash B$. A point $\mathbf{y} \in B^o$ is called an *interior point* of B.

For a set $B \subseteq \mathbb{R}^m$, its interior B^o is given by

$$B^o = \{\mathbf{y} \in \mathbb{R}^m : \mathbf{y} \operatorname{awf} (\mathbb{R}^m \backslash B)\}. \qquad (3.6.2)$$

As mentioned in Remark 3.2.3, the definition for away from (Definition 1.5.11) characterizes the key feature of open sets (Definition 3.2.2): Points in an open set have neighborhoods contained in the set. Take a look at Figure 3.6.2 again. The interior points $\mathbf{a}$ and $\mathbf{b}$ of the set B have neighborhoods contained in B.

This leads us to the following relationship with the definition for interior (Definition 3.6.3), an analogue of Lemma 3.1.2 which tells us closed sets are their closures. The proof is omitted since it follows immediately from these definitions.

E $\longmapsto$)   •   ∘···• •
            0   1   2   3 3.5 4

$\overline{E}$ $\longmapsto$ •   •···• •
            0   1   2   3 3.5 4

E^o (——)
            0   1

∂E •   •   •   •···• •
            0   1   2   3 3.5 4

E' |——| •
            0   1   3

I_E •   ∘···• •
            2   3 3.5 4

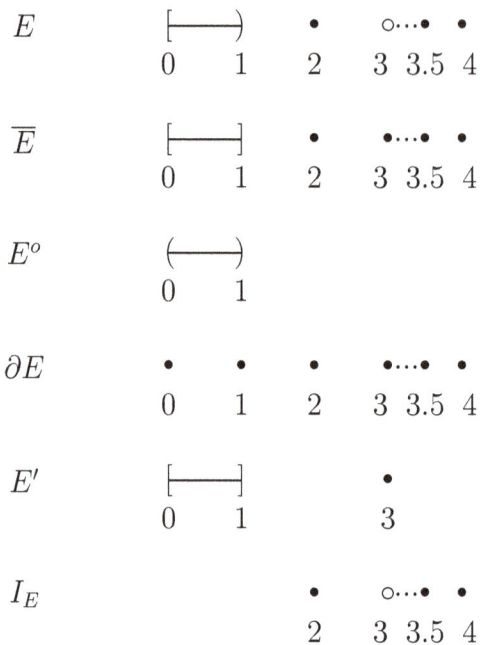

Figure 3.6.1: The set of real numbers E from Examples 3.1.12, 3.2.6, and 3.6.9 along with its closure $\overline{E}$, interior E^o, boundary ∂E, accumulation points E', and isolated points I_E. What does the exterior E^e look like? See Definitions 3.6.1 through 3.6.8.

Lemma 3.6.4: Open sets are their interiors

A set $U \subseteq \mathbb{R}^m$ is open if and only if $U = U^o$.

By turning the tables and considering the points away from a set, we get the *exterior*.

Definition 3.6.5: Exterior and exterior point

The *exterior* of a set $B \subseteq \mathbb{R}^m$, denoted by B^e, is the set of points away from B. A point $\mathbf{y} \in B^e$ is called an *exterior point* of B.

For a set $B \subseteq \mathbb{R}^m$, its exterior B^e is given by

$$B^e = \{\mathbf{y} \in \mathbb{R}^m : \mathbf{y}\,\mathrm{awf}\,B\}. \tag{3.6.3}$$

Thus, an exterior point of a set comes with a neighborhood contained in the complement of the set. In Figure 3.6.2, the exterior point $\mathbf{w}$ has a neighborhood contained in $\mathbb{R}^m \backslash B$. With this and Lemma 3.6.4 in mind, we have the following result.

Theorem 3.6.6: Interior and exterior via neighborhoods

For any $B \subseteq \mathbb{R}^m$ we have:

(i) The interior of B, B^o, is the set of points in B which have a neighborhood contained in B.

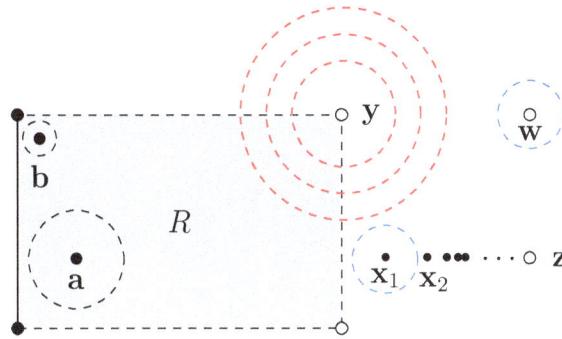

$$B = R \cup \{\mathbf{x}_n : n \in \mathbb{N}\}$$

Figure 3.6.2: A set B in the plane $\mathbb{R}^2$ which is the union of a rectangle R and the range of a convergent sequence $(\mathbf{x}_n)$. Here we have: $\mathbf{x}_1$, $\mathbf{x}_2$, $\mathbf{y}$, and $\mathbf{z}$ are boundary points of B; $\mathbf{a}$ and $\mathbf{b}$ are interior points of B; $\mathbf{w}$ is an exterior point of B; $\mathbf{a}$, $\mathbf{b}$, $\mathbf{y}$, and $\mathbf{z}$ are accumulation points of B; and $\mathbf{x}_1$ and $\mathbf{x}_2$ are isolated points of B. See Example 3.6.9 and Definitions 3.6.1 through 3.6.8.

(ii) The exterior of B, B^e, is the set of points in the complement $\mathbb{R}^m \backslash B$ which have a neighborhood in $\mathbb{R}^m \backslash B$.

Next up is an idea that characterizes the notion of a point having other points from a set nearby.

Definition 3.6.7: Accumulation point

A point $\mathbf{y} \in \mathbb{R}^m$ is an *accumulation point* of a set $B \subseteq \mathbb{R}^m$ if $\mathbf{y} \operatorname{acl}(B \backslash \{\mathbf{y}\})$, meaning $\mathbf{y}$ is arbitrarily close to other points in B. The set of accumulation points of B is denoted by B'.

For a set $B \subseteq \mathbb{R}^m$, its set of accumulation points B' is given by

$$B' = \{\mathbf{y} \in \mathbb{R}^m : \mathbf{y} \operatorname{acl}(B \backslash \{y\})\}. \tag{3.6.4}$$

The penultimate definition of this section characterizes points in a set that are not near any other points from the set.

Definition 3.6.8: Isolated point

A point $\mathbf{y} \in \mathbb{R}^m$ is an *isolated point* of a set $B \subseteq \mathbb{R}^m$ if $\mathbf{y} \in B$ but $\mathbf{y} \operatorname{awf}(B \backslash \{y\})$, meaning $\mathbf{y}$ is in B but away from the rest of B. The set of isolated points of B is denoted by I_B.

For a set $B \subseteq \mathbb{R}^m$, its set of isolated points I_B is given by

$$I_B = \{\mathbf{y} \in \mathbb{R}^m : \mathbf{y} \in B \text{ and } \mathbf{y} \operatorname{awf}(B \backslash \{y\})\}. \tag{3.6.5}$$

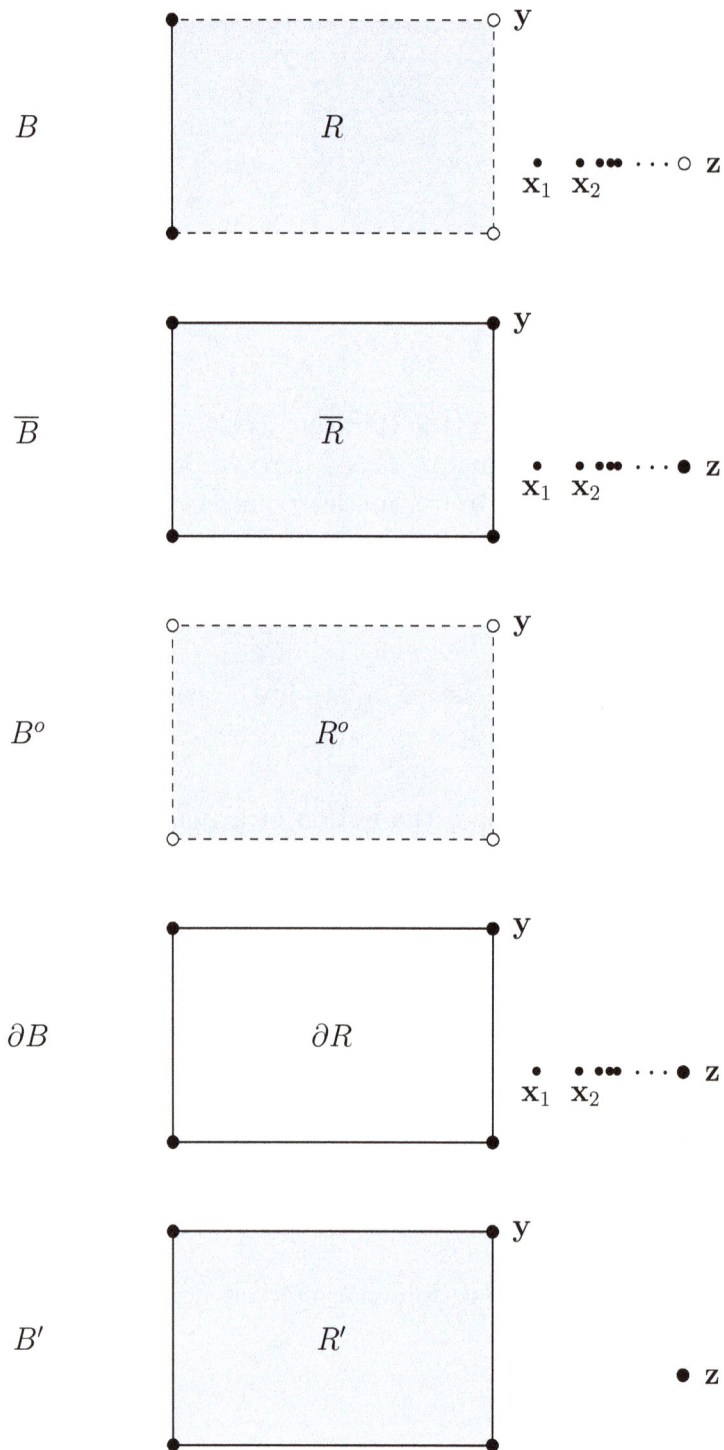

Figure 3.6.3: A set B in the plane $\mathbb{R}^2$ along with its closure $\overline{B}$, interior B^o, boundary ∂B, and accumulation points B'. Compare with Figure 3.6.2. What do the exterior B^e and isolated points I_B look like? See Definitions 3.6.1 through 3.6.8.

The following pair of examples help us explore a bunch of the definitions from this chapter in the context of the plane $\mathbb{R}^2$ and the real line $\mathbb{R}$.

Example 3.6.9: More points arbitrarily close to sets

To get a better idea of what's going on with Definitions 3.6.1 through 3.6.8, consider the subset B of the plane $\mathbb{R}^2$ in Figures 3.6.2 and 3.6.3, and for a subset of the real line $\mathbb{R}$ consider the set E in Figure 3.6.1. We have

$$B = R \cup \{\mathbf{x}_n : n \in \mathbb{N}\} \tag{3.6.6}$$

where R is a rectangle that includes its left side and two of its corners while $\{\mathbf{x}_n : n \in \mathbb{N}\}$ is the range of a convergent sequence $(\mathbf{x}_n)$ whose limit $\mathbf{z}$ is not in B.

The set E in Figure 3.6.1 is an uncountable set of real numbers which is neither an interval nor the range of a sequence, but is the union of two such sets. E also appears in Examples 3.1.12 and 3.2.6 as well as their Figures 3.1.3 and 3.2.3. Here's a summary.

$$
\begin{aligned}
\textit{original set:} \quad & E = [0,1) \cup \{2\} \cup \{3 + (1/n) : n \in \mathbb{N}\} & (3.6.7) \\
\textit{closure:} \quad & \overline{E} = [0,1] \cup \{2\} \cup \{3\} \cup \{3 + (1/n) : n \in \mathbb{N}\} & (3.6.8) \\
\textit{boundary:} \quad & \partial E = \{0,1,2,3\} \cup \{3 + (1/n) : n \in \mathbb{N}\} & (3.6.9) \\
\textit{interior:} \quad & E^o = (0,1) & (3.6.10) \\
\textit{exterior:} \quad & E^e = & (3.6.11)
\end{aligned}
$$

$$(-\infty, 0) \cup (1,2) \cup (2,3) \cup \left[\bigcup_{n=1}^{\infty} \left(3 + \frac{1}{n+1}, 3 + \frac{1}{n} \right) \right] \cup (4, \infty) \tag{3.6.12}$$

$$
\begin{aligned}
\textit{accumulation:} \quad & E' = [0,1] \cup \{3\} & (3.6.13) \\
\textit{isolated:} \quad & I_E = \{2\} \cup \{3 + (1/n) : n \in \mathbb{N}\} & (3.6.14)
\end{aligned}
$$

Theorem 2.3.1 establishes a central result which appears throughout the textbook: A point is arbitrarily close to a set if and only if there is a sequence in the set whose limit is the point. The negation of this statement is also useful: A point is away from a set if and only if no sequence in the set converges to the point.

Together with the definitions found in this section—all of which involve arbitrarily close or away from—we can immediately characterize closure, boundary, interior, exterior, accumulation points, and isolated points in terms of sequences and where they converge. See the definitions for *tails* (Definition 2.1.7), *eventually* (Definition 2.2.13) as well as Definitions 3.1.1 and 3.6.1 through 3.6.8. The characterizations are summarized in the following theorem.

Theorem 3.6.10: Limits versus sets

Let $\mathbf{y} \in \mathbb{R}^m$ and $B \subseteq \mathbb{R}^m$. Then:

(i) $\mathbf{y} \in \overline{B}$ if and only if a sequence in B converges to $\mathbf{y}$.

(ii) $\mathbf{y} \in \partial B$ if and only if a sequence in B and a sequence in $\mathbb{R}^m \backslash B$ converge to $\mathbf{y}$.

(iii) $\mathbf{y} \in B^o$ if and only if the sequences that converge to $\mathbf{y}$ have a tail contained in B (thus, they are eventually contained in B).

(iv) $\mathbf{y} \in B^e$ if and only if no sequence in B converges to $\mathbf{y}$.

(v) $\mathbf{y} \in B'$ if and only if some sequence in B excludes $\mathbf{y}$ and converges to $\mathbf{y}$.

(vi) $\mathbf{y} \in I_B$ if and only if every sequence in B that converges to $\mathbf{y}$ has a constant tail defined by the term $\mathbf{y}$ (thus, these sequences are eventually constant).

Totally disconnected sets exhibit a particular notion of sparseness.

Definition 3.6.11: Totally disconnected

A set $S \subseteq \mathbb{R}^m$ is *totally disconnected* if all of its connected subsets are singletons.

Example 3.6.12: Totally disconnected sequences

Consider the positive integers $\mathbb{N}$ and their reciprocals S given by

$$S = \frac{1}{\mathbb{N}} = \left\{ \frac{1}{n} : n \in \mathbb{N} \right\} = \left\{ 1, \frac{1}{2}, \frac{1}{3}, \dots \right\}. \tag{3.6.15}$$

$\mathbb{N}$ and S are totally disconnected. Both sets are countable, so neither contains a nontrivial open interval, which are uncountable. (See Exercise 2.8.8.) Hence, the only connected subsets of $\mathbb{N}$ and S are singletons.

The next and final definition in this section provides a characterization of sets which provide a nice approximations of the points in a given set.

Definition 3.6.13: Dense

Given two sets $A, B \subseteq \mathbb{R}^m$, we say A is *dense with respect to* B if every point in B is arbitrarily close to A. In this case, we have $B \subseteq \overline{A}$. If we also have $A \subseteq B$, then we say A is *dense in* B.

Remark 3.6.14: Density results

A nice way to interpret the meaning of density as in Definition 3.6.13 is as follows: Thinking of B as a set of points we'd like to approximate and A is dense with respect to B, then we can use the points in A as approximations for the points in B and make the approximations as close as we like.

We have already seen statements involving density. Here's a summary involving rational and irrational real numbers:

(i) $\mathbb{Q}$ is dense in $\mathbb{R}$: Every real number is approximately a rational number. See Theorem 1.4.10.

(ii) $\mathbb{R}\backslash\mathbb{Q}$ is dense in $\mathbb{R}$: Every real number is approximately an irrational number. See Corollary 1.4.13.

(iii) $\mathbb{Q}$ is dense with respect to $\mathbb{R}\backslash\mathbb{Q}$: Every rational number is approximately irrational. This follows immediately from (ii).

(iv) $\mathbb{R}\backslash\mathbb{Q}$ is dense with respect to $\mathbb{Q}$: Every irrational number is approximately rational. This follows immediately from (i).

The next chapter explores the ways in which functions *preserve closeness*. For instance, what kind of function will take a point arbitrarily close to a subset of the domain and preserve that closeness in their images? This type of question leads to parallel approaches for continuity and limits of functions.

Exercises

3.6.1. Give an example to show that the interior of a connected set is not necessarily connected.

3.6.2. Give examples of nonempty sets with the following properties:

(i) A set A where both $A^o = \varnothing$ and $A^e = \varnothing$.

(ii) A set B where $B' = \overline{B}$.

(iii) A set C where C is infinite, $C = I_C$, and $C' = \varnothing$.

(iv) A set D which is open and yet $\overline{D} \subseteq D$.

(v) A countable set E whose boundary ∂E is uncountable.

3.6.3. Prove for every $B \subseteq \mathbb{R}^m$ we have the following:

(i) $\overline{B} = B \cup B'$.

(ii) $I_B = B\backslash B'$.

(iii) $\partial B = \overline{B}\backslash B^o$.

3.6.4. Prove the interior of a given set in $\mathbb{R}^m$ is the largest open set contained in the given set in the following sense: Given a set $S \subseteq \mathbb{R}^m$, every open set which is contained in S is also contained in the interior S^o.

3.6.5. Prove the closure of a given set in $\mathbb{R}^m$ is the smallest closed set containing the given set in the following sense: Given a set $T \subseteq \mathbb{R}^m$, every closed set that contains T is also contains the closure $\overline{T}$.

3.6.6. Given any set $B \subseteq \mathbb{R}^m$, prove the trio of sets B^o, B^e, and ∂B are pairwise disjoint and we have $B^o \cup B^e \cup \partial B = \mathbb{R}^m$.

3.6.7. Prove $\mathbb{Q}$ and $\mathbb{R} \backslash \mathbb{Q}$ are totally disconnected.

3.6.8. Prove nonempty countable subsets of $\mathbb{R}^m$ are totally disconnected. In contrast, find an example of a countably infinite subset of the real line $\mathbb{R}$ whose closure is connected.

3.6.9. Recall the Cantor set C from Exercises 2.8.9 and 3.5.4. Prove the Cantor set is totally disconnected.

Chapter 4

Continuity

How do functions transform points, sets, and sequences? If a given sequence has one of the properties explored in Chapter 2 like convergence or boundedness, does its image under a function have the same property? If a set is closed, connected, or compact, or if it has another topological property from Chapter 3, can we say the same about its images?

Our focus will be on functions that map one Euclidean space to another, often the real line to the real line. And of course, arbitrarily close and away from provide fruitful perspectives to build on.

Notions such as continuity, limits, convergence, and codas defined in this chapter allow us to explore to the structure and behavior of functions. The first section builds on the formal definitions for functions and images covered in Section 1.2 by delving into some of their basic structures and adds a formal definition and various descriptions of *preimages*. (See Definition 4.1.7.)

4.1   Functions, images, and preimages

The following terminology and notation summarizes a few of the definitions involving functions and sequences given in Section 1.2 and elsewhere. Moving forward, the term *image* is used as a catch-all to refer to the outputs we get when plugging various inputs into a function.

Notation 4.1.1: Images

The following terminology and notation may be used whenever $f : A \to B$, summarizing a few of the definitions given in Section 1.2 and elsewhere:

(i) Images of points are points: When $f(x) = y$, we say x is an *input* in the domain A and y is its *output* or *image* in the range $f(A)$.

(ii) Images of sequences are sequences: For a sequence (x_n) in the domain A, its *output* or *image* is the sequence $(f(x_n))$ in the range $f(A)$.

(iii) Images of sets are sets: For a subset S of the domain A, its *image* $f(S)$ is the subset

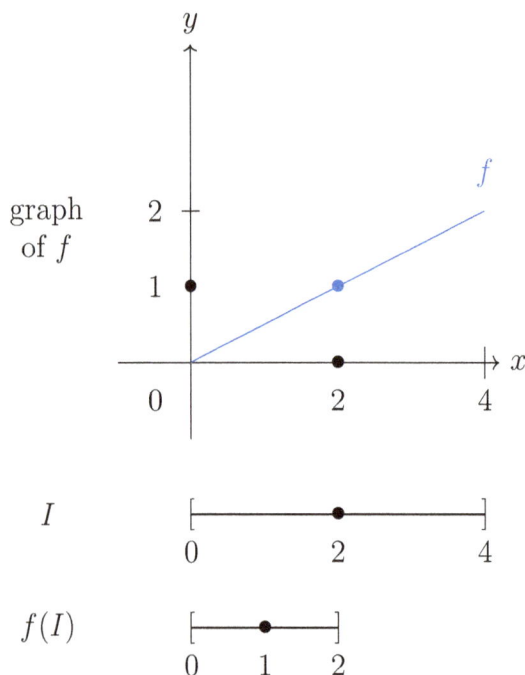

Figure 4.1.1: The function $f : \mathbb{R} \to \mathbb{R}$ in Example 4.1.2 given by $f(x) = x/2$ transforms the connected interval $I = [0, 4]$ into the connected interval $f(I) = [0, 2]$.

of the range $f(A)$ given by

$$f(S) = \{y \in B : f(x) = y \text{ for some } x \in S\} \subseteq f(A). \qquad (4.1.1)$$

For the definitions of function, domain, range, and image, see Definitions 1.2.12, 1.2.13, and 1.2.14. A formal definition for sequences is provided by Definition 2.1.1.

The examples in this section focus on images and preimages of points and sets. Sequences are explored later since we have plenty explore for now and sequences will add an interesting perspective.

For a first set of examples to get the ball rolling and revisit throughout the chapter, we consider a quartet of functions from the real line to the real line. How do these functions transform sets? What do their images look like?

Example 4.1.2: A line with slope 1/2

Consider the function $f : \mathbb{R} \to \mathbb{R}$ given by given by

$$f(x) = x/2. \qquad (4.1.2)$$

See Figure 4.1.1 which features the connected and compact interval $I = [0, 4]$ in the domain along with its image $f(I) = [0, 2]$ which is also connected and compact. Both $I = [0, 4]$

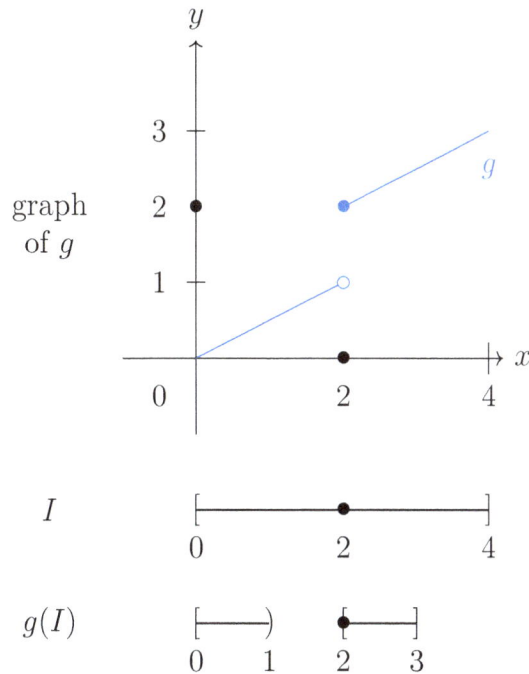

Figure 4.1.2: The function $g : \mathbb{R} \to \mathbb{R}$ in Example 4.1.3 transforms the connected interval $I = [0, 4]$ into the disconnected image $g([0, 4]) = [0, 1) \cup [2, 3]$.

and $f(I) = [0, 2]$ are intervals, so they are connected by Theorem 3.3.17. Both $I = [0, 4]$ and $f(I) = [0, 2]$ are closed and bounded, so by the Heine-Borel Theorem 3.5.1, both are compact.

Based on my intuition from calculus[a], it looks to me like f is continuous at $c = 2$. Actually, it looks like f is continuous at every $c \in I = [0, 4]$, but for now my focus is on $c = 2$ specifically.

_____

[a]From way back at the start of my senior year of high school in 1996.

Example 4.1.3: A piecewise defined function

Consider the function $g : \mathbb{R} \to \mathbb{R}$ given by

$$g(x) = \begin{cases} x/2, & \text{if } x < 2, \\ 1 + (x/2), & \text{if } x \geq 2. \end{cases} \qquad (4.1.3)$$

See Figure 4.1.2 which features the connected and compact interval $I = [0, 4]$ in the domain along with its image $g([0, 4]) = [0, 1) \cup [2, 3]$ which is neither connected nor compact. The sets $[0, 1)$ and $[2, 3]$ form a separation of $g([0, 4])$, so $g([0, 4])$ is disconnected. The real number 1 is arbitrarily close to but not in $g([0, 4])$, so $g([0, 4])$ is not closed.

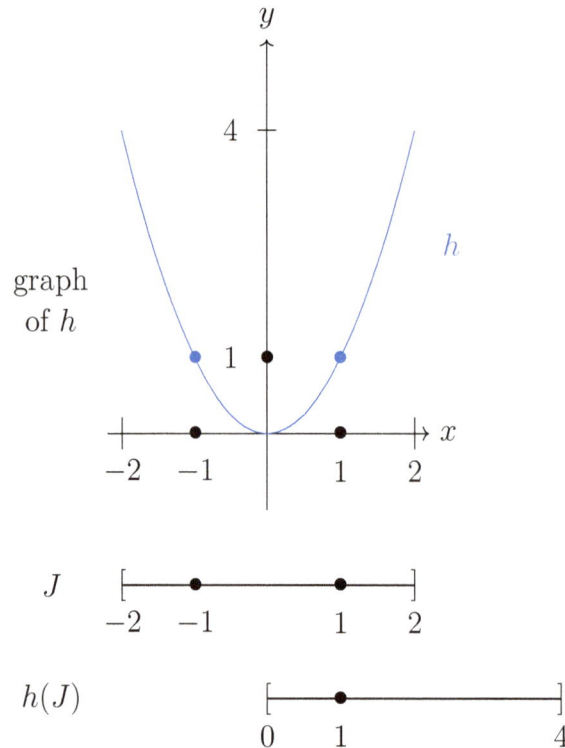

Figure 4.1.3: The graph of the square function $h : \mathbb{R} \to \mathbb{R}$ given by $h(x) = x^2$ in Example 4.1.4 transforms the connected interval $J = [-2, 2]$ into the connected interval $h(J) = [0, 4]$. The two distinct real numbers -1 and 1 in the domain map to the same image 1 in the range.

My intuition tells me g is discontinuous at $c = 2$. The graph of g (in blue) has a definitive break, a jump in the height from 2 to 3 at $c = 2$. Continuous functions should not create jumps like this, right? But how can we capture that behavior with our definitions?

Example 4.1.4: The square function

Figure 4.1.3 features the graph of the function $h : \mathbb{R} \to \mathbb{R}$ given by

$$h(x) = x^2. \tag{4.1.4}$$

Note $h(-1) = h(1) = 1$ and the range is $h([-2, 2]) = [0, 4]$. Both $J = [-2, 2]$ and $h(J) = [0, 4]$ are intervals, so they are connected by Theorem 3.3.17. Both $J = [-2, 2]$ and $h(J) = [0, 4]$ are closed and bounded, so by the Heine-Borel Theorem 3.5.1, both are compact. Also, h is not one-to-one since $h(-1) = h(1) = 1$. My intuition tells me h is continuous at every point in its domain.

Example 4.1.5: Images of sine

Consider the trigonometric function $s : \mathbb{R} \to \mathbb{R}$ and subsets of the domain T and U where

$$s(x) = \sin x, \quad T = \{\pi z : z \in \mathbb{Z}\}, \quad \text{and} \quad U = (0, 2\pi). \tag{4.1.5}$$

A couple things to note:

(i) The set $T = \{\pi z : z \in \mathbb{Z}\}$ is countable since $\mathbb{Z}$ is countable by Example 2.8.3 and its image is a singleton

$$s(T) = \{\sin \pi z : z \in \mathbb{Z}\} = \{0\}. \tag{4.1.6}$$

(ii) The interval $U = (0, 2\pi)$ is open but its image $s(U) = [-1, 1]$ is not open since 1 is in $s(U)$ but no neighborhood of 1 is contained in $s(U)$.

To get a glimpse into how functions between Euclidean spaces behave, consider the following example which transforms a disk in the plane $\mathbb{R}^2$ to a disconnected subset of the real line $\mathbb{R}$.

Example 4.1.6: A function from the plane to the real line

Consider the function $w : \mathbb{R}^2 \to \mathbb{R}$ given by

$$w(\mathbf{x}) = \begin{cases} \|\mathbf{x}\|, & \text{if } 0 \leq \|\mathbf{x}\| < 1, \\ 2, & \text{if } \|\mathbf{x}\| \geq 1. \end{cases} \tag{4.1.7}$$

The function w maps every point in the open unit disk $V_1(\mathbf{0})$ to its magnitude (also its length from the origin) and maps every point outside of $V_1(\mathbf{0})$ to 2.

In Figure 4.1.4 featuring the function w, the standard basis vectors $\mathbf{e}_1$ and $\mathbf{e}_2$ are plotted as points (•) instead of arrows. Since neither $\mathbf{e}_1$ nor $\mathbf{e}_2$ is in $V_1(\mathbf{0})$, we have $w(\mathbf{e}_1) = w(\mathbf{e}_2) = 2$.

In the domain, the unit circle C contains the unit vectors $\mathbf{e}_1$ and $\mathbf{e}_2$ which are arbitrarily close to open unit disk $V_1(\mathbf{0})$. However, in the range the images $w(\mathbf{e}_1) = w(\mathbf{e}_2) = 2$ and $w(V_1(\mathbf{0})) = [0, 1)$ are away from each other. This is dichotomy is true of every point in C and their images since

$$w(V_1(\mathbf{0})) = [0, 1) \quad \text{while} \quad w(C) = \{2\}. \tag{4.1.8}$$

Does this mean w is discontinuous?

Preimages also play an important role in the study of functions, especially when it comes to continuity. Preimages are also called *inverse images*.

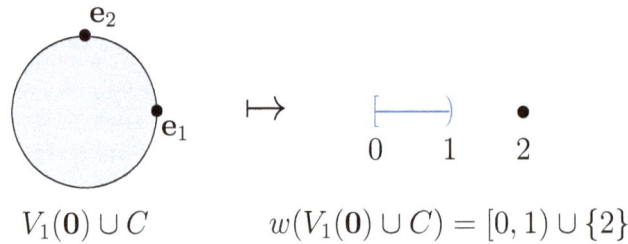

$$V_1(\mathbf{0}) \cup C \qquad\qquad w(V_1(\mathbf{0}) \cup C) = [0,1) \cup \{2\}$$

Figure 4.1.4: In Example 4.1.6, we have $\mathbf{e}_1$ and $\mathbf{e}_2$ are in the unit circle C, but neither is in the open unit disk $V_1(\mathbf{0})$. However, we have both $\mathbf{e}_1 \operatorname{acl} V_1(\mathbf{0})$ and $\mathbf{e}_2 \operatorname{acl} V_1(\mathbf{0})$ in the domain. But the function w transforms $\mathbf{e}_1$, $\mathbf{e}_2$, and $V_1(\mathbf{0})$ into images $w(\mathbf{e}_1) = w(\mathbf{e}_1) = 2$ and $w(V_1(\mathbf{0})) = [0,1)$ which are away from each other in the range.

Definition 4.1.7: Preimage

Given a function $f : A \to B$ and a subset S of the codomain B (so $S \subseteq B$), the *preimage* of S is the subset of the domain whose image is S. The preimage of S is denoted by $f^{-1}(S)$ and given by

$$f^{-1}(S) = \{x \in A : f(x) \in S \subseteq B\}. \tag{4.1.9}$$

Remark 4.1.8: Terminology of preimages

The word preimage is used in multiple contexts, namely points, sequences, and sets. This is similar to way we use the word image as in Notation 4.1.1, but we are not as definitive.

For instance, in Example 4.1.11 where $h(x) = x^2$ we can ask: What is $h^{-1}(5)$, the preimage of the real number $\sqrt{5}$? Depending on the conversation, the answers "$\sqrt{5}$", "$-\sqrt{5}$", or even "$\left\{\sqrt{5}, -\sqrt{5}\right\}$" could make sense. Of course, the first two are points and the second is a set. Sometimes, the preimage of a point could be a sequence.

Instead of providing a formal definition for the preimage of a point or a sequence, situations are handled more loosely in general but made specific by the context at hand when they come up. However, in agreement with Definition 4.1.7, preimages of sets are always sets.

Example 4.1.9: Preimages of a line

Consider the function $f : \mathbb{R} \to \mathbb{R}$ from Example 4.1.2 and subsets of the codomain E and G where

$$f(x) = x/2, \quad E = \{1\}, \quad \text{and} \quad G = (0,2). \tag{4.1.10}$$

Every real number y in the codomain $\mathbb{R}$ is the image of exactly one real number $x_y = 2y$ in the domain since

$$f(x_y) = x_y/2 = (2y)/2 = y. \tag{4.1.11}$$

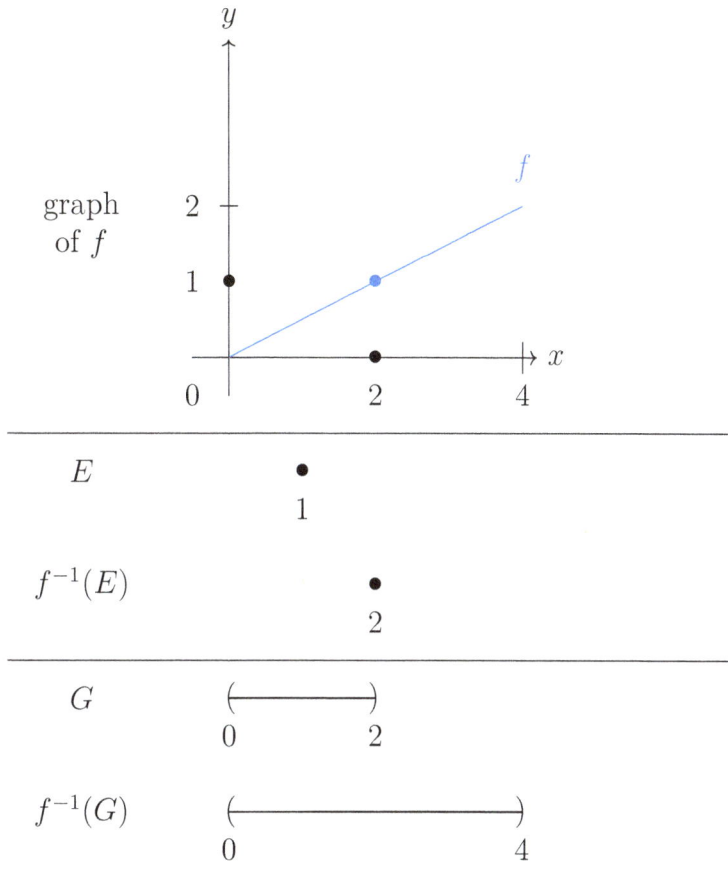

Figure 4.1.5: The preimage of the singleton $E = \{1\}$ is the singleton $f^{-1}(E) = \{2\}$ while the preimage of the open set $G = (0,2)$ is the open set $f^{-1}(G) = (0,4)$. See Example 4.1.9.

This shows f is onto and the preimages of E and G are given

$$f^{-1}(E) = \{2\}, \quad \text{and} \quad f^{-1}(G) = (0,4). \tag{4.1.12}$$

See Figure 4.1.5. A few things to note:

(i) $E = \{1\}$ is a closed singleton in the codomain and its preimage $f^{-1}(E) = \{2\}$ is a closed singleton in the domain.

(ii) $G = (0,2)$ is an open interval in the domain and $f^{-1}(G) = (0,4)$ is an open interval in the domain.

Example 4.1.10: Preimages of a piecewise defined function

Consider the function $g : \mathbb{R} \to \mathbb{R}$ from Example 4.1.3 and the subset of the codomain H

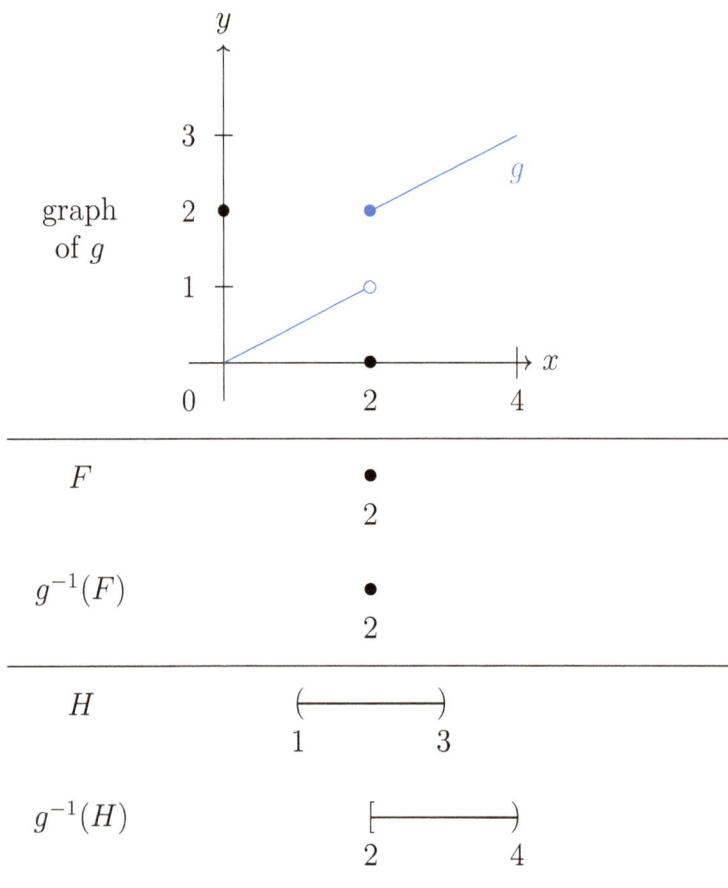

Figure 4.1.6: The preimage of the singleton $F = \{2\}$ is the singleton $g^{-1}(F) = \{2\}$. The set $H = (1,3)$ is open in the codomain, but its preimage $g^{-1}(H) = [2,4)$ is not open in the domain. See Example 4.1.10.

where

$$g(x) = \begin{cases} x/2, & \text{if } x < 2, \\ 1 + (x/2), & \text{if } x \geq 2, \end{cases} \quad \text{and} \quad H = (1,3). \tag{4.1.13}$$

See Figure 4.1.6. The interval $H = (1,3)$ is open in the codomain, but its preimage $g^{-1}(H)$ is not open in the domain: We have

$$g^{-1}(H) = [2,4) \tag{4.1.14}$$

and 2 is in $g^{-1}(H)$ but no neighborhood of 2 is contained in $g^{-1}(H)$.
Also, in the codomain $\mathbb{R}$, no real number in the interval $[1,2)$ is an output of g, so g is not onto.

Example 4.1.11: Preimages of the square function

Consider the function $h : \mathbb{R} \to \mathbb{R}$ and subsets of the codomain S, T, and U where

$$h(x) = x^2, \quad S = \{1\}, \quad T = [-4, -1], \quad \text{and} \quad U = (0, 4). \tag{4.1.15}$$

Their preimages are given by

$$h^{-1}(S) = \{-1, 1\}, \quad h^{-1}(T) = \varnothing, \quad \text{and} \quad h^{-1}(U) = (-2, 2). \tag{4.1.16}$$

See Figure 4.1.7. A few things to note:

(i) $S = \{1\}$ is a singleton but its preimage $h^{-1}(S) = \{-1, 1\}$ is not. So, h is not one-to-one.

(ii) Since every real number in $T = [-4, -1]$ is negative but all the outputs of $h(x) = x^2$ are nonnegative, no real number in the domain maps into T. So, $h^{-1}(T) = \varnothing$ and h is not onto.

(iii) $U = (0, 4)$ is open in the codomain and its preimage $h^{-1}(U) = (-2, 0) \cup (0, 2)$ is open in the domain.

Example 4.1.12: Preimages of sine

Consider the function $s : \mathbb{R} \to \mathbb{R}$ and subsets of the codomain A, B, and C where

$$s(x) = \sin x, \quad A = \{1\}, \quad B = [2, \infty), \quad \text{and} \quad C = \left(-\frac{1}{2}, \frac{1}{2}\right). \tag{4.1.17}$$

A few things to note:

(i) The preimage of $A = \{1\}$ is the range of a sequence since

$$s^{-1}(\{1\}) = \left\{\frac{\pi}{2} + 2\pi z : z \in \mathbb{Z}\right\} \tag{4.1.18}$$

and $\mathbb{Z}$ is countable by Example 2.8.3.

(ii) We have $s^{-1}(B) = \varnothing$ since every real number in the interval $B = [2, \infty)$ is at least 2 but the outputs $s(x) = \sin x$ are bounded by 1: For every $x \in \mathbb{R}$ we have

$$|s(x)| = |\sin x| \le 1 \quad \Longleftrightarrow \quad -1 \le \sin x \le 1. \tag{4.1.19}$$

(iii) The interval C and its preimage $s^{-1}(C)$ are open since open intervals are open sets and

$$s^{-1}(C) = s^{-1}\left(\left(-\frac{1}{2}, \frac{1}{2}\right)\right) \tag{4.1.20}$$

$$= \ldots \cup \left(-\frac{\pi}{6}, \frac{\pi}{6}\right) \cup \left(\frac{5\pi}{6}, \frac{7\pi}{6}\right) \cup \left(\frac{11\pi}{6}, \frac{13\pi}{6}\right) \ldots \tag{4.1.21}$$

$$= \bigcup_{z \in \mathbb{Z}} \left(\pi z - \frac{\pi}{6}, \pi z + \frac{\pi}{6}\right). \tag{4.1.22}$$

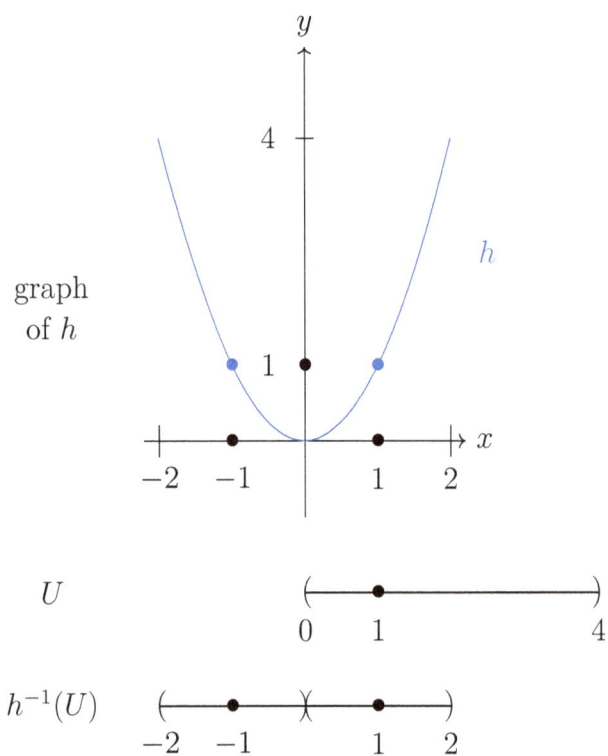

Figure 4.1.7: The set $S = \{1\}$ in the codomain is a singleton but its preimage $h^{-1}(S) = \{-1, 1\}$ in the domain is not. The set $U = (0, 4)$ is open in the codomain and its preimage $h^{-1}(U) = (-2, 0) \cup (0, 2)$ is open in the domain. See Example 4.1.11.

This section concludes with the relationship between preimages, unions, and intersections. Unions and images have a different relationship with images, as explored in the exercises.

Lemma 4.1.13: Preimages of unions and intersections

Suppose $D \subseteq \mathbb{R}^k$ and $f : D \to \mathbb{R}^m$. Then the preimages of f respect unions and intersections, as follows: For any $A, B \subseteq \mathbb{R}^m$ we have

(i) $f^{-1}(A \cup B) = f^{-1}(A) \cup f^{-1}(B)$;   and

(ii) $f^{-1}(A \cap B) = f^{-1}(A) \cap f^{-1}(B)$.

Note that A and B are subsets of the codomain $\mathbb{R}^m$.

Scratch Work 4.1.14: Follow from the definitions

Both parts of Lemma 4.1.13 follow from the definitons of preimages, unions, and intersections. Basically, points in the union of two sets in the codomain come from at least one of their preimages, while points in the intersection of two sets in the codomain must be in both preimages. And vice versa.

Proof of Lemma 4.1.13. Throughout the proof, suppose we have $D \subseteq \mathbb{R}^k, f : D \to \mathbb{R}^m$, and $A, B \subseteq \mathbb{R}^m$. By the definitions of union and intersection (Definition 1.2.5) and the definition of preimage (Definition 4.1.7), we have

$$f^{-1}(A \cup B) = \{\mathbf{x} \in D : f(\mathbf{x}) \in A \text{ or } f(\mathbf{x}) \in B\} \tag{4.1.23}$$
$$= \{\mathbf{x} \in D : f(\mathbf{x}) \in A\} \cup \{\mathbf{x} \in D : f(\mathbf{x}) \in B\} \tag{4.1.24}$$
$$= f^{-1}(A) \cup f^{-1}(B) \tag{4.1.25}$$

as well as

$$f^{-1}(A \cap B) = \{\mathbf{x} \in D : f(\mathbf{x}) \in A \text{ and } f(\mathbf{x}) \in B\} \tag{4.1.26}$$
$$= \{\mathbf{x} \in D : f(\mathbf{x}) \in A\} \cap \{\mathbf{x} \in D : f(\mathbf{x}) \in B\} \tag{4.1.27}$$
$$= f^{-1}(A) \cap f^{-1}(B). \tag{4.1.28}$$

$\square$

The next section develops my personal take on *continuity*. Continuity is when a function transforms a point and set which are arbitrarily close into images which are arbitrarily close. This version of continuity is called the *preservation of closeness*.

Exercises

4.1.1. Give an example of a function $g : U \to \mathbb{R}$ where $U \subseteq \mathbb{R}$ is open but not closed and yet its image $g(U)$ is closed.

4.1.2. Give an example of a function $t : F \to \mathbb{R}$ where $F \subseteq \mathbb{R}$ is closed but not open and yet its image $t(F)$ is open.

4.1.3. Give an example of a function $q : B \to \mathbb{R}$ where $B \subseteq \mathbb{R}$ is a bounded interval and yet its image $q(B)$ is an unbounded interval.

4.1.4. Give an example of a function $v : S \to \mathbb{R}$ where $S \subseteq \mathbb{R}$ and (x_n) is a convergent sequence in I whose image $(v(x_n))$ diverges.

4.1.5. Suppose $D \subseteq \mathbb{R}^k$, $f : D \to \mathbb{R}^m$, and $S, T \subseteq \mathbb{R}^m$. The image of a union is the union of images, but something else holds for intersections.

  (i) Prove $f(S \cup T) = f(S) \cup f(T)$.

  (ii) Prove $f(S \cap T) \subseteq f(S) \cap f(T)$.

  (iii) Find an example where $f(S \cap T) \subsetneq f(S) \cap f(T)$. (That is, the containment in (ii) can be proper.)

4.1.6. Consider the square function $h : \mathbb{R} \to \mathbb{R}$ given by $h(x) = x^2$. Fix $c \in \mathbb{R}$ and let $\delta > 0$.

  (i) Show that the image of the δ-neighborhood $h(V_\delta(c)) = h((c - \delta, c + \delta))$ is an interval.

  (ii) Find a value of $c_0 \in \mathbb{R}$ such that $h(V_\delta(c_0)) = h((c_0 - \delta, c_0 + \delta))$ is not open, regardless of the value of δ.

4.1.7. Consider the square function $h : \mathbb{R} \to \mathbb{R}$ given by $h(x) = x^2$. Fix $c \in \mathbb{R}$ and let $\varepsilon > 0$. Prove that the inverse image $h^{-1}(V_\varepsilon(h(c)) = h^{-1}((h(c) - \varepsilon, h(c) + \varepsilon))$ is an open interval.

4.2   Preserving closeness

Continuous functions preserve many properties by transforming sets and sequences while keeping much of their structures intact. For instance, continuous functions leave intervals unbroken since their images are intervals. But how does the formal definition for continuity (Definition 4.3.2) capture an ability to preserve properties? And how can we motivate this vital but notoriously difficult definition?

This section looks at how functions transform relationships between points and sets in the domain using arbitrarily close and away from (Definitions 1.5.1 and 1.5.11).

Let's revisit the functions f and g from Section 4.1 in and add the function v in Example 4.2.4 to help us along. For f, g, and v, let's consider a common collection of intervals and a common real number and see how they're transformed by these functions.

> **Example 4.2.1: A quarter of sets**
>
> Consider the following quartet of intervals, each of which is arbitrarily close to $c = 2$. See Figure 4.2.1.

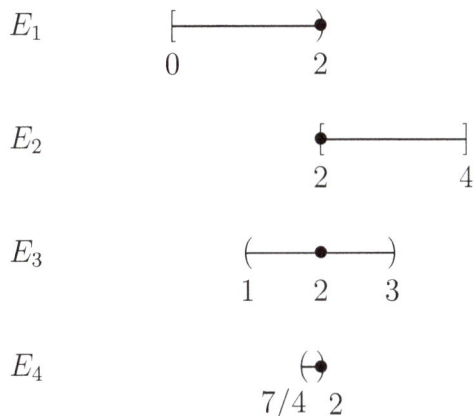

Figure 4.2.1: Various intervals in the real line $\mathbb{R}$ from Example 4.2.1. Each is connected, arbitrarily close to $c = 2$, and used throughout this section.

(i) $E_1 = [0, 2)$

(iii) $E_3 = (1, 3)$

(ii) $E_2 = [2, 4]$

(iv) $E_4 = (7/4, 2)$

Recall that for a distance $\delta > 0$, the δ-neighborhood of c in the real line $\mathbb{R}$ is the open interval

$$V_\delta(c) = (c - \delta, c + \delta). \tag{4.2.1}$$

For instance, E_3 is a δ-neighborhood of $c = 2$ with $\delta = 1$. We have

$$V_1(2) = (2 - 1, 2 + 1) = (1, 3) = E_3. \tag{4.2.2}$$

My use of the variable δ instead of ε is deeply intentional, by the way.

Example 4.2.2: Images under a function defining a line

For the function $f : \mathbb{R} \to \mathbb{R}$ from Examples 4.1.2 and 4.1.9 given by

$$f(x) = x/2, \tag{4.2.3}$$

we have the images

(i) $f(2) = 1$,

(ii) $f(E_1) = f([0, 2)) = [0, 1)$,

(iii) $f(E_2) = f([2, 4]) = [1, 2]$,

(iv) $f(E_3) = f((1, 3)) = (1/2, 3/2)$, and

(v) $f(E_4) = f((7/4, 2)) = (7/8, 1)$.

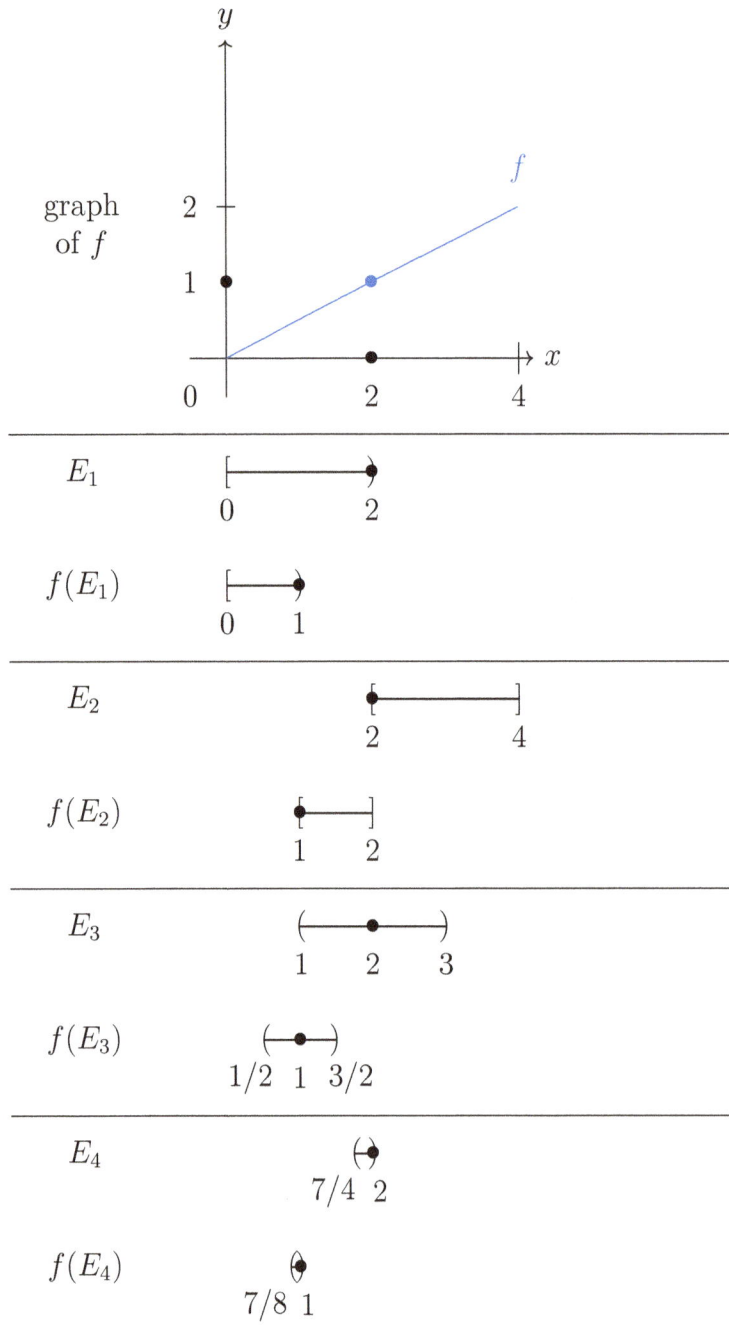

Figure 4.2.2: Images under f of the intervals from Example 4.2.1. Each interval in the domain is connected and arbitrarily close to 2, while their images are connected and arbitrarily close to $f(2) = 1$. See Example 4.2.2.

See Figure 4.2.2. In the domain, the point $c = 2$ is arbitrarily close to each of the E_k intervals. The function f maps the point 2 and the sets E_k to a point $f(2) = 1$ and sets $f(E_k)$ in the range in a way that *preserves closeness*. That is, at least for $k = 1, 2, 3, 4$, we have f ensures

$$2 \operatorname{acl} E_k \quad \Longrightarrow \quad f(2) \operatorname{acl} f(E_k). \tag{4.2.4}$$

What about g?

Example 4.2.3: Images under a split line

For the function $g : \mathbb{R} \to \mathbb{R}$ from Examples 4.1.3 and 4.1.10 given by

$$g(x) = \begin{cases} x/2, & \text{if } x < 2, \\ 1 + (x/2), & \text{if } x \geq 2, \end{cases} \tag{4.2.5}$$

we have the images $g(2) = 2$,

(i) $g(2) = 2$,

(ii) $g(E_1) = g([0, 2)) = [0, 1)$,

(iii) $g(E_2) = g([2, 4]) = [2, 3]$,

(iv) $g(E_3) = g((1, 3)) = (1/2, 1) \cup [2, 5/2)$, and

(v) $g(E_4) = g((7/4, 2)) = (7/8, 1)$.

See Figure 4.2.3. What do you notice?

Let's investigate the images of the E_k one at a time. Even though 2 is arbitrarily close to E_1 in the domain, in the range $g(2) = 2$ is away from $g(E_1) = [0, 1)$. That is,

$$2 \operatorname{acl} E_1 \quad \text{but} \quad g(2) \operatorname{awf} g(E_1). \tag{4.2.6}$$

So, g does not preserve the closeness of 2 and E_1 like f does.

On other hand, $g(2) = 2$ is in both $g(E_2)$ and $g(E_3)$, so $g(2)$ is arbitrarily close to both of these images by Lemma 1.5.4. However, in the case of $g(E_3)$, g did not preserve the connectedness of the interval $E_3 = (1, 3)$ since $g(E_3) = (1/2, 1) \cup [2, 5/2)$ is disconnected (see Definition 3.3.12).

The case for E_4 is like E_1. We have

$$2 \operatorname{acl} E_4 \quad \text{but} \quad g(2) \operatorname{awf} g(E_4). \tag{4.2.7}$$

We can prove $g(2) \operatorname{awf} g(E_1)$ and $g(2) \operatorname{awf} g(E_4)$ simultaneously using the definition of *away from* (Definition 1.5.11).

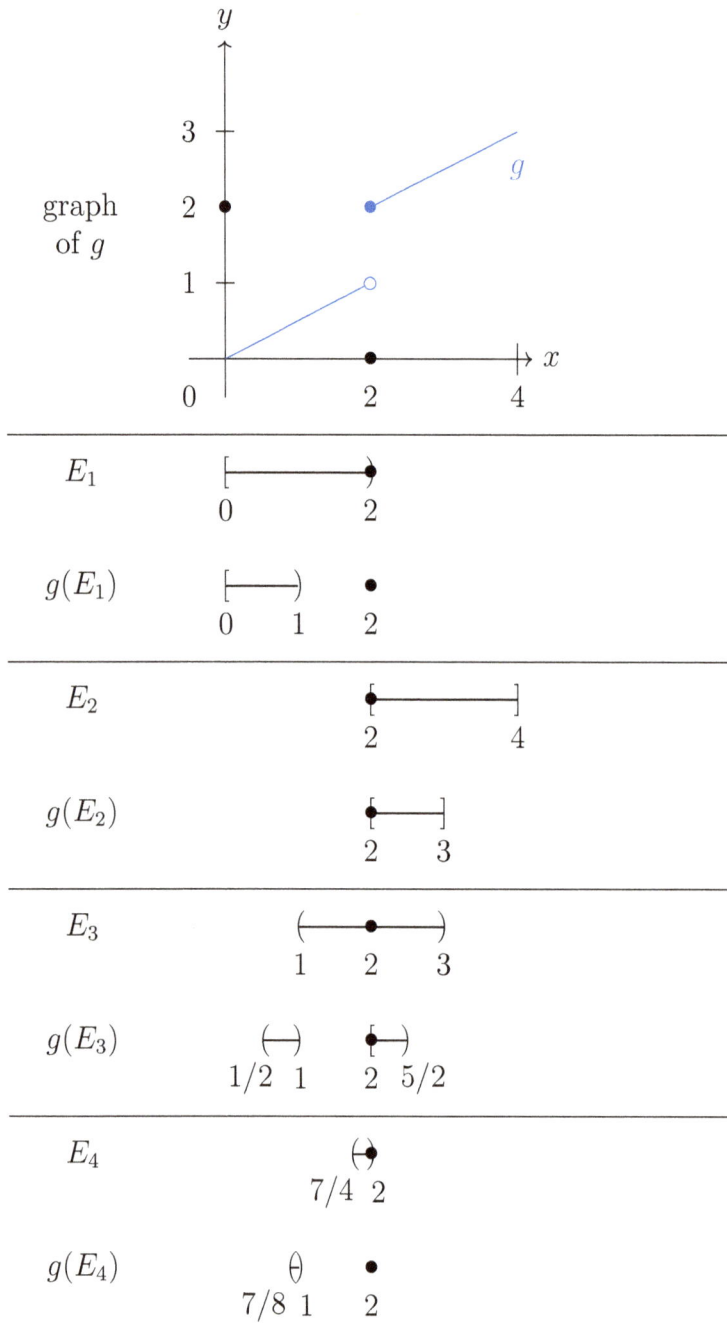

Figure 4.2.3: Images under g of the intervals from Example 4.2.1. Each interval in the domain is connected and arbitrarily close to 2, but their images have different relationships with $g(2) = 2$. See Example 4.2.3.

Proof for Example 4.2.3. Every output $g(x)$ in either $g(E_1)$ or $g(E_4)$ satisfies $g(x) < 1 \leq 2$. Hence, when $x \in E_1$ or $x \in E_4$, by the reverse triangle inequality (1.2.37) we have

$$|g(x) - g(2)| = |g(x) - 2| = 2 - g(x) > 1. \tag{4.2.8}$$

Hence, $g(2) \operatorname{awf} g(E_1)$ and $g(2) \operatorname{awf} g(E_4)$. □

The next example behaves somewhat differently at $c = 2$.

Example 4.2.4: Images under another piece-wise defined function

Consider the unbounded function $v : \mathbb{R} \to \mathbb{R}$ given by

$$v(x) = \begin{cases} \dfrac{1}{2-x}, & \text{if } x < 2, \\[2mm] 1 + \dfrac{x}{2}, & \text{if } x \geq 2. \end{cases} \tag{4.2.9}$$

For the function v we have images $v(2) = 2$ and

(i) $v(E_1) = v([0, 2)) = [1/2, \infty)$,

(ii) $v(E_2) = v([2, 4]) = [2, 3]$,

(iii) $v(E_3) = v((1, 3)) = (1, \infty)$, and

(iv) $v(E_4) = v((7/4, 2)) = (4, \infty)$.

See Figure 4.2.4. What do you notice this time?

It looks like v preserves the connectedness of all of the E_k intervals since all of the images $v(E_k)$ are intervals, too. (See Theorem 3.3.17.)

But the images $v(E_4) = (4, \infty)$ and $v(2) = 2$ in Figure 4.2.4 show us v does not preserve the closeness between $c = 2$ and $E_4 = (7/2, 2)$. We have

$$2 \operatorname{acl} E_4 \quad \text{but} \quad v(2) \operatorname{awf} v(E_4). \tag{4.2.10}$$

Once again, we can prove $v(2) \operatorname{awf} v(E_4)$ using the definition of *away from* (Definition 1.5.1).

Proof for Example 4.2.4. Every output $v(x)$ in $v(E_4)$ satisfies $v(x) > 4 > 2$. Hence, when $x \in E_4$ we have

$$|v(x) - v(2)| = |v(x) - 2| = v(x) - 2 > 4 - 2 = 2. \tag{4.2.11}$$

Therefore, $v(2) \operatorname{awf} v(E_4)$. □

But the function f is different. Example 4.2.2 and Figure 4.2.2 suggest f preserves the closeness at $c = 2$ to not only the intervals E_1, E_2, E_3, and E_4, but *any* set in the real line that's arbitrarily close to $c = 2$.

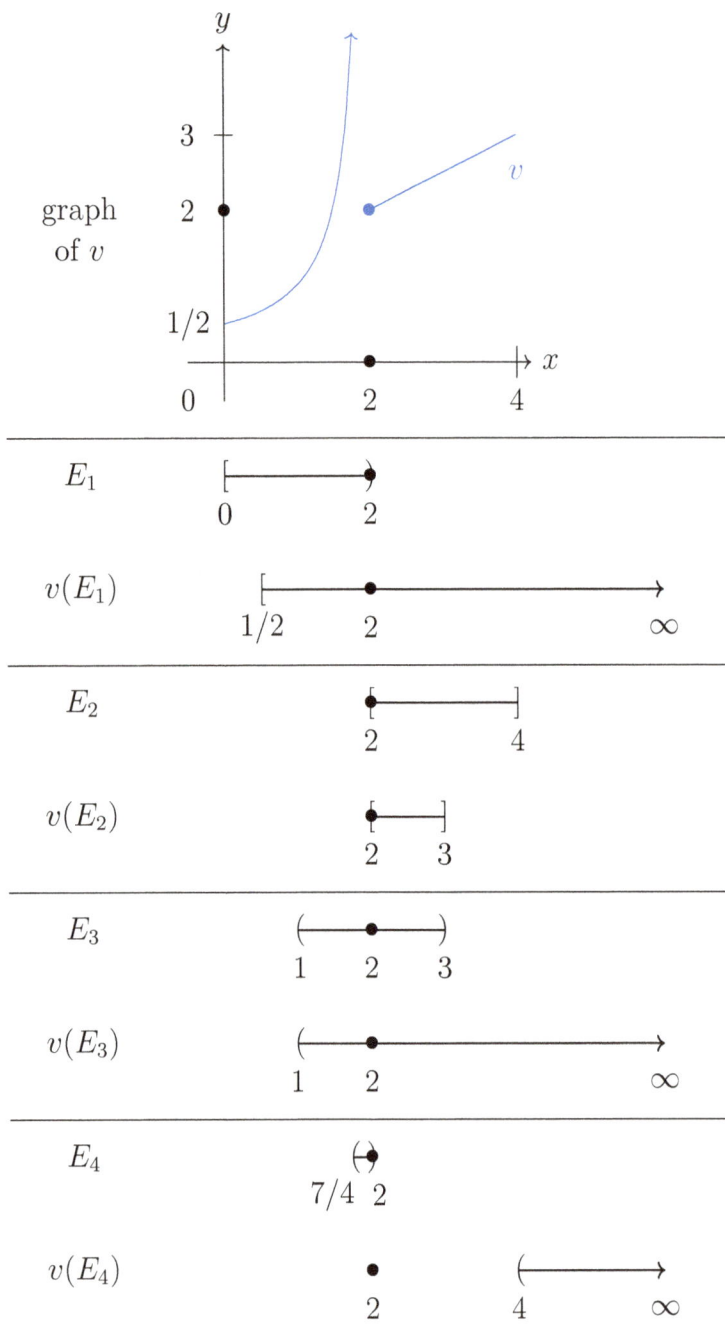

Figure 4.2.4: Images under v of the intervals from Example 4.2.1. All of the images are connected, but only $v(E_4) = (4, \infty)$ is away from $v(2) = 2$. See Example 4.2.4.

How can we prove this? We need definition for *preserving closeness* to work with.

Definition 4.2.5: Preserving closeness

Let $D \subseteq \mathbb{R}^k$, $\mathbf{c} \in D$, and $f : D \to \mathbb{R}^m$. We say f *preserves closeness at* $\mathbf{c}$ if for every subset E of the domain D we have

$$\mathbf{c} \operatorname{acl} E \quad \implies \quad f(\mathbf{c}) \operatorname{acl} f(E). \tag{4.2.12}$$

If f preserves closeness at every point in its domain D, we say f *preserves closeness*.

So, whether E is an interval, the range of a sequence, or something else, if f preserves closeness at $\mathbf{c}$, then anytime E is arbitrarily close to $\mathbf{c}$ in the domain we also have $f(E)$ is arbitrarily close to $f(\mathbf{c})$ in the range.

To show a function does not preserve closeness at $\mathbf{c}$, it suffices to find a single set in the domain arbitrarily close to $\mathbf{c}$ whose image is away from the image of $\mathbf{c}$.

Example 4.2.6: Not preserving closeness

The functions g and v from Examples 4.1.3 and 4.2.4 do not preserve closeness at $c = 2$. Recall that $E_4 = (7/4, 2)$. We have:

(i) $2 \operatorname{acl} E_4$ but $g(2) \operatorname{awf} g(E_4)$; and

(ii) $2 \operatorname{acl} E_4$ but $v(2) \operatorname{awf} v(E_4)$.

What about the case when we believe a function f preserves closeness? It is more difficult since we would need to prove that for *any* subset E of the domain where $\mathbf{c} \operatorname{acl} E$ we end up with $f(\mathbf{c}) \operatorname{acl} f(E)$. More importantly, the process of proving a function preserves closeness via Definition 4.2.5 is very similar to the process of proving a function is continuous via its Definition 4.3.2.

Example 4.2.7: The line f preserves closeness

The function $f : \mathbb{R} \to \mathbb{R}$ given by

$$f(x) = x/2 \tag{4.2.13}$$

preserves closeness at $c = 2$. See Figure 4.2.5.

Scratch Work 4.2.8: Start at the end

Given the definition of preserving closeness (Definition 4.2.5), it looks like we should consider the definition for arbitrarily close *twice*: Can we show arbitrarily close in the domain implies arbitrarily close in the range? See Definitions 1.1.8 and 1.5.1 for the definitions of arbitrarily close in the real line and Euclidean spaces, respectively.

As done throughout Chapters 1 and 2, the scratch work starts at the end. To show $f(c) \operatorname{acl} f(E)$, first consider how close we would like the images $f(E)$ and $f(c) = f(2) = 1$ to be, again using $\varepsilon > 0$ to denote an arbitrary positive distance. The goal is to end up with this: For some $f(x) \in f(E)$, we have

$$|f(x) - f(c)| = \left| \frac{x}{2} - \frac{c}{2} \right| < \varepsilon. \tag{4.2.14}$$

To ensure $f(x) \in f(E)$, we consider $x \in E$ only.

To ensure $f(x)$ is within ε of $f(c)$, we can use the assumption that E is arbitrarily close to $c = 2$ in the domain. That is, given any positive distance $\varepsilon > 0$ for the range, we want to find a distance $\delta_\varepsilon > 0$ for the domain where

$$x \in E \text{ with } |x - c| < \delta_\varepsilon \quad \Longrightarrow \quad |f(x) - f(c)| < \varepsilon. \tag{4.2.15}$$

This is a key step: Since $f(x) = x/2$, we can try to find a suitable formula for δ_ε by multiplying the inequality (4.2.14) through by 2. We get

$$|x - c| < 2\varepsilon = \delta_\varepsilon. \tag{4.2.16}$$

Since $2 \operatorname{acl} E$, there is some x_ε in E and within $\delta_\varepsilon = 2\varepsilon$ of $c = 2$. We should end up with $f(c) \operatorname{acl} f(E)$.

Time for a proof. The key step in the scratch work of finding a suitable distance δ_ε for the domain pays off right after asserting the assumptions. Please keep Figure 4.2.5 in mind as you work through it.

Proof for Example 4.2.7. Let $\varepsilon > 0$ denote an arbitrary positive distance in the range and suppose E is arbitrarily close to $c = 2$ in the domain. For the domain, choose the distance δ_ε given by

$$\delta_\varepsilon = 2\varepsilon. \tag{4.2.17}$$

Since $\delta_\varepsilon > 0$ and $c = 2$ is arbitrarily close to E, by the definition of arbitrarily close in the real line (Definition 1.1.8) there is a real number x_ε where

$$x_\varepsilon \in E \qquad \text{and} \qquad |x_\varepsilon - c| < 2\varepsilon = \delta_\varepsilon. \tag{4.2.18}$$

We have $f(x) = x/2$ for every real number x, so

$$|f(x_\varepsilon) - f(c)| = \left| \frac{x_\varepsilon}{2} - \frac{c}{2} \right| = \frac{1}{2} |x_\varepsilon - c| < \frac{1}{2}(2\varepsilon) = \varepsilon. \tag{4.2.19}$$

Also, $x_\varepsilon \in E$ implies $f(x_\varepsilon) \in f(E)$, hence $f(c) = f(2) = 1$ is arbitrarily close to $f(E)$. Therefore, f preserves closeness at $c = 2$. $\square$

Example 4.2.7 extends to the more general setting of *basic affine transformations* without much effort. They transform their inputs by a scaling followed by a translation.

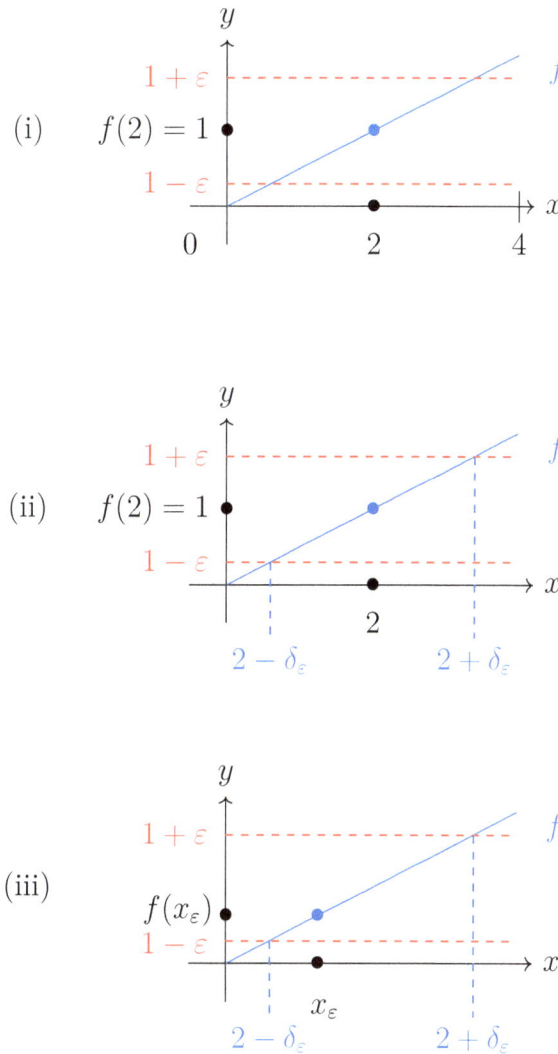

Figure 4.2.5: A three-step figure built on $E = [0, 4]$ in the domain to accompany the scratch work and proof that $f(x) = x/2$ preserves closeness at $c = 2$: (i) Start with an $\varepsilon > 0$ to give us a distance around $f(c) = f(2) = 1$; (ii) *the key step*, do scratch work to find a suitable distance $\delta_\varepsilon > 0$ around $c = 2$; (iii) use the assumption $c \operatorname{acl} E$ to prove there is some point x_ε is in E and within δ_ε of $c = 2$, leading to an output $f(x_\varepsilon)$ in $f(E)$ and within ε of $f(c) = f(2) = 1$. This shows $f(c) \operatorname{acl} f(E)$.

Definition 4.2.9: Basic affine transformation

A *basic affine transformation* is a function $f : \mathbb{R}^m \to \mathbb{R}^m$ given by a linear combination of form

$$f(\mathbf{x}) = \alpha \mathbf{x} + \mathbf{v} \tag{4.2.20}$$

where $\alpha \in \mathbb{R}$ and $\mathbf{v} \in \mathbb{R}^m$.

Theorem 4.2.10: Basic affine transformations preserve closeness

If $f : \mathbb{R}^m \to \mathbb{R}^m$ is a basic affine transformation given by $f(\mathbf{x}) = \alpha \mathbf{x} + \mathbf{v}$ where $\alpha \in \mathbb{R}$ and $\mathbf{v} \in \mathbb{R}^m$, then f preserves closeness.

Scratch Work 4.2.11: Generalize an example

Following Scratch Work 4.2.8, let's start at the end. From the definition of preserving closeness (Definition 4.2.5), we want to show $f(\mathbf{c}) \operatorname{acl} f(E)$. By the definition for arbitrarily close (Definition 1.5.1) and assuming $f(\mathbf{x}) = \alpha \mathbf{x} + \mathbf{v}$, applying the homogeneity of the Euclidean norm (1.2.33) tells us we want to end up with

$$\|f(\mathbf{x}_\varepsilon) - f(\mathbf{c})\|_m = \|\alpha \mathbf{x}_\varepsilon + \mathbf{v} - (\alpha \mathbf{c} + \mathbf{v})\|_m \tag{4.2.21}$$
$$= \|\alpha(\mathbf{x}_\varepsilon - \mathbf{c})\|_m \tag{4.2.22}$$
$$= |\alpha| \|\mathbf{x}_\varepsilon - \mathbf{c}\|_m \tag{4.2.23}$$
$$< \varepsilon \tag{4.2.24}$$

for some $\mathbf{x}_\varepsilon \in E$. To take advantage of the assumption that $\mathbf{c}$ is arbitrarily close to a set $E \subseteq \mathbb{R}^m$ in the domain, dividing by $|\alpha|$ suggests a suitable choice for a distance δ_ε to constrain the domain is given by

$$\delta_\varepsilon = \frac{\varepsilon}{|\alpha|}. \tag{4.2.25}$$

However, such a choice is only valid when $|\alpha| \neq 0$. To compensate, the proof is broken into two cases: $\alpha = 0$ and $\alpha \neq 0$. When $\alpha = 0$, the proof becomes trivial in the sense that $f(\mathbf{c}) \operatorname{acl} f(E)$ regardless of the point $\mathbf{c}$ and the set E.

Proof of Theorem 4.2.10. Throughout the proof, suppose $f : \mathbb{R}^m \to \mathbb{R}^m$ is given by $f(\mathbf{x}) = \alpha \mathbf{x} + \mathbf{v}$ for some $\alpha \in \mathbb{R}$ and $\mathbf{v} \in \mathbb{R}^m$. Also, suppose $\mathbf{c} \in \mathbb{R}^m$ and $E \subseteq \mathbb{R}^m$ where $\mathbf{c} \operatorname{acl} E$.

Case (i): Suppose $\alpha = 0$. Then for every $\mathbf{x} \in E$ we have $f(\mathbf{x}) = \alpha \mathbf{x} + \mathbf{v} = \mathbf{v}$, hence $f(E) = \{\mathbf{v}\}$. Since $f(\mathbf{c}) = \mathbf{v}$ as well and points in a set are arbitrarily close to the set (Lemma 1.5.4), we have

$$\mathbf{v} \operatorname{acl} \{\mathbf{v}\} \qquad \Longleftrightarrow \qquad f(\mathbf{c}) \operatorname{acl} f(E). \tag{4.2.26}$$

Therefore, f preserves closeness when $\alpha = 0$.

Case (ii): Now suppose $\alpha \neq 0$ and let $\varepsilon > 0$. Since $|\alpha| \neq 0$, define

$$\delta_\varepsilon = \frac{\varepsilon}{|\alpha|} \qquad (4.2.27)$$

to provide a constraint for the domain. Since $\mathbf{c}\operatorname{acl}E$ and $\delta_\varepsilon > 0$, by the definition of arbitrarily close (Definition 1.5.1), there is a point $\mathbf{x}_\varepsilon \in E$ such that

$$f(\mathbf{x}_\varepsilon) \in f(E) \qquad \text{and} \qquad \|\mathbf{x}_\varepsilon - \mathbf{c}\|_m < \delta_\varepsilon = \frac{\varepsilon}{|\alpha|}. \qquad (4.2.28)$$

Hence, by the homogeneity of the Euclidean norm (1.2.33) we have

$$\|f(\mathbf{x}_\varepsilon) - f(\mathbf{c})\|_m = \|\alpha\mathbf{x}_\varepsilon + \mathbf{v} - (\alpha\mathbf{c} + \mathbf{v})\|_m \qquad (4.2.29)$$
$$= \|\alpha(\mathbf{x}_\varepsilon - \mathbf{c})\|_m \qquad (4.2.30)$$
$$= |\alpha|\|\mathbf{x}_\varepsilon - \mathbf{c}\|_m \qquad (4.2.31)$$
$$< |\alpha| \cdot \frac{\varepsilon}{|\alpha|} \qquad (4.2.32)$$
$$= \varepsilon. \qquad (4.2.33)$$

Therefore, $f(\mathbf{c})\operatorname{acl}f(E)$ and f preserves closeness when $\alpha \neq 0$. □

Let's work through another challenging example.

Example 4.2.12: The square function preserves closeness

The function $h : \mathbb{R} \to \mathbb{R}$ given by

$$h(x) = x^2 \qquad (4.2.34)$$

preserves closeness at $c = 0$.

Scratch Work 4.2.13: Square function preserves closeness at $c = 0$

The scratch work here follows a path similar to Scratch Work 4.2.8 as laid out in Figure 4.2.5:

- Start at the end using $\varepsilon > 0$;

- *key step*, find a suitable $\delta_\varepsilon > 0$;

- use arbitrarily close in the domain to conclude we have arbitrarily close in the range.

The goal is to end up with this: Given $h(x) = x^2$, $c = 0$, $c\operatorname{acl}E$ in the domain, and a distance $\varepsilon > 0$ for the range, we want some $h(x) \in h(E)$ where

$$|h(x) - h(c)| = |x^2 - 0^2| = x^2 < \varepsilon. \qquad (4.2.35)$$

To ensure $h(x) \in h(E)$, we consider $x \in E$ only.

Next is the key step: Find a suitable distance $\delta_\varepsilon > 0$ for the domain where

$$x \in E \text{ with } |x - c| = |x| < \delta_\varepsilon \quad \Longrightarrow \quad |h(x) - h(c)| = x^2 < \varepsilon. \tag{4.2.36}$$

We can do this by taking square roots in the inequality $x^2 < \varepsilon$. Since $c = 0$, we get

$$|x - c| = |x| < \sqrt{\varepsilon} = \delta_\varepsilon. \tag{4.2.37}$$

By assumption, $c \operatorname{acl} E$. So, there is some x_ε in E and within $\delta_\varepsilon = \sqrt{\varepsilon}$ of $c = 0$. We should end up with $h(c) \operatorname{acl} h(E)$.

One again, the key step of finding a suitable distance δ_ε for the domain pays off early in the proof.

Proof for Example 4.2.12. Let $\varepsilon > 0$ denote an arbitrary positive distance in the range and suppose E is arbitrarily close to $c = 0$ in the domain. For the domain, choose the distance δ_ε given by

$$\delta_\varepsilon = \sqrt{\varepsilon}. \tag{4.2.38}$$

Since $\delta_\varepsilon > 0$ and $c = 0$ is arbitrarily close to E, by the definition of arbitrarily close in the real line (Definition 1.1.8) there is a real number x_ε where

$$x_\varepsilon \in E \quad \text{and} \quad |x_\varepsilon - c| = |x_\varepsilon| < \sqrt{\varepsilon} = \delta_\varepsilon. \tag{4.2.39}$$

We have $h(x) = x^2$ for every real number x and $0 \le y < z$ implies $y^2 < z^2$, therefore

$$|h(x_\varepsilon) - h(c)| = |x_\varepsilon^2 - 0^2| = x_\varepsilon^2 < (\sqrt{\varepsilon})^2 = \varepsilon. \tag{4.2.40}$$

Also, $x_\varepsilon \in E$ implies $h(x_\varepsilon) \in h(E)$, hence $h(c) = h(0) = 0$ is arbitrarily close to $h(E)$. So, h preserves closeness at $c = 0$. $\qquad \square$

As evidenced by the effort it took to establish the results in this section, it can be quite a challenge to prove functions preserve closeness. On the other hand, proving a function does not preserve closeness can be relatively straightforward.

Example 4.2.14: Dirichlet's function

Consider *Dirichlet's function*, the indicator function of the rationals $\mathbb{1}_\mathbb{Q} : \mathbb{R} \to \mathbb{R}$ given by

$$\mathbb{1}_\mathbb{Q}(x) = \begin{cases} 1, & \text{if } x \in \mathbb{Q}, \\ 0, & \text{if } x \in \mathbb{R} \backslash \mathbb{Q}. \end{cases} \tag{4.2.41}$$

Dirichlet's function $\mathbb{1}_\mathbb{Q}$ does not preserve closeness at any $c \in \mathbb{R}$.

The proof follows from the density of both the rationals and the irrationals in the real line (Theorem 1.4.10 and Corollary 1.4.13, respectively).

Proof for Example 4.2.14. Suppose $c \in \mathbb{Q}$ and $y \in \mathbb{R}\backslash\mathbb{Q}$. Then

$$\mathbb{1}_{\mathbb{Q}}(c) = 1 \qquad \text{and} \qquad \mathbb{1}_{\mathbb{Q}}(y) = 0. \qquad (4.2.42)$$

Hence, $\mathbb{1}_{\mathbb{Q}}(\mathbb{R}\backslash\mathbb{Q}) = \{0\}$. By the density of the irrationals in the reals (Corollary 1.4.13), we have $c \operatorname{acl} \mathbb{R}\backslash\mathbb{Q}$. However, using the distance $\varepsilon_0 = 1/2$ and the definition of away from (Definition 1.5.11), we have $1 \operatorname{awf} \{0\}$. Therefore, $\mathbb{1}_{\mathbb{Q}}$ does not preserve closeness at c since

$$c \operatorname{acl} \mathbb{R}\backslash\mathbb{Q} \qquad \text{but} \qquad \mathbb{1}_{\mathbb{Q}}(c) \operatorname{awf} \mathbb{1}_{\mathbb{Q}}(\mathbb{R}\backslash\mathbb{Q}). \qquad (4.2.43)$$

A similar argument handles the complementary situation. Suppose $c \in \mathbb{R}\backslash\mathbb{Q}$ and $x \in \mathbb{Q}$. Then

$$\mathbb{1}_{\mathbb{Q}}(c) = 0 \qquad \text{and} \qquad \mathbb{1}_{\mathbb{Q}}(x) = 1. \qquad (4.2.44)$$

Hence, $\mathbb{1}_{\mathbb{Q}}(\mathbb{R}\backslash\mathbb{Q}) = \{1\}$. By the density of the rationals in the reals (Theorem 1.4.10), we have $c \operatorname{acl} \mathbb{Q}$. However, by using the distance $\varepsilon_0 = 1/2$ again, we have $0 \operatorname{awf} \{1\}$. Thus, $\mathbb{1}_{\mathbb{Q}}$ does not preserve closeness at c since

$$c \operatorname{acl} \mathbb{Q} \qquad \text{but} \qquad \mathbb{1}_{\mathbb{Q}}(c) \operatorname{awf} \mathbb{1}_{\mathbb{Q}}(\mathbb{Q}). \qquad (4.2.45)$$

Therefore, $\mathbb{1}_{\mathbb{Q}}$ does not preserve closeness at any $c \in \mathbb{R}$. $\square$

I hope the scratch work and proofs showing functions preserve closeness have begun to make sense. Given a distance $\varepsilon > 0$ for the range, we ensure the preservation of closeness by using scratch work to find a responding distance $\delta > 0$ for the domain, then showing that points within δ of each other in the domain produce images within ε of each other in the range. This process is very similar to the work done to prove functions are *continuous*.

John Rodriguez, thank you for the conversations about proving functions preserve closeness using the definition of arbitrarily close and how it leads to continuity!

Exercises

4.2.1. Prove monomials preserve closeness at $c = 0$. That is, the functions $h_n : \mathbb{R} \to \mathbb{R}$ given by $h_n(x) = x^n$ for each $n \in \mathbb{N}$ preserve closeness at $c = 0$. (See Example 4.2.12.)

4.2.2. Give an example of a function $f : \mathbb{R} \to \mathbb{R}$ where $|f|$ preserves closeness at $c = 1$ but f does not.

4.2.3. Prove that the absolute value function $a : \mathbb{R} \to \mathbb{R}$ given by $a(x) = |x|$ preserves closeness on $\mathbb{R}$.

4.2.4. Suppose $g : \mathbb{R}^k \to \mathbb{R}^m$ preserves closeness. Prove that if $C \subseteq \mathbb{R}^k$ is connected, then its image $g(C) \subseteq \mathbb{R}^m$ is connected.

4.2.5. Suppose $D \subseteq \mathbb{R}^k, \mathbf{c} \in D$, and $f : D \to \mathbb{R}^m$. Prove the following statements are equivalent:

(i) *f preserves closeness at* $\mathbf{c}$ (Definition 4.2.5):

$$E \subseteq D \text{ with } \mathbf{c} \text{ acl } E \quad \Longrightarrow \quad f(\mathbf{c}) \text{ acl } f(E). \tag{4.2.46}$$

(ii) *f is sequentially continuous at* $\mathbf{c}$ (Definition 4.4.5):

$$(\mathbf{x}_n) \subseteq D \text{ with } \lim_{n\to\infty} \mathbf{x}_n = \mathbf{c} \quad \Longrightarrow \quad \lim_{n\to\infty} f(\mathbf{x}_n) = f(\mathbf{c}). \tag{4.2.47}$$

Hint: Use the fundamental connection between arbitrarily close and convergent sequences established in Theorem 2.3.1. (Even more is true, see Theorem 4.4.7.)

4.2.6. Suppose $f : \mathbb{R}^k \to \mathbb{R}^m$ preserves closeness and $K \subseteq \mathbb{R}^k$ is compact. Prove the image $f(K)$ is compact. Hint: Use the result of Exercise 4.2.5 and *sequential compactness* from the Heine-Borel Theorem 3.5.1.

4.2.7. Suppose $f : \mathbb{R} \to \mathbb{R}$ preserves closeness. Prove $f^{-1}([a,b])$ is closed. (That is, for such f, the inverse image of a compact interval in the codomain is a closed set in the domain.)

4.2.8. Suppose $g : \mathbb{R}^k \to \mathbb{R}^m$. Prove the following statements are equivalent:

(i) *g preserves closeness* (Definition 4.2.5): For all $\mathbf{c} \in \mathbb{R}^k$ and $E \subseteq \mathbb{R}^k$,

$$\mathbf{c} \text{ acl } E \quad \Longrightarrow \quad g(\mathbf{c}) \text{ acl } g(E). \tag{4.2.48}$$

(ii) *g is topologically continuous* in the following sense: If F is a closed subset of the codomain $\mathbb{R}^m$, the its inverse image $g^{-1}(F)$ is a closed subset of the domain $\mathbb{R}^k$. (That is, inverse images of closed sets are closed.)

4.3 Continuity

This section is designed to face the challenge of understanding *continuity* by building on the definition for preserving closeness (Definition 4.2.5). The classic "ε-δ" definitions of continuity and functional limits are perhaps the most difficult concepts for new mathematicians to understand. (See Definitions 4.3.2 and 5.1.2, respectively.) This perspective is corroborated in the works [3–5, 9, 11].

Remark 4.3.1: From preserving closeness to continuity

The scratch work and proofs showing the preservation of closeness in Examples 4.2.7 and 4.2.12 along with Theorem 4.2.10 preview the process of showing a function is continuous via Definition 4.3.2. For both preserving closeness and continuity, the scratch work first considers how close we would like outputs to be in the range, then work is done to find how close their inputs should be in the domain to compensate. In the proofs, both consider a distance $\varepsilon > 0$ for the range first, then assert a choice of a distance δ for the domain based on the scratch work. The proofs conclude by showing inputs within δ of each other in the domain yield outputs within ε of each other in the range.

Time for the classic ε-δ definition of continuity in analysis.

Definition 4.3.2: Continuity

Let $D \subseteq \mathbb{R}^k$, let $f : D \to \mathbb{R}^m$, and let $\mathbf{c} \in D$. We say f is *continuous at* $\mathbf{c}$ if for every distance $\varepsilon > 0$ there is a *threshold* $\delta > 0$ providing a distance for the domain such that

$$\mathbf{x} \in D \text{ with } \|\mathbf{x} - \mathbf{c}\|_k < \delta \quad \Longrightarrow \quad \|f(\mathbf{x}) - f(\mathbf{c})\|_m < \varepsilon. \tag{4.3.1}$$

In other words, f is *continuous at* $\mathbf{c}$ if for every error $\varepsilon > 0$ there is some threshold $\delta > 0$ where the δ-neighborhood of $\mathbf{c}$ in the domain maps into the ε-neighborhood of $f(\mathbf{c})$ in the range:

$$f(V_\delta(\mathbf{c}) \cap D) \subseteq V_\varepsilon(f(\mathbf{c})). \tag{4.3.2}$$

If f is continuous at every point in the domain D, we say f is *continuous*. If f is not continuous at some point $\mathbf{z}$, then we say f is *discontinuous at* $\mathbf{z}$.

Remark 4.3.3: Thresholds

The word "threshold" appears in the definitions of limit and convergence for sequences (Definition 2.2.1) as well as the definition of continuity (Definition 4.3.2). In both settings, the threshold—either an index $n_\varepsilon \in \mathbb{N}$ or distance $\delta > 0$—is determined in response to a given but arbitrary distance $\varepsilon > 0$ for the range.

The scratch work and proofs showing $f(x) = x/2$ preserves closeness at $c = 2$ in Example 4.2.7, more generally that basic affine transformations preserve closeness as in Theorem 4.2.10, and $h(x) = x^2$ preserves closeness at $c = 0$ in Example 4.2.12 preview the process of showing a function is continuous via Definition 4.3.2. Much like the guide for proving sequences converge in Remark 2.2.4, here's a guide for developing scratch work and proving "ε-δ" continuity by verifying Definition 4.3.2.

Remark 4.3.4: Guide for continuity proofs

Scratch work for proving f is continuous at c:

- Consider the inequality you want to end up with, typically:
$$d_m(f(\mathbf{x}), f(\mathbf{c})) = \|f(\mathbf{x}) - f(\mathbf{c})\|_m < \varepsilon. \tag{4.3.3}$$

- *Key step*: Use the previous inequality to find a formula for a threshold $\delta > 0$ as a function of ε and $\mathbf{c}$ where $\mathbf{x}$ is in the domain and
$$d_k(\mathbf{x}, \mathbf{c}) = \|\mathbf{x} - \mathbf{c}\|_k < \delta. \tag{4.3.4}$$

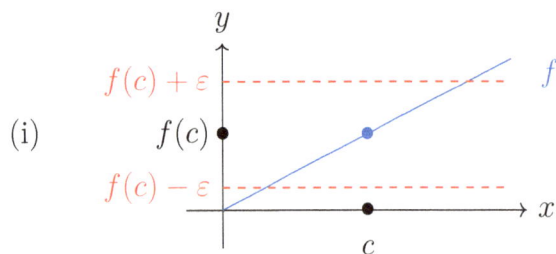

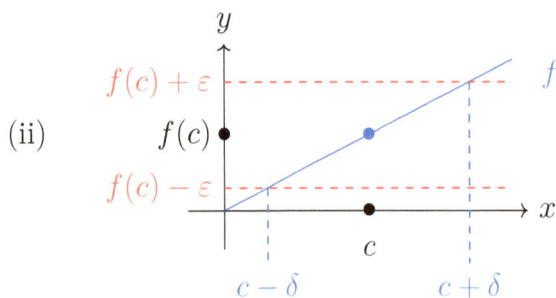

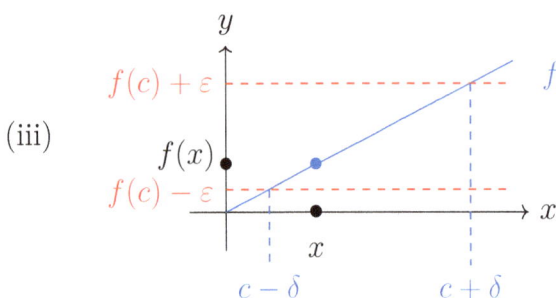

Figure 4.3.1: A three-step figure to accompany the scratch work and proof that $f(x) = x/2$ is continuous at every $c \in \mathbb{R}$ as in the guide for continuity proofs in Remark 4.3.4: (i) Start with an $\varepsilon > 0$ to give us a distance around $f(c)$ in the range; (ii) *the key step*, do scratch work to find a suitable threshold $\delta > 0$ giving us a distance around c in the domain; (iii) in the proof, show that every x in the domain within δ of c yields an output $f(x)$ within ε of $f(c)$ in the range. Also, use the QR code to play around the Desmos activity "Continuity". https://www.desmos.com/calculator/l12dxcm4ul

(i) Let $\varepsilon > 0$. $\bullet$ $\mapsto$
$\mathbf{c}$

ε

$f(\mathbf{c})$

$V_\varepsilon(f(\mathbf{c}))$

(ii) Choose $\delta > 0$. δ $\bullet$ $\mapsto$ $\bullet\, f(\mathbf{c})$
$\mathbf{c}$

$V_\delta(\mathbf{c}) \cap D$ $f(V_\delta(\mathbf{c}) \cap D)$

(iii) Verify $f(V_\delta(\mathbf{c}) \cap D) \subseteq V_\varepsilon(f(\mathbf{c}))$. $\bullet\, f(\mathbf{c})$

$f(V_\delta(\mathbf{c}) \cap D)$

$V_\varepsilon(f(\mathbf{c}))$

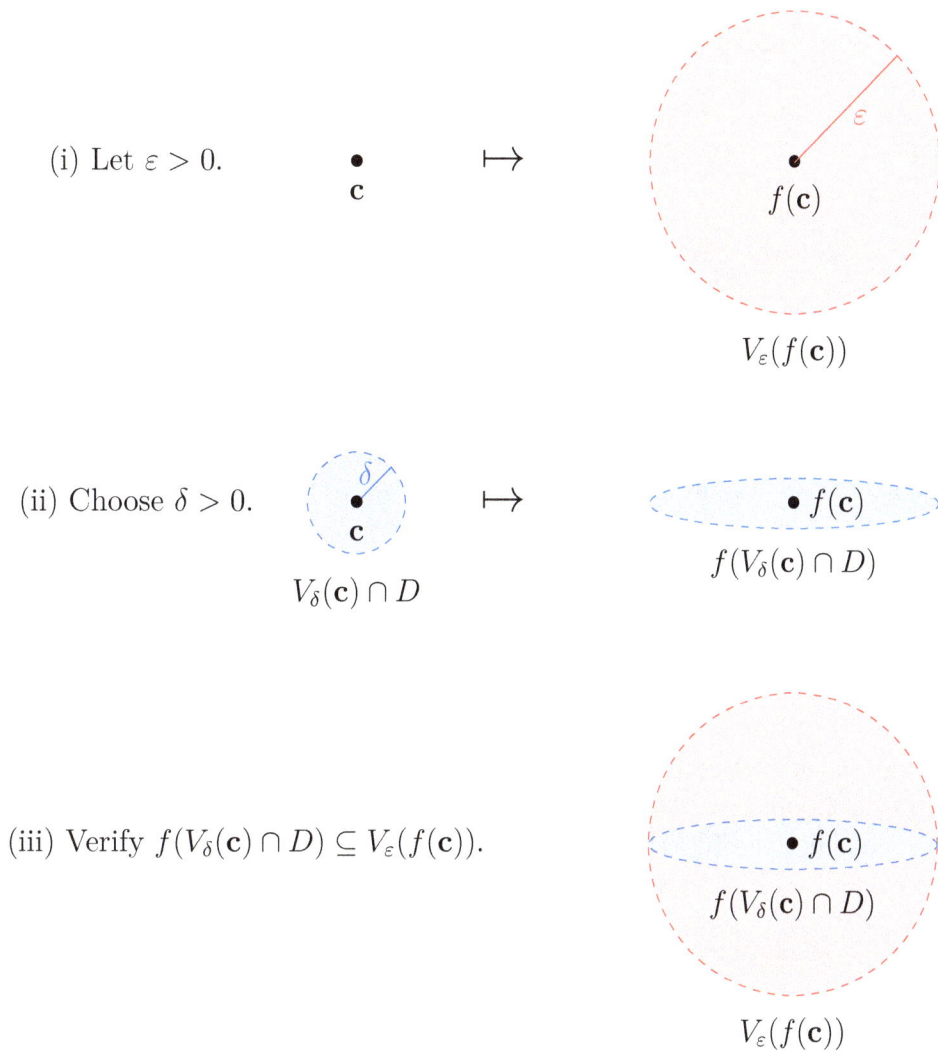

Figure 4.3.2: A visualization of the definition of continuity in the context of a function f mapping $\mathbb{R}^2$ to $\mathbb{R}^2$. Steps (i), (ii), and (iii) play the same roles as in Figure 4.3.1 to accompany Definition 4.3.2 and the guide for continuity proofs in Remark 4.3.4.

The triangle inequality as it appears in (1.2.32) and (1.2.34) is used quite often at this step in scratch work and proofs involving "ε-δ" continuity.

- Include a figure with the function f, the point $\mathbf{c}$ in the domain, and the point $f(\mathbf{c})$ in the range, perhaps as a graph. You may also want to plot the domain and range on their own.

Proving f is continuous at c:

- Early in the proof, perhaps the first step, write "Let $\varepsilon > 0$" or something similar, indicating you are accounting for *all* positive distances in the range at the same time.

- Define a candidate for the threshold $\delta > 0$ to give a distance in the domain based on your scratch work, typically in terms of a function of ε and $\mathbf{c}$.

- Verify δ is truly a threshold for the continuity of f at $\mathbf{c}$ by considering every point $\mathbf{x}$ in the domain within δ of $\mathbf{c}$, then showing

$$\mathbf{x} \in D \text{ with } \|\mathbf{x} - \mathbf{c}\|_k < \delta \quad \implies \quad \|f(\mathbf{x}) - f(\mathbf{c})\|_m < \varepsilon. \qquad (4.3.5)$$

In other words, verify the δ-neighborhood of $\mathbf{c}$ in the domain maps into the ε-neighborhood of $f(\mathbf{c})$ in the range:

$$f(V_\delta(\mathbf{c}) \cap D) \subseteq V_\varepsilon(f(\mathbf{c})). \qquad (4.3.6)$$

See Figures 4.3.1 and 4.3.2.

The remainder of the section explores a few examples and a theorem to give us practice working with the definition of continuity (Definition 4.3.2).

Once again, consider the function f from Examples 4.1.2, 4.2.2, and 4.2.7, which is shown to preserve closeness at $c = 2$. This function is continuous at every real number.

Example 4.3.5: A continuous line

The function $f : \mathbb{R} \to \mathbb{R}$ given by

$$f(x) = x/2 \qquad (4.3.7)$$

is continuous. See Figure 4.3.1.

Remark 4.3.6: Proving continuity

The scratch work, figures, and proofs showing $f(x) = x/2$ preserves closeness at $c = 2$ and f is continuous are very similar. Both situations start by considering an arbitrary distance $\varepsilon > 0$ for the range, then scratch work is done to find a suitable distance $\delta > 0$ for the domain, and finally the proofs show our choices for the δ thresholds lead to outputs within ε.

Still, there are some differences. In Example 4.2.7 shows f preserves closeness at $c = 2$ only. However, to show f is continuous in Example 4.3.5, we must consider every $c \in \mathbb{R}$. Also, when showing preservation of closeness, the scratch work and proofs consider every subset of the domain which is arbitrarily close to the point c in question. But with continuity, we eschew these subsets and consider all points in the domain within the suitable distance of c. Compare Figures 4.2.5 and 4.3.1.

Scratch Work 4.3.7: Lines are continuous

Following the guide for proving continuity in Remark 4.3.4, start with an arbitrary distance ε for the range and somehow determine a formula for a suitable threshold δ to give us

$$x \in \mathbb{R} \text{ with } |x - c| < \delta \quad \implies \quad |f(x) - f(c)| < \varepsilon. \tag{4.3.8}$$

Once we have a suitable δ, we can reorganize the information into a formal proof. See Figures 4.3.1 and 4.3.2.

Since $f(x) = x/2$, sticking everything into inequality (4.3.8) yields

$$d_{\mathbb{R}}(f(x), f(c)) = |f(x) - f(c)| = \left| \frac{x}{2} - \frac{c}{2} \right| = \frac{1}{2}|x - c| < \varepsilon. \tag{4.3.9}$$

Note the lack of a variable δ in the expression (4.3.9). The role of δ is to control the distance between inputs x and c in the domain, so we should solve for $|x - c|$ and use the resulting relationship to help us define δ.

This is a key step: We can find a suitable formula for δ by multiplying the inequality (4.3.9) through by 2. (We did the same thing when proving f preserves closeness at $c = 2$ in Example 4.2.7.) We get

$$|x - c| < 2\varepsilon = \delta. \tag{4.3.10}$$

With a choice for δ available, let's see if it is truly a threshold for the continuity of f at c.

Proof for Example 4.3.5. Let $f : \mathbb{R} \to \mathbb{R}$ be given by $f(x) = x/2$ and let $\varepsilon > 0$. Define

$$\delta = 2\varepsilon. \tag{4.3.11}$$

Note that $\delta = 2\varepsilon > 0$ since $\varepsilon > 0$. Then for every real number x in the domain where

$$d_{\mathbb{R}}(x, c) = |x - c| < \delta = 2\varepsilon, \tag{4.3.12}$$

we have

$$|f(x) - f(c)| = \left| \frac{x}{2} - \frac{c}{2} \right| = \frac{1}{2}|x - c| < \frac{1}{2}(2\varepsilon) = \varepsilon. \tag{4.3.13}$$

Therefore, $\delta = 2\varepsilon$ is a suitable threshold and f is continuous at c. Since c represents an arbitrary real number in the domain, f is continuous by Definition 4.3.2. $\square$

Figure 4.3.3: To explore the notion of discontinuity via the *floor* function defined ahead in Exercise 4.5.3, play around the Desmos activity "Floor discontinuity". In particular, note that for $c = 1$ and $\varepsilon = 1$, no value of δ can be found to serve as a threshold. Hence, the floor function is discontinous at $c = 1$. https://www.desmos.com/calculator/acc6b5b36f.

Remark 4.3.8: Threshold depends on error

The scratch work and proofs showing f preserves closenesss $c = 2$ in Example 4.2.7 and showing f is continuous at every real number c in Example 4.3.5 found and made use of the same threshold, namely

$$\delta = \delta_\varepsilon = 2\varepsilon. \tag{4.3.14}$$

Note that this common choice for the threshold explicitly depends on the distance (or error) $\varepsilon > 0$ for the range. In general, the threshold δ depends on ε, how the function f is defined, *and* the point $\mathbf{c}$ in question. Since δ depends on so many things, I decided to follow convention and leave out subscripts. Still, it is important to keep in mind the threshold δ generally depends on f, $\mathbf{c}$, and ε. See Figures 4.3.1, 4.3.2, and 4.3.4.

To contrast and explore discontinuity, play around with *floor* function using the Desmos activity "Floor discontinuity" accessed through the QR code in Figure 4.3.3. See Exercise 4.5.3 for the definitions of the *floor* function and the related *sawtooth* function.

Next, consider basic affine transformations as in Definition 4.2.9.

Theorem 4.3.9: Continuity of basic affine transformations

If $f : \mathbb{R}^m \to \mathbb{R}^m$ is a basic affine transformation given by $f(\mathbf{x}) = \alpha\mathbf{x} + \mathbf{v}$ where $\alpha \in \mathbb{R}$ and $\mathbf{v} \in \mathbb{R}^m$, then f is continuous.

Scratch Work 4.3.10: Building continuity from the preservation of closeness

Following the guide for proving continuity in Remark 4.3.4 and Scratch Work 4.2.11 which supports the proof that basic affine transformations preserve closeness, let's start at the end. From the definitions of basic affine transformation and continuity (Definitions 4.2.9 and 4.3.2) and the homogeneity of the Euclidean norm (1.2.33), we want to end up with

$$d_m(f(\mathbf{x}), f(\mathbf{c})) = \|f(\mathbf{x}) - f(\mathbf{c})\|_m \tag{4.3.15}$$
$$= \|\alpha\mathbf{x} + \mathbf{v} - (\alpha\mathbf{c} + \mathbf{v})\|_m \tag{4.3.16}$$
$$= \|\alpha(\mathbf{x} - \mathbf{c})\|_m \tag{4.3.17}$$
$$= |\alpha|\|\mathbf{x} - \mathbf{c}\|_m \tag{4.3.18}$$
$$< \varepsilon. \tag{4.3.19}$$

As in Scratch Work 4.2.11, dividing by $|\alpha|$ suggests a suitable choice for a distance δ to constrain the domain is given by

$$\delta = \frac{\varepsilon}{|\alpha|}. \tag{4.3.20}$$

Once again, such a choice is only valid when $|\alpha| \neq 0$. To compensate, the proof is broken into two cases: $\alpha = 0$ and $\alpha \neq 0$. When $\alpha = 0$, the proof becomes trivial in that any choice for a threshold $\delta > 0$ suffices.

Proof of Theorem 4.3.9. Throughout the proof, suppose $\mathbf{c} \in \mathbb{R}^m$ and $f : \mathbb{R}^m \to \mathbb{R}^m$ is a basic affine transformation of the form $f(\mathbf{x}) = \alpha\mathbf{x} + \mathbf{v}$ for some $\alpha \in \mathbb{R}$ and $\mathbf{v} \in \mathbb{R}^m$.

Case (i): Suppose $\alpha = 0$, let $\varepsilon > 0$, and define $\delta = 42 > 0$. Then for every $\mathbf{x} \in \mathbb{R}^m$ we have $f(\mathbf{x}) = \mathbf{v}$. Therefore,

$$\|\mathbf{x} - \mathbf{c}\|_m < \delta = 42 \quad \implies \quad \|f(\mathbf{x}) - f(\mathbf{c})\|_m = \|\mathbf{v} - \mathbf{v}\|_m = 0 < \varepsilon. \tag{4.3.21}$$

Since $\mathbf{c}$ is arbitrary, we have f is continuous when $\alpha = 0$.

Case (ii): Now suppose $\alpha \neq 0$ and let $\varepsilon > 0$. Since $|\alpha| \neq 0$, define

$$\delta = \frac{\varepsilon}{|\alpha|} \tag{4.3.22}$$

to provide a constraint for the domain. Suppose

$$\|\mathbf{x} - \mathbf{c}\|_m < \delta = \frac{\varepsilon}{|\alpha|}. \tag{4.3.23}$$

then by the homogeneity of the Euclidean norm (1.2.33), we have

$$\|f(\mathbf{x}) - f(\mathbf{c})\|_m = \|\alpha\mathbf{x} + \mathbf{v} - (\alpha\mathbf{c} + \mathbf{v})\|_m \tag{4.3.24}$$
$$= \|\alpha(\mathbf{x} - \mathbf{c})\|_m \tag{4.3.25}$$
$$= |\alpha|\|\mathbf{x} - \mathbf{c}\|_m \tag{4.3.26}$$
$$< |\alpha| \cdot \frac{\varepsilon}{|\alpha|} \tag{4.3.27}$$
$$= \varepsilon. \tag{4.3.28}$$

Since **c** is arbitrary, we have f is continuous when $\alpha \neq 0$.   □

The function from the real line to the real line given by taking the absolute value of the input is a continuous function.

Example 4.3.11: Absolute value is continuous

The function $g : \mathbb{R} \to \mathbb{R}$ given by $g(x) = |x|$ is continuous.

Scratch Work 4.3.12: Reverse triangle inequality

The reverse triangle inequality (1.2.38) tells us that for any $x, c \in \mathbb{R}$, we have

$$|g(x) - g(c)| = ||x| - |c|| \leq |x - c|. \tag{4.3.29}$$

So, choosing $\delta = \varepsilon$ yields the result.

Proof for Example 4.3.11. Suppose $g : \mathbb{R} \to \mathbb{R}$ given by $g(x) = |x|$ and suppose $c \in \mathbb{R}$. Let $\varepsilon > 0$ and choose $\delta = \varepsilon$. Then δ is a threshold for the continuity of g at c since

$$|x - c| < \delta = \varepsilon \tag{4.3.30}$$

immediately implies through the reverse triangle inequality (1.2.38) that

$$|g(x) - g(c)| = ||x| - |c|| \leq |x - c| < \delta = \varepsilon. \tag{4.3.31}$$

Therefore, g is continuous.   □

The next two examples show how difficult it can be to find a suitable threshold δ when the algebra defining a function is more complicated.

Example 4.3.13: Square root function is continuous

The square root function $r : [0, \infty) \to \mathbb{R}$ where

$$r(x) = \sqrt{x} \tag{4.3.32}$$

is continuous at every $c \geq 0$. See Figure 4.3.4.

Scratch Work 4.3.14: Conjugates

We want the proof to end up with

$$d_{\mathbb{R}}(r(x), r(c)) = |\sqrt{x} - \sqrt{c}| < \varepsilon. \tag{4.3.33}$$

To get there, we need a suitable threshold δ where

$$|x - c| < \delta \quad \implies \quad |\sqrt{x} - \sqrt{c}| < \varepsilon. \tag{4.3.34}$$

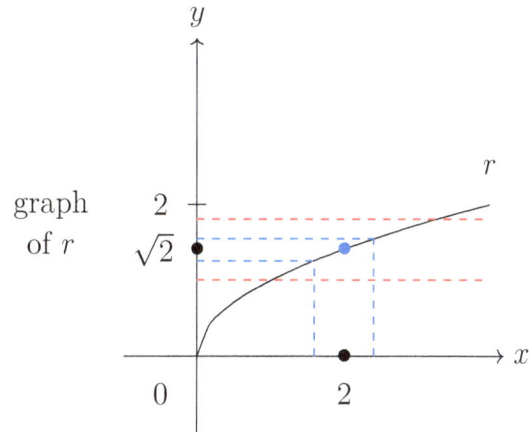

Figure 4.3.4: The graph of a function $r : D \to \mathbb{R}$ given by $D = [0, \infty)$ and $r(x) = \sqrt{x}$ to accompany the definition of continuity in terms of neighborhoods (Definition 4.3.2) and Example 4.3.13. Given a distance $\varepsilon > 0$ from $r(2) = \sqrt{2}$ in the range, there is a threshold $\delta > 0$ providing a distance from $c = 2$ in the domain ensuring $r(V_\delta(2) \cap D) \subseteq V_\varepsilon(r(2))$.

An idea from calculus or perhaps precalculus can help us here: Conjugates. As long as we consider $c > 0$ to ensure our denominators are never zero, since $\sqrt{x} \geq 0$ we have

$$|\sqrt{x} - \sqrt{c}| = |\sqrt{x} - \sqrt{c}| \left(\frac{|\sqrt{x} + \sqrt{c}|}{|\sqrt{x} + \sqrt{c}|} \right) \tag{4.3.35}$$

$$= \frac{|x - c|}{\sqrt{x} + \sqrt{c}} \tag{4.3.36}$$

$$\leq \frac{|x - c|}{\sqrt{c}}. \tag{4.3.37}$$

From here—and as long as $c > 0$—we can ensure

$$|\sqrt{x} - \sqrt{c}| \leq \frac{|x - c|}{\sqrt{c}} < \varepsilon \tag{4.3.38}$$

by choosing

$$|x - c| < \delta = \varepsilon\sqrt{c}. \tag{4.3.39}$$

But what about when $c = 0$? In this case, we want to end up with

$$|x - 0| = x < \delta \quad \Longrightarrow \quad |\sqrt{x} - \sqrt{0}| = \sqrt{x} < \varepsilon \tag{4.3.40}$$

Keeping in mind $x \geq 0$ and $0 \leq y < z$ implies $y^2 < z^2$, squaring both sides of the rightmost inequality in 4.3.40 yields

$$|x - c| = x = (\sqrt{x})^2 < \varepsilon^2. \tag{4.3.41}$$

So, let's try using $\delta = \varepsilon^2$ when $c = 0$.

Proof for Example 4.3.13. The proof is handled in two cases: $c = 0$ and $c \neq 0$.

Case (i): Let $\varepsilon > 0$, let $c = 0$, and define $\delta = \varepsilon^2 > 0$. Then for every nonnegative real number x where

$$d_{\mathbb{R}}(x, c) = |x - 0| = x < \delta = \varepsilon^2, \tag{4.3.42}$$

since $0 \leq x < \varepsilon^2$ implies $0 \leq \sqrt{x} < \sqrt{\varepsilon^2} = \varepsilon$, we have

$$d_{\mathbb{R}}(f(x), f(c)) = |\sqrt{x} - 0| = \sqrt{x} < \sqrt{\varepsilon^2} = \varepsilon. \tag{4.3.43}$$

Hence, $r(V_\delta(0) \cap D) \subseteq V_\varepsilon(r(0))$ where $D = [0, \infty)$ is the domain. Therefore, $\delta = \varepsilon^2$ is a suitable threshold and the square root function r is continuous at $c = 0$ by Definition 4.3.2.

Case (ii): Let $\varepsilon > 0$, let $c > 0$, and define $\delta = \varepsilon\sqrt{c} > 0$. Then for every nonnegative real number x where

$$d_{\mathbb{R}}(x, c) = |x - c| = x < \delta = \varepsilon\sqrt{c}, \tag{4.3.44}$$

we have

$$|\sqrt{x} - \sqrt{c}| = |\sqrt{x} - \sqrt{c}| \left(\frac{|\sqrt{x} + \sqrt{c}|}{|\sqrt{x} + \sqrt{c}|} \right) \tag{4.3.45}$$

$$= \frac{|x - c|}{\sqrt{x} + \sqrt{c}} \tag{4.3.46}$$

$$\leq \frac{|x - c|}{\sqrt{c}} \tag{4.3.47}$$

$$< \varepsilon. \tag{4.3.48}$$

Hence, $r(V_\delta(c) \cap D) \subseteq V_\varepsilon(r(c))$ where $D = [0, \infty)$ is the domain. Therefore, $\delta = \varepsilon\sqrt{c}$ is a suitable threshold and the square root function r is continuous at $c > 0$ by Definition 4.3.2.   □

Example 4.3.15: Square function is continuous

The square function $h : \mathbb{R} \to \mathbb{R}$ given by $h(x) = x^2$ is continuous.

Scratch Work 4.3.16: Factor and control

We want to end up with

$$d_\mathbb{R}(h(x), h(c)) = |x^2 - c^2| < \varepsilon. \tag{4.3.49}$$

Some algebra yields

$$|x^2 - c^2| = |x - c||x + c|. \tag{4.3.50}$$

Substituting δ_0 for $|x - c|$ in inequality (4.3.49) yields

$$d_\mathbb{R}(h(x), h(c)) = |x^2 - c^2| = |x - c||x + c| < \delta_0|x + c| < \varepsilon. \tag{4.3.51}$$

A first, naive choice for a threshold δ_0 might be

$$\delta_0 = \frac{\varepsilon}{|x + c|}. \tag{4.3.52}$$

However, this δ_0 is not an appropriate choice for a threshold. For instance, if $x = -c$, then this δ_0 would be undefined since the denominator would be zero.

More importantly, our choice for a threshold must not depend on the variable x. A suitable threshold δ imposes a condition of the value of an input x in terms of how close it should be to c in order to ensure their outputs are within ε of each other. We need to control the size of $|x+c|$ in some way, keeping in mind that we generally want x to be near c.

This may seem strange, but to get a handle on both $|x - c|$ and $|x + c|$ we are free to choose any positive real number as a preliminary version of δ and see what happens. After all, the underlying idea is to find a constraint on how close x needs to be to c, not to find a nice formula for δ.

Let's see what happens when we take $|x - c| < 0.5$. There's nothing special about 0.5, except that it is a positive number and gives us a baseline distance to build on, like this: When x is within 0.5 of c, how large is $|x + c|$? By the reverse triangle inequality (1.2.37) we have

$$|x| - |c| \leq |x - c| < 0.5. \tag{4.3.53}$$

Therefore, adding $|c|$ yields

$$|x| < |c| + 0.5. \tag{4.3.54}$$

By the (regular) triangle inequality (1.2.35), we further have

$$|x + c| \leq |x| + |c| < 2|c| + 0.5. \tag{4.3.55}$$

Under the preliminary assumption that $|x - c| < 0.5$, our scratch work looks like

$$|x^2 - c^2| = |x - c||x + c| < |x - c|(2|c| + 0.5) < \varepsilon. \tag{4.3.56}$$

Hence, choosing a threshold so that

$$|x - c| < \frac{\varepsilon}{2|c| + 0.5} \tag{4.3.57}$$

seems in order. However, inequality (4.3.57) is only valid when $|x - c| < 0.5$, so we need to keep this constraint in mind. Hence, a suitable choice for a threshold δ may well be

$$\delta = \min\left\{0.5, \frac{\varepsilon}{2|c| + 0.5}\right\}. \tag{4.3.58}$$

Play around with the "Delta Tester - continuity" Desmos activity using the QR code in Figure 4.3.5. On to the proof.

Proof for Example 4.3.15. Suppose $c \in \mathbb{R}$ and let $\varepsilon > 0$. Define a candidate for a threshold δ by

$$\delta = \min\left\{0.5, \frac{\varepsilon}{2|c| + 0.5}\right\}. \tag{4.3.59}$$

Note that $\delta > 0$. Now suppose $|x - c| < \delta$. Since

$$|x - c| < \delta \leq 0.5, \tag{4.3.60}$$

by the reverse triangle inequality (1.2.37) we have

$$|x| - |c| \leq |x - c| < 0.5 \implies |x| < |c| + 0.5. \tag{4.3.61}$$

Next, by the triangle inequality and inequality (4.3.61) we have

$$|x + c| \leq |x| + |c| < 2|c| + 0.5. \tag{4.3.62}$$

Figure 4.3.5: To accompany Example 4.3.15 and Scratch Work 4.3.16, play around with the Desmos activity "Delta Tester - continuity" accessed through the QR code. https://www.desmos.com/calculator/fgis71izbe

.

By our choice for δ given by (4.3.59) we also have

$$|x - c| < \delta \le \frac{\varepsilon}{2|c| + 0.5}. \tag{4.3.63}$$

Therefore, if $|x - c| < \delta$, then

$$|x^2 - c^2| = |x - c||x + c| < \left(\frac{\varepsilon}{2|c| + 0.5} \right)(2|c| + 0.5) = \varepsilon. \tag{4.3.64}$$

Hence, $h(x) = x^2$ is continuous on $\mathbb{R}$. $\qquad\qquad\square$

The trigonometric function sine is continuous.

Example 4.3.17: Sine is continuous

The function $f : \mathbb{R} \to \mathbb{R}$ given by $f(x) = \sin x$ is continuous.

Scratch Work 4.3.18: Use trigonometry

There are lots of ways to prove the continuity of the sine function, including identifying sine as a particular *power series*. The approach taken here relies on some trigonometric identities and inequalities, including a "sum-to-product" identity. This allows us to find a suitable threshold δ for the continuity of sine without much more effort.

Here are the relevant trigonometric properties. Their proofs are omitted, but are pretty cool nonetheless. For every $x, c \in \mathbb{R}$, we have:

$$\sin x - \sin c = 2 \cos \left(\frac{x + c}{2} \right) \sin \left(\frac{x - c}{2} \right), \tag{4.3.65}$$

$$|\sin x| \le x, \quad \text{and} \tag{4.3.66}$$

$$|\cos x| \le 1. \tag{4.3.67}$$

Now, let's see where we'd like to end up. To show sine is continuous by Definition 4.3.2, the concluding inequality would be

$$|\sin x - \sin c| < \varepsilon. \tag{4.3.68}$$

We can convert this inequality into one involving the distance $|x - c|$ in the domain using the trigonometric properties above, in the order provided. We have

$$|\sin x - \sin c| = \left|2\cos\left(\frac{x+c}{2}\right)\sin\left(\frac{x-c}{2}\right)\right| \tag{4.3.69}$$

$$\leq 2\left|\sin\left(\frac{x-c}{2}\right)\right| \tag{4.3.70}$$

$$\leq 2\left|\frac{x-c}{2}\right| \tag{4.3.71}$$

$$= |x - c|. \tag{4.3.72}$$

So, a choice of $\delta = \varepsilon$ should work.

Proof for Example 4.3.17. Let $\varepsilon > 0$ and define $\delta = \varepsilon > 0$. Suppose $|x-c| < \varepsilon$. Stringing together the trigonometric properties from Scratch Work 4.3.18 yields

$$|\sin x - \sin c| = \left|2\cos\left(\frac{x+c}{2}\right)\sin\left(\frac{x-c}{2}\right)\right| \tag{4.3.73}$$

$$\leq 2\left|\sin\left(\frac{x-c}{2}\right)\right| \tag{4.3.74}$$

$$\leq 2\left|\frac{x-c}{2}\right| \tag{4.3.75}$$

$$= |x - c| \tag{4.3.76}$$

$$< \varepsilon. \tag{4.3.77}$$

Therefore, sine is continuous by Definition 4.3.2. $\square$

As with the definition of limit and convergence for sequences (Definition 2.2.1), the threshold δ from the definition of continuity (Definition 4.3.2) gives us a way to interpret *rate of convergence for continuity.*

Remark 4.3.19: Continuity and rate of convergence

The relationship between the distances $\varepsilon > 0$ for the codomain and $\delta > 0$ for the domain D establishes a rate of convergence with the key implication

$$\mathbf{x} \in D \text{ with } \|\mathbf{x} - \mathbf{c}\|_k < \delta \quad \Longrightarrow \quad \|f(\mathbf{x}) - f(\mathbf{c})\|_m < \varepsilon. \tag{4.3.78}$$

Essentially, the rate of convergence tells us through the threshold δ how close inputs $\mathbf{x}$ need to be the point $\mathbf{c}$ should be to ensure their outputs $f(\mathbf{x})$ are within ε of the output $f(\mathbf{c})$. Typically, the smaller we take ε to be, the smaller δ needs to be.

Consider the scratch work involved with finding suitable thresholds for the functions f,

r, and h in Examples 4.3.5, 4.3.13, 4.3.15, and 4.3.17, along with the array of resulting formulas. We have

$$\delta = 2\varepsilon \quad \text{for} \quad f(x) = x/2 \text{ and every } c \in \mathbb{R}, \tag{4.3.79}$$

$$\delta = \varepsilon^2 \quad \text{for} \quad r(x) = \sqrt{x} \text{ at } c = 0, \tag{4.3.80}$$

$$\delta = \varepsilon\sqrt{c} \quad \text{for} \quad r(x) = \sqrt{x} \text{ at } c > 0, \tag{4.3.81}$$

$$\delta = \min\left\{0.5, \frac{\varepsilon}{2|c| + 0.5}\right\} \quad \text{for} \quad h(x) = x^2 \text{ and each } c \in \mathbb{R}, \quad \text{and} \tag{4.3.82}$$

$$\delta = \varepsilon \quad \text{for} \quad f(x) = \sin x \text{ and every } c \in \mathbb{R}. \tag{4.3.83}$$

Each of these can be visualized using the "Delta Tester - continuity" Desmos activity from Figure 4.3.5 by changing the formulas for the function and the threshold accordingly.

The next definition sets us up to prove continuous functions exhibit an interesting relationship with boundedness: Continuous functions are *locally bounded*, but not necessarily *bounded*. The definition of a bounded function generalizes the definition of a bounded set (Definition 1.5.20), as follows: A function is bounded if its range is a bounded set. On the other hand, a function is locally bounded at a point if its outputs are bounded when the inputs are restricted to a neighborhood of the point.

Definition 4.3.20: Bounded function

Let $D \subseteq \mathbb{R}^k$ and $f : D \to \mathbb{R}^m$. The function f is *bounded* if its range $f(D)$ is a bounded set. That is, f is bounded if there exists some $b \geq 0$ where

$$\|f(\mathbf{x})\| \leq b \quad \text{for all} \quad \mathbf{x} \in D. \tag{4.3.84}$$

Such a nonnegative real number b is called a *bound* for f and we say f is *bounded by* b. If f is not bounded, we say f is *unbounded*.

Additionally, given a point $\mathbf{c} \in D$, the function f is *locally bounded at* $\mathbf{c}$ if there exist a distance $\delta_{\mathbf{c}} > 0$ and a *local bound* $b_{\mathbf{c}} \geq 0$ where

$$\|f(\mathbf{x})\| \leq b_{\mathbf{c}} \quad \text{for all} \quad \mathbf{x} \in D \cap V_{\delta_{\mathbf{c}}}(\mathbf{c}). \tag{4.3.85}$$

The proof of the next theorem follows from the careful manipulation of the definition of continuity (Definition 4.3.2). Here, continuity is assumed and the threshold we obtain for the domain provides the neighborhood on which the function exhibits a "local" behavior.

Theorem 4.3.21: Continuous functions are locally bounded

Suppose $D \subseteq \mathbb{R}^k$, $\mathbf{c} \in D$, and $f : D \to \mathbb{R}^m$. If f is continuous at $\mathbf{c}$, then f is locally bounded at $\mathbf{c}$.

Scratch Work 4.3.22: Like boundedness of convergent sequences

The proof is very similar in spirit to the proof that convergent sequences are bounded (Theorem 2.3.15). In that context, all but a finite number of terms are within a neighborhood of the limit. So, either one of the terms outside such a neighbhorhood provides a bound, or the neighborhood of the limit does.

Here, since we are dealing with local boundedness, we only need to consider inputs within a neighborhood and bound their outputs. Continuity allows us to do just that. By picking any positive distance (say 7) for the outputs, continuity ensures the existence of a threshold that defines a distance—and therefore neighborhood—in the domain whose image is bounded.

Proof of Theorem 4.3.21. Suppose $D \subseteq \mathbb{R}^k$, $\mathbf{c} \in D$, $f : D \to \mathbb{R}^m$, and f is continuous at $\mathbf{c}$. Choose $\varepsilon_0 = 7 > 0$ as a positive distance in the range. Since f is continuous at $\mathbf{c}$, by Definition 4.3.2 there is a threshold $\delta > 0$ giving us a distance for the domain where

$$\mathbf{x} \in D \text{ with } \|\mathbf{x} - \mathbf{c}\|_k < \delta \quad \Longrightarrow \quad \|f(\mathbf{x}) - f(\mathbf{c})\|_m < 7. \tag{4.3.86}$$

So, by the reverse triangle inequality (1.2.37), for every $\mathbf{x} \in D \cap V_\delta(\mathbf{c})$ we have

$$\|f(\mathbf{x})\|_m - \|f(\mathbf{c})\|_m \leq \|f(\mathbf{x}) - f(\mathbf{c})\|_m < 7. \tag{4.3.87}$$

By adding $\|f(\mathbf{c})\|_m$ we also have

$$\|f(\mathbf{x})\|_m < \|f(\mathbf{c})\|_m + 7. \tag{4.3.88}$$

Therefore, $b_{\mathbf{c}} = \|\mathbf{c}\|_m + 7$ is a local bound and f is locally bounded at $\mathbf{c}$ (see Definition 4.3.20). $\square$

Theorem 4.3.21 gives us a glimpse look into the types of results we get with continuous functions. The ability to control how close we want outputs to be by controlling inputs is a powerful tool. However, even though continuous functions are locally bounded, they are not necessarily bounded across the whole domain.

Example 4.3.23: Lines are not bounded

The function $f : \mathbb{R} \to \mathbb{R}$ given by

$$f(x) = x/2 \tag{4.3.89}$$

is continuous but not bounded. As shown in Example 4.3.5, f is continuous. To see that f is not bounded, suppose $y \geq 0$. By the Archimedean Property (Theorem 1.4.6), there is a positive integer $n_y \in \mathbb{N}$ such that $y < n_y$. Hence,

$$f(y) = \frac{y}{2} < y < n_y. \tag{4.3.90}$$

So, y is not a bound for f. Since y represents an arbitrary positive real number, f is not bounded.

In general, thresholds can be challenging to find, whether we choose to work with preserving closeness (Definition 4.2.5), continuity (Definition 4.3.2), or limits of sequences (Definition 2.2.1). The following section defines and explores yet another meaning for continuity and provides us with alternative paths to proving and working with continuity. This time, the ideas are based on preserving the convergence of sequences, allowing us to take advantage of the results in Chapter 2.

Exercises

4.3.1. Given a threshold for the continuity of a function, any smaller positive number is also a threshold (see Definition 4.3.2). To prove this, suppose $D \subseteq \mathbb{R}^k$, $f : D \to \mathbb{R}^m$, $\mathbf{c} \in D$, and f is continuous at $\mathbf{c}$ with threshold $\delta > 0$ responding to the distance $\varepsilon > 0$. Prove that if $0 < \sigma \leq \delta$, then σ is also a threshold for the continuity of f at $\mathbf{c}$ in response to ε.

4.3.2. Suppose $f : \mathbb{R} \to \mathbb{R}$ is the exponential function given by $f(x) = e^x$. Assuming the algebraic properties of the exponential function from calculus hold, prove f is continuous.

4.3.3. Suppose $g : \mathbb{R} \to \mathbb{R}$ is the natural logarithm given by $g(x) = \ln x$. Assuming the algebraic properties of the natural logarithm from calculus hold, prove g is continuous.

4.3.4. Prove monomials are continuous. That is, the functions $h_n : \mathbb{R} \to \mathbb{R}$ given by $h_n(x) = x^n$ for each $n \in \mathbb{N}$ are continuous.

4.3.5. Prove nth roots are continuous. That is, the functions $r_n : [0, \infty) \to \mathbb{R}$ given by $r_n(x) = \sqrt[n]{x}$ for each $n \in \mathbb{N}$ are continuous.

4.3.6. Consider the function $q : \mathbb{R} \to \mathbb{R}$ given by

$$q(x) = \begin{cases} \dfrac{x^2 - 9}{x - 3}, & \text{if } x \neq 3, \\ 6, & \text{if } x = 3. \end{cases} \tag{4.3.91}$$

Prove q is continuous at $c = 3$.

4.3.7. Suppose $h : \mathbb{R} \to \mathbb{R}$ is continuous and $h(c) > 0$ for some $c \in \mathbb{R}$. Prove h is locally positive at c in the sense that there is some $\delta > 0$ such that

$$|x - c| < \delta \quad \Longrightarrow \quad f(x) > 0. \tag{4.3.92}$$

In other words, $h(x)$ is positive for all x in this δ-neighborhood of c.

4.3.8. The maximum of a finite number of real-valued continuous functions is continuous. Prove this result by completing the following steps.

(i) Suppose $a, b \in \mathbb{R}$. Prove

$$\max\{a, b\} = \frac{1}{2}[(a + b) + |a - b|]. \tag{4.3.93}$$

(ii) Suppose $D \subseteq \mathbb{R}$, $f, g : D \to \mathbb{R}$, and f and g are continuous. Define $h : D \to \mathbb{R}$ by

$$h(x) = \max\{f(x), g(x)\} \qquad \text{for all } x \in D. \tag{4.3.94}$$

Prove h is continuous.

(iii) Suppose $D \subseteq \mathbb{R}$, $n \in \mathbb{N}$, $f_1, \ldots, f_n : D \to \mathbb{R}$, and $f_1, \ldots, f_n$ are continuous. Define $q : D \to \mathbb{R}$ by

$$q(x) = \max\{f_1(x), \ldots, f_n(x)\} \qquad \text{for all } x \in D. \tag{4.3.95}$$

Prove q is continuous.

4.3.9. Suppose $f : \mathbb{R} \to \mathbb{R}$ is continuous. Prove $f^{-1}((a,b))$ is open. (That is, for such f, the inverse image of an open interval in the codomain is a open set in the domain.)

4.4   Equivalent forms of continuity

So far, continuity has been motivated by the preservation of closeness in Section 4.2 and formally defined in Section 4.3. Given the similarity of the techniques used in those sections, especially in the examples and finding suitable thresholds, it may come as no surprise that the definitions for the preservation of closeness and continuity are equivalent.

This section introduces a third version of continuity by way of convergent sequences called *sequential continuity* (Definition 4.4.5). To me, this version connects formal mathematics to the informal phrasing to describe continuity I heard in my calculus classes: $\mathbf{x}$ "approaches" $\mathbf{c}$ implies $f(\mathbf{x})$ "approaches" $f(\mathbf{c})$. Hopefully the fundamental connection between arbitrarily close and convergent sequences (Theorem 2.3.1) adds to your experience in your calculus classes and helps you develop some intuition for sequential continuity.

Given our large set of results on convergent sequences in Chapter 2, sequential continuity provides a useful and powerful perspective to build on. The key result in this section is Theorem 4.4.7 which establishes the equivalence of continuity, preserving closeness, and sequential continuity (Definitions 4.3.2, 4.2.5, and 4.4.5, respectively).

Before proving this big result, an exploration of how functions transform sequences is in order. To help, consider the following trio of sequences and how they are transformed by the functions f, g, and v originally from Examples 4.1.2, 4.1.3, and 4.2.4.

Example 4.4.1: Three sequences

Consider the following trio of sequences (x_n), (y_n), and (z_n) defined for each positive integer n by

$$x_n = 2 + \frac{2}{n}, \quad y_n = 2 - \frac{2}{n}, \quad \text{and} \quad z_n = 2 + \frac{2(-1)^n}{n}. \tag{4.4.1}$$

See Figure 4.4.1.

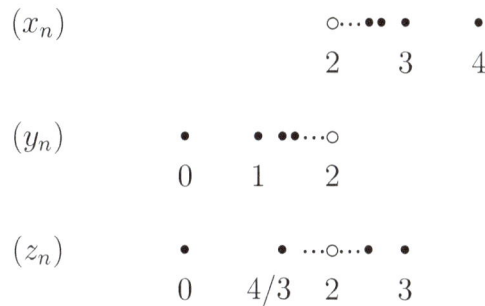

Figure 4.4.1: The sequences $(x_n), (y_n)$, and (z_n) from Example 4.4.1 are contained in the interval $I = [0, 4]$ and each converges to 2.

Each of the sequences $(x_n), (y_n)$, and (z_n) converges to 2. I'll prove the case for (z_n) but skip the scratch work for the threshold n_ε.

Proof for Example 4.4.1. Let $\varepsilon > 0$. By the Archimedean Property 1.4.6, we can choose $n_\varepsilon \in \mathbb{N}$ large enough to ensure

$$n_\varepsilon > \frac{2}{\varepsilon} \qquad \text{and therefore} \qquad \frac{2}{n_\varepsilon} < \varepsilon. \tag{4.4.2}$$

So, n_ε is a threshold since for all $n \geq n_\varepsilon$ since in this case we have

$$|z_n - 2| = \left|2 + \frac{2(-1)^n}{n} - 2\right| = \left|\frac{2(-1)^n}{n}\right| = \frac{2}{n} \leq \frac{2}{n_\varepsilon} < \varepsilon. \tag{4.4.3}$$

Hence, $\lim_{n\to\infty} z_n = 2$. □

Proofs showing $\lim_{n\to\infty} x_n = 2$ and $\lim_{n\to\infty} y_n = 2$ are nearly identical to the proof showing $\lim_{n\to\infty} z_n = 2$. In fact, we can use the same choice for a threshold in each case, namely $n_\varepsilon \in \mathbb{N}$ large enough to ensure $n_\varepsilon > 2/\varepsilon$. Try it yourself. My guess is your scratch work will lead you to the same choice for n_ε, but it'd be interest to see what you come up with.

Let's see how the functions f, g, and v from Examples 4.1.2, 4.1.3, and 4.2.4 transform the sequences $(x_n), (y_n)$, and (z_n) from Example 4.4.1.

Example 4.4.2: Images of three sequences under a line

As in Example 4.1.2, define $f : \mathbb{R} \to \mathbb{R}$ define by

$$f(x) = x/2, \tag{4.4.4}$$

and define $(x_n), (y_n)$, and (z_n) as in Example 4.4.1. For each $n \in \mathbb{N}$ we have

$$f(x_n) = 1 + \frac{1}{n}, \quad f(y_n) = 1 - \frac{1}{n}, \quad \text{and} \quad f(z_n) = 1 + \frac{1(-1)^n}{n}. \tag{4.4.5}$$

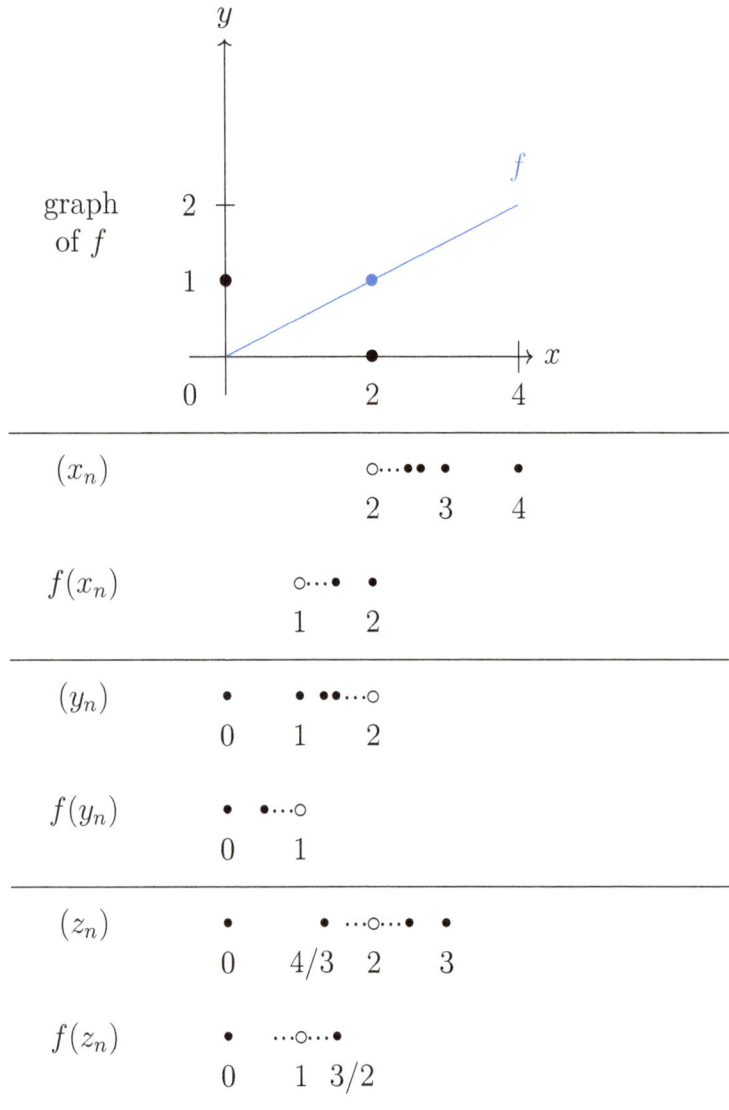

Figure 4.4.2: The images of the sequences $(x_n), (y_n)$, and (z_n) from Example 4.4.1 under the function f. See Example 4.4.2. The sequences $(f(x_n)), (f(y_n))$, and $(f(z_n))$ each converge to $f(2) = 1$. Note $f(2) = 1$ is indicated with a $\bullet$ in the graph of f but with a $\circ$ with the sequences $(f(y_n))$ and $(f(z_n))$ since it is not a term. For $(f(x_n))$, we happen to have $f(x_1) = f(4) = 2$.

See Figure 4.4.2. We have

$$\lim_{n\to\infty} x_n = 2 \quad \text{while} \quad \lim_{n\to\infty} f(x_n) = 1 = f(2), \tag{4.4.6}$$

$$\lim_{n\to\infty} y_n = 2 \quad \text{while} \quad \lim_{n\to\infty} f(y_n) = 1 = f(2), \quad \text{and} \tag{4.4.7}$$

$$\lim_{n\to\infty} z_n = 2 \quad \text{while} \quad \lim_{n\to\infty} f(z_n) = 1 = f(2). \tag{4.4.8}$$

In fact, if we wanted to prove these limits are all equal to 1, the same choice for the threshold suffices in all three cases. Namely, any positive integer n_ε where $n_\varepsilon > 1/\varepsilon$ works. Plus, this inequality establishes the rates of convergence for the output sequences $f(x_n), f(y_n)$, and $f(z_n)$.

Also, the common rate of convergence for the sequences $(x_n), (y_n)$, and (z_n) is given by the inequality for the threshold $n_\varepsilon > 2/\varepsilon$, twice that of the rate of convergence for their output sequences.

Example 4.4.3: Images of three sequences under a piecewise defined function

As in Example 4.1.3, define $g : \mathbb{R} \to \mathbb{R}$ define by

$$g(x) = \begin{cases} x/2, & \text{if } x < 2, \\ 1 + (x/2), & \text{if } x \geq 2, \end{cases} \tag{4.4.9}$$

and define $(x_n), (y_n)$, and (z_n) as in Example 4.4.1. For each $n \in \mathbb{N}$ we have

$$g(x_n) = 2 + \frac{1}{n}, \tag{4.4.10}$$

$$g(y_n) = 1 - \frac{1}{n}, \quad \text{and} \tag{4.4.11}$$

$$g(z_n) = \begin{cases} 1 - \dfrac{1}{n}, & \text{if } n \text{ is odd,} \\[2mm] 2 + \dfrac{1}{n}, & \text{if } n \text{ is even.} \end{cases} \tag{4.4.12}$$

See Figure 4.4.3. We have

$$\lim_{n\to\infty} x_n = 2 \quad \text{while} \quad \lim_{n\to\infty} g(x_n) = 2 = g(2), \tag{4.4.13}$$

$$\lim_{n\to\infty} y_n = 2 \quad \text{while} \quad \lim_{n\to\infty} g(y_n) = 1 \neq g(2), \quad \text{and} \tag{4.4.14}$$

$$\lim_{n\to\infty} z_n = 2 \quad \text{while} \quad \lim_{n\to\infty} g(z_n) \text{ does not exist.} \tag{4.4.15}$$

Therefore, g does not preserve the convergence of sequences. Specifically, (z_n) converges in the domain, but $g(z_n)$ diverges in the range. Also, (y_n) converges to $c = 2$ in the domain, but $g(y_n)$ converges to 1 in the range while $g(c) = g(2) = 2$. Play around with the GeoGebra activity "images of functions, g" found in Figure 4.4.4.

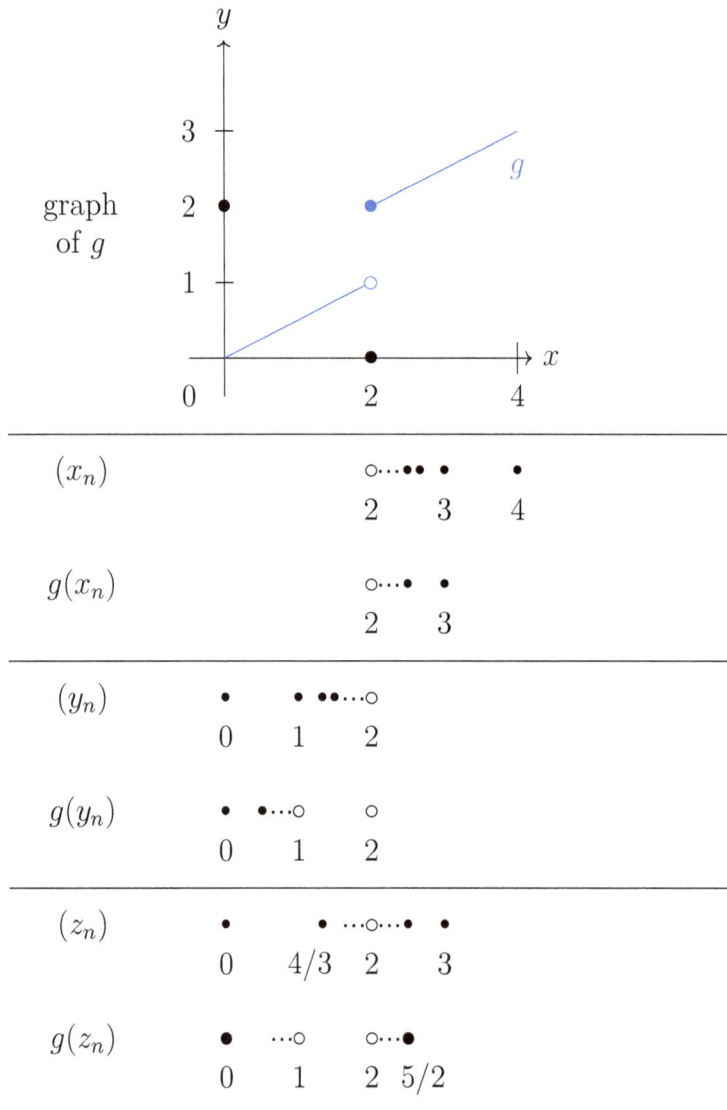

Figure 4.4.3: The images of the sequences $(x_n), (y_n)$, and (z_n) from Example 4.4.1 under the function g. See Example 4.4.3. The sequences $(g(x_n)), (g(y_n))$, and $(g(z_n))$ exhibit different limiting behavior. Note $g(2) = 2$ is indicated with a • in the graph of g but with a ∘ with the sequences $(g(x_n)), (g(y_n))$, and $(g(z_n))$ since it is not a term.

Figure 4.4.4: To accompany Example 4.4.3 and Theorem 4.4.7, play around with the GeoGebra activity "images of functions, g" accessed through this QR code. https://www.geogebra.org/m/wtmkrzsz
.

Example 4.4.4: Images of three sequences under another piecewise defined function

As in Example 4.2.4, define $v : \mathbb{R} \to \mathbb{R}$ define by

$$v(x) = \begin{cases} \dfrac{1}{2-x}, & \text{if } x < 2, \\[4mm] 1 + \dfrac{x}{2}, & \text{if } x \geq 2. \end{cases} \qquad (4.4.16)$$

and define $(x_n), (y_n)$, and (z_n) as in Example 4.4.1. For each $n \in \mathbb{N}$ we have

$$v(x_n) = 2 + \frac{1}{n}, \qquad (4.4.17)$$

$$v(y_n) = \frac{n}{2}, \qquad \text{and} \qquad (4.4.18)$$

$$v(z_n) = \begin{cases} \dfrac{n}{2}, & \text{if } n \text{ is odd}, \\[4mm] 2 + \dfrac{1}{n}, & \text{if } n \text{ is even}. \end{cases} \qquad (4.4.19)$$

See Figure 4.4.5. We have

$$\lim_{n \to \infty} x_n = 2 \quad \text{while} \quad \lim_{n \to \infty} v(x_n) = 2 = h(2), \qquad (4.4.20)$$

$$\lim_{n \to \infty} y_n = 2 \quad \text{while} \quad \lim_{n \to \infty} v(y_n) = \infty \neq v(2), \qquad \text{and} \qquad (4.4.21)$$

$$\lim_{n \to \infty} z_n = 2 \quad \text{while} \quad \lim_{n \to \infty} v(z_n) \text{ does not exist.} \qquad (4.4.22)$$

Therefore, v does not preserve the convergence of sequences. Specifically, (y_n) and (z_n) converge in the domain, but $v(y_n)$ and $v(z_n)$ diverge in the range. You might recall that $\lim_{n \to \infty} v(y_n) = \infty$ means $v(y_n)$ diverges in a particular way: The terms $v(y_n)$ get larger and larger without bound in the positive direction as the index n increases.

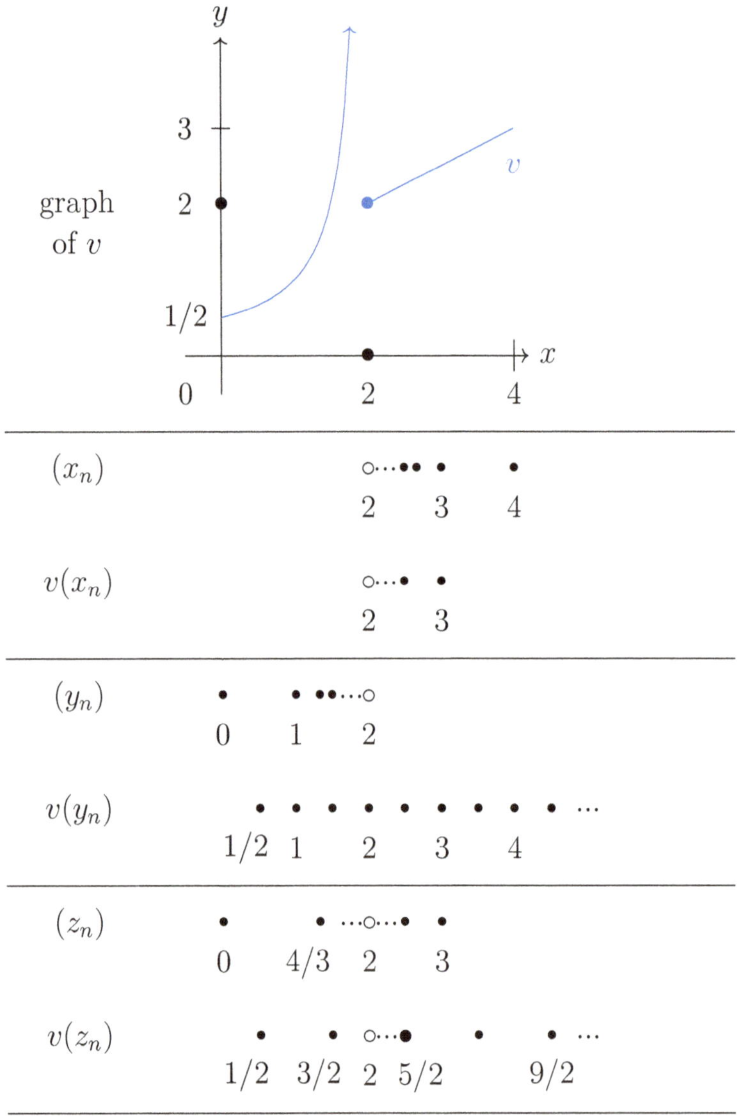

Figure 4.4.5: The images of the sequences $(x_n), (y_n)$, and (z_n) from Example 4.4.1 under the function v. See Example 4.4.4. The sequences $(v(x_n)), (v(y_n))$, and $(v(z_n))$ exhibit different limiting behavior. Note $v(2) = 2$ is indicated with a • in the graph of v but with a ∘ with the sequences $(v(x_n)), (v(y_n))$, and $(v(z_n))$.

The functions g in Examples 4.1.3, 4.2.3, and 4.4.3, and v in 4.2.4 and 4.4.4 each take the point $c = 2$ along with the set $E_4 = (7/4, 2)$ and sequence $(y_n) = (2 - (2/n))$—which are arbitrarily close in the domain—and map them to images that are away from each other in the range. In other words, the fact that g and v are discontinuous at $c = 2$ is captured by the lack of preservation of closeness as well as the lack of the preservation of the convergence of sequences.

On the other hand, function f defined in Examples 4.1.2 and 4.2.2 is proven to preserve closeness at $c = 2$ in Example 4.2.7. In Example 4.4.2, f preserves the convergence of (x_n), (y_n), and (z_n) to $c = 2$ in that their images (x_n), (y_n), and (z_n) converge to $f(c) = f(2) = 1$. But more is true: f preserves the convergence of *every* sequence that converges to $c = 2$.

The following definition for *sequential continuity* provides a formal notion for when a function preserves the convergence and limits of sequences.

> ### Definition 4.4.5: Sequential continuity
>
> Suppose $D \subseteq \mathbb{R}^k$, $\mathbf{c} \in D$, and $f : D \to \mathbb{R}^m$. We say f is *sequentially continuous at* $\mathbf{c}$ if
>
> $$(\mathbf{x}_n) \subseteq D \text{ with } \lim_{n \to \infty} \mathbf{x}_n = \mathbf{c} \qquad \Longrightarrow \qquad \lim_{n \to \infty} f(\mathbf{x}_n) = f(\mathbf{c}). \tag{4.4.23}$$
>
> If f is sequentially continuous at every point in the domain D, then f is *sequentially continuous*.

Remark 4.4.6: Move limits in and out

We can think of sequential continuity symbolically as moving the limit notation from outside f to inside:

$$\lim_{n \to \infty} f(\mathbf{x}_n) = f\left(\lim_{n \to \infty} \mathbf{x}_n \right) = f(\mathbf{c}). \tag{4.4.24}$$

Sequential continuity basically says that either way—whether we plug $(\mathbf{x}_n)$ into f first then take the limit of $f(\mathbf{x}_n)$, or we take the limit of $(\mathbf{x}_n)$ first then plug the limit $\mathbf{c}$ into f—we end up with the same value $f(\mathbf{c})$. See Figure 4.4.6.

The definitions of preserving closeness (Definition 4.2.5), sequential continuity (Definition 4.4.5), and "ε-δ" continuity (Definition 4.3.2) are equivalent. Each provides its own perspective on continuity along with its own benefits and drawbacks. Each can be used to check whether the others hold. Ultimately, when I think of continuity in analysis, I have each of these equivalent notions in mind knowing I can run with the one that seems to be the easiest to work with.

Theorem 4.4.7: Equivalent forms of continuity

Suppose $D \subseteq \mathbb{R}^k$, $\mathbf{c} \in D$, and $f : D \to \mathbb{R}^m$. Then the following statements are equivalent forms of the continuity of f at $\mathbf{c}$:

(i) f preserves closeness at $\mathbf{c}$ (Definition 4.2.5):

$$E \subseteq D \text{ with } \mathbf{c} \operatorname{acl} E \qquad \Longrightarrow \qquad f(\mathbf{c}) \operatorname{acl} f(E). \tag{4.4.25}$$

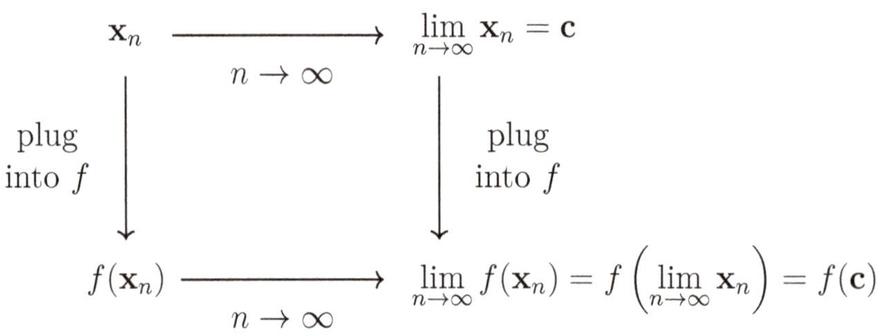

$$\mathbf{x}_n \xrightarrow{\quad n \to \infty \quad} \lim_{n\to\infty} \mathbf{x}_n = \mathbf{c}$$

$$\downarrow \text{plug into } f \qquad\qquad \downarrow \text{plug into } f$$

$$f(\mathbf{x}_n) \xrightarrow{\quad n \to \infty \quad} \lim_{n\to\infty} f(\mathbf{x}_n) = f\left(\lim_{n\to\infty} \mathbf{x}_n\right) = f(\mathbf{c})$$

Figure 4.4.6: As in Remark 4.4.6, this commutative diagram visualizes sequential continuity in that whether we take the limit as n tends to infinity or evaluate f first, we end up with the same result $f(\mathbf{c})$. Also, sequential continuity symbolically amounts to moving the "lim" symbol in and out of the function f with no effect on the resulting value.

(ii) f is sequentially continuous at $\mathbf{c}$ (Definition 4.4.5):

$$(\mathbf{x}_n) \subseteq D \text{ with } \lim_{n\to\infty} \mathbf{x}_n = \mathbf{c} \quad\Longrightarrow\quad \lim_{n\to\infty} f(\mathbf{x}_n) = f(\mathbf{c}). \tag{4.4.26}$$

(iii) f is ε-δ continuous at $\mathbf{c}$ (Definition 4.3.2):

$$\text{For every } \varepsilon > 0, \text{ there is a threshold } \delta > 0 \text{ such that}$$
$$\mathbf{x} \in D \text{ with } \|\mathbf{x} - \mathbf{c}\|_k < \delta \quad\Longrightarrow\quad \|f(\mathbf{x}) - f(\mathbf{c})\|_m < \varepsilon. \tag{4.4.27}$$

The reference to the definition of continuity as "ε-δ continuity" for Definition 4.3.2 distinguishes it from sequential continuity in Definition 4.4.5. Also, play around with the GeoGebra activity found in Figure 4.4.4.

Scratch Work 4.4.8: Contraposition

The proof argues via contraposition multiple times. So, along with the fundamental connection between arbitrarily close and convergent sequences in Theorem 2.3.1, the negations of the definitions for preserving closeness, sequential continuity, and ε-δ continuity are used throughout the proof. Their definitions are provided directly in the statement of the theorem for convenience, but you can also see Definitions 4.2.5, 4.4.5, and 4.3.2, respectively.

Proof of Theorem 4.4.7. Let's show (i) $\Longrightarrow$ (ii), (ii) $\Longrightarrow$ (iii), and (iii) $\Longrightarrow$ (i), all by contraposition.

(i) $\Longrightarrow$ (ii): Suppose f is not sequentially continuous at a point $\mathbf{c}$ in D. Then there is a sequence $(\mathbf{x}_n)$ of points in D and a point $\mathbf{c}$ in D where

$$\lim_{n\to\infty} \mathbf{x}_n = \mathbf{c} \quad \text{but} \quad \lim_{n\to\infty} f(\mathbf{x}_n) \neq f(\mathbf{c}). \tag{4.4.28}$$

Hence, there must be some $\varepsilon_0 > 0$ such that no matter which positive integer N we consider, there is a positive integer $q \geq N$ where $\mathbf{x}_q$ is in D and

$$\|f(\mathbf{x}_q) - f(\mathbf{c})\|_m \geq \varepsilon_0. \tag{4.4.29}$$

We can use the previous statement to construct a suitable subsequence $(\mathbf{x}_{n_j})$ of $(\mathbf{x}_n)$ whose image stays away from $f(\mathbf{c})$. That is, let $n_1 \geq 1$ be a positive integer that satisfies the previous statement. Proceeding inductively, for each positive integer j, there is a positive integer $n_j > n_{j-1}$ where

$$\|f(\mathbf{x}_{n_j}) - f(\mathbf{c})\|_m \geq \varepsilon_0. \tag{4.4.30}$$

Now, since $\lim_{n \to \infty} \mathbf{x}_n = \mathbf{c}$ and $(\mathbf{x}_{n_j})$ is a subsequence of $(\mathbf{x}_n)$, we have

$$\lim_{j \to \infty} \mathbf{x}_{n_j} = \mathbf{c} \tag{4.4.31}$$

as well. By Theorem 2.3.1, we have $\mathbf{c}\,\mathrm{acl}(\mathbf{x}_{n_j})$. But we also have $f(\mathbf{c})\,\mathrm{awf}(f(\mathbf{x}_{n_j}))$. Therefore, f does not preserve closeness at $\mathbf{c}$. (See Definition 4.2.5.)

(ii) $\implies$ (iii): Suppose f is not ε-δ continuous at a point $\mathbf{c}$ in D. Then there must be some $\varepsilon_0 > 0$ such that no matter which value we take for $\delta > 0$, there is a point $\mathbf{x}_\delta$ in D with

$$\|\mathbf{x}_\delta - \mathbf{c}\|_k < \delta \quad \text{and} \quad \|f(\mathbf{x}_\delta) - f(\mathbf{c})\|_m \geq \varepsilon_0. \tag{4.4.32}$$

Much as in the proof of Theorem 2.3.1, we can use the previous statement to construct a suitable sequence. For each positive integer n, there must be a point $\mathbf{x}_n$ in D where

$$\|\mathbf{x}_n - \mathbf{c}\|_k < 1/n \quad \text{and} \quad \|f(\mathbf{x}_n) - f(\mathbf{c})\|_m \geq \varepsilon_0. \tag{4.4.33}$$

(Note that $1/n$ plays the role of δ here). Hence, $(\mathbf{x}_n)$ is a sequence of points in D where

$$\lim_{n \to \infty} \mathbf{x}_n = \mathbf{c} \quad \text{but} \quad \lim_{n \to \infty} f(\mathbf{x}_n) \neq f(\mathbf{c}). \tag{4.4.34}$$

Therefore, f is not sequentially continuous at $\mathbf{c}$. (See Definition 4.4.5.)

(iii) $\implies$ (i): Finally, suppose f does not preserve closeness at a point $\mathbf{c}$ in D. Then there must be some $E \subseteq D$ where $\mathbf{c}\,\mathrm{acl}\,E$ but $f(\mathbf{c})\,\mathrm{awf}\,f(E)$. So, there is some $\varepsilon_0 > 0$ such that for every point $\mathbf{x}$ in E we have

$$\|f(\mathbf{x}) - f(\mathbf{c})\|_m \geq \varepsilon_0. \tag{4.4.35}$$

Now, let $\delta > 0$. Since $\mathbf{c}\,\mathrm{acl}\,E$, there is a point $\mathbf{y}_\delta$ in E where we have both

$$\|\mathbf{y}_\delta - \mathbf{c}\|_k < \delta \quad \text{and} \quad \|f(\mathbf{y}_\delta) - f(\mathbf{c})\|_m \geq \varepsilon_0. \tag{4.4.36}$$

Therefore, f is not ε-δ continuous at $\mathbf{c}$. (See Definition 4.3.2.) $\qquad\square$

Continuous functions have interesting effects on rates of convergence of sequences.

Remark 4.4.9: Continuity transforms rates of convergence

The functions given by $f(x) = x/2$, $h(x) = x^2$, and $r(x) = \sqrt{x}$ from Examples 4.1.2, 4.1.4, and 4.3.13 transform the rates of change of their convergent input sequences in ways that reflect the transformations of the inputs themselves. For instance, the input sequence $x_n = 1/n$ has limit 0 determined by a threshold $n_\varepsilon > 1\varepsilon$ which comes from solving the following inequality for the index n:

$$|x_n - 0| = \frac{1}{n} < \varepsilon \qquad \Longrightarrow \qquad n > \frac{1}{\varepsilon}. \tag{4.4.37}$$

The images $(f(x_n)) = 1/(2n)$, $(h(x_n)) = (1/n^2)$, and $(r(x_n)) = (1/\sqrt{n})$ have a common limit 0 as ensured by sequential continuity. A more detailed analysis of their rates of convergence comes from solving the following inequalities for the index n:

$$|f(x_n) - 0| = \frac{1}{2n} < \varepsilon \qquad \Longrightarrow \qquad n > \frac{1}{2\varepsilon}, \tag{4.4.38}$$

$$|h(x_n) - 0| = \frac{1}{n^2} < \varepsilon \qquad \Longrightarrow \qquad n > \frac{1}{\sqrt{\varepsilon}}, \qquad \text{and} \tag{4.4.39}$$

$$|r(x_n) - 0| = \frac{1}{\sqrt{n}} < \varepsilon \qquad \Longrightarrow \qquad n > \frac{1}{\varepsilon^2}. \tag{4.4.40}$$

Which output sequence converges to 0 the fastest?

The contrapositions of the implications in Theorem 4.4.7 are also useful for showing a function is discontinuous at $\mathbf{c}$. These results are summarized in Discontinuity Criteria (Corollary 4.6.13) along with negations of other results proven later on.

For now, the following example shows us how the negation of sequential continuity allows to prove the *topologist's sine curve* is discontinuous at $c = 0$.

Example 4.4.10: Topologist's sine curve

Consider the *topologist's sine curve* $t : \mathbb{R} \to \mathbb{R}$ given by

$$t(x) = \begin{cases} \sin(1/x), & \text{if } x \neq 0, \\ 0, & \text{if } x = 0. \end{cases} \tag{4.4.41}$$

The topologist's sine curve t is discontinuous at $c = 0$. This one is tricky to graph, try it yourself!

Scratch Work 4.4.11: Discontinuous via different sequential limits

The idea is to take advantage of the equivalence of ε-δ continuity and sequential continuity as well as the periodic structure of the sine function. Our goal is to find a sequence (x_n) of real numbers in the domain which converge to 0 but whose image $t(x_n)$ does not converge to 0. This will show t is not sequentially continuous at $c = 0$, so by Theorem 4.4.7, t is not continuous at $c = 0$.

The periodicity of the sine function tells us that for every index $n \in \mathbb{N}$ we have

$$\theta_n = \frac{\pi}{2} + 2\pi n \quad \implies \quad \sin \theta_n = \sin\left(\frac{\pi}{2} + 2\pi n\right) = \sin\frac{\pi}{2} = 1. \tag{4.4.42}$$

Now, motivated by the definition of t, if we take the reciprocals of the θ_n for each index n we get the sequence (x_n) where

$$x_n = \frac{1}{\frac{\pi}{2} + 2\pi n}. \tag{4.4.43}$$

In this case,

$$\lim_{n\to\infty} x_n = \lim_{n\to\infty}\left(\frac{1}{\frac{\pi}{2} + 2\pi n}\right) = 0 \quad \text{while} \tag{4.4.44}$$

$$\lim_{n\to\infty} t(x_n) = \lim_{n\to\infty} \sin\left(\frac{\pi}{2} + 2\pi n\right) = 1 \neq 0 = t(0). \tag{4.4.45}$$

The proof amounts to rearranging the scratch work.

Proof for Example 4.4.10. Define the sequence of real numbers (x_n) by

$$x_n = \frac{1}{\frac{\pi}{2} + 2\pi n} \quad \text{for each } n \in \mathbb{N}. \tag{4.4.46}$$

We have

$$\lim_{n\to\infty} x_n = \lim_{n\to\infty}\left(\frac{1}{\frac{\pi}{2} + 2\pi n}\right) = 0 \quad \text{while} \tag{4.4.47}$$

$$\lim_{n\to\infty} t(x_n) = \lim_{n\to\infty} \sin\left(\frac{\pi}{2} + 2\pi n\right) = 1 \neq 0 = t(0). \tag{4.4.48}$$

Therefore, by the negation of Definition 4.4.5, t is not sequentially continuous at $c = 0$. Therefore, by Theorem 4.4.7, t is discontinuous at $c = 0$. □

A more general recap of methods we can use to prove a function is discontinuous is provided by the result Discontinuity Criteria (Corollary 4.6.13) which incorporates some of the results in the next section.

Now that we've established Theorem 4.4.7, we have a variety of equivalent forms of continuity work with, namely the preservation of closeness, sequential continuity, and ε-δ continuity. In the next section, we use this flexibility to prove some classic results on examples and properties of continuous functions.

Exercises

4.4.1. Given a function $f : \mathbb{R} \to \mathbb{R}$, a *level set* is set of the form

$$L_k = \{x \in \mathbb{R} : f(x) = k\} \tag{4.4.49}$$

for some fixed real number k. Prove that if f is continuous, then its level sets are closed. (For the special case where $k = 0$, the level set L_0 is called the *zero set* of f and is sometimes denoted by $Z(f)$.)

4.4.2. Let $h : \mathbb{R} \to \mathbb{R}$ be given by

$$h(x) = \begin{cases} \dfrac{|x-2|}{x-2}, & \text{if } x \neq 2, \\ 0, & \text{if } x = 2. \end{cases} \tag{4.4.50}$$

(i) Prove h is continuous at every $c \neq 2$.

(ii) Prove h is discontinuous at $c = 2$.

(iii) Prove that no matter how we redefine the output $h(2)$, h is discontinuous at $c = 2$.

4.4.3. Suppose $f : \mathbb{R}^k \to \mathbb{R}^m$ is continuous and $E \subseteq \mathbb{R}^k$.

(i) Prove $f(\overline{E}) \subseteq \overline{f(E)}$.

(ii) Find an example of a function $f : \mathbb{R} \to \mathbb{R}$ and a set $A \subseteq \mathbb{R}$ where the containment in (i) is proper (that is, $f(\overline{A}) \subsetneq \overline{f(A)}$).

4.4.4. Recall Dirichlet's function $\mathbb{1}_{\mathbb{Q}}$ from Example 4.2.14 given by

$$\mathbb{1}_{\mathbb{Q}}(x) = \begin{cases} 1, & \text{if } x \in \mathbb{Q}, \\ 0, & \text{if } x \in \mathbb{R}\backslash\mathbb{Q}. \end{cases} \tag{4.4.51}$$

Prove $\mathbb{1}_{\mathbb{Q}}$ is discontinuous everywhere. That is, prove $\mathbb{1}_{\mathbb{Q}}$ is discontinuous at every $c \in \mathbb{R}$.

4.4.5. Consider the modification of Dirichlet's function $\mathbb{1}_{\mathbb{Q}}$ from Example 4.2.14 given by

$$f(x) = x\mathbb{1}_{\mathbb{Q}}(x) = \begin{cases} x, & \text{if } x \in \mathbb{Q}, \\ 0, & \text{if } x \in \mathbb{R}\backslash\mathbb{Q}. \end{cases} \tag{4.4.52}$$

Prove f is continuous at $c = 0$ but discontinuous everywhere else ($c \neq 0$).

4.4.6. *Thomae's function* $g : \mathbb{R} \to \mathbb{R}$ is a subtle modification of Dirichlet's function $\mathbb{1}_{\mathbb{Q}}$ from Example 4.2.14 given by

$$g(x) = \begin{cases} 1, & \text{if } x = 0, \\ \dfrac{1}{n}, & \text{if } x \in \mathbb{Q}\backslash\{0\} \text{ with reduced form } x = \dfrac{m}{n}, \\ 0, & \text{if } x \in \mathbb{R}\backslash\mathbb{Q}. \end{cases} \tag{4.4.53}$$

(i) Prove g is discontinuous at every $c \in \mathbb{Q}$.

(ii) Prove g is continuous at every $c \in \mathbb{R}\backslash\mathbb{Q}$.

Thomae's function g is continuous on the rationals and discontinuous on the irrationals!

4.4.7. Suppose $f, g : \mathbb{R}^k \to \mathbb{R}^m$ and H is a dense subset of $\mathbb{R}^k$. Prove that if f and g are continuous, then

$$f(\mathbf{y}) = g(\mathbf{y}) \text{ for all } \mathbf{y} \in H \qquad \Longrightarrow \qquad f(\mathbf{x}) = g(\mathbf{x}) \text{ for all } \mathbf{x} \in \mathbb{R}^k. \qquad (4.4.54)$$

That is, continuous functions between Euclidean spaces are completely determined by their behavior on any dense subset of the domain.

4.4.8. Suppose $g : \mathbb{R}^k \to \mathbb{R}^m$. Prove the following statements are equivalent:

(i) g is continuous (Definition 4.3.2).

(ii) g is *topologically continuous* in the following sense: If U is an open subset of the codomain $\mathbb{R}^m$, the its inverse image $g^{-1}(U)$ is an open subset of the domain $\mathbb{R}^k$. (That is, inverse images of open sets are open.)

4.5 Algebraic properties of continuity

Continuous functions exhibit a wealth of powerful properties, including linearity. To help motivate the upcoming results on continuity and provide a family of functions which will be revisited throughout the textbook, consider polynomials. Polynomials are defined along with monomials in 1.6.8, but a copy is given here for convenience.

Definition 4.5.1: Polynomial

A *polynomial* is a function $p : \mathbb{R} \to \mathbb{R}$ defined by a linear combination of monomials of the form

$$p(x) = \sum_{j=0}^{n} a_j x^j = a_0 + a_1 x + a_2 x^2 + \cdots + a_{n-1} x^{n-1} + a_n x^n \qquad (4.5.1)$$

where $n \in \mathbb{N} \cup \{0\}$, the coefficients $a_0, a_1, a_2, \ldots, a_{n-1}, a_n$ are real numbers, and the convention $x^0 = 1$ is used.

The proof of the next theorem is a culmination of results from this section and appears later.

Theorem 4.5.2: Polynomials are continuous

Polynomials are continuous on the real line $\mathbb{R}$.

Remark 4.5.3: Multiple perspectives on continuity

The proofs in this section make use of the three equivalent notions for continuity spelled out in Theorem 4.4.7: preserving closeness, continuity, and sequential continuity (respectively Definitions 4.2.5, 4.3.2, and 4.4.5). For proofs showing continuity based on the ε-δ form in Definition 4.3.2, the work follows the guide in Remark 4.3.4. The numerous results on convergent sequences in Chapter 2 can be paired with sequential continuity to prove the all of the results in this section.

A special case of the continuity of basic affine transformations (Theorem 4.3.9) tells us lines are continuous. The proof is omitted. Do you remember "slope-intercept form"?

Corollary 4.5.4: Lines are continuous

Suppose m and b are real numbers and let $f : \mathbb{R} \to \mathbb{R}$ be given by

$$f(x) = mx + b. \tag{4.5.2}$$

Then f is continuous.

The next result and many that follow assume we are given continuous functions to work with and show that certain combinations result in new continuous functions. As with the linearity of limits for sequences (Theorem 2.3.9), continuity is linear.

Theorem 4.5.5: Linearity of continuity

Suppose $D \subseteq \mathbb{R}^k, \mathbf{c} \in D$, and $f, g : D \to \mathbb{R}^m$. If f and g are continuous at $\mathbf{c}$, then

(i) $f + g$ is continuous at $\mathbf{c}$ *(additivity)*; and

(ii) αf is continuous at $\mathbf{c}$ for every scalar $\alpha \in \mathbb{R}$ *(homogeneity)*.

Scratch Work 4.5.6: A proof in three cases

The proof is broken down into three cases: $f + g$, $\alpha = 0$, and $\alpha \neq 0$. The $\alpha = 0$ case ends up being trivial, as in the $\alpha = 0$ case for the continuity of basic affine transformations (Theorem 4.3.9). The other two follow from the definition for continuity (Definition 4.3.2) by adapting assumed thresholds for f and g to fit the needs of the new functions αf and $f + g$, respectively. The approach reminds me of the proofs of the linearity of limits for sequences (Theorem 2.3.9).

For the sum $f + g$ in part (i), we want to end up

$$\|(f(\mathbf{x}) + g(\mathbf{x})) - (f(\mathbf{c}) + g(\mathbf{c}))\|_m < \varepsilon. \tag{4.5.3}$$

To leverage the individual continuity of f and g, we can split the distance ε in half to share between them. Each yields its own threshold (δ_f and δ_g) to ensure

$$\|\mathbf{x} - \mathbf{c}\|_k < \delta_f \quad \implies \quad \|f(\mathbf{x}) - f(\mathbf{c})\|_m < \frac{\varepsilon}{2} \qquad \text{and} \qquad (4.5.4)$$

$$\|\mathbf{x} - \mathbf{c}\|_k < \delta_g \quad \implies \quad \|g(\mathbf{x}) - g(\mathbf{c})\|_m < \frac{\varepsilon}{2}. \qquad (4.5.5)$$

We can get back to the goal inequality (4.5.3) from inequalities (4.5.4) and (4.5.5) using the triangle inequality (1.2.32), but we need to ensure both (4.5.4) and (4.5.5) are valid at the same time. To do this, we choose the threshold δ to be the smaller of individual thresholds δ_f and δ_g.

For the case where $\alpha \neq 0$ in part (ii), we want to end up with

$$\|\alpha f(\mathbf{x}) - \alpha f(\mathbf{c})\|_m = |\alpha| \|f(\mathbf{x}) - f(\mathbf{c})\|_m < \varepsilon. \qquad (4.5.6)$$

We can leverage the continuity of f at $\mathbf{c}$ by dividing by $|\alpha| > 0$ and consider

$$\|f(\mathbf{x}) - f(\mathbf{c})\|_m < \frac{\varepsilon}{|\alpha|}. \qquad (4.5.7)$$

The continuity of f at $\mathbf{c}$ allows us to choose a threshold δ for the domain in response to the distance $\varepsilon/|\alpha|$ for the range. The proof works this argument in reverse order and is similar to the proof of the continuity of basic affine transformations (Theorem 4.3.9) in the case where $\alpha \neq 0$.

Proof of Theorem 4.5.5. Throughout the proof, suppose $\alpha \in \mathbb{R}$, $D \subseteq \mathbb{R}^k$, $\mathbf{c} \in D$, and $f, g : D \to \mathbb{R}^m$ where f and g are continuous at $\mathbf{c}$.

Part (i): Suppose $\varepsilon > 0$. Since f and g are continuous at $\mathbf{c}$, there are thresholds $\delta_f > 0$ and $\delta_g > 0$ such that

$$\|\mathbf{x} - \mathbf{c}\|_k < \delta_f \quad \implies \quad \|f(\mathbf{x}) - f(\mathbf{c})\|_m < \frac{\varepsilon}{2} \qquad \text{and} \qquad (4.5.8)$$

$$\|\mathbf{x} - \mathbf{c}\|_k < \delta_g \quad \implies \quad \|g(\mathbf{x}) - g(\mathbf{c})\|_m < \frac{\varepsilon}{2}. \qquad (4.5.9)$$

Define δ to be the smaller of the thresholds δ_f and δ_g. That is, let

$$\delta = \min\{\delta_f, \delta_g\} \qquad (4.5.10)$$

and note $\delta > 0$. Now suppose

$$\|\mathbf{x} - \mathbf{c}\|_k < \delta. \qquad (4.5.11)$$

Since $\delta \leq \delta_f$ and $\delta \leq \delta_g$, both implications (4.5.8) and (4.5.9) hold. Hence, by the triangle inequality (1.2.32) we have

$$\|(f(\mathbf{x}) + g(\mathbf{x})) - (f(\mathbf{c}) + g(\mathbf{c}))\|_m = \|f(\mathbf{x}) - f(\mathbf{c}) + g(\mathbf{x}) - g(\mathbf{c})\|_m \tag{4.5.12}$$

$$\leq \|f(\mathbf{x}) - f(\mathbf{c})\|_m + \|g(\mathbf{x}) - g(\mathbf{c})\|_m \tag{4.5.13}$$

$$< \frac{\varepsilon}{2} + \frac{\varepsilon}{2} \tag{4.5.14}$$

$$= \varepsilon. \tag{4.5.15}$$

Therefore, δ is a threshold and the sum $f + g$ is continuous at $\mathbf{c}$.

Part (ii): Suppose $\alpha = 0$. Then $\alpha f(\mathbf{x}) = 0 \cdot f(\mathbf{x}) = \mathbf{0}$ for every $\mathbf{x} \in \mathbb{R}^k$. Define $\delta = 21 > 0$. Then for every $\mathbf{x} \in \mathbb{R}^k$ we have

$$\|\mathbf{x} - \mathbf{c}\|_k < \delta = 21 \quad \Longrightarrow \quad \|\alpha f(\mathbf{x}) - \alpha f(\mathbf{c})\|_m = \|\mathbf{0}\|_m = 0 < \varepsilon. \tag{4.5.16}$$

Therefore, αf is continuous when $\alpha = 0$.

Now suppose $\alpha \neq 0$ and $\varepsilon > 0$. Since f is continuous at $\mathbf{c}$, there is a threshold $\delta > 0$ such that for every $\mathbf{x} \in \mathbb{R}^k$ we have

$$\|\mathbf{x} - \mathbf{c}\|_k < \delta \quad \Longrightarrow \quad \|f(\mathbf{x}) - f(\mathbf{c})\|_m < \frac{\varepsilon}{|\alpha|}. \tag{4.5.17}$$

Hence, by the relationship between absolute value and Pythagorean distance from Corollary 1.2.30, we also have

$$\|\alpha f(\mathbf{x}) - \alpha f(\mathbf{c})\|_m = |\alpha| \|f(\mathbf{x}) - f(\mathbf{c})\|_m < |\alpha| \cdot \frac{\varepsilon}{|\alpha|} = \varepsilon. \tag{4.5.18}$$

Therefore, αf is continuous when $\alpha \neq 0$. $\square$

Once again and as mentioned in Remark 1.6.18, a corollary of the linearity of continuity holds for linear combinations. As with the proof Corollary 1.6.16 on arbitrarily close and linear combinations of sets, the proof of Corollary 4.5.7 follows from induction on linearity. So, the proof is left as an exercise.

Corollary 4.5.7: Continuity and linear combinations

Suppose $A \subseteq \mathbb{R}^\ell$, $k \in \mathbb{N}$, and for each $j = 1, \ldots, k$ we have $c_j \in \mathbb{R}$ and the functions $f_j : A \to \mathbb{R}^m$ are continuous. Then the linear combination $f : A \to \mathbb{R}^m$ given by

$$f(\mathbf{x}) = \sum_{j=1}^{k} c_j f_j(\mathbf{x}) = c_1 f_1(\mathbf{x}) + \ldots + c_k f_k(\mathbf{x}) \tag{4.5.19}$$

is continuous.

If we restrict the context to functions from the real line to the real line, then products of continuous functions are continuous. That is, continuity is multiplicative.

Theorem 4.5.8: Products of continuous functions are continuous

Suppose $D \subseteq \mathbb{R}, c \in D$, and $f, g : D \to \mathbb{R}$ where f and g are continuous at c. Then the product fg is continuous at c.

Scratch Work 4.5.9: Perspective of sequential continuity

So far in this section, the proofs follow the guide in Remark 4.3.4 to directly address the definition of continuity (Definition 4.3.2) and respond to an arbitrary distance $\varepsilon > 0$ for the range with a threshold $\delta > 0$ for the domain. The same approach could be made to work here, but to showcase the equivalent forms of continuity in Theorem 4.4.7, we work with sequential continuity (Definition 4.4.5) and the idea that for sequences, the limit of a product is the product of the limits (Theorem 2.3.17).

Proof of Theorem 4.5.8. Suppose $D \subseteq \mathbb{R}, c \in D$, and $f, g : D \to \mathbb{R}$ where f and g are continuous at c. By the implication (iii) $\implies$ (ii) from the equivalent forms of continuity given by Theorem 4.4.7, both f and g are sequentially continuous at c. So, consider an arbitrary convergent sequence of real numbers $(x_n) \subseteq D$ where

$$\lim_{n \to \infty} x_n = c. \tag{4.5.20}$$

By the definition of sequential continuity, we have

$$\lim_{n \to \infty} f(x_n) = f(c) \quad \text{and} \quad \lim_{n \to \infty} g(x_n) = g(c). \tag{4.5.21}$$

For sequences of real numbers, the limit of a product is the product of the limits (Theorem 2.3.17), so we have

$$\lim_{n \to \infty} f(x_n)g(x_n) = \left(\lim_{n \to \infty} f(x_n) \right) \left(\lim_{n \to \infty} g(x_n) \right) = f(c)g(c). \tag{4.5.22}$$

Hence, the product fg is sequentially continuous at c. By the implication (ii) $\implies$ (iii) in Theorem 4.4.7, fg is continuous at c. $\qquad \square$

Remark 4.5.10: Alternative proof

If you are interested in a proof of Theorem 4.5.8 based on the ε-δ definition for continuity (Definition 4.3.2), take a look at the proof of Theorem 2.3.17 regarding the limit of a sequence of products. That proof makes use of Theorem 2.3.15: Convergent sequences are bounded.

Since continuous functions are *locally* bounded (Theorem 4.3.21), we can make a similar argument for a pair of functions $f, g : \mathbb{R} \to \mathbb{R}$ which are continuous at c using a common local bound $b_c > 0$. Ultimately, given an arbitrary distance $\varepsilon > 0$ for the range, the distance

$$\delta = \frac{\varepsilon}{2b_c} \tag{4.5.23}$$

can be proven to serve as a threshold for the continuity of fg at c. If you like, give it a shot.

Products lead to monomials which are then continuous by induction.

Corollary 4.5.11: Monomials are continuous

Suppose $f : \mathbb{R} \to \mathbb{R}$ is a monomial given by

$$f(x) = x^n \tag{4.5.24}$$

for some $n \in \mathbb{N}$. Then f is continuous.

Proof of Corollary 4.5.11. We argue by induction. Throughout this proof, the domain and range of the functions are the real line $\mathbb{R}$.

Base case: Suppose $f_1(x) = x$. Then f_1 is continuous by Corollary 4.5.4 where $m = 1$ and $b = 0$.

Inductive case: Consider the polynomial $f_k(x) = x^k$ where $k \in \mathbb{N}$, and assume f_k is continuous. Also consider the polynomial

$$f_{k+1}(x) = x^{k+1} = x^k \cdot x. \tag{4.5.25}$$

Both $f_k(x) = x^k$ and $f_1(x) = x$ are continuous, so we have $f_{k+1}(x) = x^{k+1}$ is continuous since it is the product of continuous functions (Theorem 4.5.8).

Therefore, the monomial $f(x) = x^n$ is continuous for every $n \in \mathbb{N}$. $\qquad\square$

At this point, we can combine the results in this section with another induction argument to prove polynomials are continuous (Theorem 4.5.2).

Theorem 4.5.12: Polynomials are continuous (copy)

Polynomials are continuous on the real line $\mathbb{R}$.

Proof of Theorem 4.5.2. We proceed by induction. Throughout this proof, the domain and range of the functions are the real line $\mathbb{R}$.

Base case: Suppose p_1 is a polynomial of the form

$$p_1(x) = a_0 + a_1 x \tag{4.5.26}$$

where $a_0, a_1 \in \mathbb{R}$. Then p_1 is continuous by Corollary 4.5.4 where $m = a_1$ and $b = a_0$.

Inductive case: Suppose p_k is a polynomial of the form

$$p_k(x) = a_0 + a_1 x + a_2 x^2 + \cdots + a_{k-1} x^{k-1} + a_k x^k \tag{4.5.27}$$

where $k \in \mathbb{N}$ and the coefficients $a_0, a_1, \ldots, a_{k-1}, a_k$ are real numbers. Assume p_k is continuous. Every polynomial f_{k+1} of the form

$$f_{k+1}(x) = a_{k+1} x^{k+1} \tag{4.5.28}$$

is continuous by part (ii) of the linearity of continuity (Theorem 4.5.5) and the continuity of monomials (Corollary 4.5.11). By part (i) of the linearity of continuity (Theorem 4.5.5), the polynomial p_{k+1} given by

$$p_{k+1}(x) = p_k(x) + f_{k+1}(x) \tag{4.5.29}$$

$$= a_0 + a_1 x + a_2 x^2 + \cdots + a_k x^k + a_{k+1} x^{k+1} \tag{4.5.30}$$

is continuous since it is the sum of continuous functions.

With the induction completed, we conclude polynomials are continuous. □

The next result follows immediately from preservation of closeness (Definition 4.2.5). The hypotheses ensure the composition is well-defined.

Theorem 4.5.13: Compositions of continuous functions are continuous

Suppose $D \subseteq \mathbb{R}^k$, $E \subseteq \mathbb{R}^j$, $f : D \to \mathbb{R}^j$, and $g : E \to \mathbb{R}^m$ where $f(D) \subseteq E$. Further suppose f is continuous at $\mathbf{c} \in \mathbb{R}^k$ and g is continuous at $f(\mathbf{c})$. Then the composition $g \circ f : D \to \mathbb{R}^m$ defined by

$$g \circ f(\mathbf{x}) = g(f(\mathbf{x})) \qquad \text{for all } \mathbf{x} \in D \tag{4.5.31}$$

is continuous at $\mathbf{c}$.

Proof of Theorem 4.5.13. Suppose all the hypotheses as stated in Theorem 4.5.13 hold. Since f and g are continuous at $\mathbf{c}$ and $f(\mathbf{c})$, respectively, they preserve closeness as well by Theorem 4.4.7. Suppose $B \subseteq D$ where $\mathbf{c} \operatorname{acl} B$. By the definition of preserving closeness (Definition 4.2.5)—twice—we have

$$\mathbf{c} \operatorname{acl} B \quad \Longrightarrow \quad f(\mathbf{c}) \operatorname{acl} f(B) \quad \Longrightarrow \quad g(f(\mathbf{c})) \operatorname{acl} g(f(B)). \tag{4.5.32}$$

Therefore, the composition $g \circ f$ preserves closeness at $\mathbf{c}$. By another application of Theorem 4.4.7, $g \circ f$ is continuous at $\mathbf{c}$. □

The combination of the previous theorem with the next result leads to a nice conclusion regarding quotients of continuous functions.

Example 4.5.14: Continuity of the reciprocal

The function $f : \mathbb{R} \backslash \{0\} \to \mathbb{R}$ given by

$$f(x) = \frac{1}{x} \tag{4.5.33}$$

is continuous. See the Desmos activity "Continuity of reciprocal" in Figure 4.5.1.

Scratch Work 4.5.15: Careful algebra

Following the guide in Remark 4.3.4 for proofs using the ε-δ form of continuity in Definition 4.3.2, consider the expression

$$\left| \frac{1}{x} - \frac{1}{c} \right| < \varepsilon \tag{4.5.34}$$

where x and c are nonzero. The goal is to figure out an expression for the distance $|x - c|$ in the domain and come up with a suitable threshold δ. Using the common denominator

Figure 4.5.1: To accompany Example 4.5.14, play around with the Desmos activity "Continuity of reciprocal" accessed through this QR code. In particular, note that when we keep ε fixed but allow c to get closer to 0, the value of the threshold δ must decrease to compensate. https://www.desmos.com/calculator/clwvuzoenv

xc, we have the revised expression

$$\left|\frac{1}{x} - \frac{1}{c}\right| = \left|\frac{c-x}{xc}\right| = \frac{1}{|xc|}|x-c| < \varepsilon. \tag{4.5.35}$$

Note that when x and c are near zero, $1/|xc|$ could be so large that (4.5.35) may fail to hold. Play around with the Desmos activity "Continuity of reciprocal" in Figure 4.5.1 to see this in action. One way to compensate it to have both x and c more than a distance $|c|/2 > 0$ away from zero to keep $1/|xc|$ under control. By the reverse triangle inequality (1.2.37) we have

$$|c| - |x| \le |x-c| < \frac{|c|}{2} \quad \implies \quad \frac{|c|}{2} < |x| \quad \implies \quad \frac{2}{|c|} > \frac{1}{|x|}. \tag{4.5.36}$$

Hence, we also have

$$|x-c| < \frac{|c|}{2} \quad \implies \quad \frac{2}{c^2} > \frac{1}{|x||c|} \quad \iff \quad \frac{1}{|xc|} < \frac{2}{c^2}. \tag{4.5.37}$$

Therefore, considering both the goal (4.5.35) and (4.5.37) leads to

$$|x-c| < \frac{|c|}{2} \quad \implies \quad \left|\frac{1}{x} - \frac{1}{c}\right| = \frac{1}{|xc|}|x-c| < \frac{2}{c^2}|x-c| < \varepsilon. \tag{4.5.38}$$

Similar to Example 4.3.15, the updated goal (4.5.38) is only valid when $|x-c| < |c|/2$, so we need to keep this constraint in mind. For a threshold δ, setting

$$\delta = \min\left\{\frac{|c|}{2}, \frac{\varepsilon c^2}{2}\right\} \tag{4.5.39}$$

should work out. In fact, this can be visualized using the "Delta Tester - continuity" Desmos activity from Figure 4.3.5 by changing the formulas for the function and the threshold accordingly.

Proof for Example 4.5.14. Suppose $\varepsilon > 0$ and $c \neq 0$. Choose the distance

$$\delta = \min\left\{\frac{|c|}{2}, \frac{\varepsilon c^2}{2}\right\} \tag{4.5.40}$$

for the domain. Note $\delta > 0$ and we have both

$$|x - c| < \delta \leq \frac{|c|}{2} \qquad \text{and} \qquad |x - c| < \delta \leq \frac{\varepsilon c^2}{2}. \tag{4.5.41}$$

So as in Scratch Work 4.5.15, by the reverse triangle inequality (1.2.37) for $x \neq 0$ we have

$$|c| - |x| \leq |x - c| < \frac{|c|}{2} \implies \frac{|c|}{2} < |x| \implies \frac{2}{c^2} > \frac{1}{|xc|} \iff \frac{1}{|xc|} < \frac{2}{c^2}. \tag{4.5.42}$$

Hence, we have

$$|x - c| < \delta \implies \left|\frac{1}{x} - \frac{1}{c}\right| = \frac{1}{|xc|}|x - c| < \left(\frac{2}{c^2}\right)\left(\frac{\varepsilon c^2}{2}\right) = \varepsilon. \tag{4.5.43}$$

Therefore, f is continuous on $\mathbb{R}\backslash\{c\}$. $\qquad\square$

To conclude the section, quotients of continuous functions are continuous, within reason.

Theorem 4.5.16: Quotients of continuous functions are continuous

Suppose $D \subseteq \mathbb{R}$, $c \in D$, and $f, g : D \to \mathbb{R}$. If f and g are continuous at c and $g(x) \neq 0$ for all $x \in D$, then the quotient $f/g : D \to \mathbb{R}$ given by

$$\frac{f(x)}{g(x)} \qquad \text{for all } x \in D \tag{4.5.44}$$

is continuous at c.

We can prove Theorem 4.5.16 by combining results proven in this section.

Proof of Theorem 4.5.16. Suppose the hypotheses stated in Theorem 4.5.16 hold. The function f/g satisfies

$$\frac{f(x)}{g(x)} = f(x)\frac{1}{g(x)} \qquad \text{for all } x \in D. \tag{4.5.45}$$

Note that $1/g(x)$ is the composition of $h(x) = 1/x$ and $g(x)$ in that

$$h \circ g(x) = h(g(x)) = \frac{1}{g(x)}. \tag{4.5.46}$$

Since $g(x) \neq 0$ on D, the composition $h \circ g = 1/g$ is well-defined. Furthermore, since $h(x) = 1/x$ is continuous on $\mathbb{R}\backslash\{0\}$ by Example 4.5.14 and g is continuous at c by assumption, the composition $h \circ g = 1/g$ is continuous by Theorem 4.5.13. Finally, f is continuous at c and products of continuous functions are continuous by Theorem 4.5.8, we have f/g is continuous at c. $\qquad\square$

There is much more to do with continuous functions before moving on to limits and derivatives in the next chapter.

Exercises

4.5.1. For each statement below, find a pair of functions $f, g : \mathbb{R} \to \mathbb{R}$ that satisfy the statement.

(i) f and g are discontinuous at 0, but their sum $f + g$ is continuous at 0.

(ii) f and g are discontinuous at 0, but their product fg is continuous at 0.

(iii) f and g are discontinuous at 0, but their quotient f/g is continuous at 0.

(iv) f and g are discontinuous at 0, but both $f + g$ and fg are continuous at 0.

4.5.2. Fix $c \in \mathbb{R}$. For each $n \in \mathbb{N}$, consider the function $q_n : \mathbb{R} \to \mathbb{R}$ given by

$$q_n(x) = \begin{cases} \dfrac{x^n - c^n}{x - c}, & \text{if } x \neq c, \\ nc^{n-1}, & \text{if } x = c. \end{cases} \tag{4.5.47}$$

Prove q_n is continuous at c.

4.5.3. Every real number x has unique decomposition into an *integer part* $\lfloor x \rfloor$ and a *fractional part* $\{x\}$ where $\lfloor x \rfloor$ is the integer satisfying

$$x \leq \lfloor x \rfloor < x + 1 \quad \text{and} \quad \{x\} = x - \lfloor x \rfloor. \tag{4.5.48}$$

Let $g, h : \mathbb{R} \to \mathbb{R}$ be *floor* and *sawtooth* functions given by

$$g(x) = \lfloor x \rfloor \quad \text{and} \quad h(x) = \{x\} = x - \lfloor x \rfloor, \tag{4.5.49}$$

respectively. See the Desmos activity used to explore discontinuity in Figure 4.3.3.

(i) Draw the graphs of g and h. (Can you see how they got their names?)

(ii) Determine where g and h are discontinuous and prove your result.

(iii) Determine where g and h are continuous and prove your result.

4.5.4. Suppose $a < p < b$, $f : [a, p] \to \mathbb{R}$ is continuous, $g : [p, b] \to \mathbb{R}$ is continuous, and $f(p) = g(p)$. Consider the *glued* function $h : [a, b] \to \mathbb{R}$ given by

$$h(x) = \begin{cases} f(x), & \text{if } x \in [a, p], \\ g(x), & \text{if } x \in [p, b]. \end{cases} \tag{4.5.50}$$

Prove h is continuous.

4.5.5. Suppose $f : \mathbb{R} \to \mathbb{R}$ is continuous. Prove there are nonnegative functions[1] $g, h : \mathbb{R} \to \mathbb{R}$ such that

$$f(x) = g(x) - h(x) \quad \text{for all } x \in \mathbb{R}. \tag{4.5.51}$$

---

[1] $g(x) \geq 0$ and $h(x) \geq 0$ for all $x \in \mathbb{R}$.

4.5.6. A function $g : \mathbb{R} \to \mathbb{R}$ is *even* when $g(-x) = g(x)$ for all $x \in \mathbb{R}$. A function $h : \mathbb{R} \to \mathbb{R}$ is *odd* when $h(-x) = -h(x)$ for all $x \in \mathbb{R}$. Prove that if $f : \mathbb{R} \to \mathbb{R}$ is continuous, then there is an even function g and an odd function h where

$$f(x) = g(x) + h(x) \qquad \text{for all } x \in \mathbb{R}. \tag{4.5.52}$$

4.5.7. Let $C[a, b]$ denote the set of real-valued continuous functions on $[a, b]$. Use Lemma 1.6.7 to prove $C[a, b]$ is a vector space.

4.5.8. Suppose $c, \ell \in \mathbb{R}$ and let $C(c, \ell)$ denote the set of real-valued functions on $\mathbb{R}$ that are continuous at c with output ℓ. (Thus, for each $f \in C(c, \ell)$ we have $f(c) = \ell$.) Use Lemma 1.6.7 to prove $C(c, \ell)$ is a vector space if and only if $\ell = 0$.

4.6 More properties of continuity

This section features a flurry of results on continuous functions. They preserve properties and relationships between points, sequences, and sets. The equivalent forms of continuity established with Theorem 4.4.7 provide flexibility in the way we decide to prove the results.

First, consider a result that came out of a discussion on Discord. One night, I posted a conjecture that continuous functions preserve coupled sets. By the next day, John Rodriguez provided the proof you see here.

Lemma 4.6.1: Continuous images of coupled sets are coupled

Suppose $D \subseteq \mathbb{R}^k$, $f : D \to \mathbb{R}^m$, and $G, H \subseteq D$. If f is continuous and G and H are coupled, then $f(G)$ and $f(H)$ are coupled.

Scratch Work 4.6.2: Definitions built on arbitrarily close

John takes advantage of the fact that the definitions for coupled sets and preserving closeness (Definitions 3.3.1 and 4.2.5) are both stated in terms of arbitrarily close.

Proof of Lemma 4.6.1. Suppose f is continuous while G and H are coupled subsets of the domain. By the definition of coupled (Definition 3.3.1) and without loss of generality, there is some $\mathbf{x} \in G$ where $\mathbf{x} \operatorname{acl} H$. Since f is continuous, f preserves closeness at $\mathbf{x}$ by Theorem 4.4.7. By the definitions of preserving closeness and image (respectively Definitions 4.2.5 and 1.2.14), we have

$$\mathbf{x} \operatorname{acl} H \quad \implies \quad f(\mathbf{x}) \operatorname{acl} f(H) \quad \text{and} \tag{4.6.1}$$
$$\mathbf{x} \in G \quad \implies \quad f(\mathbf{x}) \in f(G). \tag{4.6.2}$$

Therefore, the images $f(G)$ and $f(H)$ are coupled. $\square$

The preservation of coupled sets by continuous functions leads directly to the preservation of connectedness.

Theorem 4.6.3: Continuous images of connected sets are connected

If $f : D \to \mathbb{R}^m$ is continuous and D is connected, then the range $f(D)$ is connected.

Scratch Work 4.6.4: Continuity as preservation of closeness

The following proof takes advantage of the equivalence of continuity and the preservation of closeness in Theorem 4.4.7 as well as a key property of preimages: By Lemma 4.1.13, preimages respect unions and intersections.

Proof of Theorem 4.6.3. Suppose $f : D \to \mathbb{R}^m$ is continuous where D is connected. Let $f(D) = A \cup B$ where A and B are nonempty. By the definitions of domain and range (Definition 1.2.13), every point in the domain is the preimage of a point in the range and every point in the range is the image of a point in the domain, hence $D = f^{-1}(f(D))$. So, by Lemma 4.1.13 we have

$$D = f^{-1}(f(D)) = f^{-1}(A \cup B) = f^{-1}(A) \cup f^{-1}(B). \tag{4.6.3}$$

Since A and B are nonempty, their preimages $f^{-1}(A)$ and $f^{-1}(B)$ are nonempty as well. Since D is connected (Definition 3.3.4), $f^{-1}(A)$ and $f^{-1}(B)$ are coupled sets in the domain. Since continuous functions preserve coupled sets by Lemma 4.6.1, A and B are coupled sets in the range. Finally, since A and B are arbitrary, the range $f(D)$ is connected. $\square$

A corollary of Theorem 4.6.3 is the classic *Intermediate Value Theorem*: Continuous functions from the real line to the real line map intervals to intervals. In this way, the range has no holes or jumps. (Cf. [1, Theorem 4.5.1, p.136].)

Theorem 4.6.5: Intermediate Value Theorem

Suppose $f : [a, b] \to \mathbb{R}$ is continuous. If ℓ is a real number satisfying

$$f(a) < \ell < f(b) \qquad \text{or} \qquad f(a) > \ell > f(b), \tag{4.6.4}$$

then there is some $c \in (a, b)$ such that $f(c) = \ell$.

Scratch Work 4.6.6: Intervals are the connected sets in the real line

Intervals characterize the connected subsets of the real line (see Theorem 3.3.17), and intervals themselves are characterized by inequalities like those in (4.6.4) (see Lemma 3.3.16). Combining these results with Theorem 4.6.3 yield a proof of the Intermediate Value Theorem 4.6.5.

Proof of the Intermediate Value Theorem 4.6.5. Suppose $f : [a, b] \to \mathbb{R}$ is continuous and ℓ is a real number satisfying

$$f(a) < \ell < f(b) \qquad \text{or} \qquad f(a) > \ell > f(b). \tag{4.6.5}$$

By Theorem 3.3.17, the domain $[a, b]$ is connected. Since continuous images of connected sets are connected (Theorem 4.6.3), the range $f([a, b])$ is connected. As a connected subset of the real line, the $f([a, b])$ is an interval as well. Hence, we have either

$$[f(a), f(b)] \subseteq f([a, b]) \qquad \text{or} \qquad [f(b), f(a)] \subseteq f([a, b]). \tag{4.6.6}$$

Either way, ℓ is in the range $f([a, b])$. Therefore, ℓ is the image of a point $c \in [a, b]$ and $f(c) = \ell$. Since ℓ is equal to neither $f(a)$ nor $f(b)$, we have $c \in (a, b)$. □

Continuous functions preserve compactness, too.

Theorem 4.6.7: Continuous images of compact sets are compact

If $K \subseteq \mathbb{R}^k$ and $f : K \to \mathbb{R}^m$ where f is continuous and K is compact, then the range $f(K)$ is compact.

Scratch Work 4.6.8: Pair of sequential characterizations

The proof takes advantage of the sequential characterizations of both compactness and continuity thanks to the Heine-Borel Theorem 3.5.1 and the equivalent forms of continuity in Theorem 4.4.7, respectively.

Proof of Theorem 4.6.7. Suppose $f : K \to \mathbb{R}^m$ is continuous and K is compact. By Theorem 4.4.7, f is sequentially continuous (Definition 4.4.5). By the Heine-Borel Theorem 3.5.1, K is sequentially compact (Definition 4.4.5).

So, let $(\mathbf{y}_n)$ be a sequence of points in the range $f(K)$. (The goal is to show $(\mathbf{y}_n)$ has a convergent subsequence whose limit is in $f(K)$.) By the definition of range (Definition 1.2.13), for each index $n \in \mathbb{N}$ there is a point $\mathbf{x}_n$ in the domain K where $f(\mathbf{x}_n) = \mathbf{y}_n$. Furthermore, $(\mathbf{y}_n) = (f(\mathbf{x}_n))$. Since K is sequentially compact, there is a convergent subsequence $(\mathbf{x}_{n_k})$ of the sequence $(\mathbf{x}_n)$ where

$$\lim_{k \to \infty} \mathbf{x}_{n_k} = \mathbf{c} \qquad \text{and} \qquad \mathbf{c} \in K. \tag{4.6.7}$$

Since f is sequentially continuous at every point in the domain K, we have

$$\lim_{k \to \infty} \mathbf{x}_{n_k} = \mathbf{c} \in K \qquad \Longrightarrow \qquad \lim_{k \to \infty} f(\mathbf{x}_{n_k}) = f(\mathbf{c}) \in f(K). \tag{4.6.8}$$

Hence, $(f(\mathbf{x}_{n_k}))$ is a subsequence of $(\mathbf{y}_n) = (f(\mathbf{x}_n))$ that converges to $f(\mathbf{c})$, and $f(\mathbf{c})$ is in $f(K)$. Therefore, $f(K)$ is sequentially compact and also compact by the Heine-Borel Theorem 3.5.1. □

Theorem 4.6.7 immediately yields the classic Extreme Value Theorem: Continuous functions attain their maximum and minimum values when they map compact subsets of the real line to the real line.

Corollary 4.6.9: Extreme Value Theorem

If K is a compact subset of the real line $\mathbb{R}$ and $f : K \to \mathbb{R}$ is continuous, then f attains its minimum and maximum values. That is, there are inputs $s, t \in K$ where

$$f(s) = \inf f(K) = \min f(K) \qquad \text{and} \qquad f(t) = \sup f(K) = \max f(K). \qquad (4.6.9)$$

Scratch Work 4.6.10: Pervasiveness of arbitrarily close

The proof takes advantage of the way arbitrarily close pervades the definitions and characterizations of supremum, infimum, continuity, and compactness (through closed).

Proof of the Extreme Value Theorem 4.6.9. Suppose $f : K \to \mathbb{R}$ is continuous and K is compact. By Theorem 4.6.7, $f(K)$ is compact as well. By the Heine-Borel Theorem 3.5.1, $f(K)$ is closed and bounded. As a bounded set of real numbers, both $\sup f(K)$ and $\inf f(K)$ exist by the Axiom of Completeness 1.3.8 and its mirror image for infima Theorem 1.4.1, respectively. By the definitions of supremum and infimum (Definition 1.1.14), we have both

$$(\sup f(K)) \operatorname{acl} f(K) \quad \text{and} \quad (\inf f(K)) \operatorname{acl} f(K). \qquad (4.6.10)$$

Since $f(K)$ is closed, by Definition 3.1.1 we have both

$$\sup f(K) \in f(K) \quad \text{and} \quad \inf f(K) \in f(K). \qquad (4.6.11)$$

Hence, there are inputs $s, t \in K$ where

$$f(s) = \inf f(K) \qquad \text{and} \qquad f(t) = \sup f(K). \qquad (4.6.12)$$

Since $\sup f(K)$ is an upper bound for $f(K)$, $\inf f(K)$ is a lower bound for $f(K)$, and both are in $f(K)$, both $\max f(K)$ and $\min f(K)$ exist (see Definition 1.1.2). $\qquad \square$

Another result that builds on our work with sequences is a squeeze theorem for continuity.

Theorem 4.6.11: Squeeze Theorem for continuity

Suppose $D \subseteq \mathbb{R}$, $c \in D$, and $f, g, h : D \to \mathbb{R}$. If

(i) $f(c) = h(c)$,

(ii) $f(x) \le g(x) \le h(x)$ for all $x \in D$, and

(iii) f and h are continuous at c,

then g is continuous at c as well.

Scratch Work 4.6.12: Sequential continuity and the Squeeze Theorem for sequences

We could prove the Squeeze Theorem for continuity 4.6.11 with an ε-δ argument by defining a threshold for g as the minimum of the thresholds for f and h, but a combination of sequential continuity via Theorem 4.4.7 and the Squeeze Theorem for sequences 2.4.3 makes for a short proof.

Proof of the Squeeze Theorem for continuity 4.6.11. Suppose the hypotheses of Theorem 4.6.11 hold. Since $f(c) = h(c)$ and $f(c) \le g(c) \le h(c)$, we have

$$f(c) = g(c) = h(c). \tag{4.6.13}$$

Now assume

$$(x_n) \subseteq D \quad \text{such that} \quad \lim_{n \to \infty} x_n = c. \tag{4.6.14}$$

Since f and h are continuous at c, they are sequentially continuous at c by Theorem 4.4.7. Hence,

$$\lim_{n \to \infty} f(x_n) = f(c) = h(c) = \lim_{n \to \infty} h(x_n). \tag{4.6.15}$$

Since $f(x_n) \le g(x_n) \le h(x_n)$ for every index $n \in \mathbb{N}$, the conditions of the Squeeze Theorem for Sequences 2.4.3 are met. So, since $f(c) = g(c) = h(c)$, we have

$$\lim_{n \to \infty} g(x_n) = g(c). \tag{4.6.16}$$

Therefore, g is sequentially continuous at c and also continuous at c by Theorem 4.4.7. $\quad\square$

The wealth of results developed in this chapter lead to a wide variety of criteria we can use to show a function is discontinuous without working directly with the negation of the ε-δ definition of continuity in Definition 4.3.2.

Corollary 4.6.13: Discontinuity Criteria

If $D \subseteq \mathbb{R}^k$, $\mathbf{c} \in D$, $f : D \to \mathbb{R}^m$, and f satisfies any of the following conditions, then f is discontinuous at $\mathbf{c}$:

(i) There is a set $E \subseteq D$ where

$$\mathbf{c} \operatorname{acl} E \quad \text{but} \quad f(\mathbf{c}) \operatorname{awf} f(E). \tag{4.6.17}$$

(ii) There is a sequence $(\mathbf{x}_n) \subseteq D$ where

$$\lim_{n \to \infty} \mathbf{x}_n = \mathbf{c} \quad \text{but} \quad \lim_{n \to \infty} f(\mathbf{x}_n) \neq f(\mathbf{c}). \tag{4.6.18}$$

(iii) f is not locally bounded at $\mathbf{c}$.

Additionally, f is discontinuous (on D) if either of the following conditions is satisfied:

(iv) There is a connected set $C \subseteq D$ whose image $f(C)$ is disconnected.

(v) There is a compact set $K \subseteq D$ whose image $f(K)$ is not compact.

Some of the functions explored in this chapter showcase the various notions of discontinuity described in Discontinuity Criteria 4.6.13.

Example 4.6.14: A discontinuous function from the plane to the real line

Consider the function $w : \mathbb{R}^2 \to \mathbb{R}$ in Example 4.1.6 given by

$$w(\mathbf{x}) = \begin{cases} \|\mathbf{x}\|, & \text{if } 0 \le \|\mathbf{x}\| < 1, \\ 2, & \text{if } \|\mathbf{x}\| \ge 1. \end{cases} \qquad (4.6.19)$$

See Figure 4.1.4. In the domain, the unit vector $\mathbf{e}_1$ is arbitrarily close to open unit disk $V_1(\mathbf{0})$. However, we have

$$w(\mathbf{e}_1) = 2 \quad \text{while} \quad w(V_1(\mathbf{0})) = [0, 1). \qquad (4.6.20)$$

Since $2\,\text{awf}\,[0, 1)$, w does not preserve closeness at $\mathbf{e}_1$ since

$$\mathbf{e}_1 \,\text{acl}\, V_1(\mathbf{0}) \quad \text{but} \quad w(\mathbf{e}_1)\,\text{awf}\,w(V_1(\mathbf{0})). \qquad (4.6.21)$$

By part (i) of the Discontinuity Criteria 4.6.13, w is discontinuous at $\mathbf{c} = \mathbf{e}_1$.

Example 4.6.15: A discontinuous, piecewise defined function

For the function $g : \mathbb{R} \to \mathbb{R}$ from Examples 4.1.3 and 4.2.3 given by

$$g(x) = \begin{cases} x/2, & \text{if } x < 2, \\ 1 + (x/2), & \text{if } x \ge 2, \end{cases} \qquad (4.6.22)$$

See Figure 4.2.3. The interval $E_3 = (1, 3)$ is a connected subset of the domain by Theorem 3.3.17. Its image $g(E_3) = (1/2, 1) \cup [2, 5/2)$ is disconnected since $(1/2, 1)$ and $[2, 5/2)$ form a separation (see Definition 3.3.12). By part (iv) of the Discontinuity Criteria 4.6.13, g is discontinuous.

Example 4.6.16: Another discontinuous, piecewise defined function

Consider the unbounded function $v : \mathbb{R} \to \mathbb{R}$ given by

$$v(x) = \begin{cases} \dfrac{1}{2 - x}, & \text{if } x < 2, \\[2mm] 1 + \dfrac{x}{2}, & \text{if } x \ge 2. \end{cases} \qquad (4.6.23)$$

See Figure 4.2.4. To see that v is not locally bounded at $c = 2$, note that *every δ-neighborhood* of $c = 2$ given by

$$V_\delta(2) = (2 - \delta, 2 + \delta) \tag{4.6.24}$$

is bounded, but their images are the unbounded sets given by

$$v(V_\delta(2)) = \left(\frac{1}{\delta}, \infty\right) \cup \left(2, 2 + \frac{\delta}{2}\right). \tag{4.6.25}$$

By part (iii) of the Discontinuity Criteria 4.6.13, v is discontinuous at $c = 2$.

Example 4.6.17: Dirichlet's function is discontinuous everywhere

Consider Dirichlet's function, the indicator function of the rationals $\mathbb{1}_\mathbb{Q} : \mathbb{R} \to \mathbb{R}$ given by

$$\mathbb{1}_\mathbb{Q}(x) = \begin{cases} 1, & \text{if} \quad x \in \mathbb{Q}, \\ 0, & \text{if} \quad x \in \mathbb{R}\backslash\mathbb{Q}. \end{cases} \tag{4.6.26}$$

In Example 4.2.14, it is shown that $\mathbb{1}_\mathbb{Q}$ does not preserve closeness at any $c \in \mathbb{R}$. By part (i) of the Discontinuity Criteria 4.6.13, $\mathbb{1}_\mathbb{Q}$ is discontinuous at *every $c \in \mathbb{R}$.*

Example 4.6.18: Topologist's sine curve is discontinuous at zero

Consider the *topologist's sine curve* $t : \mathbb{R} \to \mathbb{R}$ given by

$$t(x) = \begin{cases} \sin(1/x), & \text{if } x \neq 0, \\ 0, & \text{if } x = 0. \end{cases} \tag{4.6.27}$$

In Example 4.4.10, it is shown that for the sequence (x_n) given by

$$x_n = \frac{1}{\frac{\pi}{2} + 2\pi n} \quad \text{for each index} \quad n \in \mathbb{N}, \tag{4.6.28}$$

we have (x_n) is in the domain where

$$\lim_{n \to \infty} x_n = \lim_{n \to \infty} \left(\frac{1}{\frac{\pi}{2} + 2\pi n} \right) = 0, \qquad \text{but} \tag{4.6.29}$$

$$\lim_{n \to \infty} t(x_n) = \lim_{n \to \infty} \sin\left(\frac{\pi}{2} + 2\pi n\right) = 1 \neq 0 = t(0). \tag{4.6.30}$$

Hence, t is not sequentially continuous at $c = 0$. By part (ii) of the Discontinuity Criteria 4.6.13, t is discontinuous at $c = 0$. In fact, no matter how we redefine the output $t(0)$, t remains discontinuous at $c = 0$.

The next section closes the chapter with a strengthening of continuity to *uniform continuity*.

The idea being that continuous functions which have a single threshold δ that suffices for every point in the domain exhibit some additional nice properties. After that, the next chapter starts of by softening continuity to define *functional limits*, allowing for us to get into *derivatives*.

Exercises

4.6.1. Prove there is a real number x such that $x = \cos x$.

4.6.2. Suppose $f : [a, b] \to \mathbb{R}$ is continuous and one-to-one. Prove the inverse f^{-1} is continuous.

4.6.3. Suppose $g : [0, 1] \to \mathbb{R}$ where g is continuous and $g(0) = g(1)$.

(i) Prove there exist $x_2, y_2 \in [0, 1]$ such that

$$|x_2 - y_2| = \frac{1}{2} \quad \text{and} \quad g(x_2) = g(y_2). \tag{4.6.31}$$

(ii) Generalize result (i) by proving that for each $n \in \mathbb{N}$, there are $x_n, y_n \in [0, 1]$ where

$$|x_n - y_n| = \frac{1}{n} \quad \text{and} \quad g(x_n) = g(y_n). \tag{4.6.32}$$

4.6.4. Suppose $f : [0, 1] \to [0, 1]$ is continuous. Note that the codomain is the compact interval $[0, 1]$, not the real line $\mathbb{R}$. Prove f has a *fixed point* x_0 in that there is some $x_0 \in [0, 1]$ where

$$f(x_0) = x_0. \tag{4.6.33}$$

4.6.5. Suppose $h : [a, b] \to \mathbb{R}$ is strictly increasing. Prove h has at most countably many discontinuities.

4.6.6. Consider a squeezed version of the topologist's sine curve from Example 4.4.10 defined by $s : \mathbb{R} \to \mathbb{R}$ where

$$s(x) = \begin{cases} x \sin(1/x), & \text{if } x \neq 0, \\ 0, & \text{if } x = 0. \end{cases} \tag{4.6.34}$$

Prove s is continuous at $c = 0$.

4.6.7. A real-valued g has the *intermediate value property* if for every a and b in the domain with $a < b$ and for every real number ℓ satisfying

$$g(a) < \ell < g(b) \quad \text{or} \quad g(a) > \ell > g(b), \tag{4.6.35}$$

then there is some $c \in (a, b)$ such that $g(c) = \ell$. This exercise directly shows that the topologist's sine curve t from Example 4.4.10 has the intermediate value property.

(i) Prove there are positive real numbers a and b where $t(a) = -1$ and $t(b) = 1$.

(ii) Prove that for each $y \in [-1, 1]$, there is a sequence of positive real numbers (x_n) where

$$\lim_{n \to \infty} x_n = 0 \qquad \text{and} \qquad \lim_{n \to \infty} t(x_n) = y. \qquad (4.6.36)$$

(iii) Prove $t(V_\delta(0)) = [-1, 1]$ for every $\delta > 0$, no matter how small we take δ to be.

Hence, the topologist's sine curve t has the intermediate value property, despite being discontinuous at $c = 0$. Moreover, the set of points arbitrarily close to the image of *every* δ-neighborhood of 0 under t is a full interval.

4.6.8. Consider the function $h : \mathbb{R}\backslash\{0\} \to \mathbb{R}$ given by

$$h(x) = \frac{1}{x} \sin \left(\frac{1}{x} \right). \qquad (4.6.37)$$

(i) Draw a figure for h.

(ii) Find sequences of positive numbers $(x_n), (y_n)$, and (z_n) where

$$\lim_{n \to \infty} x_n = \lim_{n \to \infty} y_n = \lim_{n \to \infty} z_n = 0, \qquad \text{but} \qquad (4.6.38)$$

$$\lim_{n \to \infty} h(x_n) = 0, \quad \lim_{n \to \infty} h(y_n) = \infty, \quad \text{and} \quad \lim_{n \to \infty} h(z_n) = -\infty. \qquad (4.6.39)$$

(iii) Prove

$$\bigcap_{\delta > 0} \overline{h(V_\delta(0))} = \mathbb{R}. \qquad (4.6.40)$$

Hence, the set of points arbitrarily close to the image of *every* δ-neighborhood of 0 under h is the whole real line $\mathbb{R}$.

4.7   Uniform continuity

Uniform continuity is a strengthening of continuity where we have "Sauron's δ"[2]:

> *One threshold to rule them all.*

This section explores a number of nice results that stem from uniform continuity.

Definition 4.7.1: Uniform continuity

Let $D \subseteq \mathbb{R}^k$ and $f : D \to \mathbb{R}^m$. We say f is *uniformly continuous* if for every distance $\varepsilon > 0$ there is a *uniform threshold* $\delta > 0$ such that

$$\mathbf{x}, \mathbf{c} \in D \text{ with } \|\mathbf{x} - \mathbf{c}\|_k < \delta \implies \|f(\mathbf{x}) - f(\mathbf{c})\|_m < \varepsilon. \qquad (4.7.1)$$

---

[2]Yes, this is totally a reference to *Lord of the Rings*.

Figure 4.7.1: To accompany Remark 4.7.2 and Example 4.7.8, explore the difference between continuity and uniform continuity by playing around with the Desmos activity "Continuity of reciprocal" accessed through this QR code. In particular, note that when we keep ε fixed but allow c to get closer to 0, the value of the threshold δ must decrease to compensate. https://www.desmos.com/calculator/clwvuzoenv

Remark 4.7.2: Continuity versus uniform continuity

The definitions for continuity and uniform continuity are strikingly similar: Both start with a distance $\varepsilon > 0$ for the range and respond with a threshold $\delta > 0$ for the domain that usually depends on ε. See Definitions 4.3.2 and 4.7.1. However, there is a subtle and important difference: With continuity, a threshold δ generally depends on the point **c**. With uniform continuity, a threshold δ *is independent of* **c**. So when we do scratch work to find a suitable threshold for continuity at **c** to fit Definition 4.3.2, the formula we get for δ can depend on both ε and a particular input **c**. But a suitable threshold δ for *uniform* continuity fitting Definition 4.7.1 is independent of the inputs. To explore this contrast further, revisit the reciprocal function $f(x) = 1/x$ from Example 4.5.14 and play around with the Desmos activity "Continuity of reciprocal" found in Figure 4.7.1.

The first case of uniform continuity requires no extra effort to prove.

Corollary 4.7.3: Lines are uniformly continuous

Suppose m and b are real numbers and let $f : \mathbb{R} \to \mathbb{R}$ be given by

$$f(x) = mx + b. \tag{4.7.2}$$

Then f is uniformly continuous.

Remark 4.7.4: Exactly the same threshold

Corollaries and 4.5.4 and 4.7.3 follow immediately from the continuity of basic affine transformations (Theorem 4.3.9). In particular, the same threshold δ suffices for both proofs.

For $m \neq 0$,

$$\delta = \frac{\varepsilon}{|m|} \qquad (4.7.3)$$

is a threshold for the continuity of f at *every* real number c. Since δ is independent of c, it is also a uniform threshold and we have f is uniformly continuous.

Uniform continuity is linear.

Theorem 4.7.5: Linearity of uniform continuity

Suppose $D \subseteq \mathbb{R}^k$ and $f, g : D \to \mathbb{R}^m$. If f and g are uniformly continuous, then

(i) $f + g$ is uniformly continuous *(additivity)*; and

(ii) αf is uniformly continuous for every scalar $\alpha \in \mathbb{R}$ *(homogeneity)*.

Scratch Work 4.7.6: Nearly identical to the linearity of contintuity

A subtle modification of Scratch Work 4.5.6 for the linearity of continuity at some point $\mathbf{c}$ suffices for the linearity of uniform continuity: Interpret thresholds as uniform thresholds.

Proof of Theorem 4.7.5. Throughout the proof, suppose $\alpha \in \mathbb{R}$, $D \subseteq \mathbb{R}^k$, and $f, g : D \to \mathbb{R}^m$ where f and g are uniformly continuous.

Part (i): Suppose $\varepsilon > 0$. Since f and g are uniformly continuous, there are uniform thresholds $\delta_f > 0$ and $\delta_g > 0$ such that for all $\mathbf{x}, \mathbf{c} \in D$ we have

$$\|\mathbf{x} - \mathbf{c}\|_k < \delta_f \implies \|f(\mathbf{x}) - f(\mathbf{c})\|_m < \frac{\varepsilon}{2} \quad \text{and} \qquad (4.7.4)$$

$$\|\mathbf{x} - \mathbf{c}\|_k < \delta_g \implies \|g(\mathbf{x}) - g(\mathbf{c})\|_m < \frac{\varepsilon}{2}. \qquad (4.7.5)$$

Define δ to be the smaller of the uniform thresholds δ_f and δ_g. That is, let

$$\delta = \min\{\delta_f, \delta_g\} \qquad (4.7.6)$$

and note $\delta > 0$. Now suppose $\mathbf{x}, \mathbf{c} \in D$ where

$$\|\mathbf{x} - \mathbf{c}\|_k < \delta. \qquad (4.7.7)$$

Since $\delta \leq \delta_f$ and $\delta \leq \delta_g$, both implications (4.7.4) and (4.7.5) hold. Hence, by the triangle inequality (1.2.32) we have

$$\|(f(\mathbf{x}) + g(\mathbf{x})) - (f(\mathbf{c}) + g(\mathbf{c}))\|_m = \|f(\mathbf{x}) - f(\mathbf{c}) + g(\mathbf{x}) - g(\mathbf{c})\|_m \qquad (4.7.8)$$
$$\leq \|f(\mathbf{x}) - f(\mathbf{c})\|_m + \|g(\mathbf{x}) - g(\mathbf{c})\|_m \qquad (4.7.9)$$
$$< \frac{\varepsilon}{2} + \frac{\varepsilon}{2} \qquad (4.7.10)$$
$$= \varepsilon. \qquad (4.7.11)$$

Therefore, δ is a uniform threshold and the sum $f + g$ is uniformly continuous.

<u>Part (ii)</u>: Suppose $\alpha = 0$. Then αf is constant since

$$\alpha f(\mathbf{x}) = 0 \cdot f(\mathbf{x}) = \mathbf{0} \tag{4.7.12}$$

for every $\mathbf{x} \in \mathbb{R}^k$. Now let $\varepsilon > 0$. Then $\delta = 42$ is a uniform threshold since for any $\mathbf{x}, \mathbf{c} \in D$ we have

$$\|\mathbf{x} - \mathbf{c}\|_k < 42 \quad \Longrightarrow \quad \|\alpha f(\mathbf{x}) - \alpha f(\mathbf{c})\|_m = \|\mathbf{0}\|_m = 0 < \varepsilon. \tag{4.7.13}$$

Now suppose $\alpha \neq 0$ and $\varepsilon > 0$. Since f is uniformly continuous, there is a uniform threshold $\delta > 0$ such that for every $\mathbf{x}, \mathbf{c} \in D$ we have

$$\|\mathbf{x} - \mathbf{c}\|_k < \delta \quad \Longrightarrow \quad \|f(\mathbf{x}) - f(\mathbf{c})\|_m < \frac{\varepsilon}{|\alpha|}. \tag{4.7.14}$$

Hence, by the relationship between absolute value and Pythagorean distance from Corollary 1.2.30, we also have

$$\|\alpha f(\mathbf{x}) - \alpha f(\mathbf{c})\|_m = |\alpha| \|f(\mathbf{x}) - f(\mathbf{c})\|_m < |\alpha| \cdot \frac{\varepsilon}{|\alpha|} = \varepsilon. \tag{4.7.15}$$

Therefore, αf is uniformly continuous. $\square$

Once again and as mentioned in Remark 1.6.18 and much like Corollary 4.5.7 on pointwise continuity and linear combinations, a corollary of the linearity of uniform continuity holds for linear combinations. As with the proofs of Corollaries 1.6.16 and 4.5.7, the proof of Corollary 4.7.7 follows from induction on linearity. So, the proof is left as an exercise.

Corollary 4.7.7: Uniform continuity and linear combinations

Suppose $A \subseteq \mathbb{R}^\ell$, $k \in \mathbb{N}$, and for each $j = 1, \ldots, k$ the functions $f_j : A \to \mathbb{R}^m$ are uniformly continuous. Then the linear combination $f : A \to \mathbb{R}^m$ given by

$$f(\mathbf{x}) = \sum_{j=1}^{k} c_j f_j(\mathbf{x}) = c_1 f_1(\mathbf{x}) + \ldots + c_k f_k(\mathbf{x}) \tag{4.7.16}$$

is uniformly continuous.

In general, continuous functions are not necessarily uniformly continuous.

Example 4.7.8: Reciprocal is not uniformly continuous

The function $f : \mathbb{R} \backslash \{0\} \to \mathbb{R}$ given by

$$f(x) = \frac{1}{x} \tag{4.7.17}$$

is continuous but not uniformly continuous. The fact that f is continuous is proven in

Example 4.5.14 using the threshold

$$\delta = \min\left\{\frac{|c|}{2}, \frac{\varepsilon c^2}{2}\right\}. \tag{4.7.18}$$

This δ depends on c as can be seen by playing around with the Desmos activity "Continuity of reciprocal" from Figure 4.7.1. But to fully verify f is not uniformly continuous, we should show there is no threshold δ that suffices for every $c \in \mathbb{R}\setminus\{0\}$ at the same time. We can do this with the negation of Definition 4.7.1 by showing that for some particular $\varepsilon_0 > 0$ and *every* $\delta > 0$ there are real numbers x and c where

$$|x - c| < \delta \quad \text{but} \quad |f(x) - f(c)| = \left|\frac{1}{x} - \frac{1}{c}\right| \geq \varepsilon_0. \tag{4.7.19}$$

Scratch Work 4.7.9: The issue is at zero

Considering positive x and c close enough to 0 will suffice. In fact, we can use the pair $x = 1/n$ and $c = 1/(n+1)$ where $n \in \mathbb{N}$ since their difference can be made as small as we like (less than any $\delta > 0$) while their outputs are always a distance 1 apart. In the domain we have

$$\left|\frac{1}{n} - \frac{1}{n+1}\right| = \left|\frac{n+1-n}{n(n+1)}\right| = \frac{1}{n(n+1)} < \frac{1}{(n+1)^2}, \tag{4.7.20}$$

while in the range we have

$$\left|f\left(\frac{1}{n}\right) - f\left(\frac{1}{n+1}\right)\right| = \left|\frac{1}{1/n} - \frac{1}{1/(n+1)}\right| = |n+1-n| = 1. \tag{4.7.21}$$

Proof for Example 4.7.8. Consider the distance $\varepsilon_0 = 1/2 > 0$ for the range and let $\delta > 0$ represent an arbitrary distance for the domain. Since

$$\lim_{n \to \infty} \frac{1}{n(n+1)} = 0 \tag{4.7.22}$$

there is a threshold $n_\delta \in \mathbb{N}$ large enough so that

$$\left|\frac{1}{n_\delta} - \frac{1}{n_\delta + 1}\right| = \left|\frac{1}{n_\delta(n_\delta + 1)}\right| < \frac{1}{(n_\delta + 1)^2} < \delta. \tag{4.7.23}$$

(By the way, choosing $n_\delta > 1/\sqrt{\delta}$ suffices.) Then considering inputs $x = 1/n_\delta$ and $c = 1/(n_\delta + 1)$, their outputs satisfy

$$\left|\frac{1}{1/n_\delta} - \frac{1}{1/(n_\delta + 1)}\right| = |n_\delta + 1 - n_\delta| = 1 \geq \frac{1}{2}. \tag{4.7.24}$$

Therefore, $f(x) = 1/x$ is not uniformly continuous on $\mathbb{R}\setminus\{0\}$. $\qquad\square$

Remark 4.7.10: Continuity on a bounded domain does not imply bounded

In general, continuity on a bounded domain is not enough to ensure a function is bounded. For instance, the function $g : (0,1] \to \mathbb{R}$ given by $g(x) = 1/x$ is not bounded since $(1/n) \subseteq (0,1]$ but $g((1/n)) = \mathbb{N}$ is unbounded. However, uniform continuity suffices.

Theorem 4.7.11: Uniform continuity on a bounded domain ensures a function is bounded

If $A \subseteq \mathbb{R}^k$ is a bounded set and $f : A \to \mathbb{R}^m$ is uniformly continuous, then f is bounded.

Scratch Work 4.7.12: Bounded sets covered by a finite number of δ-neighborhoods

Given any fixed $\delta > 0$, every bounded subset of a Euclidean space can be covered a finite number of δ-neighborhoods. Given any $\varepsilon > 0$, uniform continuity (Definition 4.7.1) on a bounded domain supplies us with a uniform threshold $\delta > 0$ which can then be used to cover the domain and constrain the outputs of the function.

Proof of Theorem 4.7.11. Suppose $A \subseteq \mathbb{R}^k$ is a bounded set and $f : A \to \mathbb{R}^m$ is uniformly continuous. Let $\varepsilon = 4$. By the uniform continuity of f (Definition 4.7.1), there is a uniform threshold $\delta > 0$ such that

$$\mathbf{x}, \mathbf{c} \in A \text{ with } \|\mathbf{x} - \mathbf{c}\|_k < \delta \quad \implies \quad \|f(\mathbf{x}) - f(\mathbf{c})\|_m < 4. \tag{4.7.25}$$

Since A is bounded, a finite collection of δ-neighborhoods covers A. That is, for some $n_\delta \in \mathbb{N}$ there are δ-neighborhoods $V_\delta(\mathbf{c}_1), \ldots, V_\delta(\mathbf{c}_{n_\delta})$ with centers $\mathbf{c}_1, \ldots, \mathbf{c}_{n_\delta} \in A$ such that

$$A \subseteq \bigcup_{j=1}^{n_\delta} V_\delta(\mathbf{c}_j). \tag{4.7.26}$$

Set $b = \max\{\|\mathbf{c}_j\|_k + \delta : j = 1, \ldots, n\}$, which exists since finite sets of real numbers are bounded (Theorem 3.4.1). Since $A \subseteq \cup_{j=1}^{n_\delta} V_\delta(\mathbf{c}_j)$, each $\mathbf{x} \in A$ belongs to at least one of the δ-neighborhoods $V_\delta(\mathbf{c}_1), \ldots, V_\delta(\mathbf{c}_{n_\delta})$. Hence, for each $\mathbf{x} \in A$ there is some $j_\mathbf{x} \in \{1, \ldots, n_\delta\}$ where

$$\mathbf{x} \in V_\delta(\mathbf{c}_{j_\mathbf{x}}) \quad \Longleftrightarrow \quad \|\mathbf{x} - \mathbf{c}_{j_\mathbf{x}}\|_k < \delta. \tag{4.7.27}$$

Therefore, by adding zero and applying the triangle inequality (1.2.35) we have

$$\|f(\mathbf{x})\|_m = \|f(\mathbf{x}) \underbrace{- f(\mathbf{c}_{j_\mathbf{x}}) + f(\mathbf{c}_{j_\mathbf{x}})}_{\text{add } \mathbf{0}}\|_m \tag{4.7.28}$$

$$\leq \|f(\mathbf{x}) - f(\mathbf{c}_{j_\mathbf{x}})\|_m + \|f(\mathbf{c}_{j_\mathbf{x}})\|_m \tag{4.7.29}$$

$$< 4 + b. \tag{4.7.30}$$

Therefore, f is bounded. $\square$

Time for one more result before moving on to limits of functions and derivatives in the next chapter. This theorem plays a key role in Part I of the Fundamental Theorem of Calculus 6.1.15.

Theorem 4.7.13: Continuity on a compact set is uniform continuity

Suppose $K \subseteq \mathbb{R}^k$ and $f : K \to \mathbb{R}^m$ where K is compact and f is continuous. Then f is uniformly continuous.

Scratch Work 4.7.14: Two challenging definitions

The proof makes direct use of two of the most challenging definitions in real analysis: compactness and continuity, respectively their topological and ε-δ versions in Definitions 3.4.12 and 4.3.2. The idea is to consider an arbitrary distance $\varepsilon > 0$ for the range, then use continuity to create a δ-neighborhood at each point in the domain. The resulting collection of δ-neighborhoods creates an open cover for the domain, after some modification. Since the domain is compact, every open cover has a finite subcover. This means a finite number of the modified δ-neighborhoods covers the domain. We can use the smallest of these δ to serve as "Sauron's threshold", a threshold that satisfies the definition for uniform continuity (Definition 4.7.1).

This is a long and difficult proof with lots of notation to sift through. Hang in there.

Proof of Theorem 4.7.13. Suppose $K \subseteq \mathbb{R}^k$ and $f : K \to \mathbb{R}^m$ where f is continuous and K is compact. Let $\varepsilon > 0$. Since f is continuous, by Definition 4.3.2 we have for every point $\mathbf{c} \in K$ a threshold $\delta_{\mathbf{c}} > 0$ where

$$\mathbf{x} \in K \text{ with } \|\mathbf{x} - \mathbf{c}\|_k < \delta_{\mathbf{c}} \quad \Longrightarrow \quad \|f(\mathbf{x}) - f(\mathbf{c})\|_m < \frac{\varepsilon}{2}. \tag{4.7.31}$$

Also, note that for each $\mathbf{c} \in K$, the $(\delta_{\mathbf{c}}/2)$-neighborhood of $\mathbf{c}$ given by

$$V_{\delta_{\mathbf{c}}/2}(\mathbf{c}) = \left\{ \mathbf{x} \in \mathbb{R}^k : \|\mathbf{x} - \mathbf{c}\|_k < \frac{\delta_{\mathbf{c}}}{2} \right\} \tag{4.7.32}$$

is an open set by Lemma 3.2.7. Since we also have

$$K \subseteq \bigcup_{\mathbf{c} \in K} V_{\delta_{\mathbf{c}}/2}(\mathbf{c}), \tag{4.7.33}$$

the collection of open sets

$$\mathcal{U} = \left\{ V_{\delta_{\mathbf{c}}/2}(\mathbf{c}) : \mathbf{c} \in K \right\} \tag{4.7.34}$$

is an open cover for K. By Definition 3.4.12, since K is compact, $\mathcal{U}$ has a finite subcover $\mathcal{U}_0$. This finite subcover is indexed by a finite number of points $\mathbf{c}_1, \ldots, \mathbf{c}_{n_0} \in K$ for some $n_0 \in \mathbb{N}$. We have

$$\mathcal{U}_0 = \left\{ V_{\delta_{\mathbf{c}_1}/2}(\mathbf{c}_1), \ldots, V_{\delta_{\mathbf{c}_{n_0}}/2}(\mathbf{c}_{n_0}) \right\} \quad \text{and} \quad K \subseteq \bigcup_{j=1}^{n_0} V_{\delta_{\mathbf{c}_j}/2}(\mathbf{c}_j). \tag{4.7.35}$$

Define δ to be the smallest distance among $\delta_{\mathbf{c}_1}/2, \ldots, \delta_{\mathbf{c}_{n_0}}/2$. That is, let

$$\delta = \min \left\{ \frac{\delta_{\mathbf{c}_1}}{2}, \ldots, \frac{\delta_{\mathbf{c}_{n_0}}}{2} \right\} \tag{4.7.36}$$

so that we are sure to have

$$\delta \leq \frac{\delta_{\mathbf{c}_j}}{2} \quad \text{for each} \quad j = 1, \ldots, n_0. \tag{4.7.37}$$

To establish the uniform continuity of f, suppose

$$\mathbf{x}, \mathbf{y} \in K \quad \text{where} \quad \|\mathbf{x} - \mathbf{y}\|_k < \delta. \tag{4.7.38}$$

Since $\mathcal{U}_0$ covers K (meaning the containment in (4.7.35)), there is an index t among $1, \ldots, n_0$ where

$$\mathbf{x} \in V_{\delta_{\mathbf{c}_t}/2}(\mathbf{c}_t) \quad \Longleftrightarrow \quad \|\mathbf{x} - \mathbf{c}_t\|_k < \frac{\delta_{\mathbf{c}_t}}{2}. \tag{4.7.39}$$

Therefore, by adding zero and applying the triangle inequality ((1.2.34) and (1.2.32)), we also have

$$\|\mathbf{y} - \mathbf{c}_t\|_k = \|\mathbf{y} \underbrace{-\mathbf{x} + \mathbf{x}}_{\text{add zero}} -\mathbf{c}_t\|_k \tag{4.7.40}$$

$$\leq \|\mathbf{y} - \mathbf{x}\|_k + \|\mathbf{x} - \mathbf{c}_t\|_k \tag{4.7.41}$$

$$< \delta + \frac{\delta_{\mathbf{c}_t}}{2} \tag{4.7.42}$$

$$\leq \delta_{\mathbf{c}_t}, \tag{4.7.43}$$

where the last inequality holds since $\delta \leq \delta_{\mathbf{c}_t}/2$ by (4.7.37). So, both $\mathbf{x}$ and $\mathbf{y}$ are within the threshold $\delta_{\mathbf{c}_t}$ of $\mathbf{c}_t$. Since $\delta_{\mathbf{c}_t}$ is chosen in response to the distance $\varepsilon/2$ for the range as in (4.7.31), and by adding zero again along with another application of the triangle inequality ((1.2.34) and (1.2.32)), we have

$$\|f(\mathbf{x}) - f(\mathbf{y})\|_m = \|f(\mathbf{x}) \underbrace{-f(\mathbf{c}_t) + f(\mathbf{c}_t)}_{\text{add zero}} -f(\mathbf{y})\|_m \tag{4.7.44}$$

$$\leq \|f(\mathbf{x}) - f(\mathbf{c}_t)\|_m + \|f(\mathbf{c}_t) - f(\mathbf{y})\|_m \tag{4.7.45}$$

$$< \frac{\varepsilon}{2} + \frac{\varepsilon}{2} \tag{4.7.46}$$

$$= \varepsilon. \tag{4.7.47}$$

Therefore, f is uniformly continuous.   □

The next couple of chapters dive into some of the biggest ideas in calculus: limits of functions, derivatives, and integrals. Continuity and uniform continuity feature prominently.

Exercises

4.7.1. Prove $h(x) = x^2$ is not uniformly continuous on $\mathbb{R}$.

4.7.2. Prove $h(x) = x^2$ is uniformly continuous on every bounded interval.

4.7.3. Prove $r(x) = \sqrt{x}$ is uniformly continuous on $[0, \infty)$.

4.7.4. Prove that a composition of uniformly continuous functions is uniformly continuous.

4.7.5. Prove uniform continuity is not multiplicative. That is, find uniformly continuous real-valued functions f and g whose product fg is not uniformly continuous.

4.7.6. Despite the previous exercise, prove that if real-valued functions f and g are uniformly continuous *and bounded*, then their product fg is uniformly continuous.

4.7.7. Suppose $g : \mathbb{N} \to \mathbb{R}$ (so g defines a sequence of real numbers, see Definition 2.1.1). Prove g is uniformly continuous.

4.7.8. Suppose $t : (0, 1) \to \mathbb{R}$ is the topologist's sine curve restricted to $(0, 1)$ given by

$$t(x) = \sin\left(\frac{1}{x}\right) \quad \text{for all } x \in (0, 1). \tag{4.7.48}$$

Prove t is not uniformly continuous.

4.7.9. Suppose $f : [a, b] \to \mathbb{R}$ is continuous and $f(x) > 0$ for all $x \in [a, b]$. Prove f has a positive lower bound in that there is some $\ell > 0$ where

$$f(x) \geq \ell > 0 \quad \text{for all } x \in [a, b]. \tag{4.7.49}$$

4.7.10. Prove that if $D \subseteq \mathbb{R}^k$, $f : D \to \mathbb{R}^m$ is uniformly continuous, and $(\mathbf{x}_n) \subseteq \mathbb{R}^k$ is Cauchy, then the image sequence $(f(\mathbf{x}_n)) \subseteq \mathbb{R}^m$ is Cauchy.

4.7.11. Find an example of a function showing the previous result does not hold when uniform continuity is replaced with continuity.

4.7.12. A function $f : D \to \mathbb{R}^m$ is *Lipschitz* on $D \subseteq \mathbb{R}^k$ if there is some $b > 0$ such that for all $\mathbf{x}, \mathbf{y} \in D$ we have

$$\|f(\mathbf{x}) - f(\mathbf{y})\|_m \leq b\|\mathbf{x} - \mathbf{y}\|_k. \tag{4.7.50}$$

Prove Lipschitz functions are uniformly continuous.

4.7.13. A function $f : \mathbb{R}^m \to \mathbb{R}^m$ is a *contraction* if there is some real number r where $0 < r < 1$ and for all $\mathbf{x}, \mathbf{y} \in \mathbb{R}^m$ we have

$$\|f(\mathbf{x}) - f(\mathbf{y})\|_m \leq r\|\mathbf{x} - \mathbf{y}\|_m. \tag{4.7.51}$$

The goal of this exercise is to prove a version of the *Banach Fixed-Point Theorem* which is also called the *Contraction Mapping Principle*: Every contraction has a unique fixed point.

To that end, suppose $f : \mathbb{R}^m \to \mathbb{R}^m$ is a contraction.

(i) Prove f is uniformly continuous.

(ii) Choose a point $\mathbf{x}_1 \in \mathbb{R}^m$. Consider the recursively defined sequence $(\mathbf{x}_n) \subseteq \mathbb{R}^m$ given by

$$\mathbf{x}_2 = f(\mathbf{x}_1), \quad \mathbf{x}_3 = f(\mathbf{x}_2) = f(f(\mathbf{x}_1)), \ldots \quad \mathbf{x}_{n+1} = f(\mathbf{x}_n) = f^n(\mathbf{x}_1). \tag{4.7.52}$$

Prove $(\mathbf{x}_n)$ converges using the Cauchy criterion for sequences (Theorem 2.6.5).

(iii) Suppose $\mathbf{y} = \lim_{n \to \infty} \mathbf{x}_n$. Prove $f(\mathbf{y}) = \mathbf{y}$, so $\mathbf{y}$ is a fixed point of f.

(iv) Prove the fixed point of f is unique.

Chapter 5

Limits and Derivatives

Limits are the central concept in calculus. They empower us to determine values sequences and functions approach, whether or not the value is attained. The Fundamental Theorem of Calculus features a connection between two concepts that stem from limits: *derivatives* and *integrals*.

This chapter builds limits for functions from continuity, then derivatives and eventually *codas* from limits.

5.1 Limit of a function

If your experience in calculus was anything like mine, *limits of functions* were covered before continuity. In this book, continuity is covered first because of the way it follows from arbitrarily close and asking questions about preserving closeness. Limits of functions, also called *functional limits*, follow from continuity by allowing ourselves to ignore a single output—$f(\mathbf{c})$—and asking, how close is a function f to being continuous at $\mathbf{c}$?

To help motivate a definition for limits based on continuity, let's see what we get from older techniques. Keep calculus in mind with the following example.

Example 5.1.1: A hole in the graph

Consider the rational function $r : \mathbb{R}\backslash\{3\} \to \mathbb{R}$ given by

$$r(x) = \frac{x^2 - 9}{x - 3}. \tag{5.1.1}$$

See Figure 5.1.1. The function r is acting exactly like the polynomial $f(x) = x + 3$, except when $c = 3$ where r is not defined. Have you ever computed a limit in a way that looks

---

something like this?

$$\lim_{x \to 3} r(x) = \lim_{x \to 3} \left(\frac{x^2 - 9}{x - 3} \right) \qquad (5.1.2)$$

$$= \lim_{x \to 3} \frac{(x + 3)(x - 3)}{x - 3} \qquad (5.1.3)$$

$$= \lim_{x \to 3} (x + 3) \qquad (5.1.4)$$

$$= 6. \qquad (5.1.5)$$

Assuming this looks familiar, note that we only "take the limit" in the last step $(5.1.5)$ where we conclude $x + 3$ approaches 6 as x approaches $c = 3$.

Also, we get 6 by plugging in 3 for x in the expression $x + 3$. This process is valid because $f(x) = x + 3$ is continuous at $c = 3$, so the outputs $f(x)$—and therefore $r(x)$—really do "approach" $f(c)$ as the inputs "approach" c.

But what do we mean by "approach", exactly? What kind of mathematics can we use to codify our intuition? Well, for every number $x \neq 3$, we have

$$r(x) = \frac{x^2 - 9}{x - 3} = \frac{(x + 3)(x - 3)}{x - 3} = x + 3. \qquad (5.1.6)$$

So the computation above actually starts by replacing $r(x)$ with the continuous polynomial $f(x) = x + 3$, then we compute the limit by evaluating f at $c = 3$ to get 6. From there, the ε-δ aspect of the definition for continuity (Definition 4.3.2) captures the meaning of "approach": δ tells us how the inputs "approach" c while ε tells us how the outputs "approach" $f(c)$.

Note that the rational function r is actually discontinuous at $c = 3$ since the output $r(3)$ is not defined. By Definition 4.3.2, continuity at c requires the function to be defined at c. Modifying continuity to allow the functions to *not* defined at c is accomplished by taking c to be an accumulation point of the domain (see Definition 3.6.7). Doing so allows us to keep the way ε and δ capture the values "approached" in the range and domain, giving us the formal definition for the limit of a function.

Definition 5.1.2: Convergence, threshold, and limit of a function

Suppose $D \subseteq \mathbb{R}^k$, $f : D \to \mathbb{R}^m$, and $\mathbf{c} \operatorname{acl} D \backslash \{\mathbf{c}\}$. A point $\mathbf{y} \in \mathbb{R}^m$ is the *limit of f at $\mathbf{c}$*, if for every distance $\varepsilon > 0$ for the codomain there is a *threshold* $\delta > 0$ providing a distance for the domain such that

$$\mathbf{x} \in D \text{ with } 0 < \|\mathbf{x} - \mathbf{c}\|_k < \delta \qquad \Longrightarrow \qquad \|f(\mathbf{x}) - \mathbf{y}\|_m < \varepsilon. \qquad (5.1.7)$$

When $\mathbf{y}$ is the limit of f at $\mathbf{c}$, we say f *converges to $\mathbf{y}$ at $\mathbf{c}$* and we write

$$\lim_{\mathbf{x} \to \mathbf{c}} f(\mathbf{x}) = \mathbf{y} \quad \text{or} \quad \lim_{\mathbf{x} \to \mathbf{c}} f(\mathbf{x}) = \mathbf{y}.$$

If f does not converge, we say f *diverges* at $\mathbf{c}$ and the limit of f at $\mathbf{c}$ does not exist.

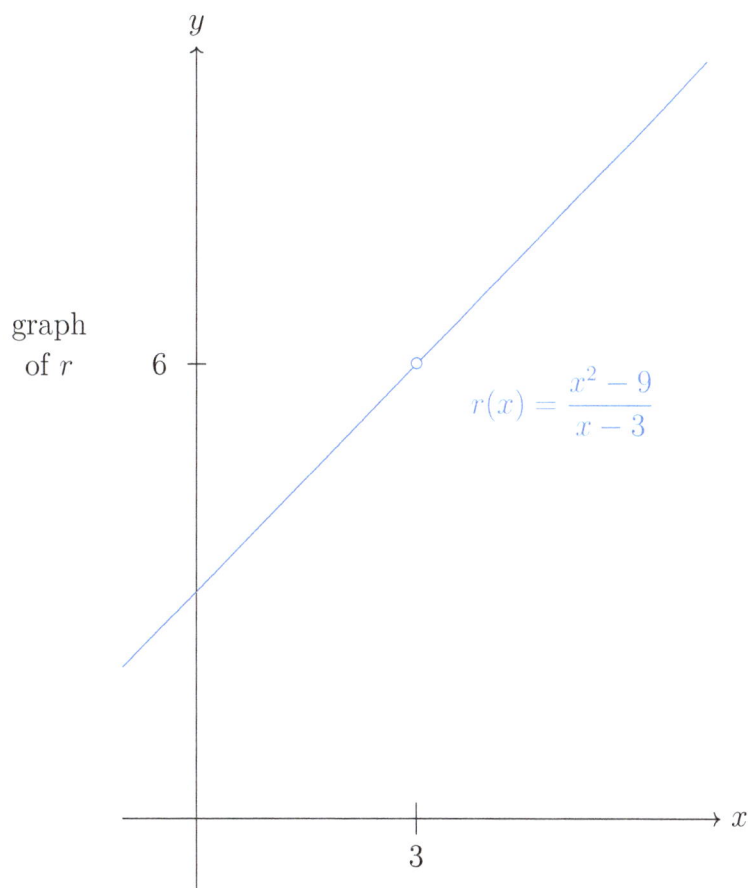

y

graph
of r   6

$$r(x) = \frac{x^2 - 9}{x - 3}$$

3

x

Figure 5.1.1: The rational function $r : \mathbb{R}\backslash\{3\} \to \mathbb{R}$ in Example 5.1.1 given by $r(x) = (x^2-9)/(x+3)$ is almost the same function as the polynomial $f(x) = x+3$. The only difference is r is not defined at $c = 3$, indicated by the hole in the graph.

Definition 5.1.2 adapts continuity (Definition 4.3.2) to fit the idea of limits. The key is to ignore the point $\mathbf{c}$ itself, whether or not it is in the domain D. To make sure our results are still meaningful and to capture the behavior of a function near $\mathbf{c}$, we consider domains where $\mathbf{c}$ is an accumulation point, hence $\mathbf{c}\,\mathrm{acl}\,D\backslash\{\mathbf{c}\}$. To give the outputs a central value to stay near, a point $\mathbf{y}$ in the codomain is used to fill the role left by $f(\mathbf{c})$. The interplay between ε and δ from the definition for continuity is kept because it provides the technical meaning for how the inputs $\mathbf{x}$ "approaching" c lead to their outputs $f(\mathbf{x})$ "approaching" the limit $\mathbf{y}$.

Remark 5.1.3: Continuity versus limit

Here is a side-by-side comparison of the definitions of continuity and functional limits:

Definition 4.3.2, continuity of f at $\mathbf{c}$	Definition 5.1.2, $\lim_{\mathbf{x}\to\mathbf{c}} f(\mathbf{x}) = \mathbf{y}$
f is continuous at $\mathbf{c}$ $\Longleftrightarrow$ $\mathbf{c}\in D$ and $\forall\,\varepsilon>0$ $\exists\,\delta>0$ such that $\mathbf{x}\in D$ with $\|\mathbf{x}-\mathbf{c}\|_k < \delta$ $\Longrightarrow$ $\|f(\mathbf{x})-f(\mathbf{c})\|_m < \varepsilon$.	f converges to $\mathbf{y}$ at $\mathbf{c}$ $\Longleftrightarrow$ $\mathbf{c}\,\mathrm{acl}\,D\backslash\{\mathbf{c}\}$ and $\forall\,\varepsilon>0$ $\exists\,\delta>0$ such that $\mathbf{x}\in D$ with $0<\|\mathbf{x}-\mathbf{c}\|_k < \delta$ $\Longrightarrow$ $\|f(\mathbf{x})-\mathbf{y}\|_m < \varepsilon$.

Note that in the Euclidean space $\mathbb{R}^k$, we have

$$\mathbf{x}=\mathbf{c} \quad\Longleftrightarrow\quad d_k(\mathbf{x},\mathbf{c}) = \|\mathbf{x}-\mathbf{c}\|_k = 0. \tag{5.1.8}$$

Hence, in the definition for the limit of f at $\mathbf{c}$, the point $\mathbf{c}$ is explicitly ignored since we only consider $\mathbf{c}$ to be an accumulation point of the domain D and

$$\mathbf{x}\in D \quad\text{with}\quad d_k(\mathbf{x},\mathbf{c}) = \|\mathbf{x}-\mathbf{c}\|_k > 0. \tag{5.1.9}$$

This means with limits we consider inputs $\mathbf{x}\in D$ where $\mathbf{x}\neq\mathbf{c}$. Still, working with the definition for limits of functions (Definition 5.1.2) is very similar to working with the definition of continuity (Definition 4.3.2).

Remark 5.1.4: Limit versus limit?

Limits have already come up in this book, specifically in Chapter 2 which deals with limits of sequences and their properties. In this chapter, the focus is on limits of functions. These concepts are deeply related with their relationship codified in Theorem 5.1.14.

Example 5.1.5: Limit via continuity

For the rational function $r : \mathbb{R} \backslash \{3\} \to \mathbb{R}$ in Example 5.1.1 given by

$$r(x) = \frac{x^2 - 9}{x - 3}, \tag{5.1.10}$$

we have

$$\lim_{x \to 3} r(x) = \lim_{x \to 3} \frac{x^2 - 9}{x - 3} = \lim_{x \to 3} (x + 3) = 6. \tag{5.1.11}$$

Scratch Work 5.1.6: Similar to showing continuity of a line

The scratch work for showing the limit of r at $c = 3$ is like Scratch Work 4.3.7 where we show $f(x) = x/2$ is continuous for Example 4.3.5. A key difference is that with the definition for limits of functions (Definition 5.1.2), we begin by ignoring $c = 3$ since we only consider inputs x where

$$0 < d_{\mathbb{R}}(x, c) = |x - c|, \tag{5.1.12}$$

which implies $x \neq c$. This allows us to replace r—which is discontinuous at $c = 3$—with a continuous polynomial, at least for this example.

From there, the scratch work is similar to dealing with continuity in Scratch Work 4.3.7, but in general *we may not know the value of the limit up front.* To deal with this, we leverage the fact that we consider only $x \neq 3$. So, we have

$$r(x) = \frac{x^2 - 9}{x - 3} = \frac{(x + 3)(x - 3)}{x - 3} = x + 3 = f(x). \tag{5.1.13}$$

For this kind of example, evaluating the continuous function f at $c = 3$ gives us a suitable candidate for the limit of r, just like we do in calculus. Specifically, let

$$y = f(c) = f(3) = 3 + 3 = 6. \tag{5.1.14}$$

So now we can start working with the concluding inequality in Definition 5.1.2, where we want to end up with

$$|r(x) - 6| = \left| \frac{x^2 - 9}{x - 3} - 6 \right| = |(x + 3) - 6| = |x - 3| < \varepsilon. \tag{5.1.15}$$

So, $\delta = \varepsilon$ seems to be the right choice for a threshold.

Proof for Example 5.1.5. Let $\varepsilon > 0$ provide a distance for the codomain and suppose $x \neq c = 3$.

We have

$$r(x) = \frac{x^2 - 9}{x - 3} = \frac{(x+3)(x-3)}{x-3} = x + 3. \tag{5.1.16}$$

Choose the distance $\delta = \varepsilon$ for the domain and let $y = 6$ serve as a candidate for the limit. We have

$$0 < |x - 3| < \delta = \varepsilon \tag{5.1.17}$$

implies

$$|r(x) - 6| = \left| \frac{x^2 - 9}{x - 3} - 6 \right| = |(x+3) - 6| = |x - 3| < \varepsilon. \tag{5.1.18}$$

Therefore, $\delta = \varepsilon$ is a threshold for the convergence of r to 6 and we have

$$\lim_{x \to 3} r(x) = 6. \tag{5.1.19}$$

$\square$

The guide for working with the ε-δ definition of continuity in Remark 4.3.4 adapts nicely to become a guide for working with the ε-δ definition for limits of functions (Definition 5.1.2). For limits, we need to make sure we are only considering inputs that are not equal to $\mathbf{c}$ and we need a candidate for the limit $\mathbf{y}$ to work with.

Remark 5.1.7: Guide for proofs of functional limits

Scratch work for proving $\lim\limits_{\mathbf{x} \to \mathbf{c}} f(\mathbf{x}) = \mathbf{y}$:

- If $f(\mathbf{c})$ is defined, ignore it.

- Find a suitable candidate for the limit $\mathbf{y}$. Consider replacing f with a function which is continuous at $\mathbf{c}$ to help find a value for $\mathbf{y}$.

- Consider the inequality you want to end up with, typically:

$$d_m(f(\mathbf{x}), \mathbf{y}) = \|f(\mathbf{x}) - \mathbf{y}\|_m < \varepsilon. \tag{5.1.20}$$

- *Key step*: Use the previous inequality to find a formula for a threshold $\delta > 0$ as a function of ε and $\mathbf{c}$ where $\mathbf{x}$ is in the domain and

$$0 < d_k(\mathbf{x}, \mathbf{c}) = \|\mathbf{x} - \mathbf{c}\|_k < \delta, \tag{5.1.21}$$

  keeping in mind we only consider $\mathbf{x} \neq \mathbf{c}$. The triangle inequality as it appears in (1.2.32) and (1.2.34) is used quite often at this step in scratch work and proofs involving "ε-δ" argument for limits.

- Include a figure with the function f, the point $\mathbf{c}$ arbitrarily close to the domain, perhaps as a graph. You may also want to plot the domain and range on their own.

Proving for proving $\lim_{\mathbf{x}\to\mathbf{c}} f(\mathbf{x}) = \mathbf{y}$:

- Early in the proof, perhaps the first step, write "Let $\varepsilon > 0$" or something similar, indicating you are accounting for *all* positive distances in the range at the same time.

- Define or choose candidates for the limit $\mathbf{y}$ in the codomain and threshold $\delta > 0$ for a distance in the domain based on your scratch work.

- Verify δ is truly a threshold for the convergence of f to $\mathbf{y}$ at $\mathbf{c}$ by considering every point $\mathbf{x}$ in the domain within δ of $\mathbf{c}$ but not equal $\mathbf{c}$, then showing

$$\mathbf{x} \in D \text{ with } 0 < \|\mathbf{x} - \mathbf{c}\|_k < \delta \quad \Longrightarrow \quad \|f(\mathbf{x}) - \mathbf{y}\|_m < \varepsilon. \tag{5.1.22}$$

In other words, verify the "deleted" δ-neighborhood of $\mathbf{c}$ in the domain maps into the ε-neighborhood of $\mathbf{y}$ in the range:

$$f(V_\delta(\mathbf{c}) \cap (D\backslash\{\mathbf{c}\})) \subseteq V_\varepsilon(\mathbf{y}). \tag{5.1.23}$$

Although they refer to working with continuity, see Figures 4.3.1, 4.3.2, and 4.3.4. The roles played by ε and δ are essentially the same for continuity and limits.

The convergence of a function is more general than continuity. The following theorem codifies this relationship tells us we can determine the value of a functional limit by evaluating a continuous function.

Theorem 5.1.8: Continuity implies convergence

Suppose $D \subseteq \mathbb{R}^k$, $\mathbf{c} \in D$, $\mathbf{c}$ acl $D\backslash\{\mathbf{c}\}$, and $f : D \to \mathbb{R}^m$. If f is continuous at $\mathbf{c}$, then f converges to $f(\mathbf{c})$ at $\mathbf{c}$ and we have

$$\lim_{\mathbf{x}\to\mathbf{c}} f(\mathbf{x}) = f(\mathbf{c}). \tag{5.1.24}$$

Scratch Work 5.1.9: The output is the limit

The proof follows immediately from the definitions where we use $f(\mathbf{c})$ to serve as a candidate for the limit $\mathbf{y}$.

Proof of Theorem 5.1.8. Suppose f is continuous at $\mathbf{c}$ which is also an accumulation point of the domain. Let $\varepsilon > 0$ and let $f(\mathbf{c})$ serve as a candidate for the limit. By the definition of continuity (Definition 4.3.2), there is a threshold $\delta > 0$ where

$$\mathbf{x} \in D \quad \text{with} \quad \|\mathbf{x} - \mathbf{c}\|_k < \delta \tag{5.1.25}$$

implies both

$$0 < \|\mathbf{x} - \mathbf{c}\|_k < \delta \quad \text{and} \quad \|f(\mathbf{x}) - f(\mathbf{c})\|_m < \varepsilon. \tag{5.1.26}$$

Hence, δ is also a threshold for the convergence of f to $f(\mathbf{c})$ at $\mathbf{c}$ (Definition 5.1.2) and

$$\lim_{\mathbf{x} \to \mathbf{c}} f(\mathbf{x}) = f(\mathbf{c}). \tag{5.1.27}$$

$\square$

A formal definition for *removable discontinuity* seems in order. It can be used to explore and take advantage of the relationship between continuity and limits of functions.

Definition 5.1.10: Removable discontinuity

Suppose $D \subseteq \mathbb{R}^k$ and $\mathbf{c}$ acl $D \backslash \{\mathbf{c}\}$. A function $g : D \to \mathbb{R}^m$ has a *removable discontinuity* at $\mathbf{c}$ if there is a function $f : D \cup \{\mathbf{c}\} \to \mathbb{R}^m$ satisfying

$$f(\mathbf{x}) = g(\mathbf{x}) \quad \text{for all} \quad \mathbf{x} \in D \backslash \{\mathbf{c}\}, \tag{5.1.28}$$

where f is continuous at $\mathbf{c}$.

Example 5.1.11: Revisiting a hole in a graph

For the rational function $r : \mathbb{R} \backslash \{3\} \to \mathbb{R}$ in Examples 5.1.1 and (5.1.5) given by

$$r(x) = \frac{x^2 - 9}{x - 3} \tag{5.1.29}$$

has a removable discontinuity at $c = 3$. To see this, take $f : \mathbb{R} \to \mathbb{R}$ to be the polynomial $f(x) = x + 3$. This gives us a continuous function (by Theorem 4.5.2) which is equal to r on $\mathbb{R} \backslash \{3\}$ by the equations in (5.1.6). See Figure 5.1.1.

In the same way continuity can be modified to define functional limits, we can modify the preservation of closeness to provide a similar condition suitable for limits. The definition of *preserving accumulation* (Definition 5.1.12) follows suit by softening the requirement of having $\mathbf{c}$ be a point in the domain to having $\mathbf{c}$ being an accumulation point and replacing the output $f(\mathbf{c})$ with a suitable point $\mathbf{y}$ in the codomain.

Definition 5.1.12: Preserve accumulation

Suppose $D \subseteq \mathbb{R}^k$, $f : D \to \mathbb{R}^m$, and $\mathbf{c}$ acl $D \backslash \{\mathbf{c}\}$. We say f *preserves accumulation at* $\mathbf{c}$ if there is a point $\mathbf{y} \in \mathbb{R}^m$ such that for every $E \subseteq D \backslash \{\mathbf{c}\}$ we have

$$\mathbf{c} \text{ acl } E \qquad \Longrightarrow \qquad \mathbf{y} \text{ acl } f(E). \tag{5.1.30}$$

In this case, we say $\mathbf{y}$ is the *limit of f at* $\mathbf{c}$.

Remark 5.1.13: Preserving closeness versus preserving accumulation

Here is a side-by-side comparison of the definitions of preserving closeness and preserving accumulation:

Definition 4.2.5 f preserves closensess at $\mathbf{c}$	Definition 5.1.12 f preserves accumulation at $\mathbf{c}$
For all $E \subseteq D$, $\mathbf{c}\operatorname{acl} E \implies f(\mathbf{c})\operatorname{acl} f(E)$.	For some $\mathbf{y} \in \mathbb{R}^m$ and for all $E \subseteq D\backslash\{\mathbf{c}\}$, $\mathbf{c}\operatorname{acl} E \implies \mathbf{y}\operatorname{acl} f(E)$.

The equivalence of the preservation of closeness, sequential continuity, and ε-δ continuity in Theorem 4.4.7 along with the strong relationship between continuity and limits of functions motivates the following theorem on equivalent forms of convergence for functions. This theorem provides flexibility in the way we choose to prove results on convergence and limits of functions throughout the rest of the book.

Theorem 5.1.14: Equivalent forms of convergence for functions

Suppose $D \subseteq \mathbb{R}^k$, $f : D \to \mathbb{R}^m$, $\mathbf{y} \in \mathbb{R}^m$, and $\mathbf{c}\operatorname{acl} D\backslash\{\mathbf{c}\}$. Then the following are equivalent forms of the convergence of f to $\mathbf{y}$ at $\mathbf{c}$:

(i) f preserves accumulation at $\mathbf{c}$ with limit $\mathbf{y}$ (Definition 5.1.12):

$$E \subseteq D\backslash\{\mathbf{c}\} \text{ with } \mathbf{c}\operatorname{acl} E \quad \implies \quad \mathbf{y}\operatorname{acl} f(E). \tag{5.1.31}$$

(ii) f transforms sequential convergence to $\mathbf{c}$ in the domain into sequential convergence to $\mathbf{y}$ in the codomain, specifically:

$$(\mathbf{x}_n) \subseteq D\backslash\{\mathbf{c}\} \text{ with } \lim_{n\to\infty} \mathbf{x}_n = \mathbf{c} \quad \implies \quad \lim_{n\to\infty} f(\mathbf{x}_n) = \mathbf{y}. \tag{5.1.32}$$

(iii) f converges to $\mathbf{y}$ at $\mathbf{c}$, so $\lim_{\mathbf{x}\to\mathbf{c}} f(\mathbf{x}) = \mathbf{y}$ (Definition 5.1.2):

$$
\begin{aligned}
&\text{For every distance } \varepsilon > 0 \text{ for the codomain,} &&(5.1.33)\\
&\text{there is a distance } \delta > 0 \text{ for the domain such that} &&(5.1.34)\\
&\mathbf{x} \in D \text{ with } 0 < \|\mathbf{x} - \mathbf{c}\|_k < \delta \quad \implies \quad \|f(\mathbf{x}) - \mathbf{y}\|_m < \varepsilon. &&(5.1.35)
\end{aligned}
$$

Scratch Work 5.1.15: Modify the continuity equivalence

The proof of Theorem 5.1.14 follows from a modification of the proof of Theorem 4.4.7: Replace the condition of having $\mathbf{c}$ in the domain with $\mathbf{c}$ being an accumulation point of the domain, and replace the output $f(\mathbf{c})$ with limit $\mathbf{y}$.

Proof of Theorem 5.1.14. Let's show (i) $\implies$ (ii), (ii) $\implies$ (iii), and (iii) $\implies$ (i), all by contraposition.

(i) $\implies$ (ii): Suppose there is a point $\mathbf{c} \in D\backslash\{\mathbf{c}\}$ and sequence $(\mathbf{x}_n)$ of points in $D\backslash\{\mathbf{c}\}$ (so

$\mathbf{x}_n \neq \mathbf{c}$) and where

$$\lim_{n\to\infty} \mathbf{x}_n = \mathbf{c} \quad \text{but} \quad \lim_{n\to\infty} f(\mathbf{x}_n) \neq \mathbf{y}. \tag{5.1.36}$$

Hence, there must be some $\varepsilon_0 > 0$ such that no matter which positive integer N we consider, there is a positive integer $q \geq N$ where $\mathbf{x}_q$ is in D and

$$\|f(\mathbf{x}_q) - \mathbf{y}\|_m \geq \varepsilon_0. \tag{5.1.37}$$

We can use the previous statement to construct a suitable subsequence $(\mathbf{x}_{n_j})$ of $(\mathbf{x}_n)$ whose image stays away from $\mathbf{y}$. That is, let $n_1 \geq 1$ be a positive integer that satisfies the previous statement. Proceeding inductively, for each positive integer j, there is a positive integer $n_j > n_{j-1}$ where

$$\|f(\mathbf{x}_{n_j}) - \mathbf{y}\|_m \geq \varepsilon_0. \tag{5.1.38}$$

That is, we have $\mathbf{y}\,\text{awf}(f(\mathbf{x}_{n_j}))$. Now, since $\lim_{n\to\infty} \mathbf{x}_n = \mathbf{c}$ and $(\mathbf{x}_{n_j})$ is a subsequence of $(\mathbf{x}_n)$, we have $\lim_{j\to\infty} \mathbf{x}_{n_j} = \mathbf{c}$ as well. By Theorem 2.3.1, we have $\mathbf{c}\,\text{acl}(\mathbf{x}_{n_j})$. Therefore, f does not preserve closeness at $\mathbf{c}$. (See Definition 5.1.12.)

(ii) $\implies$ (iii): Suppose f does not converge to $\mathbf{y}$ at a point $\mathbf{c}$ where $\mathbf{c}\,\text{acl}\,D\backslash\{\mathbf{c}\}$. Then there must be some $\varepsilon_0 > 0$ such that no matter which value we take for $\delta > 0$, there is a point $\mathbf{x}_\delta$ in D with

$$0 < \|\mathbf{x}_\delta - \mathbf{c}\|_k < \delta \quad \text{and} \quad \|f(\mathbf{x}_\delta) - \mathbf{y}\|_m \geq \varepsilon_0. \tag{5.1.39}$$

Much as in the proof of Theorem 2.3.1, we can use the previous statement to construct a suitable sequence. For each positive integer n, there must be a point $\mathbf{x}_n$ in D where

$$0 < \|\mathbf{x}_n - \mathbf{c}\|_k < 1/n \quad \text{and} \quad \|f(\mathbf{x}_n) - \mathbf{y}\|_m \geq \varepsilon_0. \tag{5.1.40}$$

(Note that $1/n$ plays the role of δ here). Hence, $(\mathbf{x}_n)$ is a sequence of points in $D\backslash\{\mathbf{c}\}$ where

$$\lim_{n\to\infty} \mathbf{x}_n = \mathbf{c} \quad \text{but} \quad \lim_{n\to\infty} f(\mathbf{x}_n) \neq \mathbf{y}. \tag{5.1.41}$$

(iii) $\implies$ (i): Finally, suppose f does not preserve accumulation to $\mathbf{y}$ at a point $\mathbf{c}$ where $\mathbf{c}\,\text{acl}\,D\backslash\{\mathbf{c}\}$. Then there must be some $E \subseteq D\backslash\{\mathbf{c}\}$ where $\mathbf{c}\,\text{acl}\,E$ but $\mathbf{y}\,\text{awf}\,f(E)$. So, there is some $\varepsilon_0 > 0$ such that for every point $\mathbf{x}$ in E we have

$$\|f(\mathbf{x}) - \mathbf{y}\|_m \geq \varepsilon_0. \tag{5.1.42}$$

Now, let $\delta > 0$. Since $\mathbf{c}\,\text{acl}\,E$, there is a point $\mathbf{y}_\delta$ in E where we have both

$$0 < \|\mathbf{y}_\delta - \mathbf{c}\|_k < \delta \quad \text{and} \quad \|f(\mathbf{y}_\delta) - \mathbf{y}\|_m \geq \varepsilon_0. \tag{5.1.43}$$

Therefore, f does not converge to $\mathbf{y}$ at $\mathbf{c}$. (See Definition 5.1.2.) $\qquad\square$

The next theorem mirrors the local boundedness of continuity established with Theorem 4.3.21. Here, convergence is assumed and the distance we obtain for the domain with a threshold provides the neighborhood on which the function is bounded.

Theorem 5.1.16: Convergence implies locally bounded

Suppose $D \subseteq \mathbb{R}^k$, $\mathbf{c} \operatorname{acl} D\backslash\{\mathbf{c}\}$, and $f : D \to \mathbb{R}^m$. If f converges at $\mathbf{c}$, then f is locally bounded at $\mathbf{c}$.

Scratch Work 5.1.17: Mirror local boundedness of continuity

The proof is very similar to the proof that continuous functions are locally bounded (Theorem 4.3.21). In both situations we deal with local boundedness, so we only need to consider inputs within a neighborhood and bound their outputs. Both continuity and convergence allow us to do so by picking any positive distance (say 14 this time) for the outputs, then convergence ensures the existence of a threshold that defines a neighborhood for the domain whose image is bounded.

Proof of Theorem 5.1.16. Suppose $D \subseteq \mathbb{R}^k$, $\mathbf{c} \operatorname{acl} D\backslash\{\mathbf{c}\}$, and $f : D \to \mathbb{R}^m$. Let $\varepsilon_0 = 14 > 0$ stand for a positive distance in the range. Since f converges at $\mathbf{c}$, by Definition 4.3.2 there is limit $\mathbf{y}$ and a threshold $\delta > 0$ defining a distance for the domain where

$$\mathbf{x} \in D \text{ with } 0 < \|\mathbf{x} - \mathbf{c}\|_k < \delta \implies \|f(\mathbf{x}) - \mathbf{y}\|_m < 14. \tag{5.1.44}$$

So, by the reverse triangle inequality (1.2.37), for every $\mathbf{x} \in D \cap V_\delta(\mathbf{c})$ we have

$$\|f(\mathbf{x})\|_m - \|\mathbf{y}\|_m \leq \|f(\mathbf{x}) - \mathbf{y}\|_m < 14. \tag{5.1.45}$$

By adding $\|\mathbf{y}\|_m$ we also have

$$\|f(\mathbf{x})\|_m < \|\mathbf{y}\|_m + 14. \tag{5.1.46}$$

To account for the possibility that $\mathbf{c} \in D$ so $f(\mathbf{c})$ is defined and would need to bounded, define

$$b_\mathbf{c} = \begin{cases} \|\mathbf{y}\|_m + 14, & \text{if } \mathbf{c} \in D, \\ \|\mathbf{y}\|_m + 14 + \|f(\mathbf{c})\|_m, & \text{if } \mathbf{c} \notin D. \end{cases} \tag{5.1.47}$$

Then $b_\mathbf{c}$ is a local bound and f is locally bounded at $\mathbf{c}$ (see Definition 4.3.20). $\square$

Another result to highlight the usefulness of multiple perspectives we can take on functional limits thanks to Theorem 5.1.14, consider order properties for sets, sequences, and functions of real numbers. The relationship between order and arbitrarily close established by Lemma 1.5.23 yields the order properties for sequential limits in $\mathbb{R}$ (Corollary 2.3.22) via Theorem 2.3.1, the fundamental connection between sequential limits and arbitrarily close. In turn, order properties for functional limits in $\mathbb{R}$ stem from the results on sequential limits part (ii) of Theorem 5.1.14 and Corollary 2.3.22.

Corollary 5.1.18: Order properties for functional limits in $\mathbb{R}$

Suppose $D \subseteq \mathbb{R}$, $f : D \to \mathbb{R}$, $c \operatorname{acl} D\backslash\{c\}$, and f converges to ℓ at c.

(i) If $f(x) \leq b$ for every $x \in D\backslash\{c\}$, then $\lim_{x \to c} f(x) = \ell \leq b$.

(ii) If $f(x) \geq a$ for every $x \in D\backslash\{c\}$, then $\lim_{x \to c} f(x) = \ell \geq a$.

Proof of Corollary 5.1.18. Throughout this proof, suppose $D \subseteq \mathbb{R}$, $f : D \to \mathbb{R}$, $c \operatorname{acl} D\backslash\{c\}$, and f converges to ℓ at c. The proofs of the two cases are quite similar.

Case (i): Suppose $f(x) \leq b$ for every $x \in D\backslash\{c\}$ and $(x_n) \subseteq D\backslash\{c\}$ where we have $\lim_{n \to \infty} x_n = c$. Then $f(x_n) \leq b$ for every index $n \in \mathbb{N}$. So, by Theorem 5.1.14 and the order properties for sequential limits in $\mathbb{R}$ (Corollary 2.3.22), we have

$$\lim_{n \to \infty} f(x_n) = \lim_{x \to c} f(x) = \ell \leq b. \tag{5.1.48}$$

Case (ii): Suppose $f(x) \geq a$ for every $x \in D\backslash\{c\}$ and $(x_n) \subseteq D\backslash\{c\}$ where we have $\lim_{n \to \infty} x_n = c$. Then $f(x_n) \geq a$ for every index $n \in \mathbb{N}$. So, by Theorem 5.1.14 and the order properties for sequential limits in $\mathbb{R}$ (Corollary 2.3.22), we have

$$\lim_{n \to \infty} f(x_n) = \lim_{x \to c} f(x) = \ell \geq a. \tag{5.1.49}$$

$\square$

One more corollary to close out this section. The statements satisfy contrapositions of sequential convergence in Theorem 5.1.14 and local boundedness of convergence in Theorem 5.1.16. The proof is omitted.

Corollary 5.1.19: Divergence Criteria for Functions

If $D \subseteq \mathbb{R}^k$, $\mathbf{c} \operatorname{acl} D\backslash\{\mathbf{c}\}$, $f : D \to \mathbb{R}^m$, and f satisfies any of the following conditions, then f diverges at $\mathbf{c}$:

(i) There are convergent sequences $(\mathbf{x}_n), (\mathbf{y}_n) \subseteq D\backslash\{\mathbf{c}\}$ where

$$\lim_{n \to \infty} \mathbf{x}_n = \lim_{n \to \infty} \mathbf{y}_n = \mathbf{c}, \qquad \text{but} \qquad \lim_{n \to \infty} f(\mathbf{x}_n) \neq \lim_{n \to \infty} f(\mathbf{y}_n). \tag{5.1.50}$$

(ii) f is not locally bounded at $\mathbf{c}$.

The next section takes further advantage of the deep similarities between continuity and functional limits explored to build properties of functional limits.

Exercises

5.1.1. Given a threshold for the convergence of a function, any smaller positive number is also a threshold (see Definition 5.1.2). To prove this, suppose $D \subseteq \mathbb{R}^k$, $f : D \to \mathbb{R}^m$, $\mathbf{c} \operatorname{acl}(D\backslash\{\mathbf{c}\})$, and f converges to $\mathbf{y}$ at $\mathbf{c}$ with threshold $\delta > 0$ responding to the distance $\varepsilon > 0$. Prove that if $0 < \sigma \leq \delta$, then σ is also a threshold for the convergence of f to $\mathbf{y}$ at $\mathbf{c}$ in response to ε.

5.1.2. Prove $\lim_{\mathbf{x} \to \mathbf{c}} \|\mathbf{x}\|_m = \|\mathbf{c}\|_m$.

5.1.3. Prove the following limits are as stated.

(i) $\lim\limits_{x\to 3}(x^2 - 5x + 6) = 0$.

(iii) $\lim\limits_{x\to -4}(x^3 + 64) = 0$.

(ii) $\lim\limits_{x\to 3}\dfrac{x^2 - 5x + 6}{x - 3} = 1$.

(iv) $\lim\limits_{x\to -4}\dfrac{x^3 + 64}{x + 4} = 48$.

5.1.4. Prove that a function has a removable discontinuity at an accumulation point if and only if the limit of the function exists at the accumulation point.

5.1.5. Prove $\lim\limits_{x\to c} x^n = c^n$ for all $c \in \mathbb{R}$.

5.1.6. Prove $\lim\limits_{x\to c} \sqrt[n]{x} = \sqrt[n]{c}$ for all $c \in [0, \infty)$.

5.1.7. Prove $\lim\limits_{x\to c}\dfrac{x^n - c^n}{x - c} = nc^{n-1}$ for all $c \in \mathbb{R}$. (Does this look familiar?)

5.1.8. Suppose $c \in \mathbb{R}$ and $h : \mathbb{R} \to \mathbb{R}$ converges to ℓ at c where $\ell > 0$. Prove h is locally positive at c—except possibly for $h(c)$—in the sense that there is some $\delta > 0$ such that

$$0 < |x - c| < \delta \qquad \Longrightarrow \qquad h(x) > 0. \tag{5.1.51}$$

In other words, $h(x)$ is positive for all x in a deleted δ-neighborhood of c.

5.2 Properties of limits

As done with the definition of continuity (Definition 4.3.2), the properties of continuity developed throughout Chapter 4 can be adapted to the context of limits. This adaptation is facilitated by replacing the condition of $\mathbf{c}$ being a point in a domain D with $\mathbf{c}$ being an accumulation point of the domain, thus "$\mathbf{c} \in D$" is replaced by "$\mathbf{c}\,\mathrm{acl}\,D\backslash\{\mathbf{c}\}$" throughout.

The final result of the section addresses a classic approach to defining continuity when limits are discussed first. It is a partial converse of Theorem 5.1.8.

Theorem 5.2.1: Continuity from limits

Suppose $D \subseteq \mathbb{R}^k$, $f : D \to \mathbb{R}^m$, and $\mathbf{c}\,\mathrm{acl}\,D\backslash\{\mathbf{c}\}$. If $\mathbf{c} \in D$ and $\lim\limits_{\mathbf{x}\to\mathbf{c}} f(\mathbf{x}) = f(\mathbf{c})$, then f is continuous at $\mathbf{c}$.

Scratch Work 5.2.2: Modify the definitions

The proof follows from the definition of functional limit (Definition 5.1.2) when we identify the limit $\mathbf{y}$ as the output $f(\mathbf{c})$, note $\|\mathbf{c}-\mathbf{c}\|_k = 0$, and take the threshold δ for the convergence of f at $\mathbf{c}$ as the threshold for the continuity of f at $\mathbf{c}$.

Proof of Theorem 5.2.1. Suppose $D \subseteq \mathbb{R}^k$, $f : D \to \mathbb{R}^m$, both $\mathbf{c}\,\mathrm{acl}\,D\backslash\{\mathbf{c}\}$ and $\mathbf{c} \in D$, and f converges to $\mathbf{y} = f(\mathbf{c})$ at $\mathbf{c}$. Let $\varepsilon > 0$. Since $\mathbf{c} \in D$, there is a threshold $\delta > 0$ for the convergence of f at $\mathbf{c}$ where we have

$$\mathbf{x} \in D \text{ with } \|\mathbf{x} - \mathbf{c}\|_k < \delta \qquad \Longrightarrow \qquad \|f(\mathbf{x}) - f(\mathbf{c})\|_m < \varepsilon. \tag{5.2.1}$$

Therefore, f is continuous at $\mathbf{c}$ by Definition 4.3.2 since δ is also a threshold for the continuity of f at $\mathbf{c}$. $\qquad\square$

Remark 5.2.3: A checklist for continuity from limits

One way to interpret Theorem 5.2.1 is the following checklist: If

  (i) $f(\mathbf{c})$ is defined,

  (ii) $\lim\limits_{\mathbf{x}\to\mathbf{c}} f(\mathbf{x})$ exists, and

  (iii) $f(\mathbf{c}) = \lim\limits_{\mathbf{x}\to\mathbf{c}} f(\mathbf{x})$ (the values of the output and the limit are the same),

then f is continuous at $\mathbf{c}$.

The contrapositions of the statements in Remark 5.2.3 provide criteria for discontinuity in addition to Discontinuity Criteria (Corollary 4.6.13).

Corollary 5.2.4: More Discontinuity Criteria

If $D \subseteq \mathbb{R}^k$, $\mathbf{c} \in D$, $f : D \to \mathbb{R}^m$, and f satisfies any of the following conditions, then f is discontinuous at $\mathbf{c}$:

  (i) $f(\mathbf{c})$ is not defined.

  (ii) $\lim\limits_{\mathbf{x}\to\mathbf{c}} f(\mathbf{x})$ does not exist.

  (iii) $f(\mathbf{c})$ is defined and $\lim\limits_{\mathbf{x}\to\mathbf{c}} f(\mathbf{x})$ exists, but

$$f(\mathbf{c}) \neq \lim\limits_{\mathbf{x}\to\mathbf{c}} f(\mathbf{x}). \tag{5.2.2}$$

Notation 5.2.5: Equivalent ways to describe the convergence of functions

Each of the following statements mean the same exact thing: Definition 5.1.2 holds.

  (i) f converges to $\mathbf{y}$ at $\mathbf{c}$.

  (ii) The limit of f at $\mathbf{c}$ is $\mathbf{y}$.

  (iii) $\lim\limits_{\mathbf{x}\to\mathbf{c}} f(\mathbf{x}) = \mathbf{y}$.

When we do not refer to the limit explicitly, the following statements are identical.

  (i) f converges at $\mathbf{c}$.

  (ii) The limit of f at $\mathbf{c}$ exists.

  (iii) $\lim\limits_{\mathbf{x}\to\mathbf{c}} f(\mathbf{x})$ exists.

Next up is the linearity of limits for functions.

Theorem 5.2.6: Linearity of functional limits

Suppose $D \subseteq \mathbb{R}^k$, $\alpha \in \mathbb{R}$, $f, g : D \to \mathbb{R}^m$, and $\mathbf{c}$ acl $D \backslash \{\mathbf{c}\}$. If f and g converge at $\mathbf{c}$, then $f + g$ and αf converge at $\mathbf{c}$ with

(i) $\lim\limits_{\mathbf{x} \to \mathbf{c}} (f(\mathbf{x}) + g(\mathbf{x})) = \lim\limits_{\mathbf{x} \to \mathbf{c}} f(\mathbf{x}) + \lim\limits_{\mathbf{x} \to \mathbf{c}} g(\mathbf{x})$ *(additivity)*; and

(ii) $\lim\limits_{\mathbf{x} \to \mathbf{c}} (\alpha f(\mathbf{x})) = \alpha \lim\limits_{\mathbf{x} \to \mathbf{c}} f(\mathbf{x})$ *(homogeneity)*.

Scratch Work 5.2.7: Nearly identical to the linearity of contintuity

Subtle yet simple modifications of Scratch Work 4.5.6 for the linearity of continuity at some point $\mathbf{c}$ suffice for the linearity of functional limits: Ignore $\mathbf{c}$ itself but consider it to be an accumulation point of the domain, keep the thresholds, and replace outputs with suitable limits.

Proof of Theorem 5.2.6. Throughout the proof, suppose $\alpha \in \mathbb{R}$, $D \subseteq \mathbb{R}^k$, $\mathbf{c}$ acl $D \backslash \{\mathbf{c}\}$, and $f, g : D \to \mathbb{R}^m$ where f and g converge at $\mathbf{c}$ to $\mathbf{y}$ and $\mathbf{z}$, respectively.

Part (i): Suppose $\varepsilon > 0$. Since f and g converge at $\mathbf{c}$, there are thresholds $\delta_f > 0$ and $\delta_g > 0$ such that

$$0 < \|\mathbf{x} - \mathbf{c}\|_k < \delta_f \quad \implies \quad \|f(\mathbf{x}) - \mathbf{y}\|_m < \frac{\varepsilon}{2} \quad \text{and} \tag{5.2.3}$$

$$0 < \|\mathbf{x} - \mathbf{c}\|_k < \delta_g \quad \implies \quad \|g(\mathbf{x}) - \mathbf{z}\|_m < \frac{\varepsilon}{2}. \tag{5.2.4}$$

Define δ to be the smaller of the thresholds δ_f and δ_g. That is, let

$$\delta = \min\{\delta_f, \delta_g\} \tag{5.2.5}$$

and note $\delta > 0$. Now suppose

$$0 < \|\mathbf{x} - \mathbf{c}\|_k < \delta. \tag{5.2.6}$$

Since $\delta \leq \delta_f$ and $\delta \leq \delta_g$, both implications (5.2.3) and (5.2.4) hold. Hence, by the triangle inequality (1.2.32) we have

$$\|(f(\mathbf{x}) + g(\mathbf{x})) - (\mathbf{y} + \mathbf{z})\|_m = \|f(\mathbf{x}) - \mathbf{y} + g(\mathbf{x}) - \mathbf{z}\|_m \tag{5.2.7}$$

$$\leq \|f(\mathbf{x}) - \mathbf{y}\|_m + \|g(\mathbf{x}) - \mathbf{z}\|_m \tag{5.2.8}$$

$$< \frac{\varepsilon}{2} + \frac{\varepsilon}{2} \tag{5.2.9}$$

$$= \varepsilon. \tag{5.2.10}$$

Therefore, δ is a threshold for the convergence of the sum $f + g$ at $\mathbf{c}$ and we have

$$\lim_{\mathbf{x} \to \mathbf{c}} (f(\mathbf{x}) + g(\mathbf{x})) = \lim_{\mathbf{x} \to \mathbf{c}} f(\mathbf{x}) + \lim_{\mathbf{x} \to \mathbf{c}} g(\mathbf{x}) = \mathbf{y} + \mathbf{z}. \tag{5.2.11}$$

<u>Part (ii)</u>: Suppose $\alpha = 0$. Then αf is constant since

$$\alpha f(\mathbf{x}) = 0 \cdot f(\mathbf{x}) = \mathbf{0} \tag{5.2.12}$$

for every $\mathbf{x} \in \mathbb{R}^k$. Now let $\varepsilon > 0$ and $\mathbf{y} = \mathbf{0}$. Then $\delta = 19$ is a threshold for the convergence of αf at any $\mathbf{c}$ to $\mathbf{0}$ since

$$0 < \|\mathbf{x} - \mathbf{c}\|_k < 19 \qquad \Longrightarrow \qquad \|\alpha f(\mathbf{x}) - \alpha \mathbf{y}\|_m = \|\mathbf{0}\|_m = 0 < \varepsilon. \tag{5.2.13}$$

Now suppose $\alpha \neq 0$ and $\varepsilon > 0$. Since f convergences to $\mathbf{y}$ at $\mathbf{c}$, there is a threshold $\delta > 0$ such that for every $\mathbf{x} \in \mathbb{R}^m$ we have

$$0 < \|\mathbf{x} - \mathbf{c}\|_k < \delta \qquad \Longrightarrow \qquad \|f(\mathbf{x}) - \mathbf{y}\|_m < \frac{\varepsilon}{|\alpha|}. \tag{5.2.14}$$

Hence, by the relationship between absolute value and Pythagorean distance from Corollary 1.2.30, we also have

$$\|\alpha f(\mathbf{x}) - \alpha \mathbf{y}\|_m = |\alpha| \|f(\mathbf{x}) - \mathbf{y}\|_m < |\alpha| \cdot \frac{\varepsilon}{|\alpha|} = \varepsilon. \tag{5.2.15}$$

Therefore, δ is a threshold for the convergence of αf at $\mathbf{c}$ and we have

$$\lim_{\mathbf{x} \to \mathbf{c}} (\alpha f(\mathbf{x})) = \alpha \lim_{\mathbf{x} \to \mathbf{c}} f(\mathbf{x}) = \alpha \mathbf{y}. \tag{5.2.16}$$

$\square$

Once again and as mentioned in Remark 1.6.18, a corollary of the linearity of limits holds for linear combinations. That is, the limit of a linear combination is the linear combination of limits. As with the proofs of Corollaries 1.6.16, 4.5.7, and 4.7.7, the proof of Corollary 5.2.8 follows from induction on linearity. So, the proof is left as an exercise.

> **Corollary 5.2.8: Linear combinations of functional limits**
>
> Suppose $A \subseteq \mathbb{R}^\ell$, $k \in \mathbb{N}$, and for each $j = 1, \ldots, k$ we have $c_j \in \mathbb{R}$ and the functions $f_j : A \to \mathbb{R}^m$ converge at $\mathbf{c}$. Then the linear combination f given by
>
> $$f(\mathbf{x}) = \sum_{j=1}^{k} c_j f_j(\mathbf{x}) = c_1 f_1(\mathbf{x}) + \ldots + c_k f_k(\mathbf{x}) \tag{5.2.17}$$
>
> converges to $\mathbf{c}$ and we have
>
> $$\lim_{\mathbf{x} \to \mathbf{c}} f(\mathbf{x}) = \lim_{\mathbf{x} \to \mathbf{c}} \left(\sum_{j=1}^{k} c_j f_j(\mathbf{x}) \right) = \sum_{j=1}^{k} \left(c_j \lim_{\mathbf{x} \to \mathbf{c}} f_j(\mathbf{x}) \right) \tag{5.2.18}$$

As with sequences, the limit of a function in a Euclidean space is unique.

Theorem 5.2.9: Functional limits in Euclidean spaces are unique

Suppose $D \subseteq \mathbb{R}^k$, $f : D \to \mathbb{R}^m$, and $\mathbf{c}$ acl $D\backslash\{\mathbf{c}\}$. If f converges at $\mathbf{c}$, its limit is unique.

Scratch Work 5.2.10: Two points arbitrarily close are the same

The idea is the same as Scratch Work 2.3.6 designed to proof sequential limits are unique in Theorem 2.3.5: Show any two limits $\mathbf{y}$ and $\mathbf{z}$ of f at $\mathbf{c}$ are arbitrarily close, so they must be the same point (see Lemma 1.5.5).

We want to end up with

$$\|\mathbf{y} - \mathbf{z}\|_m < \varepsilon, \tag{5.2.19}$$

which we get from a combination of tried and true techniques: adding zero (1.2.34) and applying the triangle inequality (1.2.32). By the definition of functional limit (Definition 5.1.2) applied to each limit $\mathbf{y}$ and $\mathbf{z}$, for every distance $\varepsilon > 0$ there are thresholds $\delta_\mathbf{y} > 0$ and $\delta_\mathbf{z} > 0$ which produce inputs whose outputs are within ε of their respective limits. To squeeze the limits together, we can find a single input $\mathbf{x}_\delta$ whose output $f(\mathbf{x}_\delta)$ is close enough to both $\mathbf{y}$ and $\mathbf{z}$. The trick is to split ε in half and define a suitable δ as the minimum of $\delta_\mathbf{y}$ and $\delta_\mathbf{z}$.

Proof of Theorem 5.2.9. Suppose $D \subseteq \mathbb{R}^k$, $f : D \to \mathbb{R}^m$, $\mathbf{c}$ acl $D\backslash\{\mathbf{c}\}$, and $\mathbf{y}, \mathbf{z} \in \mathbb{R}^m$ both satisfy

$$\mathbf{y} = \lim_{\mathbf{x} \to \mathbf{c}} f(\mathbf{x}) \qquad \text{and} \qquad \mathbf{z} = \lim_{\mathbf{x} \to \mathbf{c}} f(\mathbf{x}). \tag{5.2.20}$$

Let $\varepsilon > 0$. Then $\varepsilon/2 > 0$ and by the definition of functional limit (Definition 5.1.2), there are two thresholds $\delta_\mathbf{y}$ and $\delta_\mathbf{z}$ where

$$0 < \|\mathbf{x} - \mathbf{c}\|_k < \delta_\mathbf{y} \quad \implies \quad \|f(\mathbf{x}) - \mathbf{y}\|_m < \frac{\varepsilon}{2} \qquad \text{and} \tag{5.2.21}$$

$$0 < \|\mathbf{x} - \mathbf{c}\|_k < \delta_\mathbf{z} \quad \implies \quad \|f(\mathbf{x}) - \mathbf{z}\|_m < \frac{\varepsilon}{2}. \tag{5.2.22}$$

Now define $\delta = \min\{\delta_\mathbf{y}, \delta_\mathbf{z}\}$ and note $\delta > 0$. We have both $\delta \leq \delta_\mathbf{y}$ and $\delta \leq \delta_\mathbf{z}$. So for any $\mathbf{x}_\delta \in D$ where $0 < \|\mathbf{x}_\delta - \mathbf{c}\|_k < \delta$, both (5.2.21) and (5.2.22) holds for $\mathbf{x} = \mathbf{x}_\delta$. Therefore, by adding zero (1.2.34) and applying the triangle inequality (1.2.32), we have

$$\|\mathbf{y} - \mathbf{z}\|_m = \|\mathbf{y} \underbrace{- f(\mathbf{x}_\delta) + f(\mathbf{x}_\delta)}_{\text{add } \mathbf{0}} - \mathbf{z}\|_m \tag{5.2.23}$$

$$\leq \|\mathbf{y} - f(\mathbf{x}_\delta)\|_m + \|f(\mathbf{x}_\delta) - \mathbf{z}\|_m \tag{5.2.24}$$

$$< \frac{\varepsilon}{2} + \frac{\varepsilon}{2} \tag{5.2.25}$$

$$= \varepsilon. \tag{5.2.26}$$

Since $\varepsilon > 0$ is arbitrary, $\mathbf{y}$ acl$\{\mathbf{z}\}$. Therefore, $\mathbf{y} = \mathbf{z}$ by Lemma 1.5.5. $\qquad\square$

The idea that the limit of a product is the product of the limits holds for both sequences and functions. Again, in this context we only consider functions from the real line to the real line so products make sense.

Theorem 5.2.11: Products of functional limits

Suppose $D \subseteq \mathbb{R}$, $c \operatorname{acl} D \backslash \{c\}$, and $f, g : D \to \mathbb{R}$ where f and g converge at c. Then the product fg converges at c and we have

$$\lim_{x \to c} f(x)g(x) = \left(\lim_{x \to c} f(x)\right)\left(\lim_{x \to c} g(x)\right). \tag{5.2.27}$$

Scratch Work 5.2.12: Perspective of sequential limits

The proof of Theorem 5.2.6 follows the guide in Remark 5.1.7 to directly address the definition of functional limit (Definition 5.1.2) where we respond to an arbitrary distance $\varepsilon > 0$ for the codomain with a threshold $\delta > 0$ for the domain. The same approach could be made to work here, but to showcase the equivalent forms of functional limits in Theorem 5.1.14, we work with sequential limits (Definition 2.2.1) and the idea that for sequences of real numbers, the limit of a product is the product of the limits (Theorem 2.3.17).

Proof of Theorem 5.2.11. Suppose $D \subseteq \mathbb{R}$, $c \operatorname{acl} D \backslash \{c\}$, and $f, g : D \to \mathbb{R}$ where f and g converge at c. To take advantage of the connection with sequential limits, suppose $(x_n) \subseteq D \backslash \{c\}$ converges to c. Then by the implication (iii) $\implies$ (ii) in Theorem 5.1.14, the limit of a product is the product of the limit for sequences (Theorem 2.3.17), and the implication (ii) $\implies$ (iii) in Theorem 5.1.14, we have

$$\lim_{x \to c} (f(x)g(x)) = \lim_{n \to \infty} (f(x_n)g(x_n)) \tag{5.2.28}$$

$$= \left(\lim_{n \to \infty} f(x_n)\right)\left(\lim_{n \to \infty} g(x_n)\right) \tag{5.2.29}$$

$$= \left(\lim_{x \to c} f(x)\right)\left(\lim_{x \to c} g(x)\right). \tag{5.2.30}$$

$\square$

As with sequences, the limit of a quotient is the quotient of limits for real-valued functions.

Theorem 5.2.13: Quotients of functional limits

Suppose $D \subseteq \mathbb{R}$, $c \operatorname{acl} D \backslash \{c\}$, and $f, g : D \to \mathbb{R}$ where f and g converge at c. If $g(x) \neq 0$ for all $x \in D$ and $\lim_{x \to c} g(x) \neq 0$, then the quotient f/g converges at c and we have

$$\lim_{x \to c} \frac{f(x)}{g(x)} = \frac{\lim_{x \to c} f(x)}{\lim_{x \to c} g(x)}. \tag{5.2.31}$$

Scratch Work 5.2.14: Build on quotients of sequential limits

With a wealth of tools available, the proof follows from results on sequential limits through the equivalent form of functional limits (Theorem 5.1.14) and quotients of sequential limits (Theorem 2.3.21). It is very similar to the proof of Theorem 5.2.11.

Proof of Theorem 5.2.13. Suppose $D \subseteq \mathbb{R}$, $c \operatorname{acl} D\backslash\{c\}$, $f, g : D \to \mathbb{R}$, $g(x) \neq 0$ for all x in D, f and g converge at c, and $\lim_{x \to c} g(x) \neq 0$. To take advantage of the connection with sequential limits, suppose $(x_n) \subseteq D\backslash\{c\}$ converges to c. Then by the implication (iii) $\Longrightarrow$ (ii) in Theorem 5.1.14, quotients of sequential limits (Theorem 2.3.21), and the implication (ii) $\Longrightarrow$ (iii) in Theorem 5.1.14, we have

$$\lim_{x \to c} \frac{f(x)}{g(x)} = \lim_{n \to \infty} \frac{f(x_n)}{g(x_n)} = \frac{\lim_{n \to \infty} f(x_n)}{\lim_{n \to \infty} g(x_n)} = \frac{\lim_{x \to c} f(x)}{\lim_{x \to c} g(x)}. \tag{5.2.32}$$

$\square$

There are Squeeze Theorems wherever we have a notion of convergence in the real line. We already have a Squeeze Theorem for sequences 2.4.3 and a Squeeze Theorem for continuity 4.6.11, and now we can get one for functional limits.

Theorem 5.2.15: Squeeze Theorem for functions

Suppose $D \subseteq \mathbb{R}$, $c \operatorname{acl} D\backslash\{c\}$, $f, g, h : D \to \mathbb{R}$, and f and h converge at c. If

(i) $\lim_{x \to c} f(x) = \lim_{x \to c} h(x) = \ell$ and

(ii) $f(x) \leq g(x) \leq h(x)$ for all $x \in D\backslash\{c\}$,

then g converges at c and $\lim_{x \to c} g(x) = \ell$.

Scratch Work 5.2.16: Splitting absolute values

Thanks to the equivalent forms of convergence in Theorem 5.1.14, we could rely on the Squeeze Theorem for sequences 2.4.3. However, the plan this time mirrors Scratch Work 2.4.4 and allows to get a proof directly from the assumptions and the definition of limit and convergence for functions (Definition 5.1.2). The idea is to split the absolute values.

The convergence of f and h to ℓ at c means there are thresholds δ_f and δ_h where

$$0 < |x - c| < \delta_f \quad \Longrightarrow \quad |f(x) - \ell| < \varepsilon \quad \text{and} \tag{5.2.33}$$
$$0 < |x - c| < \delta_h \quad \Longrightarrow \quad |h(x) - \ell| < \varepsilon. \tag{5.2.34}$$

Choosing $\delta_g = \min\{\delta_f, \delta_h\}$ will work. But to achieve the goal of concluding

$$0 < |x - c| < \delta_g \quad \Longrightarrow \quad |g(x) - \ell| < \varepsilon, \tag{5.2.35}$$

we can split the absolute values thanks to Lemma 1.5.10. Also, subtracting the common limit ℓ through the assumption

$$f(x) \leq g(x) \leq h(x) \tag{5.2.36}$$

gives us

$$f(x) - \ell \leq g(x) - \ell \leq h(x) - \ell. \tag{5.2.37}$$

Tacking on some absolute values gets us here:

$$-\varepsilon < -|f(x) - \ell| \leq f(x) - \ell \leq g(x) - \ell \leq h(x) - \ell \leq |h(x) - \ell| < \varepsilon. \tag{5.2.38}$$

Keeping the portions with g and ε yields

$$-\varepsilon < g(x) - \ell < \varepsilon, \tag{5.2.39}$$

which gives us the conclusion $|g(x) - \ell| < \varepsilon$ with another application of Lemma 1.5.10.

Proof of the Squeeze Theorem for functions 5.2.15. Assume $f(x) \leq g(x) \leq h(x)$ for each $x \in D$ and

$$\lim_{x \to c} f(x) = \lim_{x \to c} h(x) = \ell. \tag{5.2.40}$$

Let $\varepsilon > 0$. Since f and g converge to ℓ at c, there are thresholds $\delta_f > 0$ and $\delta_h > 0$ where

$$0 < |x - c| < \delta_f \qquad \Longrightarrow \qquad |f(x) - \ell| < \varepsilon \qquad \text{and} \tag{5.2.41}$$
$$0 < |x - c| < \delta_h \qquad \Longrightarrow \qquad |h(x) - \ell| < \varepsilon. \tag{5.2.42}$$

Define $\delta_g = \min\{\delta_f, \delta_h\}$ and note $\delta_g > 0$. Now suppose

$$0 < |x - c| < \delta_g. \tag{5.2.43}$$

Since $\delta_g \leq \delta_f$ and $\delta_g \leq \delta_h$, both (5.2.41) and (5.2.42) hold. Hence, by splitting inequalities as in Lemma 1.5.10, subtracting ℓ from $f(x) \leq g(x) \leq h(x)$, and including absolute values from (5.2.41) and (5.2.42) gives us

$$-\varepsilon < -|f(x) - \ell| \leq f(x) - \ell \leq g(x) - \ell \leq h(x) - \ell \leq |h(x) - \ell| < \varepsilon. \tag{5.2.44}$$

In particular, we have $-\varepsilon < g(x) - \ell < \varepsilon$. So by Lemma 1.5.10 again we have

$$|g(x) - \ell| < \varepsilon. \tag{5.2.45}$$

Therefore, δ_g is a threshold for the convergence of g at c and $\lim_{x \to c} g(x) = \ell$. $\qquad \square$

The next section explores *differentiation* by building *derivatives* from limits.

Exercises

5.2.1. Prove $\lim_{x \to 2} \dfrac{|x - 2|}{x - 2}$ does not exist.

5.2.2. Consider a modified version of the topologist's sine curve from Example 4.4.10 defined by $t : \mathbb{R} \backslash \{0\} \to \mathbb{R}$ where

$$t(x) = \sin\left(\frac{1}{x}\right). \tag{5.2.46}$$

Prove $\lim_{x \to 0} t(x)$ does not exist.

5.2.3. Consider a squeezed version of the topologist's sine curve from Example 4.4.10 defined by $s : \mathbb{R} \backslash \{0\} \to \mathbb{R}$ where

$$s(x) = x \sin\left(\frac{1}{x}\right). \tag{5.2.47}$$

Prove $\lim_{x \to 0} s(x) = 0$.

5.2.4. Suppose $f : [a, b] \to \mathbb{R}$ is monotone. Prove $\lim_{x \to a} f(x)$ and $\lim_{x \to b} f(x)$ exist.

5.2.5. Recall Dirichlet's function $\mathbb{1}_{\mathbb{Q}}$ from Example 4.2.14 given by

$$\mathbb{1}_{\mathbb{Q}}(x) = \begin{cases} 1, & \text{if } x \in \mathbb{Q}, \\ 0, & \text{if } x \in \mathbb{R} \backslash \mathbb{Q}. \end{cases} \tag{5.2.48}$$

Prove $\lim_{x \to c} \mathbb{1}_{\mathbb{Q}}(x)$ does not exist at any $c \in \mathbb{R}$.

5.2.6. Consider the modification of Dirichlet's function $\mathbb{1}_{\mathbb{Q}}$ from Example 4.2.14 given by

$$f(x) = x\mathbb{1}_{\mathbb{Q}}(x) = \begin{cases} x, & \text{if } x \in \mathbb{Q}, \\ 0, & \text{if } x \in \mathbb{R} \backslash \mathbb{Q}. \end{cases} \tag{5.2.49}$$

Prove $\lim_{x \to 0} x\mathbb{1}_{\mathbb{Q}}(x) = 0$ but $\lim_{x \to c} x\mathbb{1}_{\mathbb{Q}}(x)$ does not exist for nonzero values of c.

5.2.7. *Thomae's function* $g : \mathbb{R} \to \mathbb{R}$ is given by

$$g(x) = \begin{cases} 1, & \text{if } x = 0, \\ \dfrac{1}{n}, & \text{if } x \in \mathbb{Q} \backslash \{0\} \text{ with reduced form } x = \dfrac{m}{n}, \\ 0, & \text{if } x \in \mathbb{R} \backslash \mathbb{Q}. \end{cases} \tag{5.2.50}$$

(i) Prove $\lim_{x \to c} g(x) = 0$ at each $c \in \mathbb{R} \backslash \mathbb{Q}$.

(ii) Prove $\lim_{x \to c} g(x)$ does not exist at any $c \in \mathbb{Q}$.

5.2.8. *Sided-limits* are notions from calculus that help us explore properties of functions from the real line to the real line. For instance, suppose $D \subseteq \mathbb{R}$, $c \in \text{acl}(D \backslash \{c\})$, and $f : D \to \mathbb{R}$. The *right-hand limit of f* at c is a real number r such that for every $\varepsilon > 0$ there is a threshold $\delta > 0$ where

$$x \in D \text{ with } c < x < c + \delta \qquad \Longrightarrow \qquad |f(x) - r| < \varepsilon. \tag{5.2.51}$$

(i) Carefully state a definition for the *left-hand limit of f* at *c*.

(ii) Give an example where both the left- and right-hand limits of *f* at *c* exist but the limit of *f* at *c* does not.

(iii) Prove the limit of *f* at *c* exists if and only if both the left- and right-hand limits of *f* at *c* exist.

5.2.9. Assuming concepts from trigonometry hold, show $\lim\limits_{x \to 0} \dfrac{\sin x}{x} = 1$. Hint: Check a calculus textbook.

5.2.10. Assuming concepts from trigonometry hold, show $\lim\limits_{x \to 0} \dfrac{\cos x - 1}{x} = 0$. Hint: Check a calculus textbook.

5.2.11. Suppose $c, \ell \in \mathbb{R}$ and let $L(c, \ell)$ denote the set of real-valued functions on $\mathbb{R}$ that converge to ℓ at c. (Thus, for each $f \in L(c, \ell)$ we have $\lim_{x \to c} f(x) = \ell$.) Use Lemma 1.6.7 to prove $L(c, \ell)$ is a vector space if and only if $\ell = 0$.

5.3   Differentiation

Derivatives are a motivation for and a consequence of the definition of functional limit (Definition 5.1.2). They provide a technical interpretation of ideas from calculus like "instantaneous rate of change" and "slope of the tangent line", which are not defined until a suitable notion of limit is available.

At the heart of the issue is defining the slope at a single point when slopes by their nature are determined by two distinct points. The idea is to build on the slopes of secant lines through distinct pairs of points on the graph of a function. The *difference quotient* defines the slopes of all secant lines through a particular point on the graph, then the *derivative* defines the slope of the *tangent line* as the limit of the slopes of secant lines. See Figure 5.3.1.

By the way, all of the functions considered in the development of derivatives, *integrals*, and *series* in this textbook map a subset of the real line to the real line. At present, this allows us to consider products and quotients of functions.

Definition 5.3.1: Derivative

Suppose $I \subseteq \mathbb{R}$ is an interval, $f : I \to \mathbb{R}$, and $c \in I$. The *difference quotient of f at c* is the function $q_c : I \backslash \{c\} \to \mathbb{R}$ given by

$$q_c(x) = \frac{f(x) - f(c)}{x - c}. \tag{5.3.1}$$

The *derivative of f at c* is the limit of difference quotient q_c at c given by

$$f'(c) = \lim_{x \to c} \frac{f(x) - f(c)}{x - c} = \lim_{x \to c} q_c(x), \tag{5.3.2}$$

provided this limit exists. In this case, f is said to be *differentiable at c*. If f is differentiable at every point in the domain I, we say f is *differentiable* and f' is called the *derivative* of f.

The functions that define lines are the special cases of basic affine transformations (Definition 4.2.9) that map the real line to the real line. These functions are differentiable and the derivatives are their slopes.

Lemma 5.3.2: Lines are differentiable

Suppose $m, b \in \mathbb{R}$ and $f : \mathbb{R} \to \mathbb{R}$ is given by

$$f(x) = mx + b. \tag{5.3.3}$$

Then f is differentiable with

$$f'(c) = m \tag{5.3.4}$$

for every $c \in \mathbb{R}$. In particular, if f is constant (i.e., $m = 0$), then $f'(c) = 0$ for every $c \in \mathbb{R}$.

Scratch Work 5.3.3: Simplify the difference quotient

A useful approach and starting point for developing proofs involving derivatives is to mess around with the difference quotient. For lines given by $f(x) = mx + b$, the difference quotient q_c satisfies

$$q_c(x) = \frac{mx + b - (mc + b)}{x - c} = \frac{m(x - c)}{x - c} = m. \tag{5.3.5}$$

From here, constants are continuous by Theorem 4.3.9. An application of Theorem 5.1.8, which gives us the value of a functional limit by evaluating a continuous function, yields the conclusion.

The approach of this scratch work is typical for dealing with derivatives: Manipulate difference quotients and apply properties of limits and continuity.

Proof of Lemma 5.3.2. Suppose $m, b, c \in \mathbb{R}$ and $f : \mathbb{R} \to \mathbb{R}$ is given by

$$f(x) = mx + b. \tag{5.3.6}$$

By the definition of derivative (Definition 5.3.1) and since constants are continuous by Theorem 4.3.9, applying Theorem 5.1.8 to the constant function $g(x) = m$ yields

$$f'(c) = \lim_{x \to c} q_c(x) = \lim_{x \to c} \frac{mx + b - (mc + b)}{x - c} = \lim_{x \to c} m = m. \tag{5.3.7}$$

$\square$

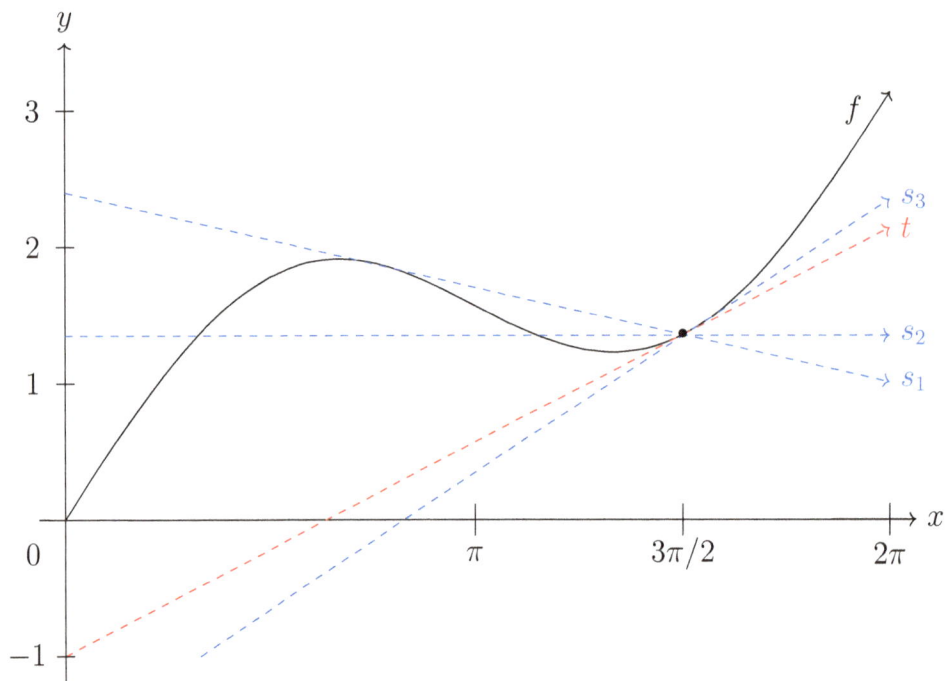

Figure 5.3.1: A differentiable function f with its tangent line t at $c = 3\pi/2$ and secant lines $s_1, s_2,$ and s_3. Also, to accompany Definition 5.3.1, play around with the Desmos activity "Derivative and difference quotient" accessed through the QR code. https://www.desmos.com/calculator/zadsgxkxca

Example 5.3.4: Derivative of the square function

The square function $h : \mathbb{R} \to \mathbb{R}$ given by $h(x) = x^2$ is differentiable with $h'(c) = 2c$ for all $c \in \mathbb{R}$.

Scratch Work 5.3.5: Again, simplify the difference quotient

The difference quotient q_c of the square function h simplifies to a polynomial thanks to thanks to the factorization of the difference of squares (1.2.39) in Lemma 1.2.32. From there, $h'(c)$ is determined by evaluating this continuous polynomial at c.

Proof for Example 5.3.4. Fix $c \in \mathbb{R}$. The difference quotient q_c for the square function h simplifies thanks to the factorization of the difference of squares (1.2.39) in Lemma 1.2.32. For all $x, c \in \mathbb{R}$ where $x \neq c$ we have

$$q_c(x) = \frac{h(x) - h(c)}{x - c} = \frac{x^2 - c^2}{x - c} = \frac{(x-c)(x+c)}{x-c} = x + c. \tag{5.3.8}$$

So $q_c(x)$ is a polynomial, at least for $x \in \mathbb{R}\setminus\{c\}$. Since polynomials are continuous (Theorem 4.5.2), the limit of $q_c(x)$ at c is found by evaluating the polynomial $x + c$ at c (Theorem 5.1.8). We have

$$h'(c) = \lim_{x \to c} \frac{x^2 - c^2}{x - c} = \lim_{x \to c} \frac{(x-c)(x+c)}{x-c} = \lim_{x \to c}(x + c) = 2c. \tag{5.3.9}$$

Since c is arbitrary, the square function h is differentiable. $\qquad\square$

Example 5.3.6: Derivative of the reciprocal

The reciprocal function $f : \mathbb{R}\setminus\{0\} \to \mathbb{R}$ given by $f(x) = 1/x$ is differentiable and for every $c \in \mathbb{R}\setminus\{0\}$ we have $f'(c) = -1/c^2$.

Scratch Work 5.3.7: Use limits of difference quotients

Once again, algebraic manipulation of the difference q_c for the reciprocal function puts us in position to apply properties of functional limits. Also, while the domain $\mathbb{R}\setminus\{0\}$ is not an interval itself, it is the disjoint union of two intervals since $\mathbb{R}\setminus\{0\} = (-\infty, 0) \cup (0, \infty)$. This is not addressed in the proof, but we could split the proof into cases where both c and x are positive or negative, if preferred.

Proof for Example 5.3.6. Suppose $f : \mathbb{R}\setminus\{0\} \to \mathbb{R}$ is given by $f(x) = 1/x$. For each $c \in \mathbb{R}\setminus\{0\}$ and all $x \in \mathbb{R}\setminus\{0, c\}$ we have $xc \neq 0$. Hence, $xc/(xc) = 1$ and we have

$$q_c(x) = \frac{\frac{1}{x} - \frac{1}{c}}{x - c} = \left(\frac{1}{x-c}\right)\left(\frac{1}{x} - \frac{1}{c}\right)\left(\frac{xc}{xc}\right) = \left(\frac{1}{x-c}\right)\left(\frac{c-x}{xc}\right) = -\frac{1}{xc}. \tag{5.3.10}$$

Now fix $c \in \mathbb{R}\backslash\{0\}$. Then by quotients and linearity of functional limits (Theorems 5.2.13 5.2.6) we have

$$f'(c) = \lim_{x \to c} \frac{\frac{1}{x} - \frac{1}{c}}{x - c} = \lim_{x \to c}\left(-\frac{1}{xc}\right) = -\frac{1}{c \lim_{x \to c} x} = -\frac{1}{c^2}. \tag{5.3.11}$$

Since $c \in \mathbb{R}\backslash\{0\}$ is arbitrary, the reciprocal function f is differentiable. $\square$

The connection between lines and differentiable functions goes much deeper than Lemma 5.3.2: They behave like lines on small scales. More specifically, differentiable functions are *locally linear* because they are nicely approximated by their *tangent lines* on suitable neighborhoods. See Figure 5.3.1 and Lemma 5.3.10 below.

Definition 5.3.8: Tangent line

Suppose $I \subseteq \mathbb{R}$ is an interval, $f : I \to \mathbb{R}$, and f is differentiable at $c \in I$. The *tangent line* to f at c is the line with point-slope form given by

$$y = f'(c)(x - c) + f(c) \tag{5.3.12}$$

for every $x \in \mathbb{R}$.

To codify idea that differentiable functions are locally linear, the definition of derivative (Definition 5.3.1) can be rearranged to show tangent lines provide arbitrarily good approximations of their functions, at least on a neighborhood.

Definition 5.3.9: Locally linear

Suppose $I \subseteq \mathbb{R}$ is an interval, $f : I \to \mathbb{R}$, and $c \in I$. Then function f is *locally linear at c* if for every $\varepsilon > 0$, there is a *threshold* $\delta > 0$ such that

$$|x - c| < \delta \text{ with } x \in I \implies |f(x) - (f'(c)(x - c) + f(c))| < \varepsilon. \tag{5.3.13}$$

Lemma 5.3.10: Differentiable implies locally linear

Suppose $I \subseteq \mathbb{R}$ is an interval, $f : I \to \mathbb{R}$, and f is differentiable at $c \in I$. Then f is locally linear at c.

Scratch Work 5.3.11: Unpack the derivative as a limit

The proof of Lemma 5.3.10 pops out of unpacking the definition of derivative (Definition 5.3.1) by considering the ε-δ definition of limit for functions (Definition 5.1.2) with the difference quotient q_c in mind. If f is differentiable at $c \in I$, then for every $\varepsilon > 0$ there is a threshold $\delta_f > 0$ such that

$$0 < |x - c| < \delta_f \text{ with } x \in I \tag{5.3.14}$$

$$\implies \left| \frac{f(x) - f(c)}{x - c} - f'(c) \right| = |q_c(x) - f'(c)| < \varepsilon. \tag{5.3.15}$$

Multiplying inequality (5.3.15) by $|x - c| > 0$ and rearranging a bit gives us

$$|f(x) - f(c) - f'(c)(x - c)| = |f(x) - (f'(c)(x - c) + f(c))| < \varepsilon |x - c|. \tag{5.3.16}$$

This is almost what we want, but the factor $|x - c|$ on the right-hand side needs to be addressed. That is, our choice for a threshold δ in the proof should accommodate both ε and $|x - c|$. This is accomplished by choosing

$$\delta = \min\{\delta_f, 1\}, \tag{5.3.17}$$

which means $|x - c| < \delta$ ensures both

$$|x - c| < \delta \leq \delta_f \quad \text{and} \quad |x - c| < \delta \leq 1. \tag{5.3.18}$$

Proof of Lemma 5.3.10. Suppose $I \subseteq \mathbb{R}$ is an interval, $f : I \to \mathbb{R}$, and f is differentiable at $c \in I$. By the definitions of functional limit and derivative (Definitions 5.1.2 and 5.3.1), for every $\varepsilon > 0$ there is a threshold $\delta_f > 0$ such that

$$0 < |x - c| < \delta_f \text{ with } x \in I \tag{5.3.19}$$

$$\implies \left| \frac{f(x) - f(c)}{x - c} - f'(c) \right| < \varepsilon. \tag{5.3.20}$$

Define $\delta = \min\{\delta_f, 1\}$ and note $\delta > 0$. Then $|x - c| < \delta$ implies both

$$|x - c| < \delta \leq \delta_f \quad \text{and} \quad |x - c| < \delta \leq 1. \tag{5.3.21}$$

Hence, we have both (5.3.20) and

$$\varepsilon |x - c| \leq \varepsilon. \tag{5.3.22}$$

So, by multiplying inequality (5.3.20) by $|x - c| > 0$ and rearranging a bit we have

$$|f(x) - f(c) - f'(c)(x - c)| = |f(x) - (f'(c)(x - c) + f(c))| < \varepsilon |x - c| \leq \varepsilon. \tag{5.3.23}$$

Therefore, δ is a threshold for the local linearity of f at c. $\qquad\square$

Remark 5.3.12: Arbitrarily close to the tangent line

The local linearity of a differentiable function f as in the implication (5.3.13) from the

Lemma 5.3.10 is equivalent to

$$x \in V_{\delta_c}(c) \cap I = (c - \delta_c, c + \delta_c) \cap I \qquad \Longrightarrow \qquad |f(x) - y| < \varepsilon, \tag{5.3.24}$$

where y is the tangent line to f at c (Definition 5.3.8):

$$y = f'(c)(x - c) + f(c). \tag{5.3.25}$$

So, in a sense, local linearity means a given function is arbitrarily close to its tangent line on a neighborhood.

Remark 5.3.13: Results on derivatives from limits

The remaining results proven in this section as well as further results from calculus explored in the exercises usually follow from algebraic manipulation of difference quotients and the numerous properties of limits and continuity developed up to this point. So when working on exercises involving derivatives, play around with difference quotients and keep properties of limits and continuity in mind.

An analog of the equivalence of the preservation of closeness, sequential continuity, and ε-δ continuity in Theorem 4.4.7 holds for derivatives as a corollary of the equivalence for functional limits in Theorem 5.1.14. Since derivatives are limits of difference quotients, the proof is omitted.

Corollary 5.3.14: Equivalent forms of derivatives

Suppose $I \subseteq \mathbb{R}$ is an interval, $f : I \to \mathbb{R}$, $c \in I$, and q_c is the difference quotient of f at c. Then the following are equivalent forms to $f'(c) = y$:

   (i) The difference quotient q_c preserves accumulation at c with y:

$$E \subseteq I \backslash \{c\} \text{ with } c \operatorname{acl} E \qquad \Longrightarrow \qquad y \operatorname{acl} q_c(E). \tag{5.3.26}$$

   (ii) The difference quotient q_c transforms sequential convergence to c in the domain into sequential convergence to y in the codomain, specifically:

$$(x_n) \subseteq I \backslash \{c\} \text{ with } \lim_{n \to \infty} x_n = c \tag{5.3.27}$$

$$\Longrightarrow \quad \lim_{n \to \infty} q_c(x_n) = \lim_{n \to \infty} \frac{f(x_n) - f(c)}{x_n - c} = y. \tag{5.3.28}$$

   (iii) f is differentiable at c with $f'(c) = y$. That is, the difference quotient q_c converges to y at c and we have

$$\lim_{x \to c} q_c(x) = \lim_{x \to c} \frac{f(x) - f(c)}{x - c} = y. \tag{5.3.29}$$

The next section begins the development of properties of differentiation and derivatives, starting with linearity.

Exercises

5.3.1. Let $f : \mathbb{R} \to \mathbb{R}$ be the cubic function $f(x) = x^3$. Prove $f'(c) = 3c^2$ for every $c \in \mathbb{R}$.

5.3.2. Give an example of a function $g : \mathbb{R} \to \mathbb{R}$ where g is one-to-one and differentiable, but $g'(0) = 0$.

5.3.3. Give an example of a function $h : \mathbb{R} \to \mathbb{R}$ where h is continuous, one-to-one, and differentiable everywhere except at $c = 0$.

5.3.4. Prove the *Power Rule* for integer powers in two parts.

  (i) For each $n \in \mathbb{N} \cup \{0\}$, let $h_n : \mathbb{R} \to \mathbb{R}$ be the monomial $h_n(x) = x^n$. Prove $h'_n(c) = nc^{n-1}$ for every $c \in \mathbb{R}$.

  (ii) For each $z \in \mathbb{Z}$, let $h_z : \mathbb{R}\backslash\{0\} \to \mathbb{R}$ be given by $h_z(x) = x^z$. Prove $h'_z(c) = zc^{z-1}$ for every $c \in \mathbb{R}\backslash\{0\}$.

5.3.5. Consider the following alternative definition for the derivative: Suppose $I \subseteq \mathbb{R}$ is an interval, $f : I \to \mathbb{R}$, and $c \in I$. The alternative derivative of f at c, also denoted $f'(c)$, is the limit

$$f'(c) = \lim_{h \to 0} \frac{f(c+h) - f(c)}{h}. \tag{5.3.30}$$

Prove this alternative definition for the derivative is equivalent to Definition 5.3.1.

5.3.6. Suppose $I \subseteq \mathbb{R}$ is an open interval, $f : I \to \mathbb{R}$, $c \in I$, and f is differentiable at c. Prove

$$\lim_{h \to 0} \frac{f(c+h) - f(c-h)}{2h} = f'(c). \tag{5.3.31}$$

Also, find an example of a function f where the above limit exists at c but f is not differentiable at c.

5.3.7. Let $r : [0, \infty) \to \mathbb{R}$ be the square root function $r(x) = \sqrt{x}$. Prove

$$r'(c) = -\frac{1}{2\sqrt{c}}. \tag{5.3.32}$$

5.4 Properties of derivatives

Differentiable functions and derivatives exhibit an interesting and sometimes surprising array of properties. For instance, differentiable functions are continuous and derivatives are linear. On the other hand, derivatives are not multiplicative. Instead, we get results like the Product Rule 5.4.6.

First up is the linearity of derivatives. This fact plays an important role in analysis and the development of topics like differential equations.

Theorem 5.4.1: Linearity of differentiation

Suppose $I \subseteq \mathbb{R}$ is an interval, $f, g : I \to \mathbb{R}$, both f and g are differentiable at $c \in I$, and $\alpha \in \mathbb{R}$. Then $f + g$ and αf are differentiable at c with

(i) $(f + g)'(c) = f'(c) + g'(c)$ *(additivity)*; and

(ii) $(\alpha f)'(c) = \alpha f'(c)$ *(homogeneity)*.

Scratch Work 5.4.2: Linearity of differentiation from linearity of functional limits

Since derivatives are defined as limits of functions, it may come as no surprise that the linearity of differentiation follows from the linearity of functional limits (Theorem 5.2.6). Algebraic manipulation of the difference quotients for $f + g$ and αf yield the linear combinations of the difference quotients for f and g, allowing us to take advantage of Theorem 5.2.6. To that end, for every $x \in I$ where $x \neq c$ we have

$$\frac{f(x) + g(x) - (f(c) + g(c))}{x - c} = \frac{f(x) - f(c)}{x - c} + \frac{g(x) - g(c)}{x - c} \tag{5.4.1}$$

as well as

$$\frac{\alpha f(x) - \alpha f(c)}{x - c} = \alpha \cdot \frac{f(x) - f(c)}{x - c}. \tag{5.4.2}$$

Taking the limit as x approaches c yields both the additivity and the homogeneity of differentiation.

Proof of Theorem 5.4.1. Suppose $I \subseteq \mathbb{R}$ is an interval, $f, g : I \to \mathbb{R}$, both f and g are differentiable at $c \in I$, and $\alpha \in \mathbb{R}$. By the definition of derivative (Definition 5.3.1), algebraic manipulation of difference quotients as in Scratch Work 5.4.2, and linearity of functional limits (Theorem 5.2.6), we have both

$$(f+g)'(c) = \lim_{x \to c} \frac{f(x) + g(x) - (f(c) + g(c))}{x - c} \tag{5.4.3}$$

$$= \lim_{x \to c} \left(\frac{f(x) - f(c)}{x - c} + \frac{g(x) - g(c)}{x - c} \right) \tag{5.4.4}$$

$$= \lim_{x \to c} \frac{f(x) - f(c)}{x - c} + \lim_{x \to c} \frac{g(x) - g(c)}{x - c} \tag{5.4.5}$$

$$= f'(c) + g'(c) \tag{5.4.6}$$

as well as

$$(\alpha f)'(c) = \lim_{x \to c} \frac{\alpha f(x) - \alpha f(c)}{x - c} \tag{5.4.7}$$

$$= \lim_{x \to c} \left(\alpha \cdot \frac{f(x) - f(c)}{x - c} \right) \tag{5.4.8}$$

$$= \alpha \lim_{x \to c} \frac{f(x) - f(c)}{x - c} \tag{5.4.9}$$

$$= \alpha f'(c). \tag{5.4.10}$$

Therefore, differentiation is linear. $\square$

Once again and as mentioned in Remark 1.6.18, a corollary of the linearity of differentiation holds for linear combinations. As with the proofs of Corollaries 1.6.16, 4.5.7, 4.7.7, and 5.2.8, the proof of Corollary 5.4.3 follows from induction on linearity and is left as an exercise.

Corollary 5.4.3: Linear combinations of derivatives

Suppose $I \subseteq \mathbb{R}$ is an interval, $c \in I$, $k \in \mathbb{N}$, and for each $j = 1, \ldots, k$ we have $a_j \in \mathbb{R}$ and the functions $f_j : I \to \mathbb{R}$ are differentiable at c. Then the linear combination f given by

$$f(x) = \sum_{j=1}^{k} a_j f_j(x) = a_1 f_1(x) + \ldots + a_k f_k(x) \tag{5.4.11}$$

is differentiable at c and with derivative given by

$$f'(c) = \sum_{j=1}^{k} a_j f_j'(c) = a_1 f_1'(c) + \ldots + a_k f_k'(c). \tag{5.4.12}$$

Differentiable functions are continuous. However, the derivative of a function is not necessarily continuous.

Theorem 5.4.4: Differentiable implies continuous

Suppose $I \subseteq \mathbb{R}$ is an interval, $f : I \to \mathbb{R}$, $c \in I$, and f is differentiable at c. Then f is continuous at c.

Scratch Work 5.4.5: Limits connect derivatives and continuity

The result follows the ideas that derivatives are limits and limits can yield continuity under the conditions of Theorem 5.2.1. Also, these conditions hold when the function is assumed to be differentiable.

We want to end up with

$$\lim_{x \to c} f(x) = f(c). \tag{5.4.13}$$

By subtracting $f(c)$ from both sides, we have (5.4.13) is equivalent to

$$\lim_{x \to c}(f(x) - f(c)) = 0 \tag{5.4.14}$$

by the linearity of functional limits (Theorem 5.2.6). Equation (5.4.14) is a better fit for the assumption that f is differentiable at c since $f(x) - f(c)$ is the numerator of the difference quotient q_c. Also, for $x \neq c$ we have

$$q_c(x)(x - c) = \frac{f(x) - f(c)}{x - c}(x - c) = f(x) - f(c). \tag{5.4.15}$$

Pairing this equation with the notion that the limit of a product is the product of limits (Theorem 5.2.11) gives us the equivalent goal (5.4.14).

Proof of Theorem 5.4.4. Suppose $I \subseteq \mathbb{R}$ is an interval, $f : I \to \mathbb{R}$, $c \in I$, and f is differentiable at c. Then the output $f(c)$ is defined, and for every input $x \in I$ where $x \neq c$ we have

$$f(x) - f(c) = \frac{f(x) - f(c)}{x - c}(x - c). \tag{5.4.16}$$

Since $\lim_{x \to c}(x - c) = 0$, the definition of derivative (Definition 5.3.1) and Theorem 5.2.11 combine to yield

$$\lim_{x \to c}(f(x) - f(c)) = \lim_{x \to c}\left(\frac{f(x) - f(c)}{x - c}(x - c) \right) \tag{5.4.17}$$

$$= \left(\lim_{x \to c} \frac{f(x) - f(c)}{x - c} \right) \left(\lim_{x \to c}(x - c) \right) \tag{5.4.18}$$

$$= f'(c) \cdot 0 \tag{5.4.19}$$

$$= 0. \tag{5.4.20}$$

By adding $f(c)$ and considering the linearity of functional limits (Theorem 5.2.6), we have

$$\lim_{x \to c}(f(x) - f(c)) = 0 \quad \implies \quad \lim_{x \to c} f(x) = f(c). \tag{5.4.21}$$

Therefore, f is continuous at c by Theorem 5.2.1. □

Unlike convergence and continuity, differentiation is not multiplicative. That is, distributing the derivative across multiplication generally does not work out. Instead, the derivative of a product of differentiable functions satisfies the *product rule*.

Theorem 5.4.6: Product Rule

Suppose $I \subseteq \mathbb{R}$ is an interval, $f, g : I \to \mathbb{R}$, and f and g are differentiable at $c \in I$. Then fg is differentiable at c with

$$(fg)'(c) = f(c)g'(c) + f'(c)g(c). \tag{5.4.22}$$

Scratch Work 5.4.7: Algebra and properties of limits

To take advantage of the differentiability of f and g separately, adding a nice version of zero allows us to split the difference quotient q_c of the product fg into a form involving the difference quotients of f and g. Note that

$$-f(x)g(c) + f(x)g(c) = 0. \tag{5.4.23}$$

So for $x \neq c$ we have

$$q_c(x) = \frac{f(x)g(x) - f(c)g(c)}{x - c} \tag{5.4.24}$$

$$= \frac{f(x)g(x) - f(x)g(c) + f(x)g(c) - f(c)g(c)}{x - c} \tag{5.4.25}$$

$$= \frac{f(x)(g(x) - g(c)) + g(c)(f(x) - f(c))}{x - c} \tag{5.4.26}$$

$$= f(x)\frac{g(x) - g(c)}{x - c} + g(c)\frac{f(x) - f(c)}{x - c}. \tag{5.4.27}$$

Properties of functional limits, continuity, and derivatives get us to the conclusion from here.

Proof the Product Rule 5.4.6. Suppose $I \subseteq \mathbb{R}$ is an interval, $f, g : I \to \mathbb{R}$, and f and g are differentiable at $c \in I$. Note that $g(c)$ is constant and by Theorem 5.4.4 we have f is continuous at c. So, by Theorem 5.2.1 we have

$$\lim_{x \to c} f(x) = f(c) \qquad \text{and} \qquad \lim_{x \to c} g(c) = g(c). \tag{5.4.28}$$

By adding the version of zero given by $-f(x)g(c) + f(x)g(c) = 0$ along with the linearity and products of functional limits (Theorems 5.2.6 and 5.2.11), we have

$$(fg)'(c) = \lim_{x \to c} \frac{f(x)g(x) - f(c)g(c)}{x - c} \tag{5.4.29}$$

$$= \lim_{x \to c} \frac{f(x)g(x) - f(x)g(c) + f(x)g(c) - f(c)g(c)}{x - c} \tag{5.4.30}$$

$$= \lim_{x \to c} \frac{f(x)(g(x) - g(c)) + g(c)(f(x) - f(c))}{x - c} \tag{5.4.31}$$

$$= \left(\lim_{x \to c} f(x)\right)\left(\lim_{x \to c} \frac{g(x) - g(c)}{x - c}\right) + \left(\lim_{x \to c} g(c)\right)\left(\lim_{x \to c} \frac{f(x) - f(c)}{x - c}\right) \tag{5.4.32}$$

$$= f(c)g'(c) + g(c)f'(c). \tag{5.4.33}$$

☐

The *chain rule* tells us how derivatives handle compositions of functions.

Theorem 5.4.8: Chain Rule

Suppose $A, B \subseteq \mathbb{R}$, $f : A \to \mathbb{R}$, and $g : B \to \mathbb{R}$ where $f(A) \subseteq B$ so the composition $g \circ f$ is defined. If f is differentiable at c and g is differentiable at $f(c)$, then $g \circ f$ is differentiable at c with

$$(g \circ f)'(c) = g'(f(c)) \cdot f'(c). \tag{5.4.34}$$

Scratch Work 5.4.9: Modify an inadequate approach

To split the difference quotient q_c for the composition $g \circ f$ into difference quotients for g and f separately, we can try multiplying q_c by the version of 1 given by dividing $f(x) - f(c)$ by itself, like this:

$$q_c(x) = \left(\frac{g(f(x)) - g(f(c))}{x - c} \right) \left(\frac{f(x) - f(c)}{f(x) - f(c)} \right) \tag{5.4.35}$$

$$= \left(\frac{g(f(x)) - g(f(c))}{f(x) - f(c)} \right) \left(\frac{f(x) - f(c)}{x - c} \right). \tag{5.4.36}$$

However, these equations are only valid if the denominator $f(x) - f(c)$ is nonzero when $f(x) \neq f(c)$. Since we do not want to impose the condition that $f(x) \neq f(c)$, consider a different difference quotient. Let $h : B \to \mathbb{R}$ be the extended difference quotient of g at $f(c)$ defined by

$$h(y) = \begin{cases} \dfrac{g(y) - g(f(c))}{y - f(c)}, & \text{if } y \neq f(c), \\ g'(f(c)), & \text{if } y = f(c). \end{cases} \tag{5.4.37}$$

The differentiability of g at $y = f(c) \in B$ ensures q is continuous at $f(c)$, once we allow limits to get involved. Additionally, for all $y \in B$ including $y = f(c)$ we have

$$h(y)(y - f(c)) = g(y) - g(f(c)). \tag{5.4.38}$$

In turn, for all $x \in A$ we have $f(x) \in B$ and so

$$h(f(x))(f(x) - f(c)) = g(f(x)) - g(f(c)). \tag{5.4.39}$$

Ultimately, for $x \in A$ where $x \neq c$, we also have

$$h(f(x)) \cdot \frac{f(x) - f(c)}{x - c} = \frac{g(f(x)) - g(f(c))}{x - c}. \tag{5.4.40}$$

Properties of continuity, functional limits, and derivatives lead to the conclusion from here.

Proof of Chain Rule 5.4.8. Suppose $A, B \subseteq \mathbb{R}$, $f : A \to \mathbb{R}$, $f(A) \subseteq B$, $g : B \to \mathbb{R}$, f is differentiable at c, and g is differentiable at $f(c)$. Define $h : B \to \mathbb{R}$ by

$$h(y) = \begin{cases} \dfrac{g(y) - g(f(c))}{y - f(c)}, & \text{if } y \neq f(c), \\ g'(f(c)), & \text{if } y = f(c). \end{cases} \tag{5.4.41}$$

Since g is differentiable at $f(c)$, we have

$$h(f(c)) = g'(f(c)) \tag{5.4.42}$$

$$= \lim_{y \to f(c)} \frac{g(y) - g(f(c))}{y - f(c)} \tag{5.4.43}$$

$$= \lim_{y \to f(c)} h(y). \tag{5.4.44}$$

By Theorem 5.2.1, h is continuous at $f(c)$. Also, for $x \in A$ where $x \neq c$, we have $f(x) \in B$ and

$$h(f(x)) \cdot \frac{f(x) - f(c)}{x - c} = \frac{g(f(x)) - g(f(c))}{x - c}. \tag{5.4.45}$$

Now, since f is differentiable at c, f is continuous at c by Theorem 5.4.4. Hence, the composition $h \circ f$ is continuous at c by Theorem 4.5.13. Therefore, by Theorems 5.1.8 and 4.5.8, the composition $g \circ f$ is differentiable at c and we have

$$(g \circ f)'(c) = \lim_{x \to c} \frac{g(f(x)) - g(f(c))}{x - c} \tag{5.4.46}$$

$$= \lim_{x \to c} \left(h(f(x)) \cdot \frac{f(x) - f(c)}{x - c} \right) \tag{5.4.47}$$

$$= \left(\lim_{x \to c} h(f(x)) \right) \left(\lim_{x \to c} \frac{f(x) - f(c)}{x - c} \right) \tag{5.4.48}$$

$$= g'(f(c)) \cdot f'(c). \tag{5.4.49}$$

$\square$

The Quotient Rule 5.4.10 from calculus can be proven from the definition of derivative (Definition 5.3.1) by manipulating the difference quotient directly and using properties of limits. However, we show the quotient rule stems from the Product Rule 5.4.6 and Chain Rule 5.4.8 via the differentiability of the reciprocal function as in Example 5.3.6.

Theorem 5.4.10: Quotient Rule

Suppose $D \subseteq \mathbb{R}$, $c \in D$, and $f, g : D \to \mathbb{R}$. If f and g are differentiable at c and $g(x) \neq 0$ for all $x \in D$, then the quotient $f/g : D \to \mathbb{R}$ given by

$$\frac{f(x)}{g(x)} \quad \text{for all } x \in D \tag{5.4.50}$$

is differentiable at c and we have

$$\left(\frac{f}{g}\right)'(c) = \frac{g(c)f'(c) - f(c)g'(c)}{(g(c))^2}. \tag{5.4.51}$$

Proof of the Quotient Rule 5.4.10. Assume the hypotheses of the Quotient Rule 5.4.10 hold. Since the reciprocal function $h(x) = 1/x$ function is differentiable on $\mathbb{R}\backslash\{0\}$ (Example 5.3.6) and g is differentiable with $g(x) \neq 0$ for all $x \in D$, the Chain Rule 5.4.8 applies and tells us the reciprocal of g given by $h \circ g(x) = 1/g(x)$ is differentiable at c with

$$\left(\frac{1}{g}\right)'(c) = -\frac{g'(c)}{(g(c))^2}. \tag{5.4.52}$$

Now, since

$$\frac{f(x)}{g(x)} = f(x)\frac{1}{g(x)} \tag{5.4.53}$$

for every $x \in D$ and f is differentiable on D, the Product Rule 5.4.6 applies and we have the quotient f/g is differentiable at c with

$$\left(\frac{f}{g}\right)'(c) = f'(c)\frac{1}{g(c)} + f(c)\left(-\frac{g'(c)}{(g(c))^2}\right) = \frac{g(c)f'(c) - f(c)g'(c)}{(g(c))^2}. \tag{5.4.54}$$

$\square$

Even though differentiable functions are continuous (Theorem 5.4.4), it is not necessarily true that a derivative is a continuous function.

Example 5.4.11: A discontinuous derivative

Consider the function $g : \mathbb{R} \to \mathbb{R}$ given by the piecewise formula

$$g(x) = \begin{cases} 0, & \text{if } x = 0, \\ x^2 \sin\left(\frac{1}{x}\right), & \text{if } x \neq 0. \end{cases} \tag{5.4.55}$$

See Figure 5.4.1. The function g is differentiable and we have a piecewise formula for the derivative g' valid for every $x \in \mathbb{R}$ given by

$$g'(x) = \begin{cases} 0, & \text{if } x = 0, \\ 2x \sin\left(\frac{1}{x}\right) - \cos\left(\frac{1}{x}\right), & \text{if } x \neq 0. \end{cases} \tag{5.4.56}$$

Moreover, g is differentiable at $c = 0$ with $g'(0) = 0$, but the derivative g' is discontinuous at $c = 0$. (Try plotting this!)

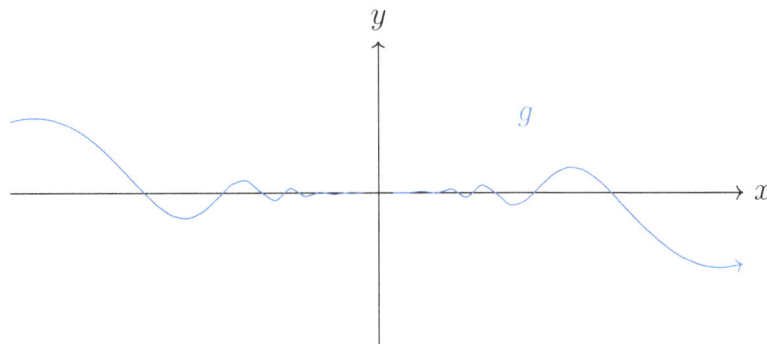

Figure 5.4.1: A differentiable function g whose derivative g' exists but is not continuous at $c = 0$.

Scratch Work 5.4.12: Combining results

There are three things to prove: $g'(0) = 0$; the formula for $g'(x)$ where $x \neq 0$; and g' is discontinuous at 0.

The Squeeze Theorem for functions 5.2.15 shows the difference quotient of g at $c = 0$ converges to 0, and so $g'(0) = 0$. The formula for $g'(x)$ for nonzero x follows from the Product Rule 5.4.6, the Chain Rule 5.4.8, the derivative of the square function in Example 5.3.4 and derivative of the reciprocal function in Example 5.3.6 along with the fact that the derivative of sine is cosine (whose proof is an exercise).

The fact that the derivative g' has a discontinuity at $c = 0$ follows from Discontinuity Criteria (Corollary 4.6.13) using a sequences of positive real numbers whose limit is 0 in the domain but whose image converges to $-1 \neq g'(0) = 0$ in the range. To that end, note that for every $n \in \mathbb{N}$ we have

$$\sin 2\pi n = 0 \quad \text{and} \quad \cos 2\pi n = 1. \tag{5.4.57}$$

So by choosing the sequence $(x_n) \subseteq \mathbb{R} \backslash \{0\}$ given by $x_n = 1/(2\pi n)$, we have

$$g'(x_n) = 2x_n \sin\left(\frac{1}{x_n}\right) - \cos\left(\frac{1}{x_n}\right) = \frac{2}{2\pi n}\sin(2\pi n) - \cos(2\pi n) = 0 - 1 = -1. \tag{5.4.58}$$

On to the proof.

Proof for Example 5.4.11. First, to show $g'(0) = 0$ let q_0 denote the difference quotient of g at $c = 0$. Then for $x \neq 0$ we have

$$q_0(x) = \frac{g(x) - g(0)}{x - 0} = \frac{x^2 \sin(1/x)}{x} = x \sin\left(\frac{1}{x}\right). \tag{5.4.59}$$

We also have

$$-1 \leq \sin\left(\frac{1}{x}\right) \leq 1 \quad \implies \quad -|x| \leq x \sin\left(\frac{1}{x}\right) \leq |x|. \tag{5.4.60}$$

Since $\lim_{x\to0}(-|x|) = \lim_{x\to0}|x| = 0$, the Squeeze Theorem for functions 5.2.15 yields

$$g'(0) = \lim_{x\to0}\frac{g(x)-g(0)}{x-0} = \lim_{x\to0}\frac{x^2\sin(1/x)}{x} = \lim_{x\to0}x\sin\left(\frac{1}{x}\right) = 0. \tag{5.4.61}$$

Hence, g is differentiable at $c = 0$ with $g'(0) = 0$.

Now suppose $c \neq 0$. By Examples 5.3.4 and 5.3.6, the square function $h(x) = x^2$ and the reciprocal function $f(x) = 1/x$ are differentiable at c. Also, the derivative of sine is cosine. So, by the Product Rule 5.4.6 and the Chain Rule 5.4.8, the function g is differentiable at c and we have

$$g'(c) = 2c\sin\left(\frac{1}{c}\right) - \cos\left(\frac{1}{c}\right). \tag{5.4.62}$$

Therefore, g is differentiable at every $c \in \mathbb{R}$.

Finally, to show the derivative g' is discontinuous at $c = 0$, consider the sequence $(x_n) \subseteq \mathbb{R}\backslash\{0\}$ given by $x_n = 1/(2\pi n)$. We have

$$g'(x_n) = 2x_n\sin\left(\frac{1}{x_n}\right) - \cos\left(\frac{1}{x_n}\right) = \frac{2}{2\pi n}\sin(2\pi n) - \cos(2\pi n) = 0 - 1 = -1. \tag{5.4.63}$$

Hence,

$$\lim_{n\to\infty}x_n = \lim_{n\to\infty}\frac{1}{2\pi n} = 0, \quad\text{but}\quad \lim_{n\to\infty}g'(x_n) = \lim_{n\to\infty}-1 = -1 \neq 0 = g'(0). \tag{5.4.64}$$

Therefore, by Discontinuity Criteria (Corollary 4.6.13), g' is discontinuous at $c = 0$. □

The next section develops further properties of derivatives.

Exercises

5.4.1. The alternative definition of derivative in Exercise 5.3.5 along with and the results of Exercises 5.2.9 and 5.2.10 help us find the derivatives of sine and cosine.

(i) Use the trigonometric identity

$$\sin(x+h) = \sin x\cos h + \cos x\sin h \tag{5.4.65}$$

to show $\dfrac{d}{dx}\sin x = \cos x$.

(ii) Use the trigonometric identity

$$\cos(x+h) = \cos x\cos h - \sin x\sin h \tag{5.4.66}$$

to show $\dfrac{d}{dx}\cos x = -\sin x$.

5.4.2. Use the results of the previous exerise to justify the formulas of the other trigonometric functions.

(i) $\dfrac{d}{dx}\tan x = \sec^2 x.$

(iii) $\dfrac{d}{dx}\cot x = -\csc^2 x.$

(ii) $\dfrac{d}{dx}\sec x = \sec x \tan x.$

(iv) $\dfrac{d}{dx}\csc x = -\csc x \cot x.$

5.4.3. Assume the general form of the Power Rule holds in that for each $p > 0$, the function $h_p : \mathbb{R} \to \mathbb{R}$ given by $h_p(x) = x^p$ is differentiable with $h_p'(c) = pc^{p-1}$ for each $c \in \mathbb{R}$. Consider a squeezed version of the topologist's sine curve from Example 4.4.10 defined by $g_p : \mathbb{R} \to \mathbb{R}$ where

$$g_p(x) = \begin{cases} x^p \sin\left(\dfrac{1}{x}\right), & \text{if } x \neq 0, \\ 0, & \text{if } x = 0. \end{cases} \tag{5.4.67}$$

(i) Find a value $p > 0$ where g_p is differentiable on $\mathbb{R}$ but g_p' is unbounded on every open interval containing 0.

(ii) Find a value $p > 0$ where g_p is differentiable on $\mathbb{R}$, g_p' is continuous on $\mathbb{R}$, but g_p' is not differentiable at $c = 0$.

5.4.4. Suppose $f : [a, b] \to \mathbb{R}$ is one-to-one and differentiable on $[a, b]$ with $f'(x) \neq 0$ for all x in $[a, b]$. Prove f^{-1} is differentiable with

$$(f^{-1})'(y) = \frac{1}{f'(x)} \quad \text{when} \quad y = f(x). \tag{5.4.68}$$

5.4.5. Prove the following *Squeeze Theorem for derivatives*[1]: Suppose $I \subseteq \mathbb{R}$ is an open interval, $c \in I$, and $f, g, h : I \to \mathbb{R}$. Further suppose f and h are differentiable at c, $f(c) = h(c)$, and

$$f(x) \leq g(x) \leq h(x) \quad \text{for all } x \in I. \tag{5.4.69}$$

Prove g is differentiable at c and

$$f'(c) = g'(c) = h'(c). \tag{5.4.70}$$

5.4.6. Suppose $p : \mathbb{R} \to \mathbb{R}$ is a polynomial. Prove that eventually (i.e., for large enough $k \in \mathbb{N}$), the kth derivative of p is identically 0.

5.4.7. Suppose $I \subseteq \mathbb{R}$ is an open interval, g is differentiable on I, and there is a bound $b > 0$ such that

$$|g'(x)| \leq b \quad \text{for all } x \in I. \tag{5.4.71}$$

(i) Prove g is uniformly continuous on I.

(ii) Find an example of a function $f : (0, 1) \to \mathbb{R}$ which is uniformly continuous and differentiable on $(0, 1)$ but the derivative f' is unbounded.

---

[1] In [7], this problem is attributed to Clyde Dubbs at New Mexico Institute of Mining and Technology.

5.4.8. Consider the modification of Dirichlet's function $\mathbb{1}_\mathbb{Q}$ from Example 4.2.14 given by

$$f(x) = x^2 \mathbb{1}_\mathbb{Q}(x) = \begin{cases} x^2, & \text{if } x \in \mathbb{Q}, \\ 0, & \text{if } x \in \mathbb{R} \backslash \mathbb{Q}. \end{cases} \tag{5.4.72}$$

Prove f is differentiable at $c = 0$ but not at nonzero values of c.

5.4.9. Let $D[a,b]$ denote the set of real-valued functions that are differentiable on $[a,b]$. Use Lemma 1.6.7 to prove $D[a,b]$ is a vector space.

5.4.10. Suppose $c, \ell \in \mathbb{R}$ and let $D(c,\ell)$ denote the set of real-valued functions on $\mathbb{R}$ that are differentiable at c with derivative ℓ. (Thus, for each $f \in D(c,\ell)$ we have $f'(c) = \ell$.) Use Lemma 1.6.7 to prove $D(c,\ell)$ is a vector space if and only if $\ell = 0$.

5.5   Mean Value Theorem

Derivatives have a variety of interesting results when it comes to the values they can attain and what they say about the behavior of functions. As seen in the previous sections, manipulation of difference quotients and properties of limits play a central role in the proofs.

First up are definitions for more local properties of functions.

Definition 5.5.1: Local maximum and local minimum

Suppose $D \subseteq \mathbb{R}$ and $f : D \to \mathbb{R}$. Then f has a *local maximum* at $c \in D$ if there is a *threshold* $\delta > 0$ such that

$$|x - c| < \delta \text{ with } x \in D \qquad \implies \qquad f(x) \leq f(c). \tag{5.5.1}$$

Similarly, f has a *local minimum* at $c \in D$ if there is a *threshold* $\delta > 0$ such that

$$|x - c| < \delta \text{ with } x \in D \qquad \implies \qquad f(c) \leq f(x). \tag{5.5.2}$$

Also, f has a *local extremum at c* when f has either a local maximum or local minimum at c.

Remark 5.5.2: The output is the extreme value

To clarify the terminology and notation a bit, if f has a local extremum at c, then the output $f(c)$ is the extreme value, not c. That is:

(i) If f has a local maximum at $c \in D$ with threshold δ, then $f(c) = \max f(V_\delta(c) \cap D)$.

(ii) If f has a local minimum at $c \in D$ with threshold δ, then $f(c) = \min f(V_\delta(c) \cap D)$.

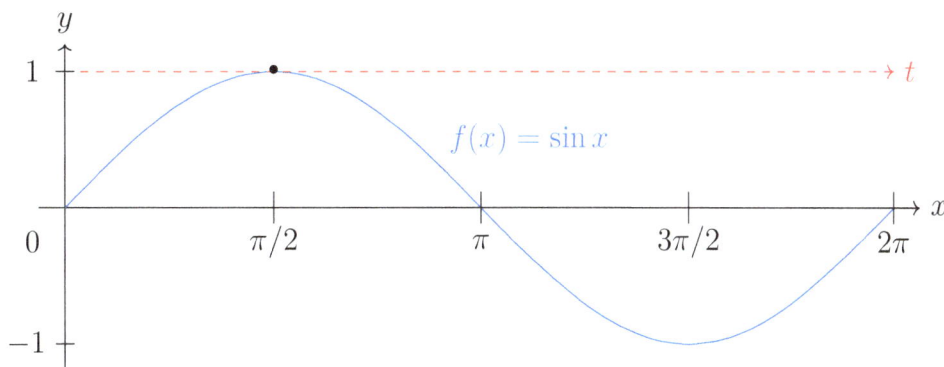

Figure 5.5.1: The sine function $f(x) = \sin x$ with its tangent line t at $c = \pi/2$ where it attains its maximum $f(\pi/2) = \max\{f([0, 2\pi])\}$. This tangent line has slope $f'(\pi/2) = \cos \pi/2 = 0$ as ensured by the Interior Extremum Theorem 5.5.3 as well as Rolle's Theorem 5.5.7 since $\sin 0 = \sin 2\pi = 0$.

The Interior Extremum Theorem 5.5.3 is the kernel of this section. It follows from the definition of derivative as a limit of difference quotients (Definition 5.3.1) and the idea that 0 is the only real number arbitrarily close to the sets of positive and negative numbers.

Theorem 5.5.3: Interior Extremum Theorem

Suppose $f : (a, b) \to \mathbb{R}$, f is differentiable at every point in the open interval (a, b), and f attains its local maximum or local minimum at some point c in (a, b). Then $f'(c) = 0$.

Scratch Work 5.5.4: Arbitrarily close to positive and negative

The approach will be to show that the derivative $f'(c)$ is arbitrarily close to positive and negative numbers, or rather nonnegative and nonpositive numbers, through a manipulation difference quotients and inequalities.

In the case where $f(c)$ is a local maximum, we have $f(x) \leq f(c)$ for any input x within the threshold of c. Hence, by subtracting $f(c)$, we have that the numerator of the difference quotient is always nonpositive since it satisfies

$$f(x) - f(c) \leq 0. \tag{5.5.3}$$

The hypothesis of having the domain be an open interval (a, b) ensures any point within it can be approached from the left and the right. By choosing inputs x and y on either side of c so that

$$x < c < y, \tag{5.5.4}$$

we can subtract c to ensure

$$x - c < 0 < y - c. \tag{5.5.5}$$

This guarantees denominators of difference quotients are either positive or negative, respectively. Combining these denomators with our nonpositive numerator allows our difference quotients, and therefore the derivative, to be arbitrarily close to both nonnegative and nonpositive real numbers. This forces the derivative to be zero by Lemma 1.5.14.

Also, only the local maximum case is proven here. The local minimum case follows from a similar argument and is left as an exercise.

Proof of the Interior Extremum Theorem 5.5.3. Suppose the hypotheses of the Interior Extremum Theorem 5.5.3 hold in the case where $f(c)$ is a local maximum with threshold $\delta > 0$. Suppose $x, y \in V_\delta(c) = (c - \delta, c + \delta)$ where δ is small enough to ensure

$$a < c - \delta < x < c < y < c + \delta < b. \tag{5.5.6}$$

Note that by Lemma 1.5.10, we have both $|x - c| < \delta$ and $|y - c| < \delta$. Also, note we have $f(c) = \max f(V_\delta(c))$. So

$$f(x) \le f(c) \quad \text{and} \quad f(y) \le f(c). \tag{5.5.7}$$

Subtracting c from (5.5.6) and $f(c)$ from (5.5.7) yields

$$x - c < 0 < y - c, \quad f(x) - f(c) \le 0, \quad \text{and} \quad f(y) - f(c) \le 0. \tag{5.5.8}$$

Let q_c denote the difference quotient of f at c. Since division by a negative number flips inequalities but division by a positive real number does not, we have

$$q_c(y) = \frac{f(y) - f(c)}{y - c} \le 0 \le \frac{f(x) - f(c)}{x - c} = q_c(x). \tag{5.5.9}$$

Therefore, by the equivalent forms of derivatives in Corollary 5.3.14, $f'(c)$ is arbitrarily close to both the set of nonnegative real numbers and the set of nonpositive real numbers. By a slight modification of Lemma 1.5.14, we have $f'(c) = 0$. $\square$

Like continuous functions, derivatives satisfy an intermediate value property. The formal statement of this result is know as *Darboux's Theorem.*

Theorem 5.5.5: Darboux's Theorem

Suppose $f : [a, b] \to \mathbb{R}$ is differentiable and $\alpha \in \mathbb{R}$ where either

$$f'(a) < \alpha < f'(b) \quad \text{or} \quad f'(b) < \alpha < f'(a). \tag{5.5.10}$$

Then there exists $c \in (a, b)$ such that

$$f'(c) = \alpha. \tag{5.5.11}$$

Scratch Work 5.5.6: Shift to fit a previous result

Without loss of generality, consider the case where $f'(a) < \alpha < f'(b)$. The other case is handled in a very similar manner. The goal is to find $c \in (a, b)$ where $f'(c) = \alpha$. The Interior Extremum Theorem 5.5.3 requires c to be where an extreme value is attained, but it yields the existence of some $c \in (a, b)$ that gives us a derivative at c equal to zero. So, our f may not apply.

Even so, we can shift the given function f to get a similar function g that will fit the hypotheses of the Interior Extremum Theorem 5.5.3. In particular, by the linearity of differentiation (Theorem 5.4.1) and the derivative of a line (Lemma 5.3.2) we have

$$g(x) = f(x) - \alpha x \qquad \Longrightarrow \qquad g'(x) = f'(x) - \alpha. \tag{5.5.12}$$

So if we can show $g'(c) = f'(c) - \alpha = 0$, we will have $f'(c) = \alpha$.

To ensure $g(c)$ is an extreme value attained with $c \in (a, b)$, we can show that neither $g(a)$ nor $g(b)$ is a local extremum by comparing the derivatives at a and b to bounds on the difference quotients.

Proof of Darboux's Theorem 5.5.5. Without loss of generality, suppose $f : [a, b] \to \mathbb{R}$ is differentiable and $\alpha \in \mathbb{R}$ where

$$f'(a) < \alpha < f'(b). \tag{5.5.13}$$

Define $g : [a, b] \to \mathbb{R}$ by

$$g(x) = f(x) - \alpha x. \tag{5.5.14}$$

By the linearity of differentiation (Theorem 5.4.1) and the derivative of a line (Lemma 5.3.2), we have g is differentiable on $[a, b]$ with

$$g'(x) = f'(x) - \alpha \tag{5.5.15}$$

for all $x \in [a, b]$. Since differentiable functions are continuous (Theorem 5.4.4), g is continuous. Since we also have $[a, b]$ is compact, the Extreme Value Theorem 4.6.9 ensures g attains both a maximum and a minimum on $[a, b]$. In particular, for some $c \in [a, b]$, we have $g(c) = \min g([a, b])$.

To ensure $c \in (a, b)$, a couple of contradiction arguments show that neither $g(a)$ nor $g(b)$ is the minimum. To that end, subtracting α across (5.5.13) yields

$$g'(a) = f'(a) - \alpha < 0 < f'(b) - \alpha = g'(b). \tag{5.5.16}$$

For the first contradiction, suppose $g(a) = \min g([a, b])$. Then for every $x \in (a, b]$ we have both

$$g(x) \geq g(a) \qquad \Longrightarrow \qquad g(x) - g(a) \geq 0 \qquad \text{and} \tag{5.5.17}$$
$$x > a \qquad \Longrightarrow \qquad x - a > 0. \tag{5.5.18}$$

Hence, the difference quotient of g at a is nonnegative since

$$q_a(x) = \frac{g(x) - g(a)}{x - a} \geq 0. \tag{5.5.19}$$

Thus, 0 is a lower bound for the image $q_a((a, b])$ and by the definition of derivative as a limit (Definition 5.3.1) and the order properties of functional limits Corollary 5.1.18, we have

$$g'(a) = \lim_{x \to a} q_a(x) = \lim_{x \to a} \frac{g(x) - g(a)}{x - a} \geq 0. \tag{5.5.20}$$

This contradicts the assertion that $g'(a) < 0$, so $g(a)$ is not the minimum of g on $[a, b]$.

For the second contradiction, suppose $g(b) = \min g([a, b])$. Then for every $x \in [a, b)$ we have both

$$g(x) \geq g(b) \quad \implies \quad g(x) - g(b) \geq 0 \quad \text{and} \tag{5.5.21}$$
$$x < b \quad \implies \quad x - b < 0. \tag{5.5.22}$$

Hence, the difference quotient of g at a is nonpositive since

$$q_b(x) = \frac{g(x) - g(b)}{x - b} \leq 0. \tag{5.5.23}$$

Thus, 0 is an upper bound for the image $q_a((a, b])$ and by the definition of derivative as a limit (Definition 5.3.1) and the order properties of functional limits Corollary 5.1.18, we have

$$g'(b) = \lim_{x \to b} q_a(x) = \lim_{x \to b} \frac{g(x) - g(b)}{x - b} \leq 0. \tag{5.5.24}$$

This contradicts the assertion that $g'(b) > 0$, so $g(b)$ is not the minimum of g on $[a, b]$.

Since the minimum of g is attained at neither a nor b, the minimum must be attained at some $c \in (a, b)$. As a minimum over the entire interval $[a, b]$, the output $g(c)$ is also a local minimum (Definition 5.5.1) with threshold given by $\delta = \min\{|a - c|, |b - c|\} > 0$, the shorter of the two distances from c to a or b. In this case we have

$$x \in V_\delta(c) = (c - \delta, c + \delta) \subseteq [a, b]. \tag{5.5.25}$$

Since $c \in (a, b)$ and $g(c)$ is a local minimum, the Interior Extremum Theorem 5.5.3 yields

$$g'(c) = 0. \tag{5.5.26}$$

Finally, plugging in c for x in (5.5.15) tells us

$$g'(c) = f'(c) - \alpha = 0. \tag{5.5.27}$$

Adding α yields the desired result,

$$f'(c) = \alpha. \tag{5.5.28}$$

$\square$

The next two theorems, Rolle's Theorem 5.5.7 and the Mean Value Theorem 5.5.9, highlight another relationship between the slopes of secant lines and tangent lines through properties of derivatives. Basically, they tell us that the slope of the secant line (a difference quotient) defined over a compact interval is equal to the slope of the tangent line (a derivative) at some point within the interval. In other words, the average rate of change is attained as an instantaneous rate of change.

Theorem 5.5.7: Rolle's Theorem

Suppose $f : [a, b] \to \mathbb{R}$ where

  (i) f is continuous on the compact interval $[a, b]$,

  (ii) f is differentiable on the open interval (a, b), and

  (iii) $f(a) = f(b)$.

Then there is a point $c \in (a, b)$ where $f'(c) = 0$.

Scratch Work 5.5.8: Apply the Interior Extremum Theorem

The conclusion of Rolle's Theorem is the same as that of the Interior Extremum Theorem 5.5.3: there is an interior point of the domain where the derivative is zero. To take advantage of this, we should again verify that f attains local extrema. The proof is broken down into a couple of cases, the first ensures f is constant and so must have derivative zero at every input, and the second allows us to apply the Interior Extremum Theorem 5.5.3.

Proof of Rolle's Theorem 5.5.7. Suppose $f : [a, b] \to \mathbb{R}$ where f is continuous on $[a, b]$, f is differentiable on (a, b), and $f(a) = f(b)$. Since f is continuous on the compact interval $[a, b]$, f attains its maximum and minimum by the Extreme Value Theorem 4.6.9.

  Case (i) Suppose both the maximum and minimum are attained at both of the endpoints. That is, assume

$$f(a) = f(b) = \max f([a, b]) = \min f([a, b]). \tag{5.5.29}$$

Then f is constant since we have

$$\max f([a, b]) = \min f([a, b]) \le f(x) \le \max f([a, b]) \tag{5.5.30}$$
$$\implies \quad f(x) = \max f([a, b]) = \min f([a, b]) \tag{5.5.31}$$

for every $x \in [a, b]$. By the derivative of lines (Lemma 5.3.2), $f'(c) = 0$ for every $c \in (a, b)$.

  Case (ii) Suppose either $\max f([a, b])$ or $\min f([a, b])$ is attained at some $c \in (a, b)$. Then by the Interior Extremum Theorem 5.5.3, $f'(c) = 0$. □

The Mean Value Theorem 5.5.9 generalizes Rolle's Theorem 5.5.7 tell us that the slope of any secant line over a compact interval is equal to the slope of the tangent line at some point within the interval. Hence, the slope of a secant line is the slope of a related tangent line.

Theorem 5.5.9: Mean Value Theorem

Suppose $f : [a, b] \to \mathbb{R}$ where

  (i) f is continuous on the compact interval $[a, b]$ and

  (ii) f is differentiable on the open interval (a, b).

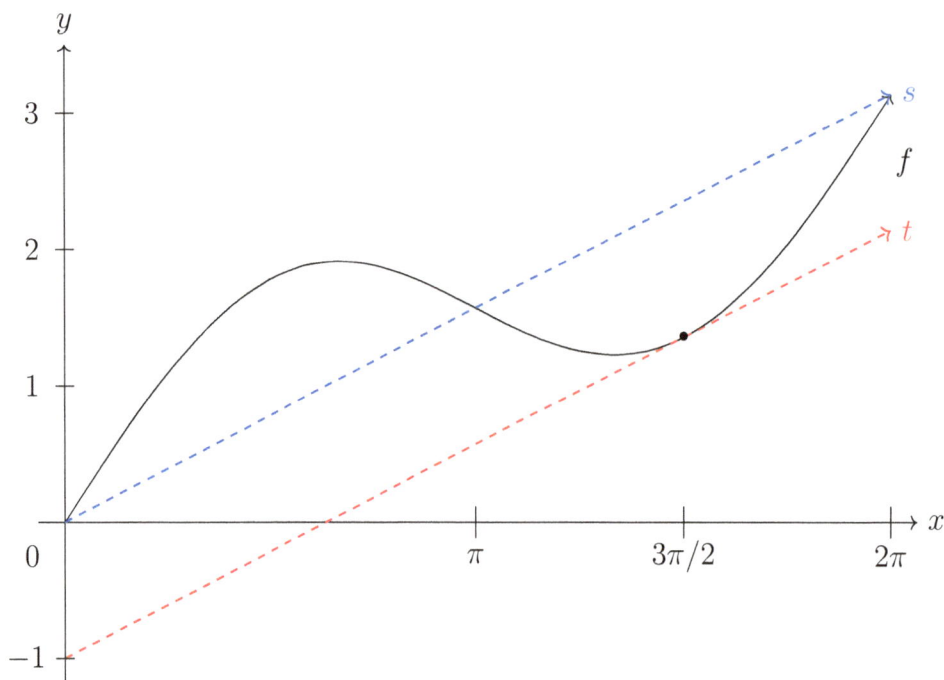

Figure 5.5.2: The differentiable function $f(x) = x/2 + \sin x$ exhibiting the Mean Value Theorem 5.5.9 with its tangent line t at $c = 3\pi/2$ and secant line s passing through the origin and $(2\pi, \pi)$, each with slope $1/2$.

Then there is a point $c \in (a, b)$ where

$$f'(c) = \frac{f(b) - f(a)}{b - a}. \tag{5.5.32}$$

Scratch Work 5.5.10: Reduce to a special case

The only difference between the hypotheses for Rolle's Theorem 5.5.7 and the Mean Value Theorem 5.5.9 is the condition that $f(a) = f(b)$. While this condition does not hold in general under the hypotheses of the Mean Value Theorem 5.5.9, we can shift such an f in as we did in the Scratch Work 5.5.6 for proof of Darboux's Theorem 5.5.5: Subtract a line from f to define a new function g satisfying the same conditions as f as well as $g(a) = g(b)$.

The secant line through the points $(a, f(a))$ and $(b, f(b))$ gives us the shift we're looking for. It is given by

$$s(x) = \left(\frac{f(b) - f(a)}{b - a} \right)(x - a) + f(a). \tag{5.5.33}$$

Then we have

$$s(a) = \left(\frac{f(b) - f(a)}{b - a}\right)(a - a) + f(a) = f(a) \qquad \text{and} \tag{5.5.34}$$

$$s(b) = \left(\frac{f(b) - f(a)}{b - a}\right)(b - a) + f(a) = f(b) - f(a) + f(a) = f(b). \tag{5.5.35}$$

So, choosing $g(x) = f(x) - s(x)$ yields

$$g(a) = f(a) - s(a) = 0 = f(b) - s(b) = g(b). \tag{5.5.36}$$

From here, Rolle's Theorem produces an interior point $c \in (a, b)$ where

$$g'(c) = 0, \quad \text{and so} \quad f'(c) = s'(c) = \frac{f(b) - f(a)}{b - a}. \tag{5.5.37}$$

Proof of the Mean Value Theorem 5.5.9. Suppose $f : [a, b] \to \mathbb{R}$ where f is continuous on $[a, b]$ and f is differentiable on (a, b). Define the secant line $s : [a, b] \to \mathbb{R}$ by

$$s(x) = \left(\frac{f(b) - f(a)}{b - a}\right)(x - a) + f(a). \tag{5.5.38}$$

By the continuity and derivatives of lines from Theorem 4.3.9 and Lemma 5.3.2, we have s is continuous on $[a, b]$, s is differentiable on (a, b),

$$s(a) = f(a), \quad s(b) = f(b), \quad \text{and} \quad s'(c) = \frac{f(b) - f(a)}{b - a}. \tag{5.5.39}$$

Next, define $g : [a, b] \to \mathbb{R}$ by

$$g(x) = f(x) - s(x). \tag{5.5.40}$$

Then we have

$$g(a) = f(a) - s(a) = 0 = f(b) - s(b) = g(b). \tag{5.5.41}$$

Furthermore, by the linearity of continuity and differentiation (Theorems 4.5.5 and 5.4.1), we have g is continuous on $[a, b]$ and differentiable on (a, b). So by Rolle's Theorem 5.5.7 applied to g, there is an interior point $c \in (a, b)$ where we have

$$g'(c) = 0 = f'(c) - s'(c) = f'(c) - \frac{f(b) - f(a)}{b - a}. \tag{5.5.42}$$

Therefore,

$$f'(c) = \frac{f(b) - f(a)}{b - a}. \tag{5.5.43}$$

$\square$

The Mean Value Theorem 5.5.9 plays a central role in calculus. In particular, it provides a key step in the proof of the Fundamental Theorem of Calculus I (Theorem 6.1.15) which tells us how evaluate integrals when we have a nice *antiderivative*. The Mean Value Theorem 5.5.9 also justifies some rules from calculus about antiderivatives, including the pair of corollaries that round out the section.

Definition 5.5.11: Antiderivative

Suppose $I \subseteq \mathbb{R}$ is an interval and $f, F : I \to \mathbb{R}$. The function F is an *antiderivative* of f if

$$F'(x) = f(x) \quad \text{for all} \quad x \in I. \tag{5.5.44}$$

An identically zero derivative implies the original function is constant. In other words, constants are the antiderivatives of the zero derivative.

Corollary 5.5.12: Zero derivative implies constant

Suppose $I \subseteq \mathbb{R}$ is an interval, $f : I \to \mathbb{R}$ is differentiable, and $f'(x) = 0$ for all $x \in I$. Then f is constant.

Proof of Corollary 5.5.12. Suppose $I \subseteq \mathbb{R}$ is an interval, $f : I \to \mathbb{R}$ is differentiable, and $f'(x) = 0$ for all $x \in I$. Without loss of generality, also suppose $a, b \in I$ where $a < b$. Then $[a, b] \subseteq I$ is a compact interval and f is differentiable on $[a, b]$. By the Mean Value Theorem 5.5.9, there is a point $c \in (a, b)$ where

$$f'(c) = 0 = \frac{f(b) - f(a)}{b - a}. \tag{5.5.45}$$

Multiplying by $b - a$ yields

$$0 = f(b) - f(a) \qquad \Longleftrightarrow \qquad f(a) = f(b). \tag{5.5.46}$$

Since a and b are arbitrary, f is constant. $\qquad\qquad\qquad\qquad\qquad\qquad\qquad\qquad\qquad\square$

Antiderivatives are not unique since any pair of functions with the same derivative across an interval are within a constant of each other on that interval. This idea justifies the need to include "$+C$" when we compute antiderivatives.

Corollary 5.5.13: Antiderivatives of a function differ by constants

Suppose $I \subseteq \mathbb{R}$ is an interval, $f, g : I \to \mathbb{R}$ are differentiable, and $f'(x) = g'(x)$ for all $x \in I$. Then for some constant $C \in \mathbb{R}$ we have

$$f(x) = g(x) + C \tag{5.5.47}$$

for all $x \in I$.

Proof of Corollary 5.5.13. Suppose $I \subseteq \mathbb{R}$ is an interval, $f, g : I \to \mathbb{R}$ are differentiable, and $f'(x) = g'(x)$ for all $x \in I$. Define $h : I \to \mathbb{R}$ by $h(x) = f(x) - g(x)$. Then by the linearity of differentiation (Theorem 5.4.1), h is differentiable and we have

$$h'(x) = f'(x) - g'(x) = 0 \tag{5.5.48}$$

for all $x \in I$. So by Corollary 5.5.12, there is some constant $C \in \mathbb{R}$ where we have

$$h(x) = f(x) - g(x) = C \tag{5.5.49}$$

for all $x \in I$. Adding C yields the desired expression $f(x) = g(x) + C$. $\qquad \square$

The next result solidifies even more notions from calculus regarding *increasing, decreasing,* and *monotone* functions.

Definition 5.5.14: Increasing, decreasing, and monotone functions

Suppose $I \subseteq \mathbb{R}$ is an interval and $f : I \to \mathbb{R}$.

(i) f is *increasing* if $x, y \in I$ with $x < y$ implies $f(x) \leq f(y)$.

(ii) f is *decreasing* if $x, y \in I$ with $x < y$ implies $f(x) \geq f(y)$.

(iii) f is *strictly increasing* if $x, y \in I$ with $x < y$ implies $f(x) < f(y)$.

(iv) f is *strictly decreasing* if $x, y \in I$ with $x < y$ implies $f(x) > f(y)$.

A function is *(strictly) monotone* if it is (strictly) increasing or (strictly) decreasing.

Corollary 5.5.15: Positive derivative implies increasing, negative derivative implies decreasing

Suppose $I \subseteq \mathbb{R}$ is an interval and $f : I \to \mathbb{R}$ is differentiable.

(i) If $f'(c) \geq 0$ for every $c \in I$, then f is increasing.

(ii) If $f'(c) \leq 0$ for every $c \in I$, then f is decreasing.

(iii) If $f'(c) > 0$ for every $c \in I$, then f is strictly increasing.

(iv) If $f'(c) < 0$ for every $c \in I$, then f is strictly decreasing.

Scratch Work 5.5.16: Get to difference quotients

Proofs of all four cases in Corollary 5.5.15 are similar to each other. Once we have one, we can modify its proof to get the others. With this in mind, the focus of this scratch work and the following proof is on the increasing case.

We would like to end up with the implication

$$x < y \qquad \Longrightarrow \qquad f(x) \le f(y), \tag{5.5.50}$$

by somehow making use of difference quotients like

$$\frac{f(y) - f(x)}{y - x} \tag{5.5.51}$$

which will connect us to the values of the derivative of f across I. Since $x < y$ if and only if $y - x > 0$, knowing the sign (positive/negative) of the difference quotient (5.5.51) will tell us the sign of the numerator $f(y) - f(x)$. For instance, we have

$$\frac{f(y) - f(x)}{y - x} \ge 0 \qquad \Longleftrightarrow \qquad f(y) - f(x) \ge 0. \tag{5.5.52}$$

Now for the subtle part: The Mean Value Theorem 5.5.9 applied to the compact interval $[x, y]$ tells us that for some $c_0 \in (x, y)$, we have

$$f'(c_0) = \frac{f(y) - f(x)}{y - x}. \tag{5.5.53}$$

So, by assuming $f'(c) \ge 0$ for all $c \in I$ and reversing the discussion here, we can show f is increasing.

Proof of Corollary 5.5.15. This proof only addresses the increasing case (i) in Corollary 5.5.15.

Suppose $I \subseteq \mathbb{R}$ is an interval, $f : I \to \mathbb{R}$ is differentiable, and $f'(c) \ge 0$ for every $c \in I$. Also, suppose $x, y \in I$ where $x < y$, and so $y - x > 0$. By the Mean Value Theorem 5.5.9 applied to the compact interval $[x, y]$, there is some $c_0 \in (x, y)$ where

$$f'(c_0) = \frac{f(y) - f(x)}{y - x} \ge 0. \tag{5.5.54}$$

Since $y - x > 0$, multiplying both sides by $y - x$ yields

$$f(y) - f(x) \ge 0 \qquad \Longleftrightarrow \qquad f(x) \le f(y). \tag{5.5.55}$$

Therefore, f is increasing. $\square$

The next chapter tackles the next big idea from calculus: *integrals.*

Exercises

5.5.1. Suppose $f : [0, 3] \to \mathbb{R}$ is given by $f(x) = \sqrt{x^2 - 3x}$. Prove f satisfies the conditions of Rolle's Theorem 5.5.7 and find $c \in (0, 3)$ where $g'(c) = 0$.

5.5.2. For each $n \in \mathbb{N}$, suppose $g_n : \mathbb{R} \to \mathbb{R}$ is given by

$$g_n(x) = \frac{1}{1 + x^{2n}}. \tag{5.5.56}$$

Prove g_n attains its maximum and determine both the maximum value and where the maximum is attained.

5.5.3. Suppose $h : [0, 4] \to \mathbb{R}$ is differentiable with $h(0) = 1$, $h(1) = 3$, and $h(4) = 3$. Prove:

  (i) there is some $c_1 \in (0, 4)$ where $h'(c_1) = 0$;

 (ii) there is some $c_2 \in (0, 4)$ where $h'(c_2) = 1/2$;   and

(iii) there is some $c_3 \in (0, 4)$ where $h'(c_3) = 1/3$.

5.5.4. Suppose $n \in \mathbb{N}$ and $0 \le x \le y$. Use the Mean Value Theorem 5.5.9 to show

$$nx^{n-1}(y - x) \le y^n - x^n \le ny^{n-1}(y - x). \tag{5.5.57}$$

5.5.5. Suppose $x > 0$. Use the Mean Value Theorem 5.5.9 to show

$$\sqrt{1 + x} < 1 + \frac{x}{2}. \tag{5.5.58}$$

5.5.6. Consider the function $g : \mathbb{R} \to \mathbb{R}$ given by

$$g(x) = \begin{cases} \dfrac{x}{2} + x^2 \sin\left(\dfrac{1}{x}\right), & \text{if } x \neq 0, \\ 0, & \text{if } x = 0. \end{cases} \tag{5.5.59}$$

  (i) Prove g is differentiable at $c = 0$ and $g'(0) > 0$.

 (ii) Prove g is differentiable on $\mathbb{R}$ and find a piecewise defined formula for g'.

(iii) In contrast to Corollary 5.5.15, prove g is not monotone on any open interval containing 0.

5.5.7. Suppose $f : (a, b) \to \mathbb{R}$ is differentiable and $f'(c) > 0$ at some $c \in (a, b)$. Prove there is some x_0 where

$$c < x_0 < b \qquad \text{and} \qquad f(x) > f(c). \tag{5.5.60}$$

Also, state and prove a similar result in the case where $f'(c) < 0$.

5.5.8. Use the Product Rule 5.4.6 to formally state and derive *Integration by Parts* for antiderivatives denoted by the "ultra-violet voodoo" formula from calculus:

$$\int u \, dv = uv - \int v \, du. \tag{5.5.61}$$

5.5.9. Use the Chain Rule 5.4.8 to formally state and derive the *Substitution* formula from calculus: If g is differentiable and $u = g(x)$, then

$$\int f(g(x))g'(x) \, dx = \int f(u) \, du. \tag{5.5.62}$$

5.5.10. Prove the following version of the *Cauchy Mean Value Theorem*: Suppose f and g are continuous on the compact interval $[a,b]$ and differentiable on the open interval (a,b). Then there is some $c \in (a,b)$ where

$$(f(b) - f(a))g'(c) = (g(b) - g(a))f'(c). \tag{5.5.63}$$

Additionally, if $g'(x) \neq 0$ for any $x \in [a,b]$, then

$$\frac{f(b) - f(a)}{g(b) - g(a)} = \frac{f'(c)}{g'(c)}. \tag{5.5.64}$$

Hint: Apply the Mean Value Theorem 5.5.9 to the function

$$h(x) = (f(b) - f(a))g(x) - (g(b) - g(a))f(x). \tag{5.5.65}$$

5.5.11. Prove the following version of the $0/0$ case of *L'Hospital's Rule*: Suppose $I \subseteq \mathbb{R}$ is an interval, $c \in I$, and $f,g : I \to \mathbb{R}$ are differentiable on I except possibly at c. Further suppose $f(c) = g(c) = 0$ and $g'(x) \neq 0$ for all $x \in I \backslash \{c\}$. Then

$$\lim_{x \to c} \frac{f'(x)}{g'(x)} = \ell \qquad \Longrightarrow \qquad \lim_{x \to c} \frac{f(x)}{g(x)} = \ell. \tag{5.5.66}$$

Chapter 6

Integration

Integrals allow us to formalize the notion of cumulative or net change, visualized by and defining "the area under a curve". Integration is another concept from calculus which can be defined using points arbitrarily close to sets.

The goal of this section is to develop a formal definition for integration based on families of linear combinations that approximate net change as closely as we like. Visually, the linear combinations represent sums of areas of rectangles which are versatile enough to be arbitrarily close to areas of more arbitrary shapes.

6.1   Defining and evaluating integrals

The formal definition for integrals (Definition 6.1.6) takes a while to build up. We need to carefully choose the terms and notation. The notion from calculus of "area under a curve" gives us a substantial visual and intuitive idea to build on, but integrals describe far more than just area.

The approach we take to defining the integral is like a squeeze theorem. At the core are approximations given by linear combinations roughly of the form

$$\text{estimate for an integral} = \sum_{k=1}^{n} \underbrace{(k\text{th height})(k\text{th width})}_{\text{area of } k\text{th rectangle}}, \tag{6.1.1}$$

where the "heights" depend on choices we make regarding outputs of a function and the "widths" depend on how we split up the domain. The value of the integral is attained by squeezing the linear combination estimates from above and below by choosing heights as upper and lower bounds on the outputs. The integral is the real number arbitrarily close to both the set of upper approximations and the set of lower approximations.

To get a more concrete idea of the approach, let's start with a rough approximation of the area of a triangle using two sets of three rectangles providing upper and lower approximations. The example itself may simple, but it is difficult and time-consuming to precisely describe the process we use to create our approximations. My hope is that by spending time working through a simple

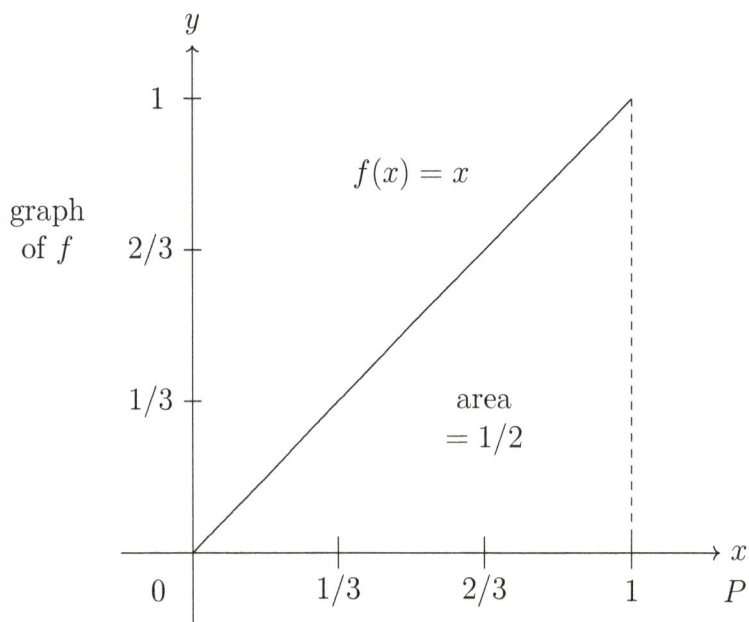

Figure 6.1.1: The triangle defined by $f(x) = x$ over the domain $[0, 1]$ has area $1/2$. See Example 6.1.1. The partition $P = \{0, 1/3, 2/3, 1\}$ splits the domain into three compact subintervals of length $1/3$ providing widths for approximations by rectangles.

example, the massive amount of notation and terminology used to define and prove results about integrals will make more sense.

Example 6.1.1: Three rectangles

Consider the function $f : [0, 1] \to \mathbb{R}$ given by

$$f(x) = x. \tag{6.1.2}$$

Geometry tells us the triangle created by the graph of f and the compact interval $[0, 1]$ on the x-axis has area $1/2$ (half the base times the height). See Figure 6.1.1.

To start building rectangles for our approximations, consider the set of endpoints

$$P = \{0, 1/3, 2/3, 1\} \subseteq [0, 1]. \tag{6.1.3}$$

The endpoints in P split the domain $[0, 1]$ into three subintervals of length $1/3$, namely

$$[0, 1/3], \quad [1/3, 2/3], \quad \text{and} \quad [2/3, 1]. \tag{6.1.4}$$

These lengths to provide widths for two approximations of the area of the triangle, one from above and another from below.

   To approximate from above, we can use the supremum of the image of each subinterval to provide the heights of the rectangles. By Definition 1.1.14, these suprema are arbitrarily close to the outputs of f and are upper bounds over their subintervals. (See Figure 6.1.2.) So, the

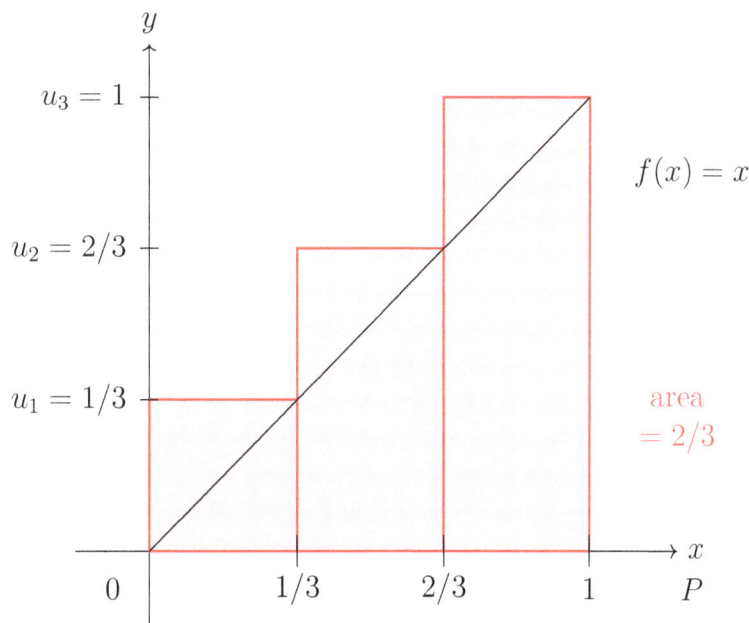

Figure 6.1.2: As in Example 6.1.1, the red rectangles use the suprema u_1, u_2, and u_3 as heights over the subintervals $[0,1]$, $[1/3,2/3]$, and $[2/3,1]$. The total area $2/3$ is an upper estimate for the area of the triangle as well as the integral of f over $[0,1]$.

suprema ensure the resulting approximation is at least pretty close to—but still an upper estimate for—the area of the triangle. We have

$$u_1 = \sup f([0,1/3]) = f(1/3) = 1/3, \tag{6.1.5}$$
$$u_2 = \sup f([1/3,2/3]) = f(2/3) = 2/3, \quad \text{and} \tag{6.1.6}$$
$$u_3 = \sup f([2/3,1]) = f(1) = 1. \tag{6.1.7}$$

Note that the supremum of the image of each subinterval is attained at the right endpoint since f is increasing. Combining these upper heights with the common width $1/3$ yields a total area given by the linear combination

$$\sum_{k=1}^{3} u_k(1/3) = (1/3)(1/3 + 2/3 + 1) = 2/3. \tag{6.1.8}$$

See Figure 6.1.2 where three red rectangles give us an upper estimate of $2/3$ for the area of the triangle.

Next, to approximate from below, we use the same endpoints and widths determined by the partition P, but this time we use the infimum of the image of each subinterval to provide the heights. (See Figure 6.1.3.) By Definition 1.1.14, these infima are arbitrarily close to the outputs of f and are lower bounds over their subintervals. We have

$$\ell_1 = \inf f([0,1/3]) = f(0) = 0, \tag{6.1.9}$$
$$\ell_2 = \inf f([1/3,2/3]) = f(1/3) = 1/3, \quad \text{and} \tag{6.1.10}$$
$$\ell_3 = \inf f([2/3,1]) = f(2/3) = 2/3. \tag{6.1.11}$$

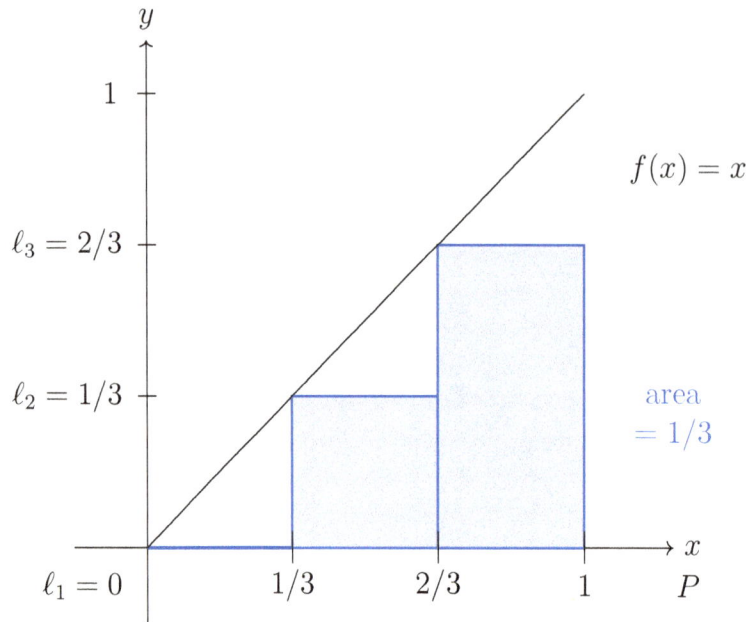

Figure 6.1.3: As in Example 6.1.1, the blue rectangles use the infima ℓ_1, ℓ_2, and ℓ_3 as heights over the subintervals $[0, 1/3], [1/3, 2/3]$, and $[2/3, 1]$. The total area $1/3$ is a lower estimate for the area of the triangle as well as the integral of f over $[0, 1]$.

Here, the infimum of the image of each subinterval is attained at the left endpoint. Combining these lower heights with the common width $1/3$ yields a total area given by the linear combination

$$\sum_{k=1}^{3} \ell_k(1/3) = (1/3)(0 + 1/3 + 2/3) = 1/3. \tag{6.1.12}$$

The three blue rectangles—well, two rectangles and one line segment—in Figure 6.1.3 show us the lower estimate of $1/3$ for the area of the triangle.

Remark 6.1.2: We can do better

In the follow-up to Example 6.1.1, we created two estimates for the area of a triangle using trios of rectangles. We ended up with an overestimate $2/3$ and an underestimate $1/3$ for the actual area $1/2$. We can do better by refining our process using more rectangles with smaller widths.

This refinement leads to a definition of the integral: Consider *all* upper and lower estimates defined in the same way as those in Example 6.1.1. There are lots of pieces to define before we put them together and nail down the integral.

Definition 6.1.3: Partitions and subintervals

For real numbers $a, b \in \mathbb{R}$ with $a < b$, a *partition* of the compact interval $[a, b]$ is a finite subset P given by

$$P = \{x_0, x_1, \ldots, x_n\} \quad \text{where} \quad a = x_0 < x_1 < \cdots < x_n = b \qquad (6.1.13)$$

for some $n \in \mathbb{N}$. Each partition of $[a, b]$ defines a finite collection of n compact subintervals given by

$$[x_{k-1}, x_k] \quad \text{for each} \quad k = 1, \ldots, n. \qquad (6.1.14)$$

Remark 6.1.4: Integrate over full intervals

Throughout this book, whenever we consider compact intervals of the form $[a, b]$ for the purposes of partitions and integrals, we work under the additional but unstated assumption that $a, b \in \mathbb{R}$ with $a < b$.

Note that each partition $P = \{x_0, x_1, \ldots, x_n\}$ of a compact interval $[a, b]$ contains the endpoints since $x_0 = a$ and $x_n = b$. Each partition also produces a set of widths of its subintervals given by

$$|x_k - x_{k-1}| = x_k - x_{k-1} \quad \text{for each} \quad k = 1, \ldots, n. \qquad (6.1.15)$$

We can drop the absolute values since the endpoints defined by the partition are taken to be increasing by default: $x_{k-1} < x_k$ for each $k = 1, \ldots, n$.

The next step is to build rectangles on these widths using upper and lower heights defined by suprema and infima. The resulting *upper* and *lower sums* are the linear combinations which represent the total areas of these rectangles:

$$\text{estimate} = \text{total area} = \sum_{k=1}^{n} \underbrace{(k\text{th height})(k\text{th width})}_{\text{area of } k\text{th rectangle}}. \qquad (6.1.16)$$

Each upper and lower sum provides an estimate for the *integral*.

To ensure the suprema and infima are there for us to work with, we only consider bounded functions. See Definition 4.3.20.

Definition 6.1.5: Upper and lower sums

Let $f : [a, b] \to \mathbb{R}$ be a bounded function and let

$$P = \{x_0, x_1, \ldots, x_n\} \qquad (6.1.17)$$

be a partition of the domain $[a, b]$. For each $k = 1, \ldots, n$, the *upper* and *lower heights*, respectively u_k and ℓ_k, are defined by

$$u_k = \sup f([x_{k-1}, x_k]) = \sup\{f(x) : x \in [x_{k-1}, x_k]\} \quad \text{and} \qquad (6.1.18)$$
$$\ell_k = \inf f([x_{k-1}, x_k]) = \inf\{f(x) : x \in [x_{k-1}, x_k]\}. \qquad (6.1.19)$$

Figure 6.1.4: To accompany Definitions 6.1.5 and 6.1.6, play around with the GeoGebra activity "Copy of upper and lower Riemann Sums" accessed through the QR code. Note that the color scheme in the GeoGebra activity differs from the figures in this section. https://www.geogebra.org/m/nz84uerx

From there, define the *upper* and *lower sums* by the linear combinations

$$u(f, P) = \sum_{k=1}^{n} u_k(x_k - x_{k-1}) \quad \text{and} \tag{6.1.20}$$

$$\ell(f, P) = \sum_{k=1}^{n} \ell_k(x_k - x_{k-1}), \quad \text{respectively.} \tag{6.1.21}$$

The sets of upper and lower sums over all partitions of $[a, b]$ are denoted by U and L, respectively. We have

$$U = \{u(f, P) : P \text{ is a partition of } [a, b]\} \quad \text{and} \tag{6.1.22}$$
$$L = \{\ell(f, P) : P \text{ is a partition of } [a, b]\}. \tag{6.1.23}$$

See Figure 6.1.4.

When it exists, the *integral* of f over $[a, b]$ is the real number which can be estimated as closely as we like by both upper and lower sums.

Definition 6.1.6: Integral

A function $f : [a, b] \to \mathbb{R}$ is *integrable* if f is bounded and there is a real number denoted by $\int_a^b f$ arbitrarily close to both the set of upper sums U and the set of lower sums L. In this case, $\int_a^b f$ is called the *integral of f over* $[a, b]$, f is called the *integrand*, and we have both

$$\int_a^b f \text{ acl } U \quad \text{and} \quad \int_a^b f \text{ acl } L.$$

See Figure 6.1.4.

Remark 6.1.7: Integral notation and terminology

We will also denote the integral of f over $[a, b]$ using a dummy variable x and its "differential" dx, as follows:

$$\int_a^b f = \int_a^b f(x)\, dx. \tag{6.1.24}$$

By the way, the elongated "s" symbols $\int$ and $\int$ both represent the integral as a type of infinite sum. There is no need to formally define the "differential" dx. Still, the notation $\int_a^b f(x)\, dx$ can roughly be thought of as an aggregate of the products of heights $f(x)$ and infinitesimally small widths dx over the interval $[a, b]$, as long as we do not rely on this analogy too much.

Since the outputs $f(x)$ can be negative, it's somewhat of a misnomer to interpret integrals as areas. On the other hand, the visualization of sums of products of real numbers as *signed* areas can lead to interesting insights and interplay: Since the product of velocity and time yields distance, we can think of distance as area.

The following definition is here for convenience.

Definition 6.1.8: Integral over a singleton, swap limits of integration

Let $f : [a, b] \to \mathbb{R}$ be an integrable function. Define

$$\int_b^a f = -\int_a^b f \quad\text{and}\quad \int_c^c f = 0 \quad\text{for } c \in [a, b]. \tag{6.1.25}$$

Let's revisit Example 6.1.1 and see if our definition for the integral (Definition 6.1.6) correctly reproduces the area $1/2$ for the triangle defined by $f(x) = x$ over $[0, 1]$.

Example 6.1.9: Recover the area of a triangle

Let $f : [0, 1] \to \mathbb{R}$ be defined by $f(x) = x$ as in Example 6.1.1. We have

$$\int_0^1 f = \int_0^1 x\, dx = \frac{1}{2}. \tag{6.1.26}$$

To prove this, it suffices to show $1/2$ is arbitrarily close to both the set of upper sums U and the set of lower sums L. We can refine the estimates following Example 6.1.1 to help us build enough upper and lower sums to show both

$$\frac{1}{2} \operatorname{acl} U \quad\text{and}\quad \frac{1}{2} \operatorname{acl} L. \tag{6.1.27}$$

The partition $P = \{0, 1/3, 2/3, 1\}$ produces upper and lower sums

$$u(f, P) = \sum_{k=1}^3 u_k \left(\frac{1}{3}\right) = \frac{2}{3} \quad\text{and}\quad \ell(f, P) = \sum_{k=1}^3 \ell_k \left(\frac{1}{3}\right) = \frac{1}{3}. \tag{6.1.28}$$

Both of these estimates differ from the true area 1/2 by 1/6. So, in order to find upper and lower sums as close as we like (arbitrarily close) to 1/2, we can create more, smaller rectangles by considering a sequence of finer and finer partitions.

The proof is long.

Proof for Example 6.1.9. Consider the partitions P_n of $[0,1]$ given by

$$P_n = \left\{0, \frac{1}{n}, \frac{2}{n}, \ldots, \frac{n-1}{n}, 1\right\} \quad \text{for each} \quad n \in \mathbb{N}. \tag{6.1.29}$$

The subintervals defined by each partition P_n are given by

$$[x_{k-1}, x_k] = \left[\frac{k-1}{n}, \frac{k}{n}\right] \quad \text{for each} \quad k = 1, \ldots, n. \tag{6.1.30}$$

Each of these subintervals has the same width, specifically

$$x_k - x_{k-1} = \frac{k}{n} - \frac{k-1}{n} = \frac{1}{n}. \tag{6.1.31}$$

From here, we can construct new sequences of upper and lower heights and sums. For now, let's focus on the lower heights and sums. The upper heights and sums are developed in a similar way.

Since $f(x) = x$ is increasing, for each $k = 1, \ldots, n$, the lower height ℓ_k is the output of the left endpoint:

$$\ell_k = \inf f([x_{k-1}, x_k]) = f(x_{k-1}) = \frac{k-1}{n}. \tag{6.1.32}$$

Hence, the lower sum $\ell(f, P_n)$ is given by

$$\ell(f, P_n) = \sum_{k=1}^{n} \ell_k(x_k - x_{k-1}) \tag{6.1.33}$$

$$= \sum_{k=1}^{n} \left(\frac{k-1}{n}\right)\left(\frac{1}{n}\right) \tag{6.1.34}$$

$$= \frac{1}{n^2} \sum_{k=1}^{n} (k-1) \tag{6.1.35}$$

$$= \frac{1}{n^2} (0 + 1 + \cdots + (n-1)). \tag{6.1.36}$$

Some classic mathematics helps us here: Carl Gauss' formula for the sum of the first n consecutive positive integers gives us a nice closed form. We have

$$\sum_{k=1}^{n} k = 1 + 2 + \cdots + n = \frac{n(n+1)}{2}. \tag{6.1.37}$$

See Figure 6.1.5 for a visual justification of this formula in the case where $n = 3$. For the lower sum $\ell(f, P_n)$, the sum of integers stops at $n - 1$. Therefore,

$$\ell(f, P_n) = \frac{1}{n^2}(1 + \cdots + (n-1)) = \frac{(n-1)n}{2n^2} = \frac{1}{2} - \frac{1}{2n}. \tag{6.1.38}$$

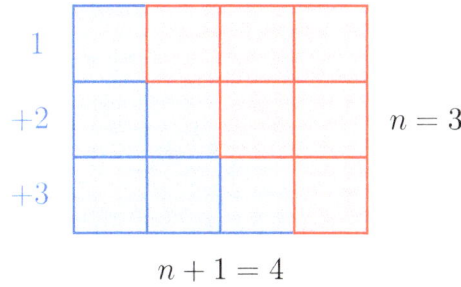

Figure 6.1.5: A visual justification of Gauss' formula for the sum of the first $n = 3$ positive integers. Two copies of the sum make up half of a rectangle whose sides are $n = 3$ and $n + 1 = 4$. So, each sum is equal to $n(n+1)/2 = 6$.

Also, since $\ell(f, P_n) \in L$ for every index $n \in \mathbb{N}$ and

$$\lim_{n \to \infty} \ell(f, P_n) = \lim_{n \to \infty} \left(\frac{1}{2} - \frac{1}{2n} \right) = \frac{1}{2}, \tag{6.1.39}$$

the fundamental connection between arbitrarily close and limits of sequences (Theorem 2.3.1) tells us $(1/2) \operatorname{acl} L$.

The argument to show $(1/2) \operatorname{acl} U$ uses the same partitions

$$P_n = \left\{ 0, \frac{1}{n}, \frac{2}{n}, \ldots, \frac{n-1}{n}, 1 \right\} \quad \text{for each} \quad n \in \mathbb{N}, \tag{6.1.40}$$

yielding the same common width

$$x_k - x_{k-1} = \frac{k}{n} - \frac{k-1}{n} = \frac{1}{n} \quad \text{for each} \quad k = 1, \ldots, n. \tag{6.1.41}$$

Now, since $f(x) = x$ is increasing, the upper height u_k is the output of the right endpoint:

$$u_k = \sup f([x_{k-1}, x_k]) = f(x_k) = \frac{k}{n}. \tag{6.1.42}$$

Hence, the upper sum $u(f, P_n)$ is given by

$$u(f, P_n) = \sum_{k=1}^{n} u_k (x_k - x_{k-1}) \tag{6.1.43}$$

$$= \sum_{k=1}^{n} \left(\frac{k}{n} \right) \left(\frac{1}{n} \right) \tag{6.1.44}$$

$$= \frac{1}{n^2} \sum_{k=1}^{n} k \tag{6.1.45}$$

$$= \frac{1}{n^2} (1 + \cdots + n). \tag{6.1.46}$$

A direct application of Carl Gauss' formula (6.1.37) for the sum of the first n consecutive positive integers gives us

$$u(f, P_n) = \frac{1}{n^2} (1 + \cdots + n) = \frac{n(n+1)}{2n^2} = \frac{1}{2} + \frac{1}{2n}. \tag{6.1.47}$$

Also, since $u(f, P_n) \in U$ for every index $n \in \mathbb{N}$ and

$$\lim_{n \to \infty} u(f, P_n) = \lim_{n \to \infty} \left(\frac{1}{2} + \frac{1}{2n} \right) = \frac{1}{2}, \tag{6.1.48}$$

the fundamental connection between arbitrarily close and limits of sequences (Theorem 2.3.1) tells us $(1/2)\,\mathrm{acl}\,U$.

Finally, since we have shown both $(1/2)\,\mathrm{acl}\,U$ and $(1/2)\,\mathrm{acl}\,L$, by the definition of the integral (Definition 6.1.6) we finally have

$$\int_0^1 f = \int_0^1 x\,dx = \frac{1}{2}. \tag{6.1.49}$$

$\square$

Working directly with the definition of the integral (Definition 6.1.6) is a tremendous challenge, even when the function under consideration is constant. The following lemma tells us the integral of a constant is the (signed) area of a rectangle.

Lemma 6.1.10: Integral of a constant

Let $f(x) = c$ on $[a, b]$ for some $c \in \mathbb{R}$. Then

$$\int_a^b f = \int_a^b c\,dx = c(b - a). \tag{6.1.50}$$

Scratch Work 6.1.11: Draw stuff

The idea for the proof follows pretty quickly from drawing a figure, which you are encouraged to do. Since f is constant, the upper and lower heights are exactly the same regardless of which partition we choose for the domain $[a, b]$. This means all upper and lower sums are the same as well.

Proof of Lemma 6.1.10. Let $f(x) = c$ on $[a, b]$ for some $c \in \mathbb{R}$ and let P be a partition of $[a, b]$ be given by

$$P = \{x_0, x_1, \ldots, x_n\} \tag{6.1.51}$$

where $n \in \mathbb{N}$, $x_0 = a$, and $x_n = b$. For each $k = 1, \ldots, n$, the upper height u_k and lower height ℓ_k are both equal to c since we have

$$u_k = \sup f([x_{k-1}, x_k]) = \sup\{c\} = c \quad \text{and} \quad \ell_k = \inf f([x_{k-1}, x_k]) = \inf\{c\} = c. \tag{6.1.52}$$

Hence, the upper sum $u(f, P)$ is given by

$$u(f, P) = \sum_{k=1}^{n} u_k(x_k - x_{k-1}) \tag{6.1.53}$$

$$= c \sum_{k=1}^{n} (x_k - x_{k-1}) \tag{6.1.54}$$

$$= c((x_1 - x_0) + (x_2 - x_1) + \cdots \tag{6.1.55}$$

$$+ (x_{n-1} - x_{n-2}) + (x_n - x_{n-1})) \tag{6.1.56}$$

$$= c(x_n - x_0) \tag{6.1.57}$$

$$= c(b - a), \tag{6.1.58}$$

after cancellation of numerous terms between lines (6.1.55) and (6.1.56). (Sums with cancellation like this are referred to as *telescoping sums*, see Remark 6.1.12.) Following a nearly identical argument, we also have

$$\ell(f, P) = \sum_{k=1}^{n} \ell_k(x_k - x_{k-1}) = c(b - a). \tag{6.1.59}$$

Hence, for any partition P of the domain $[a, b]$, we have U and L are the same singleton given by

$$U = L = \{c(b - a)\}. \tag{6.1.60}$$

By Lemma 1.5.4, we have both $c(b - a) \operatorname{acl} U$ and $c(b - a) \operatorname{acl} L$. Therefore, by the definition of the integral (Definition 6.1.6) we have

$$\int_a^b f = \int_a^b c\,dx = c(b - a). \tag{6.1.61}$$

$$\square$$

Remark 6.1.12: Telescoping sums

Telescoping sums are linear combinations whose summands mostly cancel out as in lines (6.1.55) and (6.1.56) of the proof of Lemma 6.1.10. For linear combinations of vectors in Euclidean spaces, telescoping sums behave like this:

$$\sum_{k=1}^{n} (\mathbf{x}_k - \mathbf{x}_{k-1}) = (\mathbf{x}_1 - \mathbf{x}_0) + (\mathbf{x}_2 - \mathbf{x}_1) + (\mathbf{x}_3 - \mathbf{x}_2) + \cdots \tag{6.1.62}$$

$$+ (\mathbf{x}_{n-2} - \mathbf{x}_{n-3}) + (\mathbf{x}_{n-1} - \mathbf{x}_{n-2}) + (\mathbf{x}_n - \mathbf{x}_{n-1}) \tag{6.1.63}$$

$$= \mathbf{x}_n - \mathbf{x}_0. \tag{6.1.64}$$

Can you see how most terms cancel?

Telescoping sums occur frequently in the development of results on integration and series. In addition to Lemma 6.1.10 and Example 6.2.1, telescoping sums play a central role in the evaluation half of the Fundamental Theorem of Calculus 6.1.15 which tells us how to compute integrals with antiderivatives. This role is highlighted in the proof of Corollary 6.1.13, where the Mean Value Theorem 5.5.9 applies to an antiderivative and yields telescoping sums whose common value always lies between to the upper and lower sums.

Corollary 6.1.13: Bounds for the difference of an antiderivative

Suppose $f : [a, b] \to \mathbb{R}$ is bounded and F is an antiderivative of f where $F'(x) = f(x)$ for all $x \in [a, b]$. Then for any partition P of the domain $[a, b]$ we have

$$\ell(f, P) \le F(b) - F(a) \le u(f, P). \tag{6.1.65}$$

Scratch Work 6.1.14: Mean Value Theorem yields a telescoping sum

The conclusion of the Mean Value Theorem 5.5.9 tells us

$$f(c) = F'(c) = \frac{F(b) - F(a)}{b - a}, \tag{6.1.66}$$

which we can rearrange to get

$$f(c)(b - a) = F'(c)(b - a) = F(b) - F(a). \tag{6.1.67}$$

The difference $F(b) - F(a)$ is the bound we're looking for. The product $f(c)(b-a)$ is of the "height times width" form which we can think of as an area, like the summands in an upper sum $u(f, P)$ or lower sums $\ell(f, P)$. The subtle step of applying the Mean Value Theorem 5.5.9 to each subinterval defined by the partition P produces a telescoping sum for the areas $f(c_k)(x_k - x_{k-1})$, like this

$$\sum_{k=1}^{3} f(c_k)(x_k - x_{k-1}) = \sum_{k=1}^{3} F(x_k) - F(x_{k-1}) \tag{6.1.68}$$
$$= F(x_1) - F(x_0) + F(x_2) - F(x_1) + F(x_3) - F(x_2) \tag{6.1.69}$$
$$= F(x_3) - F(x_0) \tag{6.1.70}$$
$$= F(b) - F(a). \tag{6.1.71}$$

See Figure 6.1.6.

Proof of Corollary 6.1.13. Suppose $F'(x) = f(x)$ for all $x \in [a, b]$ and let P be a partition of $[a, b]$ given by $P = \{x_0, x_1, \dots, x_n\}$ where $n \in \mathbb{N}$ and

$$a = x_0 < x_1 < \dots < x_{n-1} < x_n = b. \tag{6.1.72}$$

Then for each $k = 1, \dots, n$, the Mean Value Theorem 5.5.9 applied to the antiderivative F on the subinterval $[x_{k-1}, x_k] \subseteq [a, b]$ produces an input $c_k \in [x_{k-1}, x_k]$ such that

$$f(c_k)(x_k - x_{k-1}) = F'(c_k)(x_k - x_{k-1}) = F(x_k) - F(x_{k-1}). \tag{6.1.73}$$

Each product $f(c_k)(x_k - x_{k-1})$ can be visualized as the area of a rectangle. See Figure 6.1.6.

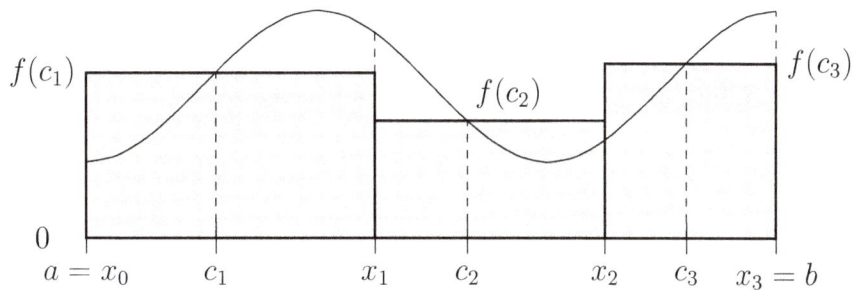

Figure 6.1.6: Three rectangles with areas given by $f(c_k)(x_k - x_{k-1})$ for $k = 1, 2, 3$. The input c_k is provided by an application of the Mean Value Theorem 5.5.9 to an antiderivative F on each subinterval $[x_{k-1}, x_k] \subseteq [a, b]$, as in the scratch work and proof for Corollary 6.1.13. The area $f(c_k)(x_k - x_{k-1})$ of the kth rectangle seems to be *exactly* the area under f between x_{k-1} and x_k. Their sum telescopes and equals $F(b) - F(a)$.

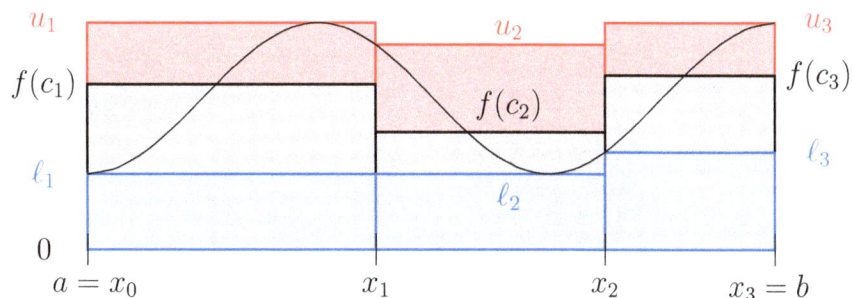

Figure 6.1.7: To accompany Corollary 6.1.13, here is a visual comparison of an upper sum, a lower sum, and a telescoping sum provided by applications of the Mean Value Theorem 5.5.9 to an antiderivative F. Since $\ell_k \leq f(c_k) \leq u_k$ on each subinterval $[x_{k-1}, x_k]$, the areas satisfy $\ell_k(x_k - x_{k-1}) \leq f(c_k)(x_k - x_{k-1}) \leq u_k(x_k - x_{k-1})$ and their sums give us $\ell(f, P) \leq F(b) - F(a) \leq u(f, P)$.

Since f is bounded, the upper height $u_k = \sup f([x_{k-1}, x_k])$ and lower height $\ell_k = \inf f([x_{k-1}, x_k])$ exist for each $k = 1, \ldots, n$. Also, the supremum is an upper bound and the infimum is a lower bound, so we have

$$\ell_k \leq f(c_k) \leq u_k. \tag{6.1.74}$$

Since $x_k - x_{k-1} > 0$, combining (6.1.73) and (6.1.74) gives us

$$\ell_k(x_k - x_{k-1}) \leq f(c_k)(x_k - x_{k-1}) = F(x_k) - F(x_{k-1}) \leq u_k(x_k - x_{k-1}). \tag{6.1.75}$$

See Figure 6.1.7 for a visual comparison as areas of rectangles. By the definition of upper and lower sums (Definition 6.1.5), taking the sum from $k = 1$ to n yields

$$\ell(f, P) = \sum_{k=1}^{n} \ell_k(x_k - x_{k-1}) \tag{6.1.76}$$

$$\leq \sum_{k=1}^{n} f(c_k)(x_k - x_{k-1}) \tag{6.1.77}$$

$$= \sum_{k=1}^{n} (F(x_k) - F(x_{k-1})) \tag{6.1.78}$$

$$\leq \sum_{k=1}^{n} u_k(x_k - x_{k-1}) \tag{6.1.79}$$

$$= u(f, P). \tag{6.1.80}$$

Again, see Figure 6.1.7. Since the sum (6.1.78) telescopes, $x_0 = a$, and $x_n = b$, we have

$$\sum_{k=1}^{n} f(c_k)(x_k - x_{k-1}) = \sum_{k=1}^{n} (F(x_k) - F(x_{k-1})) \tag{6.1.81}$$

$$= F(x_1) - F(x_0) + F(x_2) - F(x_1) + \cdots \tag{6.1.82}$$

$$+ F(x_{n-1}) - F(x_{n-2}) + F(x_n) - F(x_{n-1}) \tag{6.1.83}$$

$$= F(x_n) - F(x_0) \tag{6.1.84}$$

$$= F(b) - F(a). \tag{6.1.85}$$

Therefore,

$$\ell(f, P) \leq F(b) - F(a) \leq u(f, P). \tag{6.1.86}$$

$\square$

The Fundamental Theorem of Calculus tells us that derivatives and integrals are (roughly) inverses of each other. In this book, the theorem is stated in two parts. The Fundamental Theorem of Calculus I (Theorem 6.1.15) tells us how evaluate integrals when we have a nice antiderivative to work with. The Fundamental Theorem of Calculus II (Theorem 6.4.16) tells us that functions defined as indefinite integrals have derivatives given by the integrands, but this is proven later after establishing more results about integrals.

Theorem 6.1.15: Fundamental Theorem of Calculus I

If $f : [a, b] \to \mathbb{R}$ is integrable and F is an antiderivative of f where we have $F'(x) = f(x)$ for all $x \in [a, b]$, then

$$\int_a^b f = F(b) - F(a). \tag{6.1.87}$$

Scratch Work 6.1.16: A satisfying result

The proof of this half of the Fundamental Theorem of Calculus is a consequence of the definition of the integral in terms of arbitrarily close (Definition 6.1.6), properties of points arbitrarily close to sets, and the Mean Value Theorem 5.5.9 through Corollary 6.1.13. On a personal note, this proof was a very satisfying discovery!

Proof of the Fundamental Theorem of Calculus I (Theorem 6.1.15). Suppose f is integrable over $[a, b]$, F is an antiderivative of f on $[a, b]$, and P is a partition of $[a, b]$. By the definition of the integral (Definition 6.1.6), f is bounded on $[a, b]$ and we have both

$$\int_a^b f \text{ acl } U \quad \text{and} \quad \int_a^b f \text{ acl } L.$$

Since f is bounded, Corollary 6.1.13 yields

$$\ell(f, P) \leq F(b) - F(a) \leq u(f, P). \tag{6.1.88}$$

In particular, $F(b) - F(a) \leq u(f, P)$ for every upper sum $u(f, P) \in U$ and $\int_a^b f \text{ acl } U$ while $\ell(f, P) \leq F(b) - F(a)$ for every lower sum $\ell(f, P) \in L$ and $\int_a^b f \text{ acl } L$. Since real numbers arbitrarily close to sets respect lower and upper bounds (Lemma 1.5.23), we have

$$\int_a^b f \leq F(b) - F(a) \leq \int_a^b f. \tag{6.1.89}$$

Therefore,

$$\int_a^b f = F(b) - F(a). \tag{6.1.90}$$

$\square$

The next section develops various equivalent criteria for integrability.

Exercises

6.1.1. Suppose $f : [a, b] \to \mathbb{R}$ is bounded. Prove f is integrable over $[a, b]$ if and only if there is a real number r and sequences (P_n) and (Q_n) of partitions of $[a, b]$ where

$$\lim_{n \to \infty} u(f, P_n) = r = \lim_{n \to \infty} \ell(f, Q_n). \tag{6.1.91}$$

In this case, $\int_a^b f = r$.

6.1.2. Use Definition 6.1.6 and the *sum of squares* formula

$$\sum_{k=1}^{n} k^2 = 1^2 + 2^2 + \cdots + n^2 = \frac{n(n+1)(2n+1)}{6} \tag{6.1.92}$$

to directly prove $\int_0^1 x^2 \, dx = \frac{1}{3}$.

6.1.3. Suppose $g : [0, 1] \to \mathbb{R}$ is integrable. Define the sequence of averages (a_n) by

$$a_n = \frac{1}{n} \sum_{k=1}^{n} g\left(\frac{k}{n}\right) \qquad \text{for each } n \in \mathbb{N}. \tag{6.1.93}$$

Prove $\lim_{n \to \infty} a_n = \int_0^1 g$. Hint: Use Exercise 6.1.1 along with the Squeeze Thereom for sequences 2.4.3.

6.1.4. A function $g : \mathbb{R} \to \mathbb{R}$ is *even* when $g(-x) = g(x)$ for all $x \in \mathbb{R}$. A function $h : \mathbb{R} \to \mathbb{R}$ is *odd* when $h(-x) = -h(x)$ for all $x \in \mathbb{R}$. Even and odd functions have special properties with integrals.

Suppose $f : [-a, a] \to \mathbb{R}$ is integrable.

(i) Prove that if f is odd, then $\int_{-a}^{a} f = 0$.

(ii) Prove that if f is even, then $\int_{-a}^{a} f = 2 \int_{0}^{a} f$.

6.1.5. A function $f : \mathbb{R} \to \mathbb{R}$ is *periodic* with *period $p > 0$* if

$$f(x + p) = f(x) \qquad \text{for all } x \in \mathbb{R}. \tag{6.1.94}$$

Suppose f is periodic and integrable over every compact interval. Prove

$$\int_0^p f = \int_a^{a+p} f \qquad \text{for all } a \in \mathbb{R}. \tag{6.1.95}$$

6.1.6. Integration was nearly chosen to be the first topic explored in this book after defining arbitrarily close in the real line (Definition 1.1.8). This exercise shows that our integral can be defined without supremum and infimum.

Revise the definitions of upper and lower heights and sums in Definition 6.1.5 as follows: Let $f : [a, b] \to \mathbb{R}$ be a bounded function and let

$$P = \{x_0, x_1, \ldots, x_n\} \tag{6.1.96}$$

be a partition of the domain $[a, b]$. For each $k = 1, \ldots, n$, an *upper* * *height* is an upper bound u_k^* on the subinterval $[x_{k-1}, x_k]$. Similarly, a *lower* * *height* is a lower bound ℓ_k^* on the subinterval $[x_{k-1}, x_k]$. So, for each $k = 1, \ldots, n$ and any $x \in [x_{k-1}, x_k]$ we have

$$\ell_k^* \leq f(x) \leq u_k^*. \tag{6.1.97}$$

From there, define the *upper** and *lower** *sums* by the linear combinations

$$u^*(f, P) = \sum_{k=1}^{n} u_k^*(x_k - x_{k-1}) \quad \text{and} \tag{6.1.98}$$

$$\ell^*(f, P) = \sum_{k=1}^{n} \ell_k^*(x_k - x_{k-1}), \quad \text{respectively.} \tag{6.1.99}$$

The sets of upper* and lower* sums over all partitions of $[a, b]$ are denoted by U^* and L^*, respectively. We have

$$U^* = \{u^*(f, P) : P \text{ is a partition of } [a, b]\} \quad \text{and} \tag{6.1.100}$$
$$L^* = \{\ell^*(f, P) : P \text{ is a partition of } [a, b]\}. \tag{6.1.101}$$

Finally, replace Definition 6.1.6 as follows: A function $f : [a, b] \to \mathbb{R}$ is **-integrable* if f is bounded and there is a real number denoted by $* \int_a^b f$ arbitrarily close to both the set of upper sums U and the set of lower sums L. In this case, $* \int_a^b f$ is called the **-integral of f over* $[a, b]$, and we have both

$$* \int_a^b f \text{ acl } U \quad \text{and} \quad * \int_a^b f \text{ acl } L. \tag{6.1.102}$$

Suppose $f : [a, b] \to \mathbb{R}$ is bounded. Prove the following statements are equivalent.

(i) f is integrable over $[a, b]$ as in Definition 6.1.6.

(ii) f is *-integrable over $[a, b]$ as in the above definition.

(iii) There is a real number r and sequences (P_n) and (Q_n) of partitions of $[a, b]$ where

$$\lim_{n \to \infty} u(f, P_n) = r = \lim_{n \to \infty} \ell(f, Q_n). \tag{6.1.103}$$

(See Exercise 6.1.1.)

(iv) There is a real number r and sequences (P_n^*) and (Q_n^*) of partitions of $[a, b]$ where

$$\lim_{n \to \infty} u^*(f, P_n^*) = r = \lim_{n \to \infty} \ell^*(f, Q_n^*). \tag{6.1.104}$$

In this case, $\int_a^b f = * \int_a^b f = r$.

6.2 Criteria for integrability

When are functions integrable, exactly? This section develops equivalent criteria for integrability after building some more tools suitable for the partitions, upper sums, and lower sums that define integrals.

First, not all functions are integrable.

> ### Example 6.2.1: Dirichlet's function is not integrable
>
> Consider Dirichlet's function $\mathbb{1}_{\mathbb{Q}}$ from Example 4.2.14 given by
>
> $$\mathbb{1}_{\mathbb{Q}}(x) = \begin{cases} 1, & \text{if } x \in \mathbb{Q}, \\ 0, & \text{if } x \in \mathbb{R}\setminus\mathbb{Q}. \end{cases} \tag{6.2.1}$$
>
> Dirichlet's function is not integrable over $[0, 2]$.

> ### Scratch Work 6.2.2: Upper and lower sums are away from each other
>
> The proof follows from the density of both the rationals and the irrationals in the real line (Theorem 1.4.10 and Corollary 1.4.13). Since every interval contains rational numbers, the upper heights of $\mathbb{1}_{\mathbb{Q}}$ are all equal to 1. On the other hand, since every interval contains irrational numbers, the lower heights are all 0. This forces the sets of upper and lower sums U and L to be away from each other.

Proof for Example 6.2.1. Suppose P is a partition of $[0, 2]$ given by

$$P = \{x_0, x_1, \ldots, x_n\} \tag{6.2.2}$$

where $n \in \mathbb{N}$, $x_0 = 0$, and $x_n = 2$. Since every interval contains rational numbers (Theorem 1.4.10), for each $k = 1, \ldots, n$ the upper height u_k is given by

$$u_k = \sup \mathbb{1}_{\mathbb{Q}}([x_{k-1}x_k]) = 1. \tag{6.2.3}$$

Since all the upper heights are the same, the upper sum $u(\mathbb{1}_{\mathbb{Q}}, P)$ telescopes and we have

$$u(\mathbb{1}_{\mathbb{Q}}, P) = \sum_{k=1}^{n} u_k(x_k - x_{k-1}) \tag{6.2.4}$$

$$= \sum_{k=1}^{n} 1(x_k - x_{k-1}) \tag{6.2.5}$$

$$= (x_1 - x_0) + (x_2 - x_1) + \cdots \tag{6.2.6}$$

$$\qquad + (x_{n-1} + x_{n-2}) + (x_n - x_{n-1}) \tag{6.2.7}$$

$$= x_n - x_0 \tag{6.2.8}$$

$$= 2. \tag{6.2.9}$$

Hence, the set of upper sums is the singleton $U = \{2\}$.

Similarly, every interval contains irrational numbers (Corollary 1.4.13), so for each $k = 1, \ldots, n$ the lower height ℓ_k is given by

$$\ell_k = \inf \mathbb{1}_{\mathbb{Q}}([x_{k-1}x_k]) = 0. \tag{6.2.10}$$

Since all the lower heights are 0, the lower sum $\ell(\mathbb{1}_{\mathbb{Q}}, P)$ is also 0. We have

$$\ell(\mathbb{1}_{\mathbb{Q}}, P) = \sum_{k=1}^{n} \ell_k(x_k - x_{k-1}) = \sum_{k=1}^{n} 0(x_k - x_{k-1}) = 0. \tag{6.2.11}$$

Hence, the set of lower sums is the singleton $L = \{0\}$. So we have both

$$0 \text{ awf } U \qquad \text{and} \qquad 2 \text{ awf } L. \tag{6.2.12}$$

By Lemma 1.5.5, no real number is arbitrarily close to both U and L. Therefore, Dirichlet's function $\mathbb{1}_{\mathbb{Q}}$ is not integrable over $[0, 2]$. $\qquad\qquad\square$

To establish criteria for integrability from various perspectives, an exploration of partitions, upper sums, and lower sums will help.

Partitions of intervals are a new concept in this chapter requiring techniques that go beyond what we have seen elsewhere in the textbook. *Refinements* of partitions decompose the domain in a way we can control, allowing us to bring the linear combinations that define upper and lower sums arbitrarily close together to get the integral.

Definition 6.2.3: Refinements of partitions

Given a partition P of a compact interval $[a, b]$, a *refinement* of P is a partition Q of $[a, b]$ where

$$P \subseteq Q. \tag{6.2.13}$$

Given an pair of partitions P_1 and P_2 of $[a, b]$, a *common refinement* of P_1 and P_2 is a partition R of $[a, b]$ where

$$P_1 \cup P_2 \subseteq R. \tag{6.2.14}$$

Remark 6.2.4: A perspective on refinements

Loosely speaking, refinements split our domains into more and finer subintervals by adding endpoints, improving the estimates given by linear combinations of "heights" and "widths" that define upper and lower sums in the definition of the integral (Definition 6.1.6). Refinements also pave the way to proving a number of results on integrals from calculus in a systematic way. In particular, refinements increase lower sums and decrease upper sums (Definition 6.1.5), and brings their upper and lower estimates closer together. Compare Figures 6.2.1 and 6.2.2 which consider the same bounded function over a partition P and a refinement Q.

Lemma 6.2.5: Refinements of upper and lower sums

Suppose $f : [a, b] \to \mathbb{R}$ is bounded, P is a partition of $[a, b]$, and Q is a refinement of P. Then

$$\ell(f, P) \leq \ell(f, Q) \qquad \text{and} \qquad u(f, Q) \leq u(f, P). \tag{6.2.15}$$

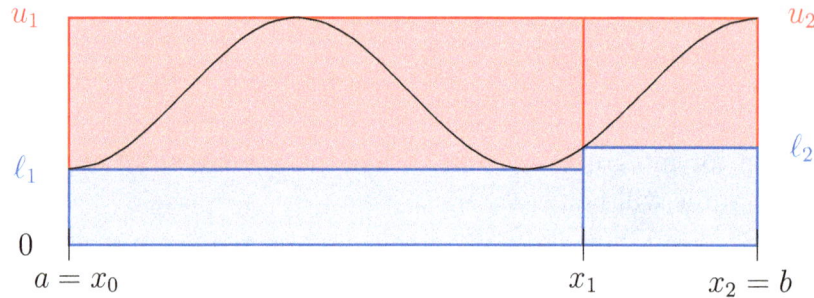

Figure 6.2.1: A bounded function on $[a, b]$ with partition $P = \{x_0, x_1, x_2\}$ along with upper and lower heights u_1, u_2, ℓ_1, and ℓ_2. This sets up a comparison with a refinement of P given by Q in Figure 6.2.2.

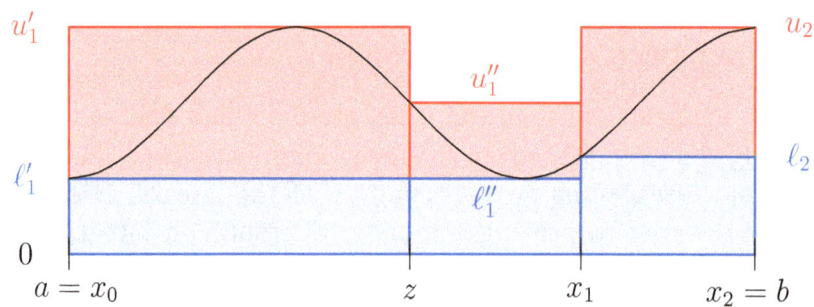

Figure 6.2.2: The same continuous function on $[a, b]$ from Figure 6.2.1, but with refinement $Q = \{x_0, z, x_1, x_2\}$ of partition $P = \{x_0, x_1, x_2\}$ along with the new upper and lower heights $u_1', u_1'', u_2, \ell_1', \ell_1''$, and ℓ_2. This exemplifies a comparison of the upper and lower sums over P and Q as in Lemma 6.2.5. Since $u_1 > u_1''$, we have $u(f, P) > u(f, Q)$. On the other hand, $\ell(f, P) = \ell(f, Q)$.

> **Scratch Work 6.2.6: Suprema and infima are the key**
>
> The order properties of suprema and infima in Corollary 1.4.4 play a key role in the proof. By adding points to get a refinement, we split our interval into *smaller* subintervals, meaning infima increase while suprema decrease. As a result, lower sums move up and upper sums move down. The proof concludes with an application of induction where the details are omitted.

Proof of Lemma 6.2.5. Suppose $f : [a, b] \to \mathbb{R}$ is bounded and P is a partition of $[a, b]$ given by

$$P = \{x_0, \ldots, x_n\}. \tag{6.2.16}$$

As an initial case, suppose Q is a refinement of P which adds a single point z where

$$x_{k-1} < z < x_k \tag{6.2.17}$$

for some $k \in \{1, \ldots, n\}$. So $Q = P \cup \{z\}$ and we have

$$[x_{k-1}, z] \subseteq [x_{k-1}, x_k] \quad \text{and} \quad [z, x_k] \subseteq [x_{k-1}, x_k]. \tag{6.2.18}$$

Since images under f respect containment, we also have

$$f([x_{k-1}, z]) \subseteq f([x_{k-1}, x_k]) \quad \text{and} \quad f([z, x_k]) \subseteq f([x_{k-1}, x_k]). \tag{6.2.19}$$

By the order properties of suprema and infima (Corollary 1.4.4), it follows that

$$u_k' = \sup f([x_{k-1}, z]) \leq \sup f([x_{k-1}, x_k]) = u_k, \tag{6.2.20}$$
$$u_k'' = \sup f([z, x_k]) \leq \sup f([x_{k-1}, x_k]) = u_k, \tag{6.2.21}$$
$$\ell_k' = \inf f([x_{k-1}, z]) \geq \inf f([x_{k-1}, x_k]) = \ell_k, \quad \text{and} \tag{6.2.22}$$
$$\ell_k'' = \inf f([z, x_k]) \geq \inf f([x_{k-1}, x_k]) = \ell_k. \tag{6.2.23}$$

Since $x_k - x_{k-1} = (x_k - z) + (z - x_{k-1}) > 0$, we have

$$u_k'(z - x_{k-1}) + u_k''(x_k - z) \leq u_k(z - x_{k-1}) + u_k(x_k - z) = u_k(x_{k-1} - x_k) \tag{6.2.24}$$
$$\implies \quad u_k'(z - x_{k-1}) + u_k''(x_k - z) - u_k(x_{k-1} - x_k) \leq 0, \tag{6.2.25}$$

as well as

$$\ell_k'(z - x_{k-1}) + \ell_k''(x_k - z) \geq \ell_k(z - x_{k-1}) + \ell_k(x_k - z) = \ell_k(x_{k-1} - x_k) \tag{6.2.26}$$
$$\implies \quad \ell_k'(z - x_{k-1}) + \ell_k''(x_k - z) - \ell_k(x_{k-1} - x_k) \geq 0. \tag{6.2.27}$$

Hence, replacing the kth term in the upper sum with the nonpositive value (6.2.25) yields

$$u(f, P) = \sum_{j=1}^{n} u_j(x_j - x_{j-1}) \tag{6.2.28}$$

$$\geq \sum_{j=1}^{n} u_j(x_j - x_{j-1}) \tag{6.2.29}$$

$$+ u_k'(z - x_{k-1}) + u_k''(x_k - z) - u_k(x_{k-1} - x_k) \tag{6.2.30}$$
$$= u(f, Q). \tag{6.2.31}$$

Similarly, replacing the kth term in the lower sum with the nonnegative value (6.2.27) yields

$$\ell(f, P) = \sum_{j=1}^{n} \ell_j(x_j - x_{j-1}) \tag{6.2.32}$$

$$\leq \sum_{j=1}^{n} \ell_j(x_j - x_{j-1}) \tag{6.2.33}$$

$$+ \ell_k'(z - x_{k-1}) + \ell_k''(x_k - z) - \ell_k(x_{k-1} - x_k) \tag{6.2.34}$$

$$= \ell(f, Q). \tag{6.2.35}$$

Therefore, the desired result (6.2.15) holds in the case where Q refines P by including a single new point. Since refinements are partitions themselves, they only add a finite number of new points to consider. Hence, the general case follows from an induction argument. □

More tools regarding upper and lower sums will prove to be helpful. Recall from Definition 6.1.5 that L represents the set of lower sums and U represents the set of upper sums.

Theorem 6.2.7: Lower sums are always below upper sums

Suppose P_1 and P_2 are partitions of $[a, b]$ and $f : [a, b] \to \mathbb{R}$ is bounded. Then

$$\ell(f, P_1) \leq u(f, P_2). \tag{6.2.36}$$

Furthermore, we have

$$\ell(f, P_1) \leq \sup L \leq \inf U \leq u(f, P_2). \tag{6.2.37}$$

Scratch Work 6.2.8: Make a common refinement

The common refinement of two partitions orders the upper and lower sums thanks to Lemma 6.2.5 to generate the desired result. It helps to consider the case where $P_1 = P_2$ first, then build the general result from there.

Once inequality (6.2.36) is established in the general case, the last three inequalities hold since an infimum is the greatest lower bound and a supremum is the least upper bound. See Theorems 1.4.3 and 1.3.10, respectively. Also, the Axiom of Completeness 1.3.8 and Theorem 1.4.1 ensure $\sup L$ and $\inf U$ exist.

Proof of Theorem 6.2.7. Suppose $f : [a, b] \to \mathbb{R}$ is bounded P_1 and P_2 are the same partition P of $[a, b]$ where

$$P_1 = P_2 = \{x_0, \ldots, x_n\} = P. \tag{6.2.38}$$

Since infima are lower bounds and suprema are upper bounds (Definition 1.1.14), for each $k = 1, \ldots, n$ and each $x \in [x_{k-1}, x_k]$ we have

$$\ell_k = \inf f([x_{k-1}, x_k]) \leq f(x) \leq \sup f([x_{k-1}, x_k]) = u_k. \tag{6.2.39}$$

So, comparing upper and lower sums (Definition 6.1.5) gives us

$$\ell(f, P) = \sum_{k=1}^{n} \ell_k(x_k - x_{k-1}) \leq \sum_{k=1}^{n} u_k(x_k - x_{k-1}) = u(f, P). \qquad (6.2.40)$$

Now suppose P_1 and P_2 are distinct partitions of $[a, b]$, so $P_1 \neq P_2$. Let Q be their common refinement given by $Q = P_1 \cup P_2$. Since $P_1 \subseteq Q$ and $P_2 \subseteq Q$, by Lemma 6.2.5 and inequality (6.2.40) we have

$$\ell(f, P_1) \leq \ell(f, Q) \leq u(f, Q) \leq u(f, P_2). \qquad (6.2.41)$$

Therefore, inequality (6.2.36) holds for any pair of partitions P_1 and P_2 of $[a, b]$.

Next, by temporarily fixing P_1, ignoring Q, and allowing P_2 to stand for any partition of $[a, b]$, inequality (6.2.41) tells us $\ell(f, P_1)$ is a lower bound for U, the set of upper sums of f over $[a, b]$. By Theorem 1.4.1, $\inf U$ exists. Since an infimum is the *greatest* lower bound by Theorem 1.4.3, we have

$$\ell(f, P_1) \leq \inf U = \inf \left\{ u(f, P_2) : P_2 \text{ is a partition of } [a, b] \right\}. \qquad (6.2.42)$$

Similarly, by temporarily fixing P_2, ignoring Q, and allowing P_1 to stand for any partition of $[a, b]$, inequality (6.2.41) tells us $u(f, P_2)$ is an upper bound for L, the set of lower sums of f over $[a, b]$. By the Axiom of Completeness 1.3.8, $\sup U$ exists. Since a supremum is the *least* upper bound by Theorem 1.3.10, we have

$$\sup L = \sup \left\{ \ell(f, P_1) : P_1 \text{ is a partition of } [a, b] \right\} \leq u(f, P_2). \qquad (6.2.43)$$

By allowing P_1 to stand for any partition of $[a, b]$, inequality (6.2.42) tells us $\inf U$ is an upper bound of L. Since $\sup L$ is the *least* upper bound by Theorem 1.3.10, we have

$$\sup L \leq \inf U. \qquad (6.2.44)$$

Finally, since a supremum is an upper bound and an infimum is a lower bound, we have the desired inequality

$$\ell(f, P_1) \leq \sup L \leq \inf U \leq u(f, P_2). \qquad (6.2.45)$$

$\square$

The following corollary of Theorem 6.2.7 is handy.

Corollary 6.2.9: Refinements improve estimates

Suppose P_1 and P_2 are partitions of $[a, b]$, Q is a refinement where $P_1 \subseteq Q$ and $P_2 \subseteq Q$, and $f : [a, b] \to \mathbb{R}$ is bounded. Then

$$0 \leq u(f, Q) - \ell(f, Q) \leq u(f, P_1) - \ell(f, P_2). \qquad (6.2.46)$$

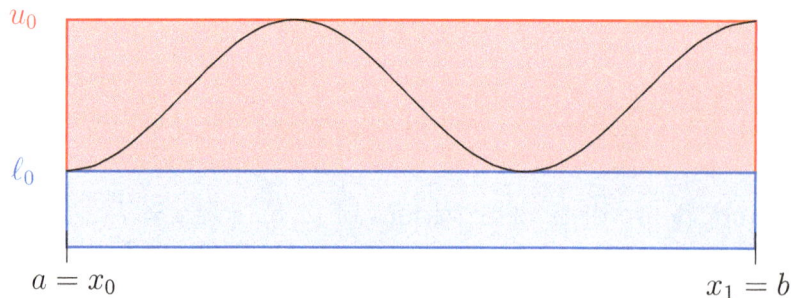

Figure 6.2.3: A bounded function on $[a, b]$ with the simplest partition $P_0 = \{x_0, x_1\}$ along with upper and lower heights u_0 and ℓ_0. All upper and lower sums of this function represent areas between the rectangles shown here. See Example 6.2.10.

Proof of Corollary 6.2.9. Suppose P_1 and P_2 are partitions of $[a, b]$, Q is a refinement where $P_1 \subseteq Q$ and $P_2 \subseteq Q$, and $f : [a, b] \to \mathbb{R}$ is bounded. By Theorem 6.2.7, we have

$$\ell(f, P_2) \leq \ell(f, Q) \leq u(f, Q) \leq u(f, P_1). \tag{6.2.47}$$

So by reorganizing as in property (vi) of Theorem 1.3.2 and Figure 1.3.1, and noting the differences are nonnegative, we get the desired result

$$0 \leq |u(f, Q) - \ell(f, Q)| \tag{6.2.48}$$
$$= u(f, Q) - \ell(f, Q) \tag{6.2.49}$$
$$\leq u(f, P_1) - \ell(f, P_2) \tag{6.2.50}$$
$$= |u(f, P_1) - \ell(f, P_2)|. \tag{6.2.51}$$

$\square$

A special case of Theorem 6.2.7 tells us that upper and lower sums of a bounded function are between rectangles.

Example 6.2.10: Between rectangles

Given a bounded function $f : [a, b] \to \mathbb{R}$, consider the partition P_0 given by

$$P_0 = \{x_0, x_1\} = \{a, b\}. \tag{6.2.52}$$

By Definition 6.1.3, every partition P of $[a, b]$ is a refinement of P_0 since we have $P_0 = \{a, b\} \subseteq P$. Now let $\ell_0 = \inf f([a, b])$ and $u_0 = \sup f([a, b])$. By Theorem 6.2.7, we have

$$\ell_0(b - a) = \ell(f, P_0) \leq \ell(f, P) \leq u(f, P) \leq u(f, P_0) = u_0(b - a). \tag{6.2.53}$$

See Figure 6.2.3.

The definition of integral in Definition 6.1.6 is not a classic one, but it is equivalent to ones that are such as the *Darboux integral.* Theorem 6.2.7 ensures the definition is sound.

Definition 6.2.11: Upper, lower, and Darboux integrals

Given a bounded function $f : [a, b] \to \mathbb{R}$, the *upper integral* $u(f)$ and *lower integral* $\ell(f)$ of f over $[a, b]$ are respectively defined by

$$u(f) = \inf U = \inf \{u(f, P) : P \text{ is a partition of } [a, b]\} \qquad \text{and} \qquad (6.2.54)$$

$$\ell(f) = \sup L = \sup \{\ell(f, P) : P \text{ is a partition of } [a, b]\}. \qquad (6.2.55)$$

The function f is *Darboux integrable* when $u(f) = \ell(f)$. In this case, the common value is called the *Darboux integral* of f.

The key result of the section provides four equivalent notions of integrability. (None of which happen to be the one originally developed by Bernhard Riemann.)

Theorem 6.2.12: Equivalent forms of integration

Suppose $f : [a, b] \to \mathbb{R}$ is bounded. Then the following are equivalent:

(i) f is integrable over $[a, b]$ as in Definition 6.1.6.

(ii) f is Darboux integrable over $[a, b]$, so $u(f) = \ell(f)$ as in Definition 6.2.11.

(iii) f is *sequentially integrable* over $[a, b]$:

$$\text{There is sequence } (P_n) \text{ of partitions of } [a, b] \text{ such that} \qquad (6.2.56)$$

$$\lim_{n \to \infty} (u(f, P_n) - \ell(f, P_n)) = 0. \qquad (6.2.57)$$

(iv) f satisfies the following *Cauchy criterion for integrability*:

$$\text{For every error } \varepsilon > 0, \qquad (6.2.58)$$

$$\text{there is a partition } P_\varepsilon \text{ of } [a, b] \text{ such that} \qquad (6.2.59)$$

$$u(f, P_\varepsilon) - \ell(f, P_\varepsilon) < \varepsilon. \qquad (6.2.60)$$

If any of these statements hold, then they all do and we have

$$\int_a^b f = u(f) = \ell(f). \qquad (6.2.61)$$

Remark 6.2.13: Comparing the different forms of integration

The definition of sequential limit (Definition 2.2.1) assumes we have a candidate for the limit $\mathbf{y}$ in mind. On the other hand, the definition of a Cauchy sequence (Definition 2.6.1) does not mention a candidate for the limit at all. Even so, the Cauchy Criterion (Theorem 2.6.5) tells us a sequence in a Euclidean space converges if and only if the sequence is Cauchy.

A similar comparison holds here. The notions of integrability in (i) Definition 6.1.6 and (ii) Darboux integrability in Definition 6.2.11 explicitly mention values of the integral as $\int_a^b f$ or $u(f)$ and $\ell(f)$, respectively. On the other hand, neither (iii) sequential integrability nor (iv) Cauchy criterion for integrability mentions a value of the integral, providing some useful flexibility. Still, the four notions do have something in common. Each codifies integrability in that upper and lower sums can be brought arbitrarily close together.

Scratch Work 6.2.14: A culmination of results and perspectives

It was fun to think about all the different ways to prove Theorem 6.2.12. Ultimately, the decision to run with the string of implications

$$\text{(i)} \implies \text{(iii)} \implies \text{(iv)} \implies \text{(ii)} \implies \text{(i)} \tag{6.2.62}$$

is due to the way results from throughout the book come together.

Proof of Theorem 6.2.12. Throughout the proof, suppose $f : [a, b] \to \mathbb{R}$ is bounded.

(i) $\implies$ (iii): Suppose $f : [a, b] \to \mathbb{R}$ integrable as in Definition 6.1.6. Recall from Definition 6.1.5 that the sets of upper sums U and lower sums L of f are given by

$$U = \{u(f, P) : P \text{ is a partition of } [a, b]\} \quad \text{and} \tag{6.2.63}$$
$$L = \{\ell(f, P) : P \text{ is a partition of } [a, b]\}. \tag{6.2.64}$$

Then the integral $\int_a^b f \in \mathbb{R}$ exists where both $\int_a^b f \operatorname{acl} L$ and $\int_a^b f \operatorname{acl} U$. By the fundamental connection between arbitrarily close and convergence (Theorem 2.3.1), there are sequences of partitions (P_n') and (P_n'') such that

$$\lim_{n \to \infty} u(f, P_n') = \int_a^b f \quad \text{and} \quad \lim_{n \to \infty} \ell(f, P_n'') = \int_a^b f. \tag{6.2.65}$$

For each index $n \in \mathbb{N}$, define Q_n to be the common refinement of P_n' and P_n'' so that $Q_n = P_n' \cup P_n''$. Since lower sums are below upper sums as in Lemma 6.2.5 and Theorem 6.2.7, for each $n \in \mathbb{N}$ we have

$$\ell(f, P_n'') \le \ell(f, Q_n) \le u(f, Q_n) \le u(f, P_n'). \tag{6.2.66}$$

By the Squeeze Theorem for sequences 2.4.3 (twice), we have

$$\lim_{n \to \infty} \ell(f, Q_n) = \int_a^b f = \lim_{n \to \infty} u(f, Q_n). \tag{6.2.67}$$

So by the linearity of sequential limits (Theorem 2.3.9), we have

$$\lim_{n \to \infty} (u(f, Q_n) - \ell(f, Q_n)) = \lim_{n \to \infty} u(f, Q_n) - \lim_{n \to \infty} \ell(f, Q_n) = \int_a^b f - \int_a^b f = 0. \tag{6.2.68}$$

Therefore, (i) $\implies$ (iii).

(iii) $\implies$ (iv): Suppose f is sequentially integrable in that there is a sequence (P_n) of partitions of $[a, b]$ such that

$$\lim_{n \to \infty} \left(u(f, Q_n) - \ell(f, Q_n) \right) = 0. \tag{6.2.69}$$

Since lower sums are below upper sums as in Corollary 6.2.9, for each index $n \in \mathbb{N}$ we have

$$0 \le u(f, P_n) - \ell(f, P_n) \implies |u(f, P_n) - \ell(f, P_n)| = u(f, P_n) - \ell(f, P_n). \tag{6.2.70}$$

Now let $\varepsilon > 0$. By the definition of sequential limit (Definition 2.2.1), there is a threshold $n_\varepsilon \in \mathbb{N}$ such that

$$u(f, P_{n_\varepsilon}) - \ell(f, P_{n_\varepsilon}) = |u(f, P_{n_\varepsilon}) - \ell(f, P_{n_\varepsilon})| < \varepsilon. \tag{6.2.71}$$

Therefore, (iii) $\implies$ (iv).

(iv) $\implies$ (ii): Suppose f satisfies the Cauchy criterion for integrability in that for every $\varepsilon > 0$, there is a partition P_ε of $[a, b]$ where

$$u(f, P_\varepsilon) - \ell(f, P_\varepsilon) < \varepsilon. \tag{6.2.72}$$

Since f is bounded on $[a, b]$, the proof of Theorem 6.2.7 ensures the existence of

$$u(f) = \inf U = \inf\{u(f, P) : P \text{ is a partition of } [a, b]\} \quad \text{and} \tag{6.2.73}$$
$$\ell(f) = \sup L = \sup\{\ell(f, P) : P \text{ is a partition of } [a, b]\}. \tag{6.2.74}$$

Also by Theorem 6.2.7, we have

$$\ell(f, P_\varepsilon) \le \ell(f) \le u(f) \le u(f, P_\varepsilon). \tag{6.2.75}$$

Manipulating these inequalities as in part (vi) of Theorem 1.3.2, and in particular noting $u(f, P_\varepsilon) - \ell(f, P_\varepsilon)$ is nonnegative, yields

$$u(f) - \ell(f) \le u(f, P_\varepsilon) - \ell(f, P_\varepsilon) = |u(f, P_\varepsilon) - \ell(f, P_\varepsilon)| < \varepsilon. \tag{6.2.76}$$

So by the definition of arbitrarily close (Definition 1.5.1), we have

$$u(f) \operatorname{acl}\{\ell(f)\}. \tag{6.2.77}$$

Since two points arbitrarily close must be the same point (Lemma 1.5.5), we have

$$u(f) = \ell(f). \tag{6.2.78}$$

Therefore, (iv) $\implies$ (ii).

(ii) $\implies$ (i): Suppose f is Darboux integrable as in Definition 6.2.11 and let v denote the common value of the upper and lower integrals so that

$$v = u(f) = \ell(f). \tag{6.2.79}$$

Since suprema and infima are arbitrarily close to their sets by Definition 1.1.14, we have

$$v \operatorname{acl} U \qquad \text{and} \qquad v \operatorname{acl} L. \tag{6.2.80}$$

Hence, by Definition 6.1.6, f is integrable over $[a, b]$ and

$$\int_a^b f = v = u(f) = \ell(f). \tag{6.2.81}$$

Therefore, we have (ii) $\Longrightarrow$ (i).

In conclusion, we have shown

$$\text{(i)} \implies \text{(iii)} \implies \text{(iv)} \implies \text{(ii)} \implies \text{(i)}. \tag{6.2.82}$$

Therefore, statements (i) through (iv) provide equivalent forms integrability. □

In Theorem 6.2.12, sequential integrability (iii) and the Cauchy criterion for integrability (iv) consider the difference between the upper and lower sum stemming from a common partition, like this:

$$u(f, P) - \ell(f, P). \tag{6.2.83}$$

The situation arises often enough to merit a lemma.

Lemma 6.2.15: Difference of upper and lower sums

Suppose $f : [a, b] \to \mathbb{R}$ is bounded. If P is a partition of $[a, b]$ where

$$P = \{x_0, x_1, \ldots, x_n\}, \tag{6.2.84}$$

then we have

$$u(f, P) - \ell(f, P) = \sum_{k=1}^n (u_k - \ell_k)(x_k - x_{k-1}). \tag{6.2.85}$$

Proof of Lemma 6.2.15. Suppose $f : [a, b] \to \mathbb{R}$ is bounded and let P be a partition of $[a, b]$ where

$$P = \{x_0, x_1, \ldots, x_n\}. \tag{6.2.86}$$

By the definition of upper and lower sums (Definition 6.1.5) along with the commutative, associative, and distributive properties of addition, we have

$$u(f, P) - \ell(f, P) = \sum_{k=1}^n u_k(x_k - x_{k-1}) - \sum_{k=1}^n \ell_k(x_k - x_{k-1}) \tag{6.2.87}$$

$$= \sum_{k=1}^n \left(u_k(x_k - x_{k-1}) - \ell_k(x_k - x_{k-1}) \right) \tag{6.2.88}$$

$$= \sum_{k=1}^n (u_k - \ell_k)(x_k - x_{k-1}). \tag{6.2.89}$$

□

When a function is integrable, its integral is between any lower sum and any upper sum.

Corollary 6.2.16: Integrals are between upper and lower sums

If $f : [a, b] \to \mathbb{R}$ is integrable, then for any pair of partitions P_1 and P_2 of $[a, b]$ we have

$$\ell(f, P_1) \le \int_a^b f \le u(f, P_2). \tag{6.2.90}$$

Proof of Corollary 6.2.16. Suppose $f : [a, b] \to \mathbb{R}$ is integrable and P_1 and P_2 are partitions of $[a, b]$. By Theorem 6.2.12, f is Darboux integrable with integral equal to the upper and lower integrals. So by Theorem 6.2.7, we have

$$\ell(f, P_1) \le \ell(f) = \int_a^b f = u(f) \le u(f, P_2). \tag{6.2.91}$$

$\square$

The next theorem establishes one of the fundamental connections between integrals and continuity.

Theorem 6.2.17: Continuous functions are integrable

If $f : [a, b] \to \mathbb{R}$ is continuous, then f is integrable over $[a, b]$.

Scratch Work 6.2.18: Continuity and the size of partitions

The Cauchy criterion for integrability ((iv) in Theorem 6.2.12) provides a perspective on integrability that pairs especially well with continuity. Given $\varepsilon > 0$, we only need to find a single partition P_ε whose upper and lower sums are within ε of each other. Suitable partitions stem from continuity.

Since the domain $[a, b]$ is compact and continuous functions on compact intervals are actually *uniformly continuous* by Theorem 4.7.13, we have a uniform threshold δ which tells us how close inputs should be—across all of $[a, b]$—to ensure their outputs are as close together as we like. This control on both the domain and the range allows us to control the widths and heights defining upper and lower sums. See Figure 6.2.4 for a visualization of this approach and to motivate our choice of δ in the following proof.

Proof of Theorem 6.2.17. Suppose $f : [a, b] \to \mathbb{R}$ is continuous. Since $[a, b]$ is compact, Theorem 4.7.13 tells us f is uniformly continuous. So let $\varepsilon > 0$ and choose a uniform threshold $\delta > 0$ (Definition 4.7.1) where

$$x, y \in [a, b] \text{ with } |x - y| < \delta \quad \implies \quad |f(x) - f(y)| < \frac{\varepsilon}{b - a}. \tag{6.2.92}$$

Next, define a partition P_ε of $[a, b]$ where $P_\varepsilon = \{x_0, x_1, \ldots, x_n\}$ whose subintervals have length less than δ. That is, choose P_ε so that for each $k = 1, \ldots, n$ we have

$$x_k - x_{k-1} = |x_k - x_{k-1}| < \delta. \tag{6.2.93}$$

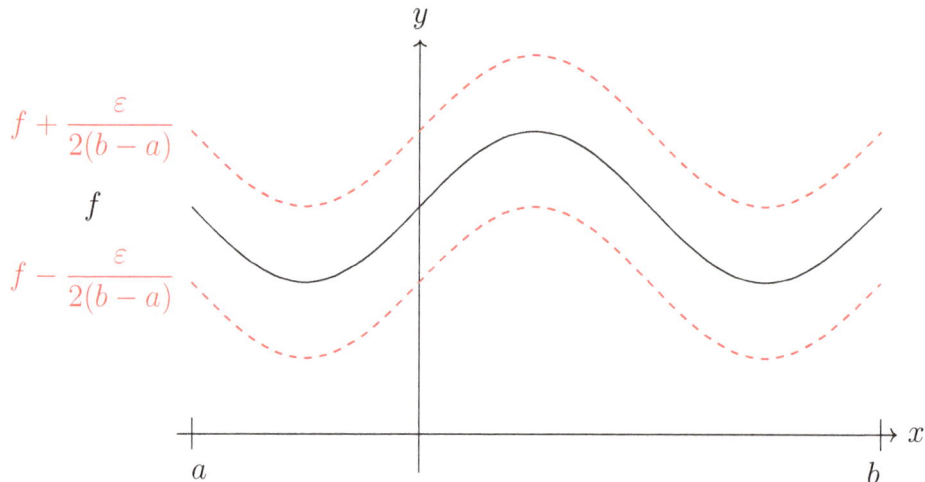

Figure 6.2.4: The squiggly area between the curves $f+\dfrac{\varepsilon}{2(b-a)}$ and $f-\dfrac{\varepsilon}{2(b-a)}$ is exactly ε, the same as a rectangle of height $\varepsilon/(b-a)$ and width $b-a$. (This follows from the linearity of integration, see Theorem 6.3.6.) In the proof of Theorem 6.2.17, the integrability of f is assured since, by uniform continuity, we can choose a partition P_ε where the upper and lower heights fit within the squiggly area. When that happens, the difference between upper and lower sums is less than ε, so the Cauchy criterion for integrability (iv) in Theorem 6.2.12 is satisfied by f.

The Extreme Value Theorem 4.6.9 applies to each compact subinterval $[x_{k-1}, x_k]$ and for each $k = 1, \ldots, n$ and yields inputs $s_k, t_k \in [x_{k-1}, x_k]$ where

$$f(s_k) = \ell_k = \inf f([x_{k-1}, x_k]) \qquad \text{and} \qquad f(t_k) = u_k = \sup f([x_{k-1}, x_k]). \tag{6.2.94}$$

Since $s_k, t_k \in [x_{k-1}, x_k]$, we have

$$|t_k - s_k| \le |x_k - x_{k-1}| < \delta. \tag{6.2.95}$$

Suprema are larger than infima on the same set, so we have

$$f(s_k) = \ell_k \le f(t_k) = u_k \qquad \Longleftrightarrow \qquad f(t_k) - f(s_k) = u_k - \ell_k \ge 0. \tag{6.2.96}$$

Hence, the extreme values of f on the kth subinterval satisfy

$$|f(t_k) - f(s_k)| = f(t_k) - f(s_k) = u_k - \ell_k < \frac{\varepsilon}{b-a}. \tag{6.2.97}$$

To establish the hypothesis of Cauchy criterion for integrability ((iv) in Theorem 6.2.12) and finish the proof, the next step shows that the difference of upper and lower sums over the partition P_ε is less than a nice telescoping sum, which in turn is less than ε. See Remark 6.1.12 where it is shown that

$$\sum_{k=1}^{n}(x_k - x_{k-1}) = b - a. \tag{6.2.98}$$

By Lemma 6.2.15, inequality (6.2.97), and the telescoping sum (6.2.98), we have

$$u(f, P_\varepsilon) - \ell(f, P_\varepsilon) = \sum_{k=1}^{n} (u_k - \ell_k)(x_k - x_{k-1}) \tag{6.2.99}$$

$$< \sum_{k=1}^{n} \left(\frac{\varepsilon}{b-a} (x_k - x_{k-1}) \right) \tag{6.2.100}$$

$$= \frac{\varepsilon}{b-a} \sum_{k=1}^{n} (x_k - x_{k-1}) \tag{6.2.101}$$

$$= \frac{\varepsilon}{b-a} (b-a) \tag{6.2.102}$$

$$= \varepsilon. \tag{6.2.103}$$

Therefore, by the Cauchy criterion for integrability ((iv) in Theorem 6.2.12), f is integrable over $[a, b]$. $\square$

The next section develops the linearity of integration. See Theorem 6.3.6.

Exercises

6.2.1. Prove the *Power Rule for definite integrals*: For every $n \in \mathbb{N}$ and every compact interval $[a, b]$, we have

$$\int_a^b x^n \, dx = \frac{b^{n+1} - a^{n+1}}{n+1}. \tag{6.2.104}$$

6.2.2. Suppose $f : [a, b] \to \mathbb{R}$ is continuous and $f(x) \geq 0$ for all $x \in [a, b]$. Prove

$$\int_a^b f = 0 \qquad \Longrightarrow \qquad f(x) = 0 \text{ for all } x \in [a, b]. \tag{6.2.105}$$

6.2.3. Suppose $f : [a, b] \to \mathbb{R}$ is monotone. Prove f is integrable.

6.2.4. Prove the *Substitution* formula for definite integrals: Suppose $g : [a, b] \to \mathbb{R}$ is differentiable where g' is continuous and $g([a, b]) = [c, d]$ with $g(a) = c$ and $g(b) = d$. Further suppose $f : [c, d] \to \mathbb{R}$ is continuous. Then

$$\int_a^b f(g(x))g'(x) \, dx = \int_c^d f(u) \, du. \tag{6.2.106}$$

6.2.5. Prove the area of the unit circle is π. Specifically, use "trig substitution" by pairing the Substitution formula in the previous exercise with the trigonometric identity

$$1 - \sin^2 \theta = \cos^2 \theta \qquad \text{for all } \theta \in \mathbb{R} \tag{6.2.107}$$

to show that

$$\int_0^1 \sqrt{1 - x^2} \, dx = \frac{\pi}{4}. \tag{6.2.108}$$

6.3   Linearity of integration

The linearity of integration is another beautiful property we can prove. Our approach makes use of a few more results regarding suprema and infima of functions from the real line to the real line.

Lemma 6.3.1: Suprema and infima of sums of functions

Suppose $D \subseteq \mathbb{R}$ where $f, g : D \to \mathbb{R}$ are bounded. Then

$$\sup (f + g)(D) \leq \sup f(D) + \sup g(D) \qquad \text{and} \qquad (6.3.1)$$
$$\inf (f + g)(D) \geq \inf f(D) + \inf g(D). \qquad (6.3.2)$$

Scratch Work 6.3.2: Same input for both summands

The sum $f + g$ is defined by plugging in the same input x into f and g separately, the taking the sum of their outputs $f(x)$ and $g(x)$ to define the output $(f + g)(x)$. This means the supremum of $f + g$ is limited to the collection of outputs of f and g evaluated at the same time. See Figure 6.3.1 to get a sense of how this works.

The proof of Lemma 6.3.1 makes a pointwise argument to generate the result. However, only the supremum case is proven below. The infimum case follows from a similar argument and its proof is omitted.

Proof of Lemma 6.3.1. Suppose $D \subseteq \mathbb{R}$ where $f, g : D \to \mathbb{R}$ are bounded. So by the Axiom of Completeness 1.3.8, $\sup f(D)$ and $\sup g(D)$ exist. Since suprema are upper bounds (Definition 1.1.14), for all $x \in D$ we have

$$(f + g)(x) = f(x) + g(x) \leq \sup f(D) + \sup g(D). \qquad (6.3.3)$$

Therefore, $\sup f(D) + \sup g(D)$ is an upper bound of the range $(f + g)(D)$. Since a supremum is the *least* upper bound by Theorem 1.3.10, we have

$$\sup(f + g)(D) \leq \sup f(D) + \sup g(D). \qquad (6.3.4)$$

$\square$

The following example shows us that the inequalities in Lemma 6.3.1 are as good as we can expect.

Example 6.3.3: Inequalities can be strict

The First Derivative Test from calculus can be used to verify the following conclusions: If $f, g : [0, 2\pi] \to \mathbb{R}$ are given by $f(x) = \sin x$ and $g(x) = \cos x$, then $f + g : [0, 2\pi] \to \mathbb{R}$ is

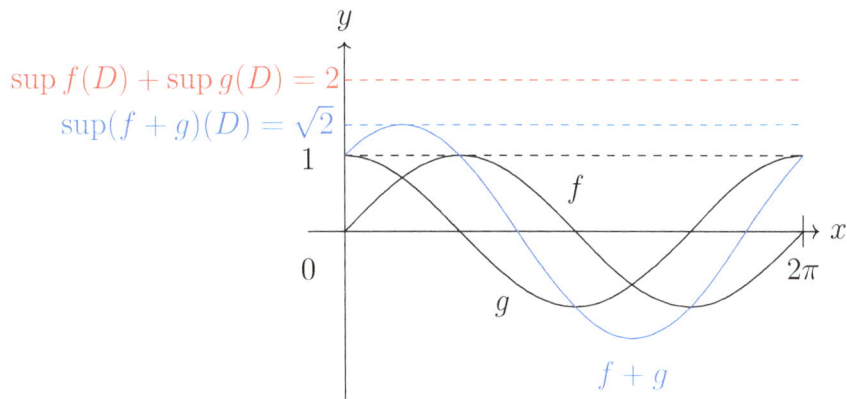

Figure 6.3.1: The functions $f(x) = \sin x$ and $g(x) = \cos x$ on the interval $D = [0, 2\pi]$ attain their suprema 1 at different inputs, while their sum $f + g$ has supremum $\sqrt{2}$ (which is strictly less 2). See Lemma 6 and Example 6.3.3.

given by $(f + g)(x) = \sin x + \cos x$ and

$$\sup f([0, 2\pi]) = \sup g([0, 2\pi]) = 1, \tag{6.3.5}$$
$$\inf f([0, 2\pi]) = \inf g([0, 2\pi]) = -1, \quad \text{but} \tag{6.3.6}$$
$$\sup(f + g)([0, 2\pi]) = \sqrt{2} \quad \text{and} \tag{6.3.7}$$
$$\inf(f + g)([0, 2\pi]) = -\sqrt{2}. \tag{6.3.8}$$

Therefore, Lemma 6.3.1 results in strict inequalities since

$$\sup(f + g)([0, 2\pi]) = \sqrt{2} < 2 = \sup f([0, 2\pi]) + \sup g([0, 2\pi]) \quad \text{and} \tag{6.3.9}$$
$$\inf(f + g)([0, 2\pi]) = -\sqrt{2} > -2 = \inf f([0, 2\pi]) + \inf g([0, 2\pi]). \tag{6.3.10}$$

See Figure 6.3.1.

Lemma 6.3.4: Suprema and infima of scaled functions

Suppose $D \subseteq \mathbb{R}$ where $f : D \to \mathbb{R}$ is bounded and $\alpha \in \mathbb{R}$.

(i) If $\alpha \geq 0$, then $\sup(\alpha f(D)) = \alpha \sup f(D)$ and $\inf(\alpha f(D)) = \alpha \inf f(D)$.

(ii) If $\alpha < 0$, then $\sup(\alpha f(D)) = \alpha \inf f(D)$ and $\inf(\alpha f(D)) = \alpha \sup f(D)$.

Scratch Work 6.3.5: Parity of the scalar matters

Suprema and infima are bounds, so the parity of a scalar as positive or negative determines whether the corresponding inequalities flip. In particular, when the scalar is negative, lower bounds become upper bounds and vice versa. This idea is use in the proof of Theorem 1.4.1, which takes the existence of a supremum for sets bounded above from Axiom of

Completeness 1.3.8 and generates a mirrored result for infima. A similar approach is taken here.

The proof addresses just two of the four results in Lemma 6.3.4, namely

$$\alpha \geq 0 \quad \Longrightarrow \quad \sup\left(\alpha f(D)\right) = \alpha \sup f(D) \quad \text{and} \tag{6.3.11}$$
$$\alpha < 0 \quad \Longrightarrow \quad \inf\left(\alpha f(D)\right) = \alpha \sup f(D) \tag{6.3.12}$$

The other cases follow from similar arguments, so their proofs are omitted. After some initial arguments, the proof splits into three cases: $\alpha > 0$, $\alpha < 0$, and $\alpha = 0$.

Proof of Lemma 6.3.4. Suppose $D \subseteq \mathbb{R}$ where $f : D \to \mathbb{R}$ is bounded and $\alpha \in \mathbb{R}$. Since f is bounded, $\sup f(D)$ exists by the Axiom of Completeness 1.3.8. Also, since a supremum is an upper bound (Definition 1.1.14), for all $x \in D$ we have

$$f(x) \leq \sup f(D). \tag{6.3.13}$$

Now let $\varepsilon > 0$ and assume $\alpha \neq 0$. Then $\varepsilon/|\alpha| > 0$ and since $(\sup f(D))\, \mathrm{acl}\, f(D)$ by the definition of supremum (Definition 1.1.14), there is some $x_\varepsilon \in D$ such that

$$|f(x_\varepsilon) - \sup f(D)| < \frac{\varepsilon}{|\alpha|}. \tag{6.3.14}$$

Since $|\alpha| > 0$, we have

$$|(\alpha f)(x_\varepsilon) - \alpha \sup f(D)| = |\alpha||f(x_\varepsilon) - \sup f(D)| < |\alpha|\frac{\varepsilon}{|\alpha|} = \varepsilon. \tag{6.3.15}$$

Hence, $(\alpha \sup f(D))\, \mathrm{acl}\, (\alpha f)(D)$.
From here, the proof is split into three cases: $\alpha > 0$, $\alpha < 0$, and $\alpha = 0$.

Case $\alpha > 0$: Assume $\alpha > 0$. Then for all $x \in D$ we have

$$(\alpha f)(x) = \alpha f(x) \leq \alpha \sup f(D). \tag{6.3.16}$$

Hence, $\alpha \sup f(D)$ is an upper bound for αf. Therefore, since $(\alpha \sup f(D))\, \mathrm{acl}\, (\alpha f)(D)$, by the definition of supremum (Definition 1.1.14) we have

$$\sup(\alpha f)(D) = \alpha \sup f(D). \tag{6.3.17}$$

Case $\alpha < 0$: Assume $\alpha < 0$. Then for all $x \in D$ we have

$$(\alpha f)(x) = \alpha f(x) \geq \alpha \sup f(D). \tag{6.3.18}$$

Hence, $\alpha \sup f(D)$ is a lower bound for αf. Therefore, since $(\alpha \sup f(D))\, \mathrm{acl}\, (\alpha f)(D)$, by the definition of infimum (Definition 1.1.14) we have

$$\inf(\alpha f)(D) = \alpha \sup f(D). \tag{6.3.19}$$

Case $\alpha = 0$: Assume $\alpha = 0$. Then for all $x \in D$ we have $(\alpha f)(x) = \alpha f(x) = 0$. Therefore,

$$\sup(\alpha f)(D) = \sup\{0\} = 0 = \alpha \sup f(D). \tag{6.3.20}$$

$\square$

We are now prepared to prove the linearity of integration.

Theorem 6.3.6: Linearity of integration

Suppose $f, g : [a, b] \to \mathbb{R}$, both f and g are integrable, and $\alpha \in \mathbb{R}$. Then $f + g$ and αf are integrable with

(i) $\displaystyle \int_a^b (f + g) = \int_a^b f + \int_a^b g$ *(additivity);*   and

(ii) $\displaystyle \int_a^b (\alpha f) = \alpha \int_a^b f$ *(homogeneity).*

With all the tools at our disposal such as the equivalent forms of integration in Theorem 6.2.12, the lemmas in this section, numerous properties on suprema and infima, and linearity in various forms, there are lots of ways to approach the proof of the linearity of integration (Theorem 6.3.6). It is split in two—additivity and homogeneity—since the arguments are a bit long.

Scratch Work 6.3.7: Perspective of arbitrarily close

The proof of additivity focuses on upper sums since the argument for lower sums is so similar. The approach here through the definitions of integration (Definition 6.1.6) and supremum (Definition 1.1.14) in terms of arbitrarily close in the real line (Definition 1.1.8).

Proof of additivity in Theorem 6.3.6. Suppose $f, g : [a, b] \to \mathbb{R}$ where both f and g are integrable. Let P be a partition of $[a, b]$ where $P = \{x_0, x_1, \ldots, x_n\}$.

To compare the upper sums of f, g, and $f + g$, for each $k = 1, \ldots, n$ let

$$u_k = \sup(f + g)([x_{k-1}, x_k]), \tag{6.3.21}$$

$$u_k' = \sup f([x_{k-1}, x_k]), \quad \text{and} \tag{6.3.22}$$

$$u_k'' = \sup g([x_{k-1}, x_k]). \tag{6.3.23}$$

By Lemma 6.3.1, for each $k = 1, \ldots, n$ we have

$$u_k = \sup(f + g)([x_{k-1}, x_k]) \tag{6.3.24}$$

$$\leq \sup f([x_{k-1}, x_k]) + \sup g([x_{k-1}, x_k]) \tag{6.3.25}$$

$$= u_k' + u_k''. \tag{6.3.26}$$

Since $x_k - x_{k-1} > 0$ for each $k = 1, \ldots, n$, taking sums gives us

$$u(f + g, P) = \sum_{k=1}^n u_k(x_k - x_{k-1}) \tag{6.3.27}$$

$$\leq \sum_{k=1}^n (u_k' + u_k'')(x_k - x_{k-1}) \tag{6.3.28}$$

$$= \sum_{k=1}^n u_k'(x_k - x_{k-1}) + \sum_{k=1}^n u_k''(x_k - x_{k-1}) \tag{6.3.29}$$

$$= u(f, P) + u(g, P). \tag{6.3.30}$$

By a similar argument for the lower sums, we also have

$$\ell(f+g, P) \geq \ell(f, P) + \ell(g, P). \tag{6.3.31}$$

Denote the sets of upper and lower sums of f, g, and $f + g$ by U_f, L_f, U_g, L_g, U_{f+g}, and L_{f+g}, respectively. Since f and g are integrable, we have $\int_a^b f \operatorname{acl} U_f$, $\int_a^b f \operatorname{acl} L_f$, $\int_a^b g \operatorname{acl} U_g$ and $\int_a^b g \operatorname{acl} L_g$.

Now let $\varepsilon > 0$. By the definitions of integral (Definition 6.1.6), upper and lower sums (Definition 6.1.5), and arbitrarily close in the real line (Definition 1.1.8) as well as splitting absolute values, there are partitions P_1, P_2, P_3, and P_4 of $[a, b]$ where

$$\left| u(f, P_1) - \int_a^b f \right| < \frac{\varepsilon}{2} \quad \Longrightarrow \quad u(f, P_1) < \int_a^b f + \frac{\varepsilon}{2}, \tag{6.3.32}$$

$$\left| \ell(f, P_2) - \int_a^b f \right| < \frac{\varepsilon}{2} \quad \Longrightarrow \quad \int_a^b f - \frac{\varepsilon}{2} < \ell(f, P_2), \tag{6.3.33}$$

$$\left| u(g, P_3) - \int_a^b g \right| < \frac{\varepsilon}{2} \quad \Longrightarrow \quad u(g, P_3) < \int_a^b g + \frac{\varepsilon}{2}, \quad \text{and} \tag{6.3.34}$$

$$\left| \ell(g, P_4) - \int_a^b g \right| < \frac{\varepsilon}{2} \quad \Longrightarrow \quad \int_a^b g - \frac{\varepsilon}{2} < \ell(g, P_4). \tag{6.3.35}$$

Let Q be the common refinement $Q = P_1 \cup P_2 \cup P_3 \cup P_4$. Since lower sums are below upper sums by Theorem 6.2.7, the previous string of inequalities yields

$$\left(\int_a^b f - \frac{\varepsilon}{2} \right) + \left(\int_a^b g - \frac{\varepsilon}{2} \right) < \ell(f, Q) + \ell(g, Q) \tag{6.3.36}$$

$$\leq \ell(f + g, Q) \tag{6.3.37}$$

$$\leq u(f + g, Q) \tag{6.3.38}$$

$$\leq u(f, Q) + u(g, Q) \tag{6.3.39}$$

$$< \left(\int_a^b f + \frac{\varepsilon}{2} \right) + \left(\int_a^b g + \frac{\varepsilon}{2} \right). \tag{6.3.40}$$

Subtracting the sum $\int_a^b f + \int_a^b g$ and focusing on $f + g$ yields

$$-\varepsilon < \ell(f + g, Q) - \left(\int_a^b f + \int_a^b g \right) \leq u(f + g, Q) - \left(\int_a^b f + \int_a^b g \right) < \varepsilon. \tag{6.3.41}$$

Considering the lower and upper sums separately, we have both

$$\left| \ell(f + g, Q) - \left(\int_a^b f + \int_a^b g \right) \right| < \varepsilon \quad \text{and} \tag{6.3.42}$$

$$\left| u(f + g, Q) - \left(\int_a^b f + \int_a^b g \right) \right| < \varepsilon. \tag{6.3.43}$$

By the definition of arbitrarily close (Definition 1.5.1), we have both

$$\left(\int_a^b f + \int_a^b g \right) \operatorname{acl} L_{f+g} \quad \text{and} \quad \left(\int_a^b f + \int_a^b g \right) \operatorname{acl} U_{f+g}. \tag{6.3.44}$$

Therefore, by the definition of integration (Definition 6.1.6), $f + g$ is integrable and

$$\int_a^b (f + g) = \int_a^b f + \int_a^b g. \tag{6.3.45}$$

$\square$

Scratch Work 6.3.8: Parity of the scalar matters

As in Lemma 6.3.4 in dealing with the suprema and infima of the ranges of scaled functions, the parity of the scalar α as either nonnegative or negative impacts the proof of the homogeneity of integration. That said, the proof below focuses on the lower sums of αf in the case where $\alpha < 0$. The results for all the other cases follow from similar arguments, so their proofs are omitted.

Proof of homogeneity in Theorem 6.3.6. Suppose $f : [a, b] \to \mathbb{R}$ where f is integrable and $\alpha \in \mathbb{R}$ such that $\alpha < 0$. Then both f and αf are bounded and their suprema and infima exist over all compact subintervals of $[a, b]$.

Let P be a partition of $[a, b]$ where $P = \{x_0, x_1, \dots, x_n\}$. Then by Lemma 6.3.4, for each $k = 1, \dots, n$ we have

$$\ell_k^\alpha = \inf(\alpha f)([x_{k-1}, x_k]) = \alpha \sup f([x_{k-1}, x_k]) = \alpha u_k. \tag{6.3.46}$$

So taking the sum from $k = 1$ to n yields

$$\ell(\alpha f, P) = \sum_{k=1}^n \ell_k^\alpha (x_k - x_{k-1}) = \sum_{k=1}^n \alpha u_k (x_k - x_{k-1}) = \alpha u(f, P). \tag{6.3.47}$$

Denote the set of lower sums of αf denoted by $L_{\alpha f}$ and denote the set of upper sums of f by U_f. Since (6.3.47) holds for an arbitrary partition P of $[a, b]$, we have

$$L_{\alpha f} = \{\ell(\alpha f, P) : P \text{ is a partition of } [a, b]\} \tag{6.3.48}$$

$$= \{\alpha u(f, P) : P \text{ is a partition of } [a, b]\} \tag{6.3.49}$$

$$= \alpha U_f. \tag{6.3.50}$$

Now let $\varepsilon > 0$. Since $\alpha < 0$, we have $\varepsilon/|\alpha| > 0$. By the definition of integral (Definition 6.1.6), we have $\int_a^b f$ acl U_f. So by the definition of arbitrarily close in the real line (Definition 1.1.8), there is a partition P_ε of $[a, b]$ such that

$$\left| u(f, P_\varepsilon) - \int_a^b f \right| < \frac{\varepsilon}{|\alpha|}. \tag{6.3.51}$$

So for the particular partition P_ε we have

$$\left| \ell(\alpha f, P_\varepsilon) - \alpha \int_a^b f \right| = \left| \alpha u(f, P_\varepsilon) - \alpha \int_a^b f \right| \tag{6.3.52}$$

$$= |\alpha| \left| u(f, P_\varepsilon) - \int_a^b f \right| \tag{6.3.53}$$

$$< |\alpha| \frac{\varepsilon}{|\alpha|} \tag{6.3.54}$$

$$= \varepsilon. \tag{6.3.55}$$

Hence, $(\alpha \int_a^b f) \operatorname{acl} L_{\alpha f}$. By a similar argument for the set $U_{\alpha f}$ of upper sums of αf, we also have $(\alpha \int_a^b f) \operatorname{acl} U_{\alpha f}$. Therefore, αf is integrable and we have

$$\int_a^b (\alpha f) = \alpha \int_a^b f. \tag{6.3.56}$$

A similar set of arguments hold for the case where $\alpha > 0$. The case where $\alpha = 0$ holds by Lemma 6.1.10 since $\alpha f = 0$ is a constant function. $\square$

Yet again and as mentioned in Remark 1.6.18, a corollary of the linearity of integration holds for linear combinations:

The integral of a linear combination is the linear combination of integrals.

As with the proofs of previous corollaries regarding linearity, the proof of Corollary 6.3.9 follows from induction on the linearity of integration in Theorem 6.3.6. So, the proof is left as an exercise.

Corollary 6.3.9: Linear combinations of integrals

Suppose $k \in \mathbb{N}$ and for each $j = 1, \ldots, k$ we have $c_j \in \mathbb{R}$ and the functions $f_j : [a, b] \to \mathbb{R}$ are integrable. Then the linear combination f given by

$$f(x) = \sum_{j=1}^{k} c_j f_j(x) = c_1 f_1(x) + \ldots + c_k f_k(x) \tag{6.3.57}$$

is integrable and with integral given by

$$\int_a^b f = \sum_{j=1}^{k} \left(c_j \int_a^b f_j \right) = c_1 \int_a^b f_1 + \ldots + c_k \int_a^b f_k. \tag{6.3.58}$$

Another corollary of the linearity of integration (Theorem 6.3.6) tells us how integrals respect order.

Corollary 6.3.10: Order property of integration

If f and g are integrable over $[a, b]$ where $f(x) \le g(x)$ for all $x \in [a, b]$, then

$$\int_a^b f \le \int_a^b g. \tag{6.3.59}$$

Proof of Corollary 6.3.10. Suppose f and g are integrable over $[a, b]$ with $f(x) \le g(x)$ for all $x \in [a, b]$. Hence, 0 is a lower bound for the difference $g - f$ since

$$0 \le g(x) - f(x) \qquad \text{for all } x \in [a, b]. \tag{6.3.60}$$

Since an infimum is the greatest lower bound, we have

$$0 \le \inf(g - f)([a, b]). \tag{6.3.61}$$

By the linearity of integration (Theorem 6.3.6), the difference $g - f$ is integrable over $[a, b]$ and

$$\int_a^b (g - f) = \int_a^b g - \int_a^b f. \tag{6.3.62}$$

Since integrals are between rectangles as in Example 6.2.10, we have

$$0 \leq \underbrace{(b - a)}_{\text{width}} \underbrace{\inf(g - f)([a, b])}_{\text{height}} \leq \int_a^b (g - f) = \int_a^b g - \int_a^b f. \tag{6.3.63}$$

Therefore,

$$\int_a^b f \leq \int_a^b g. \tag{6.3.64}$$

$\square$

The next section develops even more properties of integrals, starting with an example of a bounded function which is not integrable.

Exercises

6.3.1. Let $R[a, b]$ denote the set of integrable functions over $[a, b]$. Use Lemma 1.6.7 to prove $R[a, b]$ is a vector space.

6.3.2. Use the Product Rule 5.4.6 and the Fundamental Theorem of Calculus I 6.1.15 to prove the *Integration by Parts* formula for definite integrals: Suppose f and g are integrable functions on $[a, b]$ with antiderivatives F and G, respectively (i.e., $F' = f$ and $G' = g$ on $[a, b]$). Then

$$\int_a^b Fg = F(b)G(b) - F(a)G(a) - \int_a^b Gf. \tag{6.3.65}$$

6.4   Properties of integration

There are many more properties of integrals to explore such as integrability when compositions are in play and a triangle inequality for integrals. The first result of the section gives us another tool to control upper and lower sums through the distance of outputs.

Lemma 6.4.1: Reach of the range

Suppose $D \subseteq \mathbb{R}$ and $f : D \to \mathbb{R}$ is bounded by $q \geq 0$. Then for every $x, y \in D$ we have

$$|f(x) - f(y)| \leq \sup f(D) - \inf f(D) \leq 2q. \tag{6.4.1}$$

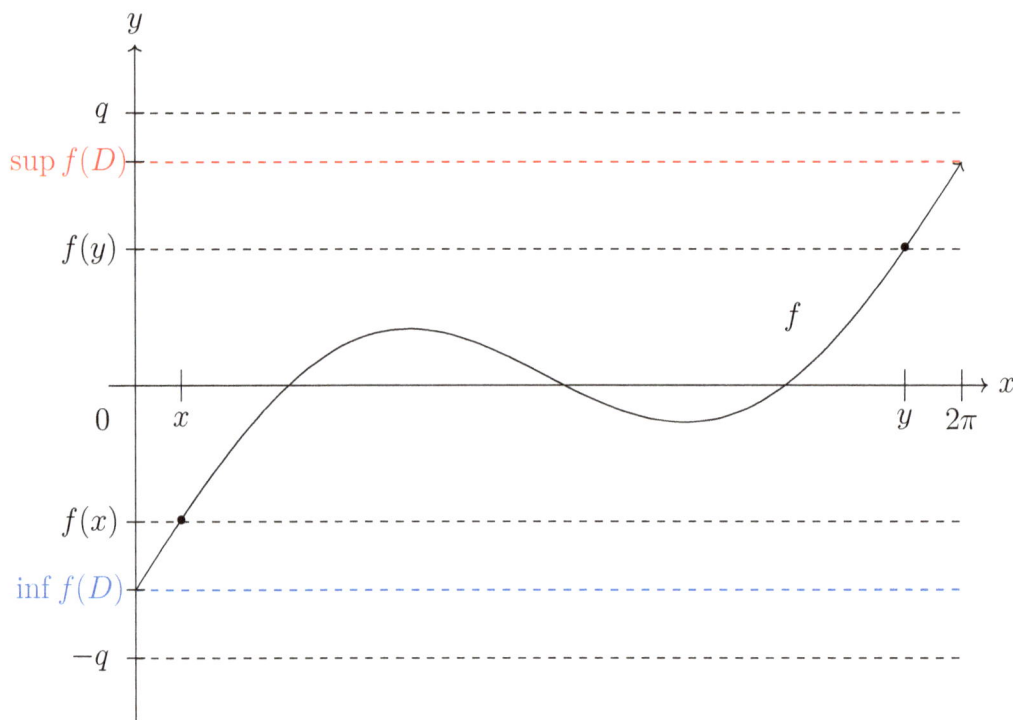

Figure 6.4.1: A function f on $D = [0, 2\pi]$ bounded by $q \geq 0$ showing that the distance between any pair of outputs is no larger than the difference between its supremum and infimum and no larger than twice the bound. See Lemma 6.4.1.

Scratch Work 6.4.2: Draw stuff

The reach of the range for a bounded function from the real line to the real line is visualized in Figure 6.4.1. Comparing the distances

$$|f(x) - f(y)|, \quad \sup f(D) - \inf f(D), \quad \text{and} \quad |q - (-q)| = 2q \qquad (6.4.2)$$

in the figure provides some nice evidence to support the concluding inequality of Lemma 6.4.1.

Proof 6.4.1. Suppose $D \subseteq \mathbb{R}$ and $f : D \to \mathbb{R}$ is bounded by $q \geq 0$. Then q is an upper bound for the $f(D)$ and $-q$ is a lower bound. Without loss of generality, let $x, y \in D$ where $f(x) \leq f(y)$. Since a supremum is both an upper bound and the *least* upper bound (Theorem 1.3.10) and an infimum is both a lower bound and the *greatest* lower bound (Theorem 1.4.3), we have

$$-q \leq \inf f(D) \leq f(x) \leq f(y) \leq \sup f(D) \leq q. \qquad (6.4.3)$$

By tracking which differences are nonnegative and rearranging the inequalities as in (vi) in Theorem 1.3.2 we have

$$|f(x) - f(y)| \leq \sup f(D) - \inf f(D) \leq 2q. \qquad (6.4.4)$$

$\square$

An application of Lemma 6.4.1 cleans up the proof of the following theorem.

Theorem 6.4.3: Integrable compositions

Suppose $K \subseteq \mathbb{R}$ is compact, $f : [a, b] \to K$ is integrable, and $g : K \to \mathbb{R}$ is continuous. Then the composition $g \circ f : [a, b] \to \mathbb{R}$ is integrable.

Scratch Work 6.4.4: A difficult argument

The following proof is an expansion of the proof of the corresponding result Theorem 6.11 in [10]. The idea is to show the composition $g \circ f$ satisfies the Cauchy criterion for integrability ((vi) in Theorem 6.2.12), but the path is not an easy one to follow. In particular, choice and usage of the uniform threshold δ in the proof is unlike anything we have come across so far.

Take your time.

Proof of Theorem 6.4.3. Suppose $K \subseteq \mathbb{R}$ is compact, $f : [a, b] \to K$ is integrable, and $g : K \to \mathbb{R}$ is continuous. Since g is continuous on a compact set, the Extreme Value Theorem 4.6.9 tells us g is bounded. Hence, $|g|$ is bounded as well and so by the Axiom of Completeness 1.3.8,

$$v = \sup\{|g(t)| : t \in K\} = \sup |g|(K). \tag{6.4.5}$$

exists.

Now let $\varepsilon > 0$. By Theorem 4.7.13, g is uniformly continuous. So by Definition 4.7.1, there is a uniform threshold $\delta > 0$ chosen so that we have both

$$s, t \in K \text{ with } |s - t| < \delta \quad \implies \quad |g(s) - g(t)| < \frac{\varepsilon}{b - a + 2v} \tag{6.4.6}$$

as well as

$$\delta < \frac{\varepsilon}{b - a + 2v}. \tag{6.4.7}$$

Since f is integrable, it is bounded and satisfies the Cauchy criterion for integrability ((vi) in Theorem 6.2.12) which tells us there is a partition P of $[a, b]$ such that

$$P = \{x_0, x_1, \ldots, x_n\} \quad \text{and} \quad u(f, P) - \ell(f, P) < \delta^2. \tag{6.4.8}$$

Since g is bounded, $g \circ f$ is bounded as well. So for each $k = 1, \ldots, n$, let

$$u_k = \sup f([x_{k-1}, x_k]), \quad \ell_k = \inf f([x_{k-1}, x_k]), \tag{6.4.9}$$
$$u_k^* = \sup (g \circ f)([x_{k-1}, x_k]), \quad \text{and} \quad \ell_k^* = \inf (g \circ f)([x_{k-1}, x_k]). \tag{6.4.10}$$

(Each exists by either the by the Axiom of Completeness 1.3.8 or Theorem 1.4.1.) Also, since $u_k^* \leq v$ and $-v \leq \ell_k^*$, we have

$$u_k^* - \ell_k^* \leq v + v = 2v. \tag{6.4.11}$$

As a new and subtle step, split indices $k = 1, \ldots, n$ of the partition P into a disjoint union $A \cup B$ where

$$A = \{k : u_k - \ell_k < \delta\} \qquad \text{and} \qquad B = \{k : u_k - \ell_k \geq \delta\}. \tag{6.4.12}$$

Hence, by implication (6.4.6) applied to each $k \in A$, our choice for δ tells us that if $x, y \in [x_{k-1}, x_k]$, we have

$$|x - y| \leq u_k - \ell_k < \delta \quad \implies \quad |g(f(x)) - g(f(y))| < \frac{\varepsilon}{b - a + 2v}. \tag{6.4.13}$$

Since x and y are arbitrary, an application of Lemma 6.4.1 to the composition $g \circ f$ yields

$$u_k^* - \ell_k^* \leq \frac{\varepsilon}{b - a + 2v}. \tag{6.4.14}$$

By the difference of upper and lower sums in Lemma 6.2.15 applied to f with partition P, the definition of B giving $u_k - \ell_k \geq \delta$ for each $k \in B$, and $u_k - \ell_k \geq 0$ for all k, we have

$$\delta \sum_{k \in B} (x_k - x_{k-1}) \leq \sum_{k \in B} (u_k - \ell_k)(x_k - x_{k-1}) \tag{6.4.15}$$

$$\leq \sum_{k=1}^{n} (u_k - \ell_k)(x_k - x_{k-1}) \tag{6.4.16}$$

$$= u(f, P) - \ell(f, P) \tag{6.4.17}$$

$$< \delta^2. \tag{6.4.18}$$

Since $\delta > 0$, dividing by δ gives us

$$\sum_{k \in B} (x_k - x_{k-1}) < \delta. \tag{6.4.19}$$

Hence, by the difference of upper and lower sums in Lemma 6.2.15 applied to $g \circ f$ with partition P, we have

$$u(g \circ f, P) - \ell(g \circ f, P) = \sum_{k=1}^{n} (u_k^* - \ell_k^*)(x_k - x_{k-1}) \tag{6.4.20}$$

$$= \sum_{k \in A} (u_k^* - \ell_k^*)(x_k - x_{k-1}) + \sum_{k \in B} (u_k^* - \ell_k^*)(x_k - x_{k-1}) \tag{6.4.21}$$

$$\leq \sum_{k \in A} \left(\frac{\varepsilon}{b - a + 2v} \right) (x_k - x_{k-1}) + \sum_{k \in B} 2v(x_k - x_{k-1}) \tag{6.4.22}$$

$$\leq \left(\frac{\varepsilon}{b - a + 2v} \right) (b - a) + 2v\delta \tag{6.4.23}$$

$$< \left(\frac{\varepsilon}{b - a + 2v} \right) (b - a) + 2v \left(\frac{\varepsilon}{b - a + 2v} \right) \tag{6.4.24}$$

$$= \varepsilon. \tag{6.4.25}$$

Therefore, $g \circ f$ satisfies the Cauchy criterion for integrability ((vi) in Theorem 6.2.12), meaning $g \circ f$ is integrable. $\qquad \square$

Integrals have their own version of a triangle inequality.

Corollary 6.4.5: Integral triangle inequality

If $f : [a, b] \to \mathbb{R}$ is integrable, then $|f|$ is integrable as well and we have

$$\left| \int_a^b f \right| \leq \int_a^b |f|. \tag{6.4.26}$$

Scratch Work 6.4.6: Squeezing upper and lower sums together

An intuitive idea for this inequality comes from noting that the integral $\left| \int_a^b f \right|$ allows for negative area to cancel positive area before taking absolute value. On the other hand, $\int_a^b |f|$ makes all outputs positive before they contribute to the area with the integral, so no such cancellation happens here.

The proof of the integrability of $|f|$ follows from Theorem 6.4.3 and the fact that the absolute value function is continuous (Example 4.3.11, modified to the restriction on a compact set as needed). The proof of the integral triangle inequality (6.4.26) follows from applying the order property of integration (Corollary 6.3.10) along with the homogeneity of integrals (Theorem 6.3.6) to $-|f|$ and $|f|$.

Proof of Corollary 6.4.5. Suppose $f : [a, b] \to \mathbb{R}$ is integrable and $K \subseteq \mathbb{R}$ is a compact set containing $f([a, b])$. Since the absolute value function $g : K \to \mathbb{R}$ given by $g(x) = |x|$ is continuous, the composition $g \circ f = |f|$ is integrable by Theorem 6.4.3.

Now, by the homogeneity of integrals (Theorem 6.3.6), we have $-|f|$ is integrable as well. Since $-|f(x)| \leq f(x) \leq |f(x)|$ for every $x \in [a, b]$, by the order property of integration (Corollary 6.3.10) applied twice and the homogeneity of integrals (Theorem 6.3.6) applied one more time, we have

$$-\int_a^b |f| \leq \int_a^b f \leq \int_a^b |f|. \tag{6.4.27}$$

Therefore, by a property of absolue value we have

$$\left| \int_a^b f \right| \leq \int_a^b |f|. \tag{6.4.28}$$

$\square$

Continuity plays a significant role in the development of properties of integrals. Even so, integration can handle a certain amount of discontinuity. The next example suggests that a single point of discontinuity does not break integrability. However, a complete proof for the next example result is omitted since it can get pretty technical. Instead, consider the scratch work and Figure 6.4.2.

Example 6.4.7: Integrating over a single discontinuity

Consider the function $g : [0, 1] \to \mathbb{R}$ given by

$$g(x) = \begin{cases} 1, & \text{if } x = 1/2, \\ 0, & \text{if } x \neq 1/2. \end{cases} \tag{6.4.29}$$

We have

$$\int_0^1 g = \int_0^1 0 = 0. \tag{6.4.30}$$

The function g is almost identical to the zero function, except it has a removable discontinuity at $c = 1/2$ where the output jumps to $g(1/2) = 1$. This discontinuity does not impact the integrability of g, nor does it change the value integral since, roughly speaking, the area over a single point is zero.

Scratch Work 6.4.8: A simple partition works

To establish the integrability of g, we can use Definition 6.1.6 since we have a good candidate for the integral, namely zero. The trick is to show zero is arbitrarily close to the sets of upper sums and lower sums of g.

Notice $\ell(g, P) = 0$ for any partition $P = \{0, \ldots, n\}$ since $\ell_k = \inf g([x_{k-1}, x_k]) = 0$ for every $k = 1, \ldots, n$. Hence, the set of lower sums L is a singleton: $L = \{0\}$. Therefore, $0 \operatorname{acl} L$ by Lemma 1.5.4.

To show $0 \operatorname{acl} U$ for the set of upper sums U, it suffices to find a single partition P_ε in response to a given $\varepsilon > 0$ where

$$|u(g, P_\varepsilon) - 0| = u(g, P_\varepsilon) < \varepsilon. \tag{6.4.31}$$

Since g is zero except for $g(1/2) = 1$, the partition

$$P_\varepsilon = \{x_0, x_1, x_2, x_3\} = \left\{0, \frac{1}{2} - \frac{\varepsilon}{3}, \frac{1}{2} + \frac{\varepsilon}{3}, 1\right\} \tag{6.4.32}$$

will get the job done (for small enough ε). See Figure 6.4.2. Then we have

$$|u(g, P_\varepsilon) - 0| = u(g, P_\varepsilon) \tag{6.4.33}$$

$$= \sum_{k=1}^{3} u_k(x_k - x_{k-1}) \tag{6.4.34}$$

$$= 0 + 1\left(\frac{1}{2} + \frac{\varepsilon}{3} - \left(\frac{1}{2} - \frac{\varepsilon}{3}\right)\right) + 0 \tag{6.4.35}$$

$$= \frac{2}{3}\varepsilon \tag{6.4.36}$$

$$< \varepsilon. \tag{6.4.37}$$

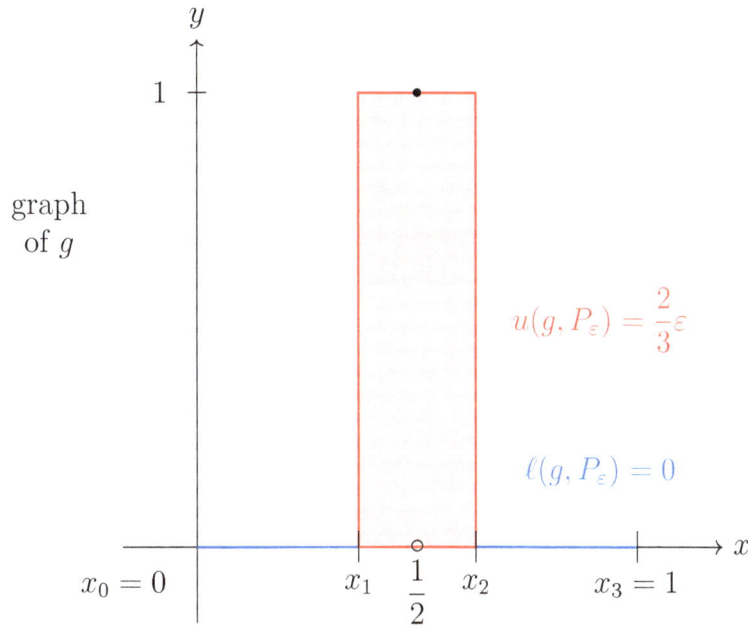

Figure 6.4.2: The function g from Example 6.4.7 is zero except at $c = 1/2$ where it jumps to 1. This discontinuity does not impact the integrability of g since every lower sum is zero and, given ε, we can always choose a partition P_ε whose upper sum $u(g, P_\varepsilon)$ is strictly less than ε. Here, P_ε comprises four endpoints where x_1 and x_2 have c between them and are $2\varepsilon/3$ apart, meaning the upper sum $u(g, P_\varepsilon)$ is just $2\varepsilon/3$.

Since both $0 \operatorname{acl} L$ and $0 \operatorname{acl} U$, we have

$$\int_0^1 g = \int_0^1 0 = 0. \tag{6.4.38}$$

The following lemma will help us split a partition of an interval $[a, b]$ into partitions of subintervals $[a, p]$ and $[p, b]$ in a useful way and allows us to recover another result from calculus. (See Theorem 6.4.11.) Also, by a mild abuse of notation to mitigate how complicated the details can become, f stands for the function f over $[a, b]$ as well its restrictions to the subintervals $[a, p]$ and $[p, b]$.

Lemma 6.4.9: Splitting sums and partitions in two

Suppose $f : [a, b] \to \mathbb{R}$ is bounded, $t \in (a, b)$, and P is a partition of $[a, b]$ containing t so that

$$P = \{x_0 = a, \ldots, x_j = t, \ldots, x_n = b\} \tag{6.4.39}$$

for some $j, n \in \mathbb{N}$ where $0 < j < n$. If $P_1 = P \cap [a, t]$ and $P_2 = P \cap [t, b]$, then

$$u(f, P) = u(f, P_1) + u(f, P_2) \quad \text{and} \tag{6.4.40}$$
$$\ell(f, P) = \ell(f, P_1) + \ell(f, P_2). \tag{6.4.41}$$

Scratch Work 6.4.10: Control partitions around the split

The proof focuses on splitting an upper sum around the point t. The trick is to carefully deal with the indices in the resulting pair of sums. The proof for the lower sums is similar, so it is omitted.

Proof of Lemma 6.4.9. Suppose the hypotheses of Lemma 6.4.9 hold. By the definition of upper sums (Definition 6.1.5) as well as the commutativity of addition we have

$$u(f, P) = \sum_{k=1}^{n} u_k(x_k - x_{k-1}) \tag{6.4.42}$$

$$= \sum_{k=1}^{j} u_k(x_k - x_{k-1}) + \sum_{k=j+1}^{n} u_k(x_k - x_{k-1}) \tag{6.4.43}$$

$$= u(f, P_1) + u(f, P_2). \tag{6.4.44}$$

Similary, $\ell(f, P) = \ell(f, P_1) + \ell(f, P_2)$.   □

Time for a classic result from calculus which, even with all the tools we have available, takes a while to prove.

Theorem 6.4.11: Splitting integrals in two

Suppose $f : [a, b] \to \mathbb{R}$ is bounded and $t \in (a, b)$. Then f is integrable over $[a, b]$ if and only if f is integrable over both $[a, t]$ and $[t, b]$. In this case, we have

$$\int_a^b f = \int_a^t f + \int_t^b f. \tag{6.4.45}$$

Scratch Work 6.4.12: Control intgrals around the split

The focus of the proof is on controlling the difference between the upper and lower sums around the point t. The Cauchy criterion for integrability ((iv) of Theorem along with a number of properties of upper sums and lower sums are involved. Also, the proof has three unnamed parts: the forward implication regarding integrability, the backward implication, and the equation with the integrals.

Proof of Theorem 6.4.11. Throughout the proof, suppose $f : [a, b] \to \mathbb{R}$ is bounded and $t \in (a, b)$.

To prove the forward implication regarding integrability, assume f is integrable over $[a, b]$ and let $\varepsilon > 0$. Then f satisfies the Cauchy criterion for integrability ((iv) of Theorem 6.2.12), meaning there is a partition P_ε of $[a, b]$ where

$$u(f, P_\varepsilon) - \ell(f, P_\varepsilon) < \varepsilon. \tag{6.4.46}$$

To ensure we can split $[a, b]$ around $t \in (a, b)$, let $Q_\varepsilon = P_\varepsilon \cup \{t\}$. Then for some $j, n \in \mathbb{N}$ where $0 < j < n$ we have

$$Q_\varepsilon = \{x_0 = a, \dots, x_j = t, \dots, x_n = b\}. \tag{6.4.47}$$

Now let $P_1 = Q_\varepsilon \cap [a,t]$ and $P_2 = Q_\varepsilon \cap [t,b]$. By carefully splitting sums at the index j as in Lemma 6.4.9, we have

$$u(f, Q_\varepsilon) - \ell(f, Q_\varepsilon) = (u(f, P_1) + u(f, P_2)) - (\ell(f, P_1)) + \ell(f, P_2)) \qquad (6.4.48)$$
$$= u(f, P_1) - \ell(f, P_1) + u(f, P_2) - \ell(f, P_2). \qquad (6.4.49)$$

So by Corollary 6.2.9 applied to the above split, we have

$$u(f, P_1) - \ell(f, P_1) + u(f, P_2) - \ell(f, P_2) = u(f, Q_\varepsilon) - \ell(f, Q_\varepsilon) \qquad (6.4.50)$$
$$\leq u(f, P_\varepsilon) - \ell(f, P_\varepsilon) \qquad (6.4.51)$$
$$< \varepsilon. \qquad (6.4.52)$$

Again by Corollary 6.2.9, we have both

$$u(f, P_1) - \ell(f, P_1) \geq 0 \qquad \text{and} \qquad u(f, P_2) - \ell(f, P_2) \geq 0. \qquad (6.4.53)$$

Hence, we have both

$$u(f, P_1) - \ell(f, P_1) \leq u(f, Q_\varepsilon) - \ell(f, Q_\varepsilon) < \varepsilon \qquad \text{and} \qquad (6.4.54)$$
$$u(f, P_2) - \ell(f, P_2) \leq u(f, Q_\varepsilon) - \ell(f, Q_\varepsilon) < \varepsilon. \qquad (6.4.55)$$

By the Cauchy criterion for integrability ((iv) in Theorem 6.2.12), f is integrable over both $[a,t]$ and $[t,b]$.

To show the backward implication regarding integrability, assume f is integrable over both $[a,t]$ and $[t,b]$, and let $\varepsilon > 0$. By the Cauchy criterion for integrability ((iv) in Theorem 6.2.12), there are partitions R_1 of $[a,t]$ and R_2 of $[t,b]$ such that

$$u(f, R_1) - \ell(f, R_1) < \frac{\varepsilon}{2} \quad \text{and} \quad u(f, R_2) - \ell(f, R_2) < \frac{\varepsilon}{2}. \qquad (6.4.56)$$

Define R_ε to the partition of $[a,b]$ given by $R_\varepsilon = R_1 \cup R_2$. Then by splitting upper and lower sums according to Lemma 6.4.9, we have

$$u(f, R_\varepsilon) - \ell(f, R_\varepsilon) = u(f, R_1) - \ell(f, R_1) + u(f, R_2) - \ell(f, R_2) \qquad (6.4.57)$$
$$< \frac{\varepsilon}{2} + \frac{\varepsilon}{2} \qquad (6.4.58)$$
$$< \varepsilon. \qquad (6.4.59)$$

Therefore, by the Cauchy criterion for integrability ((iv) in Theorem 6.2.12), we have f is integrable over $[a,b]$.

At this point we have shown f is integrable over $[a,b]$ if and only if f is integrable over both $[a,t]$ and $[t,b]$. It remains to show equation (6.4.45) holds. To that end, suppose f is integrable over $[a,b]$ so that the integrals $\int_a^b f$, $\int_a^t f$, and $\int_t^b f$ exist.

Once again, let $\varepsilon > 0$ and let $P_\varepsilon, Q_\varepsilon, P_1$, and P_2 satisfy the same relationships as in the proof of the forward implication above. In particular, we have

$$u(f, Q_\varepsilon) - \ell(f, Q_\varepsilon) < \varepsilon. \qquad (6.4.60)$$

Since integrals are between their upper and lower sums by Corollary 6.2.16 and the lower sum over Q_ε splits at p into P_1 and P_2 as in Lemma 6.4.9, we have

$$\int_a^b f \leq u(f, Q_\varepsilon) \tag{6.4.61}$$

$$< \ell(f, Q_\varepsilon) + \varepsilon \tag{6.4.62}$$

$$= \ell(f, P_1) + \ell(f, P_2) + \varepsilon \tag{6.4.63}$$

$$\leq \int_a^t f + \int_t^b f + \varepsilon. \tag{6.4.64}$$

Since ε is arbitrary, Lemma 1.5.22 implies

$$\int_a^b f \leq \int_a^t f + \int_t^b f. \tag{6.4.65}$$

For the other inequality, and again since integrals are between their upper and lower sums by Corollary 6.2.16 while the lower sum over Q_ε splits at t into P_1 and P_2 as in Lemma 6.4.9, we have

$$\int_a^t f + \int_t^b f \leq u(f, P_1) + u(f, P_2) \tag{6.4.66}$$

$$< \ell(f, P_1) + \ell(f, P_2) + \varepsilon \tag{6.4.67}$$

$$= \ell(f, Q_\varepsilon) + \varepsilon \tag{6.4.68}$$

$$\leq \int_a^b f + \varepsilon. \tag{6.4.69}$$

Since ε is arbitrary, Lemma 1.5.22 once again implies

$$\int_a^t f + \int_t^b f \leq \int_a^b f. \tag{6.4.70}$$

Therefore,

$$\int_a^b f = \int_a^t f + \int_t^b f. \tag{6.4.71}$$

□

Next, consider *definite integrals* in the sense of functions defined as integrals where one limit of integration is fixed and the other is variable.

Definition 6.4.13: Definite integral

Suppose $f : [a, b] \to \mathbb{R}$ is integrable. The *definite integral of f* over $[a, b]$ is the function $g : [a, b] \to \mathbb{R}$ defined for each $x \in [a, b]$ by

$$g(x) = \int_a^x f. \tag{6.4.72}$$

Definite integrals are nice.

Theorem 6.4.14: Definite integrals are uniformly continuous

Suppose $f : [a,b] \to \mathbb{R}$ is integrable and g is the definite integral of f over $[a,b]$ given by $g(x) = \int_a^x f$. Then g is uniformly continuous.

Scratch Work 6.4.15: Linearity and several properties of integration

Following the guide in Remark 4.3.4 for proofs of continuity, but keeping in mind we want a *uniform* threshold δ which is independent of inputs, we would like to end up with

$$|x - y| < \delta \quad \implies \quad |g(x) - g(y)| = \left| \int_a^x f - \int_a^y f \right| < \varepsilon. \tag{6.4.73}$$

Stringing together properties of integrals allows us to streamline to expression considerably. For any bound $q > 0$ of $|f|$, we have

$$|g(x) - g(y)| = \left| \int_a^x f - \int_a^y f \right| = \left| \int_y^x f \right| \leq \int_y^x |f| \leq q|x - y| < \varepsilon. \tag{6.4.74}$$

So a reasonable choice for a uniform threshold is

$$\delta = \frac{\varepsilon}{q}. \tag{6.4.75}$$

On to the proof.

Proof of Theorem 6.4.14. Suppose $f : [a,b] \to \mathbb{R}$ is integrable and g is the definite integral of f over $[a,b]$ given by $g(x) = \int_a^x f$. Since integrable functions are bounded and the absolute value of a bounded function is also bounded, let $q > 0$ be a bound for $|f|$.

Suppose $x, y \in [a,b]$ where, without loss of generality, we have $y < x$. Then by splitting the definite integral at y as in Theorem 6.4.11, we have

$$\int_a^x f = \int_a^y f + \int_y^x f \quad \Longleftrightarrow \quad \int_a^x f - \int_a^y f = \int_y^x f. \tag{6.4.76}$$

Also, as a constant, q is integrable over $[y, x]$ by Lemma 6.1.10 with

$$\int_y^x q = q(x - y) = q|x - y|. \tag{6.4.77}$$

Now let $\varepsilon > 0$ and choose $\delta = \varepsilon/q$. Suppose

$$|x - y| < \delta = \frac{\varepsilon}{q}. \tag{6.4.78}$$

Then by the integral triangle inequality and the order or integration (Corollaries 6.4.5 and 6.3.10), we have

$$|g(x) - g(y)| = \left| \int_a^x f - \int_a^y f \right| = \left| \int_y^x f \right| \leq \int_y^x |f| \leq \int_y^x q = q|x - y| < \varepsilon. \tag{6.4.79}$$

Therefore, the definite integral g is uniformly continuous. $\qquad \square$

Modifying the hypothesis of integrability of the integrand in Theorem 6.4.14 with continat a point yields the second half of the Fundamental Theorem of Calculus.

Theorem 6.4.16: Fundamental Theorem of Calculus II

Suppose $f : [a, b] \to \mathbb{R}$ is integrable over $[a, b]$ and continuous at $c \in [a, b]$. Let g be the definite integral of f over $[a, b]$ given by $g(x) = \int_a^x f$. Then g is differentiable at c and we have

$$g'(c) = \left(\int_a^x f \right)'(c) = f(c). \tag{6.4.80}$$

Scratch Work 6.4.17: Simplify the difference quotient

Since the goal is show differentiability, consider the difference quotient of the definite integral g at c. We have

$$\frac{g(x) - g(c)}{x - c} = \frac{1}{x - c}\left(\int_a^x f - \int_a^c f \right) = \frac{1}{x - c}\left(\int_c^x f \right). \tag{6.4.81}$$

So, the goal of the proof is to show the limit of this difference quotient is $f(c)$. So, we would like to show for a suitable threshold δ that

$$0 < |x - c| < \delta \quad \Longrightarrow \quad \left| \frac{1}{x - c}\left(\int_c^x f \right) - f(c) \right| < \varepsilon. \tag{6.4.82}$$

Properties on integrals allow us to show a threshold δ from the continuity of f at c allows us to accomplish this goal.

Along the way and as a clever trick, it will help to keep in mind that $f(c)$ is a constant, so it is integrable over either $[c, x]$ or $[x, c]$ by Lemma 6.1.10 and we have

$$\int_c^x f(c) = f(c)(x - c) \quad \Longrightarrow \quad \frac{1}{x - c}\left(\int_c^x f(c) \right) = f(c). \tag{6.4.83}$$

Linearity of integration (Theorem 6.3.6) helps as well.

Proof of the Fundemental Theorem of Calculus II 6.4.16. Suppose $f : [a, b] \to \mathbb{R}$ is integrable over $[a, b]$ and continuous at $c \in [a, b]$, and let g be the definite integral of f over $[a, b]$ given by $g(x) = \int_a^x f$. Note that $f(c)$ is a constant, so it is integrable by Lemma 6.1.10 and we have

$$\int_c^x f(c) = f(c)(x - c) \quad \Longrightarrow \quad \frac{1}{x - c}\left(\int_c^x f(c) \right) = f(c). \tag{6.4.84}$$

Now let $\varepsilon > 0$. Since f is continuous at c, there is a threshold $\delta > 0$ such that

$$t \in [a, b] \quad \text{with} \quad |t - c| < \delta \quad \Longrightarrow \quad |f(t) - f(c)| < \frac{\varepsilon}{2}. \tag{6.4.85}$$

Next, to show the limit of the difference of g at c is $f(c)$, suppose

$$x \in [a, b] \quad \text{with} \quad |x - c| < \delta. \tag{6.4.86}$$

At this point, it helps to reintroduce the dummy variable t and dt. Note $t \in [c, x]$ and $t \in [x, c]$ imply

$$|t - c| \leq |x - c| < \delta. \tag{6.4.87}$$

From here, considering $c < x$ without loss of generality, splitting the definite integral at c (Theorem 6.4.11), the linearity of integration (Theorem 6.3.6) and a number of other integral properties yield

$$\left| \frac{g(x) - g(c)}{x - c} - f(c) \right| = \left| \frac{1}{x - c} \left(\int_a^x f - \int_a^c f \right) - \frac{1}{x - c} \left(\int_c^x f(c) \right) \right| \tag{6.4.88}$$

$$= \left| \frac{1}{x - c} \left(\int_c^x f \right) - \frac{1}{x - c} \left(\int_c^x f(c) \right) \right| \tag{6.4.89}$$

$$= \frac{1}{|x - c|} \left| \int_c^x f(t) - f(c)\, dt \right| \tag{6.4.90}$$

$$\leq \frac{1}{x - c} \int_c^x |f(t) - f(c)|\, dt \tag{6.4.91}$$

$$\leq \frac{1}{x - c} \int_c^x \frac{\varepsilon}{2}\, dt \tag{6.4.92}$$

$$= \frac{1}{x - c} \left(\frac{\varepsilon}{2}(x - c) \right) \tag{6.4.93}$$

$$= \frac{\varepsilon}{2} \tag{6.4.94}$$

$$< \varepsilon. \tag{6.4.95}$$

Hence, the limit of the difference quotient of g at c is $f(c)$. Therefore, g is differentiable at c and $g'(c) = f(c)$. $\qquad\square$

Exercises

6.4.1. Suppose f is integrable over $[a, b]$ and $f(x) \geq 0$ for all $x \in [a, b]$. Prove $\sqrt{f}$ is integrable over $[a, b]$.

6.4.2. Suppose f and g are integrable over $[a, b]$. Define $h : [a, b] \to \mathbb{R}$ by

$$h(x) = \max\{f(x), g(x)\} \qquad \text{for all } x \in [a, b]. \tag{6.4.96}$$

Prove h is integrable over $[a, b]$

6.4.3. Suppose f and g are continuous on $[a, b]$ and $\int_a^b f = \int_a^b g$. Prove there is some $c \in [a, b]$ such that $f(c) = g(c)$.

6.4.4. Suppose $f : [a, b] \to \mathbb{R}$ is continuous except at the finite number of points $x_1, \ldots, x_n \in [a, b]$. Prove that f is bounded, then f is integrable over $[a, b]$.

6.4.5. Suppose $g : [a, b] \to \mathbb{R}$ is continuous except on a convergent sequence $(x_n) \subseteq [a, b]$. Prove that if g is bounded, then g is integrable over $[a, b]$.

6.4.6. Suppose $h : [0, 1] \to \mathbb{R}$ is continuous except on the Cantor set $C \subseteq [0, 1]$ from Exercises 2.8.9, 3.5.4, and 3.6.9. Prove that if h is bounded, then h is integrable over $[0, 1]$. Hint: According to part (ii) of Exercise 3.5.4, the Cantor set C is arbitrarily small. (This exercise shows that a function can be integrable even if it is discontinuous on an uncountable set.)

Chapter 7

Pointwise and Uniform Convergence

One of the big pictures for continuity is their preservation of nice properties between points, sequences, and sets. They preserve closeness and convergence of sequences (Theorem 4.4.7), connectedness (Theorem 4.6.3), and compactness (Theorem 4.6.7). Continuity is also linear (Theorem 4.5.5), respecting scalar multiplication and sums.

Sequences and *series* (Chapter 8) of functions have the potential to preserve nice properties like continuity, differentiability, and integrability. In this short chapter, a pair of related limit processes for sequences of functions provide a framework to prove such preservation properties.

7.1 Sequences of functions

Pointwise convergence is and extension componentwise convergence. A reminder of componentwise convergence might help.

Example 7.1.1: A divergent component

Suppose $(\mathbf{x}_n)$ is a sequence of points in $\mathbb{R}^4$ where for each index n we have

$$\mathbf{x}_n = \begin{bmatrix} x_{1,n} \\ x_{2,n} \\ x_{3,n} \\ x_{4,n} \end{bmatrix} = \begin{bmatrix} 1 \\ 1/\sqrt{2^n} \\ (-1)^n/\sqrt{2^n} \\ (-1)^n \end{bmatrix}. \tag{7.1.1}$$

See Figure 7.1.1. The first three component sequences of $(\mathbf{x}_n)$ converge since we have

$$\lim_{n\to\infty} x_{1,n} = \lim_{n\to\infty} 1 = 1, \tag{7.1.2}$$

$$\lim_{n\to\infty} x_{2,n} = \lim_{n\to\infty} \frac{1}{\sqrt{2^n}} = 0, \qquad \text{and} \tag{7.1.3}$$

$$\lim_{n\to\infty} x_{3,n} = \lim_{n\to\infty} \frac{(-1)^n}{\sqrt{2^n}} = 0, \tag{7.1.4}$$

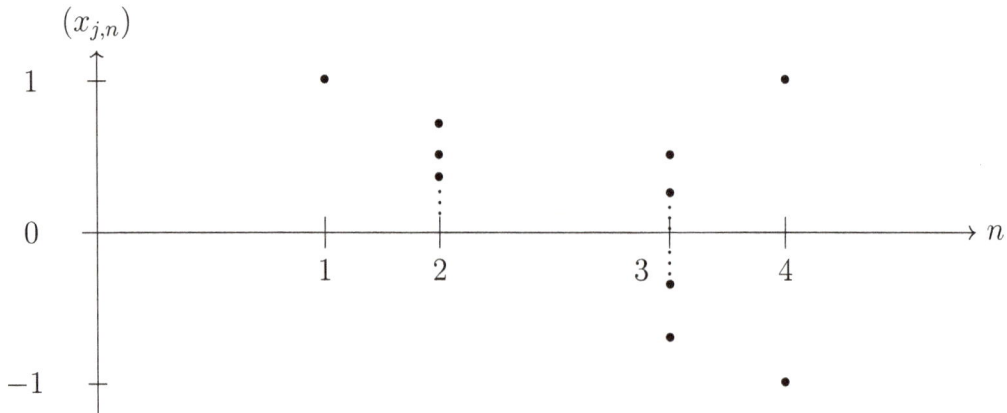

Figure 7.1.1: The sequence $(\mathbf{x}_n) \subseteq \mathbb{R}^4$ from Example 7.1.1 split into its component sequences $(x_{1,n}), (x_{2,n}), (x_{3,n}), (x_{4,n}) \subseteq \mathbb{R}$. The sequence $(\mathbf{x}_n)$ diverges since the component $(x_{4,n}) = ((-1)^n)$ alternates between -1 and 1, meaning it diverges by Divergence Criteria for Sequences 2.6.9.

where the limits for $(x_{2,n})$ and $(x_{3,n})$ are determined by Corollary 2.4.19. On the other hand, $(x_{4,n}) = ((-1)^n)$ has two subsequences converging to different limits. That is,

$$\lim_{k \to \infty} x_{2k} = 1 \neq -1 = \lim_{k \to \infty} x_{2k-1}. \qquad (7.1.5)$$

So, $(x_{4,n})$ diverges by Divergence Criteria for Sequences 2.6.9. Furthermore, $(\mathbf{x}_n)$ diverges by Theorem 2.4.11 since one of its components diverges.

Remark 7.1.2: From componentwise to pointwise

The leap from componentwise convergence of a sequence of vectors in a Euclidean space $\mathbb{R}^m$ to a the *pointwise convergence* of a sequence of functions comes from considering more sequences. Every sequence $(\mathbf{x}_n) \subseteq \mathbb{R}^m$ can be split into m component sequences $(x_{1,n}), \ldots, (x_{m,n}) \subseteq \mathbb{R}$, and $(\mathbf{x}_n)$ converges if and only if all of its component sequences converge Theorem 2.4.11.

In the setting of a sequence of functions (f_n) from the real line to the real line, the finite number of indices $1, \ldots, m$ for vectors are replaced with uncountably many inputs c from a common domain D where we write $(f_n(c))$ to denote the sequence of outputs associated with c. Then much like componentwise convergence, *pointwise convergence* occurs when each of the *output sequences* converges.

Definition 7.1.3: Output sequence

A sequence of functions (f_n) where $f_n : D \to \mathbb{R}$ for some $D \subseteq \mathbb{R}$ and every index $n \in \mathbb{N}$ is called a *sequence of functions on D*. For each $c \in D$, the *output sequence at c* is the sequence $(f_n(c))$.

Example 7.1.4: Sequence of monomials

Consider the sequence of monomials (f_n) on the compact interval $[0, 1]$ given by

$$f_n(x) = x^n \quad \text{for all} \quad x \in [0, 1] \text{ and } n \in \mathbb{N}. \tag{7.1.6}$$

See Figure 7.1.2. Each output sequence $(f_n(c)) = (c^n)$ exhibits its own behavior, depending on the input $c \in [0, 1]$. For instance, when $c = 0$ or $c = 1$, the output sequence is constant. On the other hand, inputs $c = 1/2$ and $c = 0.9$ yield strictly decreasing output sequences that converge to 0. For each index $n \in \mathbb{N}$ we have $f_n(0) = 0$, $f_n(1/2) = 1/2^n$, $f_n(0.9) = (0.9)^n$, and $f_n(1) = 1$. See Figure 7.1.2. Since $0 \leq 1/2 < 0.9 < 1$, by Corollary 2.4.19 we also have

$$\lim_{n \to \infty} f_n(0) = \lim_{n \to \infty} f_n(1/2) = \lim_{n \to \infty} f_n(0.9) = 0, \quad \text{while} \tag{7.1.7}$$

$$\lim_{n \to \infty} f_n(1) = 1. \tag{7.1.8}$$

Moreover, by Corollary 2.4.19, whenever $0 \leq c < 1$ we have

$$\lim_{n \to \infty} f_n(c) = c^n = 0. \tag{7.1.9}$$

A sequence of functions that produces convergent output sequences for each of its inputs is said to be *pointwise convergent*.

Definition 7.1.5: Pointwise convergence

Suppose $D \subseteq \mathbb{R}$, $c \in D$, and $f_n : D \to \mathbb{R}$ for every index $n \in \mathbb{N}$. The sequence of functions (f_n) *converges at* c if the output sequence $(f_n(c))$ converges. The sequence (f_n) *converges pointwise* (on D) if every output sequence converges. In this case, the function $f : D \to \mathbb{R}$ defined by

$$f(x) = \lim_{n \to \infty} f_n(x) \quad \text{for all} \quad x \in D \tag{7.1.10}$$

is called the *pointwise limit* of (f_n). See Figure 7.1.2.

Pointwise convergence can break continuity.

Example 7.1.6: Pointwise limit of monomials

For the sequence of functions (f_n) on $[0, 1]$ in Example 7.1.4 and Figure 7.1.2 given by

$$f_n(x) = x^n \quad \text{for all} \quad x \in [0, 1] \quad \text{and} \quad n \in \mathbb{N}, \tag{7.1.11}$$

the pointwise limit of (f_n) is the function $f : [0, 1] \to \mathbb{R}$ where

$$f(x) = \lim_{n \to \infty} f_n(x) = \begin{cases} 0, & \text{if } 0 \leq x < 1, \\ 1, & \text{if } x = 1. \end{cases} \tag{7.1.12}$$

Moreover, even though each f_n is continuous, the pointwise limit f is not.

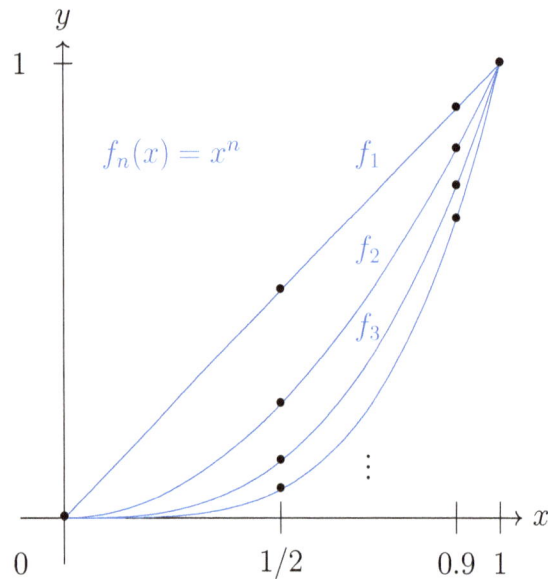

Figure 7.1.2: The sequence of monomials $(f_n(x)) = (x^n)$ on $[0,1]$ along with output sequences for $c = 1/2$ and $c = 0.9$ from Example 7.1.4. Play around with the Desmos activity "Pointwise convergence of monomials" accessed through the QR code. https://www.desmos.com/calculator/4zehljtc2q

Proof for Example 7.1.6. The formula for the pointwise limit f in (7.1.12) follows immediately from Example 7.1.4 and the definition for pointwise limit (Definition 7.1.5). Since polynomials are continuous (Theorem 4.5.2), each function f_n is continuous. However, the pointwise limit f is discontinuous at $c = 1$ because

$$\lim_{n\to\infty}\left(1 - \frac{1}{2n}\right) = 1 \tag{7.1.13}$$

in the domain $[0, 1]$ while

$$\lim_{n\to\infty} f\left(1 - \frac{1}{2n}\right) = 0 \neq 1 = f(1) \tag{7.1.14}$$

in the range (see Discontinuity Criteria 4.6.13). □

Pointwise convergence can break anticipated values of integrals, too.

Example 7.1.7: A sequence of blocks

Consider the sequence of *blocks* (b_n) on $[0, 2]$ given by

$$b_n(x) = \begin{cases} 0, & \text{when } x = 0 \text{ or } 1/n \leq x \leq 2, \\ n, & \text{when } 0 < x < 1/n. \end{cases} \tag{7.1.15}$$

See Figure 7.1.3. Even though the sequence (b_n) converges pointwise to 0 on $[0, 2]$, the integrals satisfy $\int_0^2 b_n = 1$ for each $n \in \mathbb{N}$. Hence,

$$\lim_{n\to\infty} \int_0^2 b_n = 1 \neq 0 = \int_0^2 0 = \int_0^2 \left(\lim_{n\to\infty} b_n\right). \tag{7.1.16}$$

So, the limit of the integrals is not the integral of the pointwise limit.

Proof for Example 7.1.7. For each index $n \in \mathbb{N}$, the function b_n is continuous on $[0, 2]$ except at the real numbers $c = 0$ and $c = 1/n$. By splitting the integral at $1/n$ in the domain (Theorem 6.4.11 and taking the integrals of constants (Lemma 6.1.10), we have

$$\int_0^2 b_n = \int_0^{1/n} n + \int_{1/n}^2 0 = n(1/n) + 0(2 - (1/n)) = 1. \tag{7.1.17}$$

Therefore,

$$\lim_{n\to\infty} \int_0^2 b_n = 1. \tag{7.1.18}$$

To establish that the pointwise limit of (b_n) is 0, first note $b_n(0) = 0$ for every index $n \in \mathbb{N}$. Hence,

$$\lim_{n\to\infty} b_n(0) = 0. \tag{7.1.19}$$

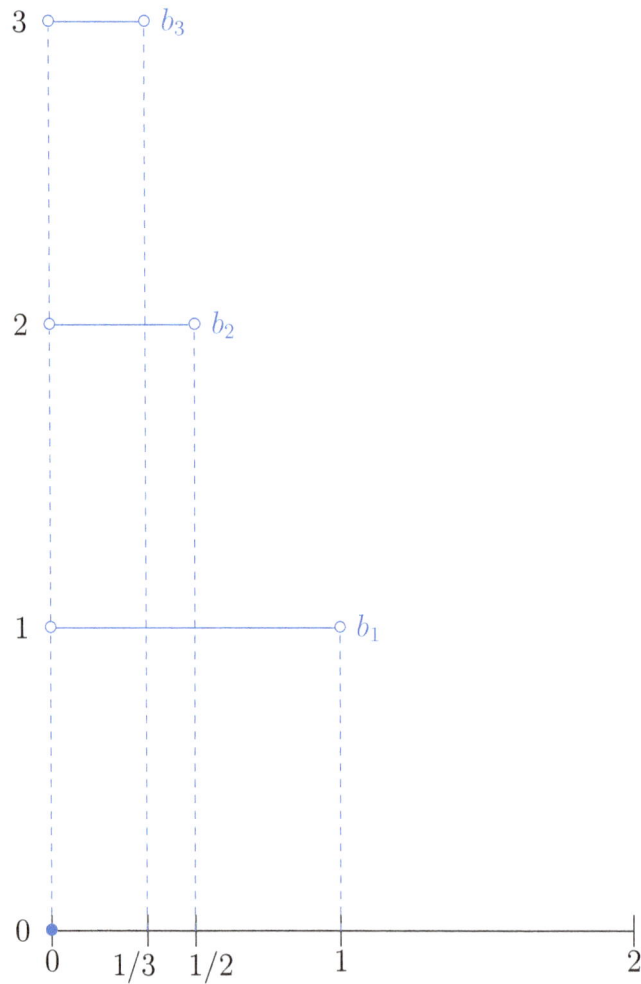

Figure 7.1.3: The sequence of blocks (b_n) on $[0, 2]$ converges pointwise to 0 even though their integrals are all equal to 1. In a way, pointwise convergence can break integration. See Example 7.1.7.

Now suppose $0 < x \leq 2$. Then the output sequence $b_n(x)$ is eventually the constant 0. To see this, note that by the corollary of the Archimedean Property (Corollary 1.4.8), there is an index $n_x \in \mathbb{N}$ such that

$$0 < 1/n_x < x. \tag{7.1.20}$$

Hence, for each index $n \geq n_x$ we have $b_n(x) = 0$. Furthermore, for every $\varepsilon > 0$, the index n_x is a threshold for the convergence of the output sequence $(b_n(x))$ to 0 since, for every $n \geq n_x$, we have

$$|b_n(x) - 0| = |0 - 0| = 0 < \varepsilon. \tag{7.1.21}$$

Therefore, 0 is the pointwise limit of (b_n) on $[0, 2]$ since

$$\lim_{n \to \infty} b_n(x) = 0 \quad \text{for every } x \in [0, 2]. \tag{7.1.22}$$

$\square$

Some sequences of functions fail to converge pointwise.

Example 7.1.8: Powers of sine

Consider the sequence of functions (g_n) on $[0, 2\pi]$ where

$$g_n(x) = (\sin x)^n = \sin^n x \quad \text{for all } x \in [0, 2\pi] \text{ and } n \in \mathbb{N}. \tag{7.1.23}$$

See Figure 7.1.4 as well as Example 7.1.1 and its Figure 7.1.1. For $c = 3\pi/2$ and every $n \in \mathbb{N}$, we have

$$g_n(3\pi/2) = \sin^n(3\pi/2) = (-1)^n = \begin{cases} -1, & \text{when } n \text{ is odd,} \\ 1, & \text{when } n \text{ is even.} \end{cases} \tag{7.1.24}$$

So, by Divergence Criteria for Sequences 2.6.9, the output sequence $(g_n(3\pi/2))$ diverges since two of its subsequences converge to different limits:

$$\lim_{k \to \infty} g_{2k-1}(3\pi/2) = -1 \quad \text{while} \quad \lim_{k \to \infty} g_{2k}(3\pi/2) = 1. \tag{7.1.25}$$

Therefore, (g_n) does not converge pointwise on $[0, 2\pi]$.

The other inputs $c \in [0, 2\pi] \setminus \{3\pi/2\}$ have convergent output sequences exhibiting a variety of behaviors. For instance,

$$g_n(\pi/2) = \sin^n(\pi/2) = 1, \tag{7.1.26}$$

$$g_n(3\pi/4) = \sin^n(3\pi/4) = \left(\frac{\sqrt{2}}{2}\right)^n = \frac{1}{\sqrt{2^n}}, \quad \text{and} \tag{7.1.27}$$

$$g_n(5\pi/4) = \sin^n(5\pi/4) = \left(-\frac{\sqrt{2}}{2}\right)^n = \frac{(-1)^n}{\sqrt{2^n}} \tag{7.1.28}$$

for every index $n \in \mathbb{N}$. So,

$$\lim_{k \to \infty} g_n(\pi/2) = 1 \quad \text{and} \quad \lim_{k \to \infty} g_n(3\pi/4) = \lim_{k \to \infty} g_n(5\pi/4) = 0, \tag{7.1.29}$$

where the latter two follow from Corollary 2.4.19.

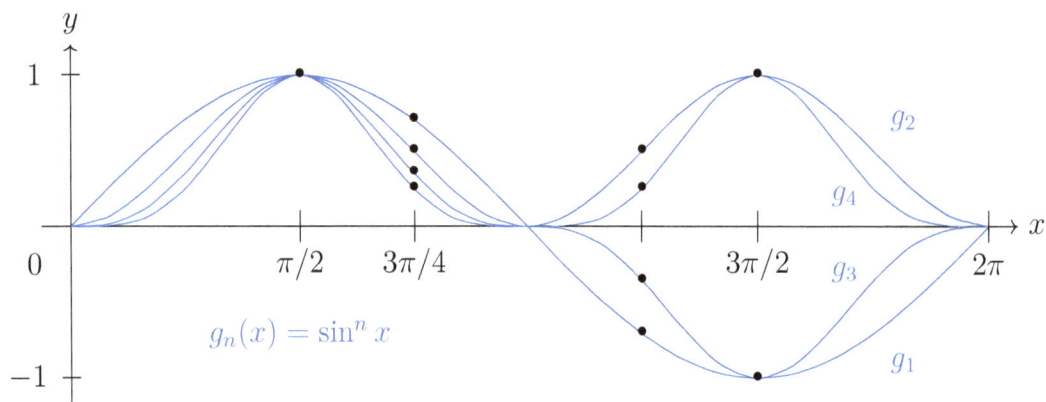

Figure 7.1.4: The sequence of powers of the sine function $(g_n(x)) = (\sin^n x)$ on $[0, 2\pi]$ along with output sequences for $c = \pi/2, 3\pi/4, 5\pi/4$, and $3\pi/2$, although $5\pi/4$ is not labeled on the x-axis. See Examples 7.1.8 and 7.1.1 as well as Figure 7.1.1. Also, play around with the Desmos activity "Sequence of powers of sine" accessed through the QR code. https://www.desmos.com/calculator/ofe6otbfyu

We can dive a bit deeper into pointwise convergence, or lack thereof, by considering the *pointwise thresholds* and *rates of convergence* for convergent output sequences. See the definition for thresholds in Definition 2.2.1. Pointwise thresholds determine the rate of convergent of their output sequences.

Definition 7.1.9: Pointwise thresholds

Let (f_n) be a pointwise convergent sequence of functions where $f_n : D \to \mathbb{R}$ for some $D \subseteq \mathbb{R}$ and every index $n \in \mathbb{N}$. For each $c \in D$, a *pointwise threshold at c* is an index $n_c \in \mathbb{N}$ which is a threshold for the convergence of the output sequence $(f_n(c))$.

More specifically, the index $n_c \in \mathbb{N}$ is a pointwise threshold at c if for every distance $\varepsilon > 0$ and some $y_c \in \mathbb{R}$ we have

$$n \geq n_c \quad \text{with} \quad n \in \mathbb{N} \quad \Longrightarrow \quad |f_n(c) - y_c| < \varepsilon. \tag{7.1.30}$$

Finding pointwise thresholds follows the guide for working with limits of sequences in Remark 2.2.4.

Example 7.1.10: Pointwise thresholds for the sequence of monomials

Consider the sequence $f_n = x^n$ on $[0, 1]$ from Example 7.1.4. The index $n_0 = n_1 = 1$ is a pointwise threshold for both $c = 0$ and $c = 1$ since for every distance $\varepsilon > 0$ and every index $n \in \mathbb{N}$ we have

$$|f_n(0) - 0| = 0 < \varepsilon \quad \text{and} \quad |f_n(1) - 1| = |1^n - 1| = 0 < \varepsilon. \tag{7.1.31}$$

For $c = 1/2$, a bit more work is involved. We want to end up with

$$|f_n(1/2) - 0| = 1/2^n < \varepsilon. \tag{7.1.32}$$

Following the guide in Remark 2.2.4, solving for n looks like this:

$$1/2^n < \varepsilon \quad \Longrightarrow \quad n \ln(1/2) < \ln \varepsilon \quad \Longrightarrow \quad n > \frac{\ln \varepsilon}{\ln(1/2)}, \tag{7.1.33}$$

where the last implication holds since $\ln(1/2) < 0$. Hence, any index

$$n_{1/2} > \frac{\ln \varepsilon}{\ln(1/2)} \tag{7.1.34}$$

serves a pointwise threshold for the sequence $(f_n(1/2)) = (1/2^n)$. Very similarly, for the input $c = 0.9$ we want to end up with

$$|f_n(0.9) - 0| = (0.9)^n < \varepsilon. \tag{7.1.35}$$

This leads to

$$(0.9)^n < \varepsilon \quad \Longrightarrow \quad n \ln(0.9) < \ln \varepsilon \quad \Longrightarrow \quad n > \frac{\ln \varepsilon}{\ln(0.9)}, \tag{7.1.36}$$

where the last implication holds since $\ln(0.9) < 0$. Hence, any index

$$n_{0.9} > \frac{\ln \varepsilon}{\ln(0.9)} \tag{7.1.37}$$

serves as a pointwise threshold for the sequence $(f_n(0.9)) = ((0.9)^n)$.

Now, instead of proving the limit of both $(f_n(1/2)) = (1/2^n)$ and $(f_n(0.9)) = ((0.9)^n)$ is 0 (which is redundant in light of Example 7.1.4), compare their rates of convergence. For the distance $\varepsilon_1 = 1/10$, the formulas (7.1.34) and (7.1.37) for the pointwise thresholds at $c = 1/2$ at $c = 0.9$ yield

$$n_{1/2} > \frac{\ln(1/10)}{\ln(1/2)} > 3.321 \quad \text{and} \tag{7.1.38}$$

$$n_{0.9} > \frac{\ln(1/10)}{\ln(0.9)} > 22.854. \tag{7.1.39}$$

So, $n_{1/2} = 4$ suffices for the distance $\varepsilon_1 = 1/10$ and the output sequence $(f_n(1/2)) = (1/2^n)$. However, $n_{0.9} = 4$ does not suffice for the same distance $\varepsilon_1 = 1/10$ the output sequence $(f_n(0.9)) = ((0.9)^n)$ since

$$|f_4(0.9) - 0| = (0.9)^4 = 0.6561 \geq 0.1 = 1/10 = \varepsilon_1. \tag{7.1.40}$$

In other words and to summarize, the output sequences $(f_n(0)) = (0^n)$, $(f_n(1/2)) = (1/2^n)$, and $(f_n(0.9)) = ((0.9)^n)$ all converge to 0, but they do so at different rates: For the distance $\varepsilon_1 = 1/10$, the *smallest* pointwise thresholds for $c = 0, 1/2$, and 0.9 are respectively given by

$$n_0 = 1, \quad n_{1/2} = 4, \quad \text{and} \quad n_{0.9} = 23. \tag{7.1.41}$$

So, $(f_n(0.9)) = ((0.9)^n)$ converges to 0 slower than $(f_n(0)) = (0^n)$ and $(f_n(1/2)) = (1/2^n)$ since $n_{0.9} = 23$ is the largest of the three minimal pointwise thresholds.

To recap some of the results in this section, sequences of functions can exhibit all sorts of behaviors.

- In Example 7.1.8, the sequence $(g_n(x)) = (\sin^n x)$ comprises continuous functions but fails to converge pointwise on $[0, 2\pi]$. The issue occurs with the divergent output sequence at $c = 3\pi/2$ given by $(g_n(3\pi/2)) = ((-1)^n)$.

- In Examples 7.1.4 and 7.1.6, the sequence $(f_n(x)) = (x^n)$ converges pointwise on $[0, 1]$ to the function

$$f(x) = \begin{cases} 0, & \text{if } 0 \leq x < 1, \\ 1, & \text{if } x = 1. \end{cases} \tag{7.1.42}$$

However, the continuity of the polynomials $f_n(x) = x^n$ is broken by the pointwise limit

f which is discontinuous at $c = 1$. So, the pointwise limit of continuous functions is not necessarily continuous.

- In Example 7.1.7, the sequence of functions (b_n) converges pointwise to 0, but the integral of each box b_n is 1. So, the limit of integrals is not necessarily the integral of the pointwise limit.

This begs the question: When does a sequence of functions preserve nice properties? The next section provides an answer with the idea of *uniform convergence.*

Exercises

7.1.1. Construct a function $g : \mathbb{R} \to \mathbb{R}$ satisfying the following properties and prove your result:

(i) g is increasing.

(ii) g is continuous at every $c \in \mathbb{R} \backslash \mathbb{N}$.

(iii) g is discontinuous at every $c \in \mathbb{N}$.

7.1.2. Let $f_n : \mathbb{R} \to \mathbb{R}$ be given by

$$f_n(x) = \frac{(x^2 + 1)}{n^2} \tag{7.1.43}$$

for each $n \in \mathbb{N}$ and $x \in \mathbb{R}$. Prove (f_n) converges pointwise to $f(x) = 0$ on $\mathbb{R}$. Do the pointwise thresholds depend on the inputs?

7.1.3. Let $g_n : \mathbb{R} \to \mathbb{R}$ be given by

$$g_n(x) = \frac{\sin x}{n} + x + \cos x \tag{7.1.44}$$

for each $n \in \mathbb{N}$ and $x \in \mathbb{R}$. Prove (g_n) converges pointwise to $g(x) = x + \cos x$ on $\mathbb{R}$. Do the pointwise thresholds depend on the inputs?

7.1.4. Let $h_n : \mathbb{R} \to \mathbb{R}$ be given by

$$h_n(x) = \frac{x}{1 + nx^2} \tag{7.1.45}$$

for each $n \in \mathbb{N}$ and $x \in \mathbb{R}$. Prove (h_n) converges pointwise on $\mathbb{R}$. Do the pointwise thresholds depend on the inputs?

7.1.5. Find an example of a sequence of bounded functions defined on $(0, 1)$ that converges pointwise to an unbounded limit on $(0, 1)$.

Figure 7.2.1: The QR code takes you to the Desmos activity "Uniform convergence" which is designed to accompany Definition 7.2.1. It allows for an exploration of a sequence of functions that converge uniformly on the real line. https://www.desmos.com/calculator/o3qsnaagez

7.2   Uniform convergence

Uniform convergence for sequences of functions is an analog of uniform continuity (Definition 4.7.1) in that we have

> *One threshold to rule them all.*

In this case, the *uniform threshold* goes by "Sauron's Nani"[1] and is a single threshold which serves as the pointwise threshold for every output sequence.

Definition 7.2.1: Uniform convergence

Suppose $D \subseteq \mathbb{R}$ and $f_n : D \to \mathbb{R}$ for every index $n \in \mathbb{N}$. The sequence of functions (f_n) *converges uniformly to f* (on D) if for every distance $\varepsilon > 0$ there is an index $n_\varepsilon \in \mathbb{N}$ such that for every input $c \in D$ we have

$$n \geq n_\varepsilon \quad \text{with} \quad n \in \mathbb{N} \quad \implies \quad |f_n(c) - f(c)| < \varepsilon. \tag{7.2.1}$$

In this case, the index n_ε is called a *uniform threshold* and f is called the *uniform limit* of (f_n). See Figure 7.2.1.

In Example 7.1.6, the sequence of functions (x^n) is shown to converge pointwise on the compact interval $[0, 1]$. By the restricting the domain to the smaller interval $[0, 0.9]$, the modified version of this sequence converges uniformly.

---

[1]Thank you, Nikita Campos!

Example 7.2.2: Revising the domain for monomials yields uniform convergence

Consider the revised sequence of functions (f_n) on the compact interval $[0, 0.9]$ given by

$$f_n(x) = x^n \quad \text{for all} \quad x \in [0, 0.9] \text{ and } n \in \mathbb{N}. \tag{7.2.2}$$

The sequence (f_n) converges uniformly to 0 on $[0, 0.9]$. See Figure 7.2.2.

Scratch Work 7.2.3: Similar scratch for pointwise and uniform convergence

The scratch work for showing uniform convergence and pointwise convergence are very similar since both involve finding suitable thresholds. The key difference is that pointwise convergence is satisfied when each input has its own pointwise threshold, but uniform convergence requires a single threshold to serve as the pointwise threshold for every input at the same time. So, to guide the determination of suitable choice for a uniform threshold, see Figure 7.2.2. Once the output $f_n(0.9) = (0.9)^n$ is within ε of 0, all of the outputs $f_n(c) = (c)^n$ for every c in the domain $[0, 0.9]$ are within ε of 0 as well.

If there is an output sequence with the slowest rate of convergence, its pointwise threshold will suffice for all the output sequences. In the case of Example 7.2.2, the output sequence $f_n(0.9) = (0.9)^n$ has the slowest rate of convergence and its pointwise threshold $n_{0.9} \in \mathbb{N}$ serves as the uniform threshold. The formula describing $n_{0.9}$ is found in Example 7.1.10, stemming from the implications

$$(0.9)^n < \varepsilon \quad \implies \quad n \ln(0.9) < \ln \varepsilon \quad \implies \quad n > \frac{\ln \varepsilon}{\ln(0.9)}, \tag{7.2.3}$$

where the last inequality holds since $\ln(0.9) < 0$. So, any index $n_\varepsilon \in \mathbb{N}$ such that

$$n_\varepsilon = n_{0.9} > \frac{\ln \varepsilon}{\ln(0.9)} \tag{7.2.4}$$

should serve as a *uniform* threshold for the uniform convergence of the sequence $(f_n(x)) = (x^n)$ to $f(x) = 0$.

Proof for Example 7.2.2. For any $c \in [0, 0.9]$ and any index $n \in \mathbb{N}$ we have

$$0 \le c \le 0.9 \quad \implies \quad 0 \le c^n \le 0.9^n \tag{7.2.5}$$

since each positive power of x is increasing. Now let $\varepsilon > 0$. Choose an index $n_\varepsilon \in \mathbb{N}$ such that

$$n_\varepsilon > \frac{\ln \varepsilon}{\ln(0.9)} \quad \Longleftrightarrow \quad 0.9^{n_\varepsilon} < \varepsilon. \tag{7.2.6}$$

Then for every input $c \in [0, 0.9]$ and index $n \in \mathbb{N}$ we have

$$n \ge n_\varepsilon \quad \implies \quad |f_n(c) - 0| = c^n \le 0.9^n \le 0.9^{n_\varepsilon} < \varepsilon \tag{7.2.7}$$

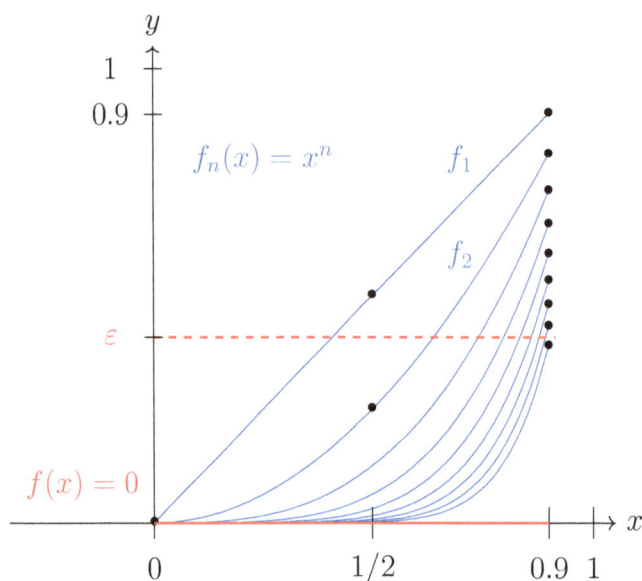

Figure 7.2.2: The revised sequence of functions $(f_n(x)) = (x^n)$ on $[0, 0.9]$ along with output sequences for $c = 1/2$ and $c = 0.9$, and a distance $\varepsilon > 0$ away from the pointwise limit $f(x) = 0$. See Example 7.2.2. The output sequence $(f_n(0.9)) = ((0.9)^n)$ has the slowest rate of convergence, so its threshold serves as the uniform threshold. Play around with the Desmos activity "Pointwise convergence of monomials" accessed through the QR code. https://www.desmos.com/calculator/4zehljtc2q

since $0 < 0.9 < 1$. Therefore, by Definition 7.2.1, n_ε is a uniform threshold and (f_n) converges uniformly to 0 on $[0, 0.9]$. □

The proof of Example 7.2.2 motivates the following lemma. If we can find a sequence of nonnegative real numbers that converge to zero and act as bounds on the difference of the outputs of a sequence of functions and another function, then this other function is the uniform limit.

Lemma 7.2.4: Sequence of bounds implies uniform convergence

Suppose $D \subseteq \mathbb{R}$, $f : D \to \mathbb{R}$, (f_n) is a sequence of real-valued functions on D, and (a_n) is a sequence of nonnegative real numbers where both

$$\lim_{n \to \infty} a_n = 0 \qquad \text{and} \qquad |f_n(x) - f(x)| \le a_n \tag{7.2.8}$$

for each index $n \in \mathbb{N}$ and each input $x \in D$. Then (f_n) converges uniformly to f.

Scratch Work 7.2.5: Uniform threshold from the convergence of real numbers

The convergence of the sequence of bounds (a_n) to zero provides a threshold via Definition 2.2.1 which immediately serves as the uniform threshold for the uniform convergence of (f_n) to f as in Definition 7.2.1.

Proof of Lemma 7.2.4. Suppose $D \subseteq \mathbb{R}$, $f : D \to \mathbb{R}$, (f_n) is a sequence of real-valued functions on D, and (a_n) is a sequence of nonnegative real numbers where (a_n) converges to zero. Further suppose that for each $n \in \mathbb{N}$ and each $x \in D$ we have

$$|f_n(x) - f(x)| \le a_n. \tag{7.2.9}$$

Now let $\varepsilon > 0$. By the definition of sequential limit (Definition 2.2.1) and inequality (7.2.9), there is a threshold $n_\varepsilon \in \mathbb{N}$ such that

$$n \ge n_\varepsilon \qquad \implies \qquad |f_n(x) - f(x)| \le a_n = |a_n - 0| < \varepsilon. \tag{7.2.10}$$

Therefore, n_ε is a uniform threshold and (f_n) converges uniformly to f as in Definition 7.2.1. □

Uniform convergence is a strengthening of pointwise convergence.

Lemma 7.2.6: Uniform convergence implies pointwise convergence

If a sequence of functions (f_n) converges uniformly to f on $D \subseteq \mathbb{R}$, then (f_n) converges pointwise to f on D.

Scratch Work 7.2.7: "One threshold to rule them all"

Any uniform threshold serves as the pointwise threshold for every output sequence. The proof follows directly from the definitions.

Proof of Lemma 7.2.6. Suppose (f_n) converges uniformly to f on D and $\varepsilon > 0$. By Definition 7.2.1, there is a uniform threshold $n_\varepsilon \in \mathbb{N}$ such that for every input $c \in D$ we have

$$n \geq n_\varepsilon \quad \text{with} \quad n \in \mathbb{N} \quad \Longrightarrow \quad |f_n(c) - f(c)| < \varepsilon. \tag{7.2.11}$$

So by Definitions 7.1.3, 7.1.5, and 7.1.9, for every $c \in D$, the uniform threshold n_ε is the pointwise threshold for the output sequence $(f_n(c))$ and the uniform limit f is the pointwise limit.   □

Uniform convergence is another property that exhibits linearity.

Theorem 7.2.8: Linearity of uniform convergence

Suppose (f_n) and (g_n) converge uniformly on D to f and g, respectively, and suppose $\alpha \in \mathbb{R}$. Then $(f_n + g_n)$ converges uniformly to $f + g$ on D, (αf_n) converges uniformly to αf on D, and for every $x \in D$ we have

  (i) $\displaystyle \lim_{n \to \infty} (f_n(x) + g_n(x)) = \lim_{n \to \infty} f_n(x) + \lim_{n \to \infty} g_n(x) = f(x) + g(x)$ *(additivity);*   and

  (ii) $\displaystyle \lim_{n \to \infty} (\alpha f_n(x)) = \alpha \lim_{n \to \infty} f_n(x) = \alpha f(x)$ *(homogeneity).*

Scratch Work 7.2.9: Almost identical to linearity of sequential limits

The scratch work and proof of Theorem 7.2.8 is essentially the same as Scratch Work 2.3.11 and 2.3.12 for the proof of Theorem 2.3.9.

For both additivity and homogeneity, we can verify the definition of uniform convergence (Definition 7.2.1) holds by considering an arbitrary $\varepsilon > 0$ and finding a suitable uniform thresholds which serves as pointwise threshold for every input. These threshold can be shown to ensure the outputs with indices large enough are within ε of the proposed uniform limit across the whole common domain.

The proof is split in two to address additivity and homogeneity separately.

Proof of additivity (i) in Theorem 7.2.8. Assume (f_n) and (g_n) converge uniformly on D to f and g, respectively, and let $\varepsilon > 0$. Then $\varepsilon/2 > 0$ and by the definition of uniform convergence (Definition 7.2.1) there are uniform thresholds $j_{\varepsilon/2}$ and $k_{\varepsilon/2}$ such that for every $x \in D$ we have

$$n \geq j_{\varepsilon/2} \implies |f_n(x) - f(x)| < \frac{\varepsilon}{2} \quad \text{and} \tag{7.2.12}$$

$$n \geq k_{\varepsilon/2} \implies |g_n(x) - g(x)| < \frac{\varepsilon}{2}. \tag{7.2.13}$$

Define n_ε to be the larger of the two uniform thresholds (so $n_\varepsilon = \max\{j_{\varepsilon/2}, k_{\varepsilon/2}\}$). Then every index n where $n \geq n_\varepsilon$ is large enough to give us both $n \geq j_{\varepsilon/2}$ and $n \geq k_{\varepsilon/2}$. So by the triangle inequality (1.2.35), (7.2.12), and (7.2.13), $n \geq n_\varepsilon$ implies that for every $x \in D$ we have

$$|(f_n(x) + g_n(x)) - (f(x) + g(x))| = |f_n(x) - f(x) + g_n(x) - g(x)| \qquad (7.2.14)$$

$$\leq |f_n(x) - f(x)| + |g_n(x) - g(x)| \qquad (7.2.15)$$

$$< \frac{\varepsilon}{2} + \frac{\varepsilon}{2} \qquad (7.2.16)$$

$$= \varepsilon. \qquad (7.2.17)$$

Therefore, n_ε is a uniform threshold, $(f_n + g_n)$ converges uniformly to $f + g$ on D, and for every $x \in D$ we have

$$\lim_{n \to \infty} (f_n(x) + g_n(x)) = \lim_{n \to \infty} f_n(x) + \lim_{n \to \infty} g_n(x) = f(x) + g(x). \qquad (7.2.18)$$

Hence, uniform convergence is additive. $\qquad \square$

Proof of homogeneity (ii) in Theorem 7.2.8. Assume (f_n) converges uniformly on D to f and assume $\alpha \in \mathbb{R}$. This proof has two cases from here: (i) $\alpha = 0$ and (ii) $\alpha \neq 0$.

$\quad$ Case (i): Suppose $\alpha = 0$ and $\varepsilon > 0$. Define $n_\varepsilon = 21$. Then for every index $n \geq n_\varepsilon = 21$ and every $x \in D$ we have

$$|\alpha f_n(x) - \alpha f(x)| = |0 - 0| = 0 < \varepsilon. \qquad (7.2.19)$$

Therefore, $n_\varepsilon = 21$ is a uniform threshold, (αf_n) converges uniformly to αf on D, and for every $x \in D$ we have

$$\lim_{n \to \infty} (\alpha f_n(x)) = \alpha \lim_{n \to \infty} f_n(x) = \alpha f(x). \qquad (7.2.20)$$

$\quad$ Case (ii): Assume $\alpha \neq 0$ and $\varepsilon > 0$. Then $\varepsilon/|\alpha| > 0$ and by the definition of uniform convergence (Definition 7.2.1) there is a uniform threshold $n_{\varepsilon/|\alpha|}$ such that for every $x \in D$

$$n \geq n_{\varepsilon/|\alpha|} \quad \implies \quad |\alpha f_n(x) - \alpha f(x)| < \frac{\varepsilon}{|\alpha|}. \qquad (7.2.21)$$

By (1.2.33) and (7.2.21), for every indx $n \geq n_\varepsilon$ and every $x \in D$ we have

$$|\alpha f_n(x) - \alpha f(x)| = |\alpha||f_n(x) - f(x)| \qquad (7.2.22)$$

$$< |\alpha| \frac{\varepsilon}{|\alpha|} \qquad (7.2.23)$$

$$= \varepsilon. \qquad (7.2.24)$$

Therefore, $n_{\varepsilon/|\alpha|}$ is a uniform threshold, (αf_n) converges uniformly to αf on D, and for every $x \in D$ we have

$$\lim_{n \to \infty} (\alpha f_n(x)) = \alpha \lim_{n \to \infty} f_n(x) = \alpha f(x). \qquad (7.2.25)$$

Hence, uniform convergence is homogeneous. $\qquad \square$

$\quad$ As mentioned in Remark 1.6.18, a corollary of the linearity of sequential limits holds for linear combinations. As with the proof Corollary 1.6.16 on arbitrarily close and linear combinations of sets and the proof of Corollary 2.3.13 on linear combinations and sequential limits, the proof of Corollary 7.2.10 follows from induction. So, it is left as an exercise. Here, the notation $f_{j,n}$ indicates the nth function of the jth sequence.

Corollary 7.2.10: Linear combinations and uniform convergence

Suppose $k \in \mathbb{N}$ and for each $j = 1, \ldots, k$ we have $\alpha_j \in \mathbb{R}$ and the sequence $(f_{j,n})$ converges uniformly to f_j on D. Then the sequence of linear combinations $\sum_{j=1}^{k} \alpha_j f_{j,n}$ converges uniformly to the linear combination $\sum_{j=1}^{k} \alpha_j f_j$ on D and we have

$$\lim_{n \to \infty} \left(\sum_{j=1}^{k} \alpha_j f_{j,n}(x) \right) = \sum_{j=1}^{k} \left(\alpha_j \lim_{n \to \infty} f_{j,n}(x) \right) = \sum_{j=1}^{k} \left(\alpha_j f_j(x) \right). \qquad (7.2.26)$$

The section concludes with a *Cauchy criterion for uniform convergence*. As with the other Cauchy criteria, we get uniform convergence without having a candidate for the limit in mind.

Theorem 7.2.11: Cauchy criterion for uniform convergence

A sequence of functions (f_n) converges uniformly on a set D if and only if for every $\varepsilon > 0$ there is a uniform threshold $n_\varepsilon \in \mathbb{N}$ such that for all $x \in D$ we have

$$j, k \in \mathbb{N} \text{ with } j, k \geq n_\varepsilon \qquad \Longrightarrow \qquad |f_j(x) - f_k(x)| < \varepsilon. \qquad (7.2.27)$$

Scratch Work 7.2.12: Cauchy criteria and the triangle inequality

The forward implication follows from a standard triangle inequality argument. The backwards implication is more subtle since a candidate for the uniform limit is needed. It comes from applying the Cauchy criterion for sequences (Theorem 2.6.5) to the output sequences, ensuring each output sequence converges. The uniform convergence to the candidate then follows from the implication (7.2.27) and properties of sequential limits.

Proof of Theorem 7.2.11. To prove the forward implication, suppose (f_n) converges uniformly to f on a set D. Let $\varepsilon > 0$. By the definition of uniform convergence (Definition 7.2.1), there is a uniform threshold $n_{\varepsilon/2} \in \mathbb{N}$ such that for all $x \in D$,

$$n \in \mathbb{N} \text{ with } n \geq n_{\varepsilon/2} \qquad \Longrightarrow \qquad |f_n(x) - f(x)| < \frac{\varepsilon}{2}. \qquad (7.2.28)$$

Now suppose $j, k \in \mathbb{N}$ where $j, k \geq n_{\varepsilon/2}$. Then by the triangle inequality (1.2.35) we have

$$|f_j(x) - f_k(x)| = |f_j(x) \underbrace{- f(x) + f(x)}_{\text{add zero}} + f_k(x)| \qquad (7.2.29)$$

$$\leq |f_j(x) - f(x)| + |f(x) - f_k(x)| \qquad (7.2.30)$$

$$< \frac{\varepsilon}{2} + \frac{\varepsilon}{2} \qquad (7.2.31)$$

$$= \varepsilon \qquad (7.2.32)$$

for every $x \in D$. Therefore, (f_n) satisfies the Cauchy criterion for uniform convergence (7.2.27).

To prove the backward implication, suppose (f_n) satisfies the Cauchy criterion for uniform convergence (7.2.27). Then every output sequence $(f_n(x))$ is a Cauchy sequence (Definition 2.6.1).

By the Cauchy criterion for sequences (Theorem 2.6.5), $(f_n(x))$ converges. As a candidate for a uniform limit, define $f : D \to \mathbb{R}$ by

$$f(x) = \lim_{n \to \infty} f_n(x) \tag{7.2.33}$$

for each $x \in D$. To show f is the uniform limit of (f_n), let $\varepsilon > 0$. By (7.2.27), there is a threshold $n_{\varepsilon/2} \in \mathbb{N}$ such that for every $x \in D$ we have

$$j, k \in \mathbb{N} \text{ with } j, k \geq n_{\varepsilon/2} \quad \Longrightarrow \quad |f_j(x) - f_k(x)| < \frac{\varepsilon}{2}. \tag{7.2.34}$$

By fixing an index $j \in \mathbb{N}$ with $j \geq n_{\varepsilon/2}$ and letting k tend to infinity, the linearity and order properties of sequential limits (Theorem 2.3.9 and Corollary 2.3.22) imply

$$|f_j(x) - f(x)| = \lim_{k \to \infty} |f_j(x) - f_k(x)| \leq \frac{\varepsilon}{2} < \varepsilon. \tag{7.2.35}$$

for every $x \in D$. Since j and x are arbitrary, the index $n_{\varepsilon/2}$ is a uniform threshold as in Definition 7.2.1. Therefore, (f_n) converges to f uniformly on D. $\qquad\square$

The next section explores the relationship between continuity, differentiability, integrability, and uniform convergence.

Exercises

7.2.1. Prove that for every $0 < b < 1$, the sequence of monomials $f_n : [0, b] \to \mathbb{R}$ where

$$f_n(x) = x^n \quad \text{for each } n \in \mathbb{N} \tag{7.2.36}$$

converges uniformly to 0 on $[0, b]$.

7.2.2. Let $f_n : \mathbb{R} \to \mathbb{R}$ be the functions from Exercise 7.1.2 given by

$$f_n(x) = \frac{(x^2 + 1)}{n^2} \tag{7.2.37}$$

for each $n \in \mathbb{N}$ and $x \in \mathbb{R}$.

(i) Explain why (f_n) does not converge uniformly on $\mathbb{R}$.

(ii) Prove that if the domain is restricted to a compact interval $[a, b]$, then (f_n) converges uniformly on $[a, b]$.

7.2.3. Let $g_n : \mathbb{R} \to \mathbb{R}$ be the functions from Exercise 7.1.3 given by

$$g_n(x) = \frac{\sin x}{n} + x + \cos x \tag{7.2.38}$$

for each $n \in \mathbb{N}$ and $x \in \mathbb{R}$. Prove $g(x_n)$ converges uniformly to $g(x) = x + \cos x$ on $\mathbb{R}$.

7.2.4. Let $h_n : \mathbb{R} \to \mathbb{R}$ be the functions from Exercise 7.1.4 given by

$$h_n(x) = \frac{x}{1 + nx^2} \tag{7.2.39}$$

for each $n \in \mathbb{N}$ and $x \in \mathbb{R}$. Does (h_n) converge uniformly on $\mathbb{R}$? (For a partial spoiler, see Example 7.3.6.)

7.2.5. Suppose $q_n : (0, \infty) \to \mathbb{R}$ is given by

$$q_n(x) = \frac{nx}{1 + nx^2} \tag{7.2.40}$$

for each $n \in \mathbb{N}$ and $x \in (0, \infty)$.

(i) Prove (q_n) converges pointwise on $(0, \infty)$

(ii) Does (q_n) converges uniformly on $(0, \infty)$?

(iii) Does (q_n) converges uniformly on $(0, 1)$?

(iv) Does (q_n) converges uniformly on $(1, \infty)$?

7.2.6. Give an example of sequences of functions (f_n) and (g_n) that converge uniformly on $D \subseteq \mathbb{R}$ whose sequence of products $(f_n g_n)$ does not converge uniformly.

7.2.7. Suppose sequences of functions (f_n) and (g_n) converge uniformly on $D \subseteq \mathbb{R}$ and suppose $b > 0$ is a uniform bound on both (f_n) and (g_n) in that

$$|f_n(x)| \le b \quad \text{and} \quad |g_n(x)| \le b \tag{7.2.41}$$

for each $n \in \mathbb{N}$ and $x \in D$. Prove the sequence of products $(f_n g_n)$ converges uniformly.

7.2.8. Suppose $D \subseteq \mathbb{R}$. For every bounded function $g : D \to \mathbb{R}$, define the *supremum norm* $\|g\|_\infty$ by

$$\|g\|_\infty = \sup |g(D)| = \sup\{|g(x)| : x \in D\}. \tag{7.2.42}$$

Suppose $f_n : D \to \mathbb{R}$ is bounded for each $n \in \mathbb{N}$ and $f : D \to \mathbb{R}$. Prove (f_n) converges uniformly to f on D if and only if

$$\lim_{n \to \infty} \|f_n - f\|_\infty = 0. \tag{7.2.43}$$

7.2.9. Suppose $D \subseteq \mathbb{R}$ and let $U(D)$ denote the set of uniformly convergent sequences of real-valued functions on D. Use Lemma 1.6.7 to prove $U(D)$ is a vector space.

7.3 Calculus and uniform convergence

This section develops results connecting uniform convergence to three concepts in calculus: continuity, integration, and differentiation. The proofs are a culmination of results developed earlier in the book.

For starters, recall that Example 7.1.6 shows us the sequence of continuous monomials (x^n) converges pointwise on the compact interval $[0, 1]$, but the pointwise limit is not continuous. Uniform convergence ensures the pointwise limit of a sequence of continuous functions is continuous.

Theorem 7.3.1: Uniform convergence preserves continuity

If (f_n) converges uniformly to f on D and f_n is continuous on D for each $n \in \mathbb{N}$, then the uniform limit f is continuous on D.

Scratch Work 7.3.2: Triangle inequality and uniform threshold

Following the guide in Remark 4.3.4 for working with the definition of continuity (Definition 4.3.2), the goal is to end up with

$$|f(x) - f(c)| < \varepsilon. \tag{7.3.1}$$

Thanks to the uniform convergence of (f_n) to f, the outputs $f_n(x)$ and $f_n(c)$ can be made as close as we like to their limits $f(x)$ and $f(c)$ using the same threshold, a uniform threshold n_ε. The continuity of f_{n_ε} allows us to make $f_{n_\varepsilon}(x)$ and $f_{n_\varepsilon}(c)$ close as we like. Stitching distances together by adding zero and using the triangle inequality allows to show that since $f(x)$ is close to $f_{n_\varepsilon}(x)$, which is close to $f_{n_\varepsilon}(c)$, which is close to $f(c)$, we have $f(x)$ and $f(c)$ are close.

Proof of Theorem 7.3.1. Suppose (f_n) converges uniformly to f on D and f_n is continuous on D for each $n \in \mathbb{N}$. Let $c \in D$ and $\varepsilon > 0$. By the definition of uniform convergence (Definition 7.2.1), there is a uniform threshold $n_{\varepsilon/3} \in \mathbb{N}$ such that

$$|f_{n_{\varepsilon/3}}(x) - f(x)| < \varepsilon/3 \quad \text{for all } x \in D. \tag{7.3.2}$$

Since $f_{n_{\varepsilon/3}}$ is continuous at c, by Definition 4.3.2 there is a threshold $\delta_{\varepsilon/3} > 0$ such that

$$x \in D \text{ with } |x - c| < \delta_{\varepsilon/3} \quad \implies \quad |f_{n_{\varepsilon/3}}(x) - f_{n_{\varepsilon/3}}(c)| < \frac{\varepsilon}{3}. \tag{7.3.3}$$

Therefore, by adding zero (1.2.34) and the triangle inequality (1.2.32), for all $x \in D$ where $|x - c| < \delta_{\varepsilon/3}$ we have

$$|f(x) - f(c)| = |f(x) \underbrace{- f_{n_{\varepsilon/3}}(x) + f_{n_{\varepsilon/3}}(x)}_{\text{add zero}} \underbrace{- f_{n_{\varepsilon/3}}(c) + f_{n_{\varepsilon/3}}(c)}_{\text{add zero}} - f(c)| \tag{7.3.4}$$

$$\leq \underbrace{|f(x) - f_{n_{\varepsilon/3}}(x)|}_{\text{unif. conv.}} + \underbrace{|f_{n_{\varepsilon/3}}(x) - f_{n_{\varepsilon/3}}(c)|}_{\text{cont. of } f_{n_{\varepsilon/3}}} + \underbrace{|f_{n_{\varepsilon/3}}(c) - f(c)|}_{\text{unif. conv.}} \tag{7.3.5}$$

$$< \frac{\varepsilon}{3} + \frac{\varepsilon}{3} + \frac{\varepsilon}{3} \tag{7.3.6}$$

$$= \varepsilon. \tag{7.3.7}$$

Hence, $\delta_{\varepsilon/3}$ is a threshold for the continuity of f at c. Since c is arbitrary, f is continuous on D. $\qquad \square$

Example 7.3.3: Pointwise does not imply integrable

Consider the sequence of functions (g_n) defined on $[0,2]$ by

$$g_n(x) = \begin{cases} 0, & \text{when } x = 0, \\ n, & \text{when } x \in (0, 1/n), \\ 1/x, & \text{when } x \in [1/n, 2]. \end{cases} \tag{7.3.8}$$

Then (g_n) converges pointwise to the unbounded function $g : [0,2] \to \mathbb{R}$ given by

$$g(x) = \begin{cases} 0, & \text{when } x = 0, \\ 1/x, & \text{when } x \in (0, 2]. \end{cases} \tag{7.3.9}$$

For each $n \in \mathbb{N}$, g_n is continuous on $(0,2]$ by Theorem 5.2.1 since

$$g_n\left(\frac{1}{n}\right) = \lim_{x \to 1/n} g_n(x) = n. \tag{7.3.10}$$

Also, each g_n is integrable since it continuous except at single point, $c = 0$, much like Example 6.4.7. However, the pointwise limit g is not bounded, so g is not integrable (Definition 6.1.6). The proofs of these claims are omitted, but drawing figures is suggested.

When a sequence of integrable functions converges uniformly, the limit of their integrals is the integral of the uniform limit.

Theorem 7.3.4: Uniform convergence respects integration

If (f_n) converges uniformly to f on the compact interval $[a, b]$ and f_n is integrable over $[a, b]$ for each $n \in \mathbb{N}$, then the uniform limit f is integrable over $[a, b]$ and

$$\lim_{n \to \infty} \int_a^b f_n(x)\, dx = \int_a^b \left(\lim_{n \to \infty} f_n(x) \right) dx = \int_a^b f(x)\, dx. \tag{7.3.11}$$

Scratch Work 7.3.5: Squeeze upper and lower sums together

As with so many of the proofs involving integration in Chapter 6, the approach here is to find a partition that will bring the upper and lower sums of the uniform limit arbitrarily close. The uniform convergence of (f_n) means a particular f_{n_ε} can be brought as close to the uniform limit f as we like across the whole domain, and the Cauchy criterion for integrability ((vi) in Theorem 6.2.12) allows us to squeeze the upper and lower sums of f_{n_ε} and f together.

Proof of Theorem 7.3.4. Suppose $a < b$, (f_n) converges uniformly to f on $[a, b]$, and f_n is integrable over $[a, b]$ for each $n \in \mathbb{N}$. Let $\varepsilon > 0$ and $x \in [a, b]$. By the definition of uniform convergence

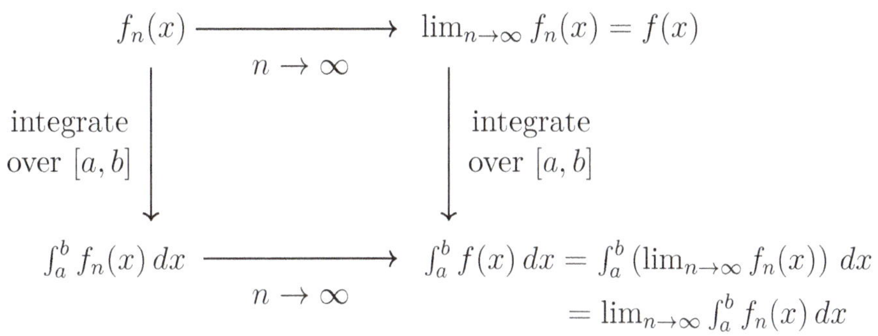

$$f_n(x) \xrightarrow[\quad n \to \infty \quad]{} \lim_{n \to \infty} f_n(x) = f(x)$$

$$\begin{array}{c} \text{integrate} \\ \text{over } [a,b] \end{array} \Bigg\downarrow \qquad\qquad\qquad \Bigg\downarrow \begin{array}{c} \text{integrate} \\ \text{over } [a,b] \end{array}$$

$$\int_a^b f_n(x)\,dx \xrightarrow[\quad n \to \infty \quad]{} \begin{aligned} \int_a^b f(x)\,dx &= \int_a^b \left(\lim_{n \to \infty} f_n(x) \right) dx \\ &= \lim_{n \to \infty} \int_a^b f_n(x)\,dx \end{aligned}$$

Figure 7.3.1: To pair with Theorem 7.3.4, this commutative diagram shows that whether we integrate or take the limit first, the uniform convergence of a sequence of integrable functions (f_n) to a function f on $[a,b]$ ensures the limit of the integrals of f_n is equal to the integral of the uniform limit f.

(Definition 7.2.1) and splitting the absolute value, there is a uniform threshold n_ε such that

$$|f_{n_\varepsilon}(x) - f(x)| < \frac{\varepsilon}{3(b-a)} \tag{7.3.12}$$

$$\Longleftrightarrow \quad -\frac{\varepsilon}{3(b-a)} < f(x) - f_{n_\varepsilon}(x) < \frac{\varepsilon}{3(b-a)}. \tag{7.3.13}$$

Adding $f_{n_\varepsilon}(x)$ gives us

$$f_{n_\varepsilon}(x) - \frac{\varepsilon}{3(b-a)} < f(x) < f_{n_\varepsilon}(x) + \frac{\varepsilon}{3(b-a)} \tag{7.3.14}$$

for all $x \in [a,b]$. So, since f_{n_ε} is bounded, f is bounded as well. Also, since f_{n_ε} is integrable, by Theorem 6.2.12 there is a partition $P_{\varepsilon/3} = \{x_0, x_1, \ldots, x_j\}$ such that

$$u(f_{n_\varepsilon}, P_{\varepsilon/3}) - \ell(f_{n_\varepsilon}, P_{\varepsilon/3}) < \frac{\varepsilon}{3}. \tag{7.3.15}$$

Now, for each $k = 1, \ldots, j$ let

$$u_k = \sup f([x_{k-1}, x_k]), \quad \ell_k = \sup f([x_{k-1}, x_k]), \tag{7.3.16}$$
$$u_k' = \sup f_{n_\varepsilon}([x_{k-1}, x_k]), \quad \text{and} \quad \ell_k' = \sup f_{n_\varepsilon}([x_{k-1}, x_k]). \tag{7.3.17}$$

Since a supremum is the least upper bound (Theorem 1.3.10), an infimum is the greatest lower bound (Theorem 1.4.3), and the supremum is greater than the infimum over the same set, we have

$$-\frac{\varepsilon}{3(b-a)} + \ell_k' \leq \ell_k \leq u_k \leq u_k' + \frac{\varepsilon}{3(b-a)}. \tag{7.3.18}$$

Since we have $x_k - x_{k-1} > 0$ for each index $k = 1, \ldots, j$, multiplying through (7.3.18) by $(x_k - x_{k-1})$, noting the telescoping sum $\sum_{k=1}^{j}(x_k - x_{k-1}) = b - a$ is in play, and taking sums yields

$$-\frac{\varepsilon}{3} + \ell(f_{n_\varepsilon}, P_{\varepsilon/3}) \leq \ell(f, P_{\varepsilon/3}) \leq u(f, P_{\varepsilon/3}) \leq u(f_{n_\varepsilon}, P_{\varepsilon/3}) + \frac{\varepsilon}{3}. \tag{7.3.19}$$

By rearranging these inequalities as in property (vi) of Theorem 1.3.2 and including inequality (7.3.15), we get

$$u(f, P_{\varepsilon/3}) - \ell(f, P_{\varepsilon/3}) \leq u(f_{n_\varepsilon}, P_{\varepsilon/3}) - \ell(f_{n_\varepsilon}, P_{\varepsilon/3}) + \frac{2\varepsilon}{3} < \frac{\varepsilon}{3} + \frac{2\varepsilon}{3} = \varepsilon. \tag{7.3.20}$$

Therefore, the uniform limit f satisfies the Cauchy criterion for integrability, so f is integrable over $[a, b]$ by Theorem 6.2.12.

Next, the goal is to show equations (7.3.11) hold. To that end, let $\varepsilon > 0$ again. By the definition of uniform convergence (Definition 7.2.1), there is a uniform threshold q_ε such that for all $x \in [a, b]$ we have

$$n \geq q_\varepsilon \text{ with } n \in \mathbb{N} \quad \implies \quad |f_n(x) - f(x)| < \frac{\varepsilon}{2(b-a)}. \tag{7.3.21}$$

Since the uniform limit f and each f_n is integrable over $[a, b]$, applications of the linearity of integration (Theorem 6.3.6), the integral triangle inequality (Corollary 6.4.5), the order property of integration (Corollary 6.3.10), and the integral of a constant (Lemma 6.1.10) combine to yield

$$\left| \int_a^b f_n - \int_a^b f \right| = \left| \int_a^b (f_n - f) \right| \tag{7.3.22}$$

$$\leq \int_a^b |f_n - f| \tag{7.3.23}$$

$$\leq \int_a^b \frac{\varepsilon}{2(b-a)} \tag{7.3.24}$$

$$= \frac{\varepsilon}{2(b-a)}(b-a) \tag{7.3.25}$$

$$= \frac{\varepsilon}{2} \tag{7.3.26}$$

$$< \varepsilon. \tag{7.3.27}$$

By the definition of sequential limit (Definition 2.2.1), the sequence of integrals $\left(\int_a^b f_n \right)$ converges to $\int_a^b f$ and we have

$$\lim_{n \to \infty} \int_a^b f_n(x)\, dx = \int_a^b \left(\lim_{n \to \infty} f_n(x) \right) dx = \int_a^b f(x)\, dx. \tag{7.3.28}$$

$\square$

Uniform convergence has a more complicated relationship with derivatives than it does with continuity and integration (see Theorems 7.3.1 and 7.3.4). For instance, the following example shows that the derivative of a uniform limit is not necessarily the limit of derivatives. More examples showing interesting behavior with derivatives despite uniform convergence are explored in the exercises.

Example 7.3.6: Uniform convergence with broken derivatives

Consider the sequence of functions (h_n) on $[-2, 2]$ given by

$$h_n(x) = \frac{x}{1 + nx^2}. \tag{7.3.29}$$

We have h_n is differentiable on $[-2, 2]$ for each index $n \in \mathbb{N}$, (h_n) converges uniformly to the zero function $h(x) = 0$ on $[-2, 2]$, but also

$$\lim_{n \to \infty} h'_n(0) = 1 \neq 0 = h'(0). \tag{7.3.30}$$

Scratch Work 7.3.7: First derivative test

Since each h_n is a rational function whose denominator is never zero, each h_n is differentiable by the Quotient Rule 5.4.10 with derivative given by

$$h'_n(x) = \frac{(1 + nx^2)(1) - x(2nx)}{(1 + nx^2)^2} = \frac{1 - nx^2}{(1 + nx^2)^2}. \tag{7.3.31}$$

So at $x = 0$, we have $h'_n(0) = 1$.

The hard part is showing (h_n) converges uniformly to the zero function on $[-2, 2]$. A uniform threshold comes from an application of the First Derivative Test from calculus: Solving the zeros of each h'_n leads to

$$h'_n(x) = \frac{1 - nx^2}{(1 + nx^2)^2} = 0 \iff 1 - nx^2 = 0 \iff x = \pm\frac{1}{\sqrt{n}}. \tag{7.3.32}$$

It turns out that for each index $n \in \mathbb{N}$, these inputs yield the maximum and minimum values of h_n. Hence, we can convert these into a sequence of bounds on h_n and get uniform convergence from Lemma 7.2.4.

Proof of Example 7.3.6. Consider the sequence of functions (h_n) on $I = [-2, 2]$ defined for each $n \in \mathbb{N}$ by

$$h_n(x) = \frac{x}{1 + nx^2}. \tag{7.3.33}$$

By the Quotient Rule 5.4.10, each h_n is differentiable on $[-2, 2]$ with derivative given by

$$h'_n(x) = \frac{(1 + nx^2)(1) - x(2nx)}{(1 + nx^2)^2} = \frac{1 - nx^2}{(1 + nx^2)^2}. \tag{7.3.34}$$

So at $x = 0$ we have $h'_n(0) = 1$. Also, motivated by the First Derivative Test, we have

$$h'_n(x) = \frac{1 - nx^2}{(1 + nx^2)^2} = 0 \iff 1 - nx^2 = 0 \iff x = \pm\frac{1}{\sqrt{n}}. \tag{7.3.35}$$

Since the denominator of h_n is positive for every $x \in [-2, 2]$, we have

$$
\begin{aligned}
x \in [-2, -1/\sqrt{n}) &\implies 1 - nx^2 < 0 &&\implies h_n'(x) < 0, &&& (7.3.36) \\
x \in (-1/\sqrt{n}, 1/\sqrt{n}) &\implies 1 - nx^2 > 0 &&\implies h_n'(x) > 0, \quad \text{and} &&& (7.3.37) \\
x \in (1/\sqrt{n}, 2] &\implies 1 - nx^2 < 0 &&\implies h_n'(x) < 0. &&& (7.3.38)
\end{aligned}
$$

By Corollary 5.5.15 applied for each $n \in \mathbb{N}$, h_n is decreasing on $[-2, -1/\sqrt{n})$, increasing on $(-1/\sqrt{n}, 1/\sqrt{n})$, then decreasing again on $(1/\sqrt{n}, 2]$. Hence, the mimimum of h_n is attained at $-1/\sqrt{n}$ and the maximum is attained at $1/\sqrt{n}$. Therefore, we have

$$
h_n\left(-\frac{1}{\sqrt{n}}\right) = -\frac{1}{2\sqrt{n}} \leq h_n(x) \leq \frac{1}{2\sqrt{n}} = h_n\left(\frac{1}{\sqrt{n}}\right) \tag{7.3.39}
$$

for all $x \in [-2, 2]$. Rearranging the inequalities yields

$$
|h_n(x)| = |h_n(x) - 0| \leq \frac{1}{2\sqrt{n}}. \tag{7.3.40}
$$

Since $(1/(2\sqrt{n}))$ is a sequence of nonnegative bounds that converges to zero, Lemma 7.2.4 tells us (h_n) converges uniformly to the zero function $h(x) = 0$ on $[-2, 2]$. $\square$

Despite Example 7.3.6, uniform convergence can preserve differentiability under suitable conditions.

Theorem 7.3.8: Uniform convergence and derivatives

Suppose (f_n) is a sequence of differentiable functions on $[a, b]$ where the sequence of derivatives (f_n') converges uniformly to a function g on $[a, b]$. If there is some $x_0 \in [a, b]$ whose output sequence $(f_n(x_0))$ converges, then there is a function $f : [a, b] \to \mathbb{R}$ such that

(i) (f_n) converges uniformly to f on $[a, b]$ with

$$
f(x) = \lim_{n\to\infty} f_n(x) \tag{7.3.41}
$$

for every $x \in [a, b]$; and

(ii) f is differentiable on $[a, b]$ and f is an antiderivative of g with

$$
f'(x) = \lim_{n\to\infty} f_n'(x) = g(x) \tag{7.3.42}
$$

for every $x \in [a, b]$.

Remark 7.3.9: Not like the others

The relationships uniform convergence has with continuity and integration are quite different from its relationship with differentiation. For instance, the hypotheses of Theorems 7.3.1 and 7.3.4 include the assumption that a sequence of functions converges uniformly to a given function. On the other hand, in Theorem 7.3.8 the hypotheses include the assumption that a sequence of *derivatives* converges uniformly to a given function while the conclusions include the uniform convergence of the antiderivatives. There are other differences as well, including the hypothesis of the convergence of a single output sequence in Theorem 7.3.8.

The proof is exceptionally long and brings together a wide variety of results to get the pair of conclusions.

Proof of Theorem 7.3.8. Suppose the hypotheses of Theorem 7.3.8 hold and let $\varepsilon > 0$. Since the output sequence $(f_n(x_0))$ converges, it is also Cauchy by the Cauchy criterion for sequences 2.6.5. So, there is a threshold $j_\varepsilon \in \mathbb{N}$ such that

$$j, k \in \mathbb{N} \text{ with } j, k \geq j_\varepsilon \quad \implies \quad |f_j(x_0) - f_k(x_0)| < \frac{\varepsilon}{2}. \qquad (7.3.43)$$

Also, by the Cauchy criterion for uniform convergence 7.2.11, there is a threshold $k_\varepsilon \in \mathbb{N}$ such that for all $t \in [a, b]$ we have

$$j, k \in \mathbb{N} \text{ with } j, k \geq k_\varepsilon \quad \implies \quad |f_j'(t) - f_k'(t)| < \frac{\varepsilon}{2(b - a)}. \qquad (7.3.44)$$

Define $n_\varepsilon = \max\{j_\varepsilon, k_\varepsilon\}$ and assume $j, k \in \mathbb{N}$ with $j, k \geq n_\varepsilon$ so that inequalities (7.3.43) and (7.3.44) both hold. By the linearity of derivatives (Theorem 5.4.1), we have $f_j - f_k$ is differentiable on $[a, b]$ with

$$(f_j - f_k)'(t) = f_j'(t) - f_k'(t) \qquad (7.3.45)$$

for all $t \in [a, b]$. Now, without loss of generality, for any pair $u, v \in [a, b]$ where $u < v$ the Mean Value Theorem 5.5.9 applies to $f_j - f_k$ on $[u, v]$. Hence, there is some $c \in (u, v) \subseteq [a, b]$ such that

$$(f_j - f_k)'(c) = f_j'(c) - f_k'(c) = \frac{f_j(v) - f_k(v) - (f_j(u) - f_k(u))}{v - u}. \qquad (7.3.46)$$

Equivalently, by multiplying by $v - u$ we have

$$(f_j'(c) - f_k'(c))(v - u) = f_j(v) - f_j(u) + f_k(u) - f_k(v). \qquad (7.3.47)$$

So, by adding zero, the triangle inequality (1.2.35), equation 7.3.47, and inequalities (7.3.43) and (7.3.44), we have for all $j, k \in \mathbb{N}$ with $j, k \geq n_\varepsilon$, all $x \in [a, b]$ and some $c_0 \in [a, b]$ we have

$$|f_j(x) - f_k(x)| = |f_j(x) - f_k(x) \underbrace{- f_j(x_0) + f_k(x_0) + f_j(x_0) - f_k(x_0)}_{\text{add zero}}| \qquad (7.3.48)$$

$$\leq |f_j(x) - f_k(x) - f_j(x_0) + f_k(x_0)| + |f_j(x_0) - f_k(x_0)| \qquad (7.3.49)$$

$$= |(f_j'(c_0) - f_k'(c_0))(x - x_0)| + |f_j(x_0) - f_k(x_0)| \qquad (7.3.50)$$

$$< \frac{\varepsilon}{2(b - a)}|x - x_0| + \frac{\varepsilon}{2} \qquad (7.3.51)$$

$$\leq \frac{\varepsilon}{2(b - a)}(b - a) + \frac{\varepsilon}{2} \qquad (7.3.52)$$

$$= \varepsilon. \qquad (7.3.53)$$

Hence, (f_n) satisfies the Cauchy criterion for uniform convergence (Theorem 7.2.11). So there is a function $f : [a, b] \to \mathbb{R}$ such that (f_n) converges to f uniformly and

$$f(x) = \lim_{n \to \infty} f_n(x) \tag{7.3.54}$$

for every $x \in [a, b]$. Therefore, part (i) in Theorem 7.3.8 holds.

To prove part (ii) in Theorem 7.3.8 holds from here, fix $x \in [a, b]$ and for each $n \in \mathbb{N}$ let q_n denote the extended difference quotient of f_n at x given by

$$q_n(y) = \begin{cases} \dfrac{f_n(y) - f_n(x)}{y - x}, & y \in [a, b] \backslash \{x\}, \\ f_n'(x), & y = x. \end{cases} \tag{7.3.55}$$

Since each f_n is differentiable on $[a, b]$, by Definition 5.3.1 we have

$$f_n'(x) = \lim_{y \to x} \frac{f_n(y) - f_n(x)}{y - x} = \lim_{y \to x} q_n(y). \tag{7.3.56}$$

Hence, q_n is continuous at x by Theorem 5.2.1.

Now, with the goal of showing f is differentiable at x with $f'(x) = g(x)$, define $q : [a, b] \to \mathbb{R}$ to be the extended difference quotient of f at x given by

$$q(y) = \begin{cases} \dfrac{f(y) - f(x)}{y - x}, & y \in [a, b] \backslash \{x\} \\ g(x), & y = x. \end{cases} \tag{7.3.57}$$

It turns out q is the pointwise limit of (q_n) since, by the linearity of sequential limits (Theorem 2.3.9) and other properties, for all $y \in [a, b] \backslash \{x\}$ we have

$$\lim_{n \to \infty} q_n(y) = \lim_{n \to \infty} \frac{f_n(y) - f_n(x)}{y - x} = \frac{\lim\limits_{n \to \infty} f_n(y) - \lim\limits_{n \to \infty} f_n(x)}{\lim\limits_{n \to \infty} (y - x)} = \frac{f(y) - f(x)}{y - x}. \tag{7.3.58}$$

Also, for $y = x$ we have

$$\lim_{n \to \infty} q_n(x) = f_n'(x) = g(x) = q(x). \tag{7.3.59}$$

Note that by Theorem 5.1.8 and the definition of derivative as the limit of the difference quotient (Definition 5.3.1), the differentiability of f at x will be ensured once we show its extended difference quotient q is continuous at x. This can be proven via Theorem 7.3.1 by showing the sequence (q_n) of continuous difference quotients converges to q uniformly.

To invoke the Cauchy criterion for uniform convergence (Theorem 7.2.11), let $\varepsilon > 0$. Since the sequence of derivatives (f_n') converges uniformly to g on $[a, b]$, there is a threshold $z_\varepsilon \in \mathbb{N}$ such that for all $t \in [a, b]$ we have

$$j, k \in \mathbb{N} \text{ with } j, k \geq z_\varepsilon \quad \Longrightarrow \quad |f_j'(t) - f_k'(t)| < \frac{\varepsilon}{2}. \tag{7.3.60}$$

In particular, we have

$$|q_j(x) - q_k(x)| = |f_j'(x) - f_k'(x)| < \frac{\varepsilon}{2}. \tag{7.3.61}$$

Also, for all $j, k \in \mathbb{N}$ with $j, k \geq z_\varepsilon$ and all $y \in [a,b]\setminus\{x\}$, by another argument involving the Mean Value Theorem 5.5.9 as in the proof of part (i) above, for some c_1 between x and y we have

$$|q_j(y) - q_k(y)| = \left| \frac{f_j(y) - f_j(x)}{y - x} - \frac{f_k(y) - f_k(x)}{y - x} \right| \tag{7.3.62}$$

$$= |f_j(y) - f_j(x) - f_k(x) + f_k(y)| \left| \frac{1}{y - x} \right| \tag{7.3.63}$$

$$= |(f_j'(c_1) - f_j'(c_1))(x - y)| \left| \frac{1}{y - x} \right| \tag{7.3.64}$$

$$\leq \frac{\varepsilon}{2}|x - y| \left| \frac{1}{y - x} \right| \tag{7.3.65}$$

$$< \varepsilon. \tag{7.3.66}$$

Therefore, since y is arbitrary, the Cauchy criterion for uniform convergence 7.2.11 tells us the sequence of difference quotients (q_n) converges uniformly on $[a,b]$ to its pointwise limit q. Since each q_n is continuous, Theorem 7.3.1 tells us the (now) uniform limit q is continuous. By Theorem 5.1.8 and the definition of derivative as the limit of difference quotients (Definition 5.3.1), we have f is differentiable at every $x \in [a,b]$ with

$$f'(x) = \lim_{t \to x} q(t) = q(x) = g(x) = \lim_{n \to \infty} f_n'(x). \tag{7.3.67}$$

Therefore, part (ii) in Theorem 7.3.8 holds. $\qquad\square$

The next section develops one of the pinnacles of real analysis: the Weierstrass Approximation Theorem 7.4.7.

Exercises

7.3.1. Suppose $0 < b < 1$ and let (f_n) be the sequence of monomials $f_n : [0,b] \to \mathbb{R}$ from Exercise 7.2.1 given by $f_n(x) = x^n$ for each $n \in \mathbb{N}$ and $x \in [0,b]$.

(i) Prove $\displaystyle\lim_{n \to \infty} \int_0^b f_n = \int_0^b \left(\lim_{n \to \infty} f_n \right) = \int_0^b 0 = 0$.

(ii) Extend the domain of (f_n) to $[0,1]$. See Example 7.1.4. Prove

$$\lim_{n \to \infty} \int_0^1 f_n = \int_0^1 \left(\lim_{n \to \infty} f_n \right) = \int_0^1 f = 0, \tag{7.3.68}$$

where f is the pointwise limit of (f_n) on $[0,1]$.

(iii) For each $n \in \mathbb{N}$, determine the sequence of derivatives (f_n') on $[0,1]$. Does (f_n') converge pointwise or uniformly on $[0,1]$? On $[0,b]$ with $0 < b < 1$?

7.3.2. Let $f_n : \mathbb{R} \to \mathbb{R}$ be the functions from Exercises 7.1.2 and 7.2.2 given by

$$f_n(x) = \frac{(x^2 + 1)}{n^2} \tag{7.3.69}$$

for each $n \in \mathbb{N}$ and $x \in \mathbb{R}$. Determine the sequence of derivatives (f_n') on $\mathbb{R}$. Does (f_n') converge uniformly on $\mathbb{R}$?

7.3.3. Let $g_n : \mathbb{R} \to \mathbb{R}$ be the functions from Exercises 7.1.3 and 7.2.3 given by

$$g_n(x) = \frac{\sin x}{n} + x + \cos x \tag{7.3.70}$$

for each $n \in \mathbb{N}$ and $x \in \mathbb{R}$. Determine the sequence of derivatives (g_n') on $\mathbb{R}$. Does (g_n') converge uniformly on $\mathbb{R}$?

7.3.4. Consider the sequence of *spikes* (t_n) on $[0,2]$ given by

$$t_n(x) = \begin{cases} n^2 x, & \text{when } 0 \le x \le 1/n, \\ -n^2 x + 2n, & \text{when } 1/n < x < 2/n, \\ 0, & \text{when } 2/n \le x \le 2. \end{cases} \tag{7.3.71}$$

See Figure 7.3.2. Prove t_n is continuous on $[0,2]$ for each index $n \in \mathbb{N}$. Furthermore, determine the pointwise limit t on $[0,2]$ and prove

$$\lim_{n \to \infty} \int_0^2 t_n \ne \int_0^2 t. \tag{7.3.72}$$

Does (t_n) converge to t uniformly on $[0,2]$?

7.3.5. Suppose (r_k) is an enumeration of $\mathbb{Q} \cap [0,2]$ where each rational number appears exactly once as a term. Define $f_n : [0,2] \to \mathbb{R}$ for each $n \in \mathbb{N}$ by

$$f_n(x) = \begin{cases} 0, & \text{if } x \in [0,2]\backslash\mathbb{Q}, \\ 1, & \text{if } x = r_k \text{ and } k \le n, \\ 0, & \text{if } x = r_k \text{ and } k > n. \end{cases} \tag{7.3.73}$$

(i) Prove each f_n is integrable over $[0,2]$.

(ii) Prove (f_n) converges pointwise on $[0,2]$ to a bounded function. What is the pointwise limit?

(iii) Prove the pointwise limit of (f_n) is not integrable. So, even if a sequence of integrable functions converges pointwise to a bounded function, the pointwise limit may not be integrable.

7.3.6. Prove *Dini's Theorem*, a partial converse to Theorem 7.3.1: Suppose $K \subseteq \mathbb{R}$ is compact and (f_n) converges pointwise to f on K such that the output sequence $(f_n(x))$ is increasing for each $x \in K$. Then (f_n) converges uniformly to f on K.

Hint: For a fixed $\varepsilon > 0$ and each $n \in \mathbb{N}$, let

$$K_n = \{x \in K : f(x) - f_n(x) \ge \varepsilon\}. \tag{7.3.74}$$

First prove $K_1 \supseteq K_2 \supseteq K_3 \supseteq \cdots$, then use this result to finish the proof of Dini's Theorem.

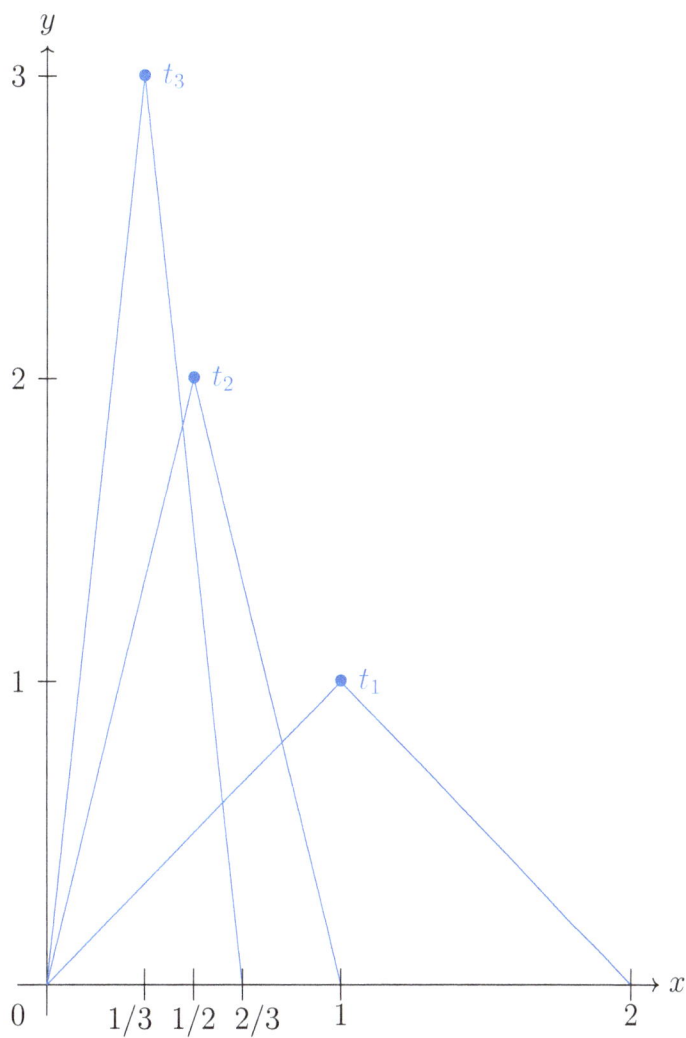

Figure 7.3.2: The sequence of spikes (t_n) on $[0, 2]$ in Exercise 7.3.4. Each t_n is continuous and has integral 1, but the sequence (t_n) converges pointwise to 0.

7.4　The Weierstrass Approximation Theorem

Polynomials are one of the most important and ubiquitous functions in mathematics. Moreover,

Every continuous function on a compact interval
is arbitrarily close to the set of polynomials.

The formal version of this statement is the Weierstrass Approximation Theorem 7.4.7. Our approach to proving it is thanks to Ukrainian mathematician Sergei Natanovich Bernstein and the polynomials that bear his name. While Bernstein's approach was probabilistic in nature, we also use a perspective of combinatorics stemming from the factorial function and binomial coefficients. Following Bernstein, we first prove a special case of the Weierstrass Approximation Theorem that holds on the unit interval (Theorem 7.4.5).

Definition 7.4.1: Bernstein basis polynomial

Given a nonnegative integer $n \in \mathbb{N} \cup \{0\}$ and $k = 0, 1, \ldots, n$, the *Bernstein basis polynomial* $b_{n,k} : [0,1] \to \mathbb{R}$ is given by

$$b_{n,k}(x) = \binom{n}{k} x^k (1-x)^{n-k}. \tag{7.4.1}$$

Example 7.4.2: Some Bernstein basis polynomials

Here are the first six Bernstein basis polynomials for $n = 0, 1, 2$:

$$b_{0,0}(x) = 1, \tag{7.4.2}$$
$$b_{1,0}(x) = (1-x), \quad b_{1,1}(x) = x, \tag{7.4.3}$$
$$b_{2,0}(x) = (1-x)^2, \quad b_{2,1}(x) = 2x(1-x), \quad \text{and} \quad b_{2,2}(x) = x^2. \tag{7.4.4}$$

The Bernstein basis polynomials exhibit some very interesting probabilistic properties and play a central role in approximating continuous functions over the compact interval $[0,1]$. The proof of the following lemma is left as a collection of exercises.

Lemma 7.4.3: Probability and Bernstein basis polynomials

Suppose $x \in [0,1]$ and $n \in \mathbb{N} \cup \{0\}$. Then

(i) $b_{n,k}(x) \geq 0$　　(*nonnegativity*);

(ii) $\displaystyle\sum_{k=0}^{n} b_{n,k}(x) = \sum_{k=0}^{n} \binom{n}{k} x^k (1-x)^{n-k} = 1$　　(*probability, partition of unity*);

(iii) $\displaystyle\sum_{k=0}^{n} \frac{k}{n} b_{n,k}(x) = \sum_{k=0}^{n} \frac{k}{n} \binom{n}{k} x^k (1-x)^{n-k} = x$　　(*mean*);　　and

(iv) $\displaystyle\sum_{k=0}^{n} \left(x - \frac{k}{n}\right)^2 b_{n,k}(x) = \sum_{k=0}^{n} \left(x - \frac{k}{n}\right)^2 \binom{n}{k} x^k (1-x)^{n-k} = \frac{x(1-x)}{n}$　　(*variance*).

Figure 7.4.1: The QR code takes you to the Desmos activity "Bernstein polynomial" designed to accompany Definition 7.4.4. It allows for an exploration of the Bernstein polynomials defined by a sine function. https://www.desmos.com/calculator/gwxwh9nrow

Bernstein polynomials approximate all continuous functions over the compact interval $[0, 1]$ as closely as we like. Each is defined as a linear combination of Bernstein basis polynomials (Definition 7.4.1) with weights (or scalars) given by an interpolation type of process using a finite set of outputs from a given continuous function.

Definition 7.4.4: Bernstein polynomial

Given a continuous function $f : [0, 1] \to \mathbb{R}$ and $n \in \mathbb{N} \cup \{0\}$, the nth *Bernstein polynomial* of f is the linear combination of Bernstein basis polynomials given by

$$B_n(f)(x) = \sum_{k=0}^{n} f\left(\frac{k}{n}\right) b_{n,k}(x) = \sum_{k=0}^{n} f\left(\frac{k}{n}\right)\binom{n}{k} x^k (1-x)^{n-k}. \qquad (7.4.5)$$

See Figure 7.4.1.

The following version of the *Weierstrass Approximation Theorem* tells us real-valued continuous functions on the unit interval $[0, 1]$ are arbitrarily close to their set of Bernstein polynomials.

Theorem 7.4.5: Weierstrass Approximation Theorem on $[0, 1]$

If $f : [0, 1] \to \mathbb{R}$ is continuous, then for every $\varepsilon > 0$ there is an index $n_\varepsilon \in \mathbb{N}$ such that for all $x \in [0, 1]$ we have

$$|B_{n_\varepsilon}(f)(x) - f(x)| < \varepsilon. \qquad (7.4.6)$$

Scratch Work 7.4.6: Split the difference

The proof of Theorem 7.4.5 comes from a careful decomposition of the difference

$$|B_n(f)(x) - f(x)|. \qquad (7.4.7)$$

The uniform continuity of f on $[0, 1]$ and the triangle inequality (1.2.35) allows us to squeeze the decomposed difference to be less than any given $\varepsilon > 0$.

Proof of Theorem 7.4.5. Suppose $f : [0, 1] \to \mathbb{R}$ is continuous. Since $[0, 1]$ is compact, the Extreme Value Theorem 4.6.9 tells us f is bounded. So, let

$$u = \max |f|([0, 1]) = \max\{|f(x)| : x \in [0, 1]\}. \tag{7.4.8}$$

Also, f is uniformly continuous by Theorem 4.7.13.

Now let $\varepsilon > 0$. By Definition 4.7.1, there is a uniform threshold $\delta > 0$ such that

$$x, y \in [0, 1] \text{ with } |x - y| < \delta \quad \Longrightarrow \quad |f(x) - f(y)| < \frac{\varepsilon}{2}. \tag{7.4.9}$$

Next, choose an index $n_\varepsilon \in \mathbb{N}$ large enough to satisfy

$$n_\varepsilon > \frac{4u}{\delta^2 \varepsilon} \quad \Longleftrightarrow \quad \frac{2u}{n_\varepsilon \delta^2} < \frac{\varepsilon}{2}. \tag{7.4.10}$$

The remainder of the proof shows that n_ε gives us the desired Bernstein polynomial B_{n_ε}.

By part (ii) of Lemma 7.4.3, for every $n \in \mathbb{N} \cup \{0\}$ and $x \in [0, 1]$ we have

$$f(x) = f(x) \sum_{k=0}^{n} b_{n,k}(x) = \sum_{k=0}^{n} f(x) b_{n,k}(x). \tag{7.4.11}$$

Hence, by the definition of Bernstein polynomials (Definition 7.4.4), combining like terms, the nonnegativity of the Bernstein basis polynomials (part (i) of Lemma 7.4.3), and repeated use of the triangle inequality (1.2.35), we have

$$|B_n(f)(x) - f(x)| = \left| \sum_{k=0}^{n} f\left(\frac{k}{n}\right) b_{n,k}(x) - \sum_{k=0}^{n} f(x) b_{n,k}(x) \right| \tag{7.4.12}$$

$$= \left| \sum_{k=0}^{n} \left(f\left(\frac{k}{n}\right) - f(x) \right) b_{n,k}(x) \right| \tag{7.4.13}$$

$$\leq \sum_{k=0}^{n} \left| f\left(\frac{k}{n}\right) - f(x) \right| b_{n,k}(x) \tag{7.4.14}$$

The next step is focus on the index n_ε to split (7.4.14) over the indices $k = 0, 1, \ldots, n_\varepsilon$ into those where the ratio k/n_ε is within δ of x and those that aren't. To that end, we have

$$\sum_{k=0}^{n_\varepsilon} \left| f\left(\frac{k}{n_\varepsilon}\right) - f(x) \right| b_{n_\varepsilon,k}(x) = \tag{7.4.15}$$

$$\sum_{\left| \frac{k}{n_\varepsilon} - x \right| < \delta} \left| f\left(\frac{k}{n_\varepsilon}\right) - f(x) \right| b_{n_\varepsilon,k}(x) \quad + \sum_{\left| \frac{k}{n_\varepsilon} - x \right| \geq \delta} \left| f\left(\frac{k}{n_\varepsilon}\right) - f(x) \right| b_{n_\varepsilon,k}(x) \tag{7.4.16}$$

For the sum over k where k/n_ε is within δ of x, by (7.4.14) and parts (i) and (ii) of Lemma 7.4.3 we have

$$\sum_{\left|\frac{k}{n_\varepsilon}-x\right|<\delta} \left| f\left(\frac{k}{n_\varepsilon}\right) - f(x) \right| b_{n_\varepsilon,k}(x) < \frac{\varepsilon}{2}\left(\sum_{\left|\frac{k}{n_\varepsilon}-x\right|<\delta} b_{n_\varepsilon,k}(x) \right) \le \frac{\varepsilon}{2}. \tag{7.4.17}$$

For the sum over k where k/n_ε is at least δ away from x, by another application of the triangle inequality (1.2.35) we have

$$\sum_{\left|\frac{k}{n_\varepsilon}-x\right|\ge\delta} \left| f\left(\frac{k}{n_\varepsilon}\right) - f(x) \right| b_{n_\varepsilon,k}(x) \le \sum_{\left|\frac{k}{n_\varepsilon}-x\right|\ge\delta} \left(\left| f\left(\frac{k}{n_\varepsilon}\right) \right| + |f(x)| \right) b_{n_\varepsilon,k}(x) \tag{7.4.18}$$

$$\le \sum_{\left|\frac{k}{n_\varepsilon}-x\right|\ge\delta} 2u\, b_{n_\varepsilon,k}(x) \tag{7.4.19}$$

$$= 2u\left(\sum_{\left|\frac{k}{n_\varepsilon}-x\right|\ge\delta} b_{n_\varepsilon,k}(x) \right). \tag{7.4.20}$$

Also, since the sum is over k where k/n_ε is at least δ away from x, multiplying a nice version of 1 and applying part (iv) of Lemma 7.4.3 yields

$$\sum_{\left|\frac{k}{n_\varepsilon}-x\right|\ge\delta} b_{n_\varepsilon,k}(x) = \sum_{\left|\frac{k}{n_\varepsilon}-x\right|\ge\delta} \frac{\left|\frac{k}{n_\varepsilon}-x\right|^2}{\left|\frac{k}{n_\varepsilon}-x\right|^2} b_{n_\varepsilon,k}(x) \tag{7.4.21}$$

$$\le \sum_{\left|\frac{k}{n_\varepsilon}-x\right|\ge\delta} \frac{\left|\frac{k}{n_\varepsilon}-x\right|^2}{\delta^2} b_{n_\varepsilon,k}(x) \tag{7.4.22}$$

$$= \frac{1}{\delta^2} \sum_{\left|\frac{k}{n_\varepsilon}-x\right|\ge\delta} \left(\frac{k}{n_\varepsilon}-x\right)^2 b_{n_\varepsilon,k}(x) \tag{7.4.23}$$

$$= \frac{x(1-x)}{n_\varepsilon\delta^2}. \tag{7.4.24}$$

Since $0 \le x \le 1$, we also have $0 \le 1-x \le 1$ as well as $0 \le x(1-x) \le 1$. Hence, by starting from (7.4.14) with index $n = n_\varepsilon$, splitting the resulting sum as in (7.4.16), then applying the bounds (7.4.17), (7.4.20), and (7.4.24) we have

$$|B_{n_\varepsilon}(f)(x) - f(x)| \le \sum_{k=0}^{n_\varepsilon} \left| f\left(\frac{k}{n_\varepsilon}\right) - f(x) \right| b_{n_\varepsilon,k}(x) \tag{7.4.25}$$

$$< \frac{\varepsilon}{2} + 2u\frac{x(1-x)}{n_\varepsilon\delta^2} \tag{7.4.26}$$

$$\le \frac{\varepsilon}{2} + \frac{2u}{n_\varepsilon\delta^2} \tag{7.4.27}$$

$$< \varepsilon. \tag{7.4.28}$$

Therefore, the Weierstrass Approximation Theorem holds on $[0,1]$. $\qquad\square$

Figure 7.4.2: The QR code takes you to the Desmos activity "Weierstrass Approximation Theorem" designed to accompany Theorem 7.4.7. It allows for an exploration of generalized Bernstein polynomials which uniformly approximate a given continuous function on some compact interval. https://www.desmos.com/calculator/p0ac1ozo3u

The full statement of the Weierstrass Approximation Theorem holds for real-valued continuous functions on compact intervals.

Theorem 7.4.7: Weierstrass Approximation Theorem

If $a < b$ and $g : [a, b] \to \mathbb{R}$ is continuous, then for every $\varepsilon > 0$ there is a polynomial p_ε such that for all $t \in [a, b]$ we have

$$|p_\varepsilon(t) - g(t)| < \varepsilon. \tag{7.4.29}$$

See Figure 7.4.2.

Scratch Work 7.4.8: Composition with a line

The proof of the Weierstrass Approximation Theorem 7.4.7 follows from the special case of the theorem on the unit interval $[0, 1]$ as in Theorem 7.4.5. The idea is to replace the given function g which is defined on $[a, b]$ with an equivalent function f defined on $[0, 1]$ so that Theorem 7.4.5 applies and gives us a suitable Bernstein polynomial. The trick is to compose a line with g to generate f through a change of variables. Since lines are continuous, g is assumed to be continuous, and compositions of continuous functions are continuous, f will be continuous as well.

The line for the job is $h : [0, 1] \to [a, b]$ and its inverse $h^{-1} : [a, b] \to [0, 1]$ given by

$$h(x) = t = (b - a)x + a \qquad \Longleftrightarrow \qquad h^{-1}(t) = x = \frac{t - a}{b - a}, \tag{7.4.30}$$

and the desired polynomial p_ε will stem from a Bernstein polynomial defined by the composition $g \circ h$.

Proof of the Weierstrass Approximation Theorem 7.4.7. Suppose $a < b$ and $g : [a, b] \to \mathbb{R}$ is continuous. Since $b - a > 0$, the line $h : [0, 1] \to [a, b]$ and its inverse $h^{-1} : [a, b] \to [0, 1]$ given by

$$h(x) = t = (b - a)x + a \qquad \Longleftrightarrow \qquad h^{-1}(t) = x = \frac{t - a}{b - a} \tag{7.4.31}$$

are both well-defined (and each is a bijection). Also, the compositions $f = g \circ h$ and $g = f \circ h^{-1}$ are well-defined with

$$f(x) = g(h(x)) = g((b - a)x + a) \quad \text{and} \quad g(t) = f(h^{-1}(t)) = f\left(\frac{t - a}{b - a}\right) \tag{7.4.32}$$

for all $x \in [0, 1]$ and all $t \in [a, b]$, respectively.

Next, since lines are continuous as a special case of the basic affine transformations in Theorem 4.3.9, g is continuous on $[a, b]$, and compositions of continuous functions are continuous (Theorem 4.5.13), the composition $f = g \circ h$ is continuous on $[0, 1]$.

Now let $\varepsilon > 0$. By the special case of the Weierstrass Approximation Theorem 7.4.5 that holds on $[0, 1]$, there is an index $n_\varepsilon \in \mathbb{N}$ whose Bernstein polynomial $B_{n_\varepsilon}(f)(x)$ is arbitrarily close to $f(x)$ in that for every $x \in [0, 1]$ we have

$$|B_{n_\varepsilon}(f)(x) - f(x)| < \varepsilon. \tag{7.4.33}$$

Define $p_\varepsilon : [a, b] \to \mathbb{R}$ by

$$p_\varepsilon(t) = B_{n_\varepsilon}(f)(h^{-1}(t)) = B_{n_\varepsilon}(f)\left(\frac{t - a}{b - a}\right). \tag{7.4.34}$$

Since $g(t) = f(x)$ and $x = (t - a)/(b - a)$, we have

$$|p_\varepsilon(t) - g(t)| = \left|B_{n_\varepsilon}(f)\left(\frac{t - a}{b - a}\right) - g(t)\right| \tag{7.4.35}$$

$$= |B_{n_\varepsilon}(f)(x) - f(x)| < \varepsilon. \tag{7.4.36}$$

To see that p_ε is indeed a polynomial, note that for every $t \in [a, b]$, the definition of Bernstein polynomials (Definition 7.4.4) along with (7.4.31) and (7.4.32) yields

$$p_\varepsilon(t) = B_{n_\varepsilon}(f)\left(\frac{t - a}{b - a}\right) \tag{7.4.37}$$

$$= \sum_{k=0}^{n_\varepsilon} f\left(\frac{k}{n_\varepsilon}\right) b_{n_\varepsilon, k}\left(\frac{t - a}{b - a}\right) \tag{7.4.38}$$

$$= \sum_{k=0}^{n_\varepsilon} f\left(\frac{k}{n_\varepsilon}\right) \left(\frac{t - a}{b - a}\right)^k \left(1 - \frac{t - a}{b - a}\right)^{n_\varepsilon - k} \tag{7.4.39}$$

$$= \sum_{k=0}^{n_\varepsilon} g\left((b - a)\frac{k}{n_\varepsilon} + a\right) \left(\frac{t - a}{b - a}\right)^k \left(1 - \frac{t - a}{b - a}\right)^{n_\varepsilon - k}. \tag{7.4.40}$$

The last two sums above show that p_ε is a polynomial. $\qquad\qquad\square$

An analog of the fundamental theorem of arbitrarily close (Theorem 2.3.1) tells us every continuous function on a compact interval is the uniform limit of a sequence of polynomials.

Corollary 7.4.9: A continuous function is the uniform limit of a sequence of polynomials

If $a < b$ and $g : [a, b] \to \mathbb{R}$ is continuous, then there is a sequence of polynomials (p_n) that converges uniformly to g on $[a, b]$ where

$$\lim_{n \to \infty} p_n(t) = g(t) \qquad (7.4.41)$$

for all $t \in [a, b]$.

Scratch Work 7.4.10: A fundamental argument

The Weierstrass Approximation Theorem 7.4.7 is truly a statement about objects that are arbitrarily close to sets. In fact, the scratch work and proof follow in much the same way as Scratch Work 2.3.2 and Theorem 2.3.1 on the fundamental connection between arbitrarily close and convergence: An object arbitrarily close to a set is the limit of a sequence of objects from the set. Here, the objects are continuous functions on compact intervals and the sets are the sets of polynomials defined on those compact intervals.

The argument follows by taking advantage of the idea that $\varepsilon > 0$ is arbitrary, so a sequence of positive real numbers which tend to zero yields the desired sequence of polynomials through the Weierstrass Approximation Theorem 7.4.7.

Proof of Corollary 7.4.9. Suppose $a < b$ and $g : [a, b] \to \mathbb{R}$ is continuous, and let $\varepsilon > 0$. Consider the sequence of positive real numbers (ε_n) given by

$$\varepsilon_n = \frac{1}{n} > 0 \quad \text{for all} \quad n \in \mathbb{N}. \qquad (7.4.42)$$

By the Weierstrass Approximation Theorem 7.4.7, there is a sequence of polynomials (p_n) such that for each $n \in \mathbb{N}$ we have

$$|p_n(t) - g(t)| < \varepsilon_n = \frac{1}{n} \quad \text{for all} \quad t \in [a, b]. \qquad (7.4.43)$$

Now choose $n_\varepsilon \in \mathbb{N}$ such that

$$n_\varepsilon > \frac{1}{\varepsilon} \qquad \Longleftrightarrow \qquad \frac{1}{n_\varepsilon} < \varepsilon. \qquad (7.4.44)$$

Then for every $n \in \mathbb{N}$ where $n \geq n_\varepsilon$ and every $t \in [a, b]$ we have

$$|p_n(t) - g(t)| < \frac{1}{n} \leq \frac{1}{n_\varepsilon} < \varepsilon. \qquad (7.4.45)$$

Therefore, by Definition 7.2.1, n_ε is a uniform threshold and (p_n) converges uniformly to g on $[a, b]$. $\qquad \square$

Here's a fun way to finish the chapter.

Remark 7.4.11: Weierstrass Approximation Theorem haiku

> Continuity,
> arbitrarily close to
> polynomial.

The next chapter turns our attention to *series*.

Exercises

7.4.1. Use mathematical software (Desmos, GeoGebra, etc.) to explore the Bernstein polynomials of the following functions defined on $[0, 1]$.

(i) $f_1(x) = 2x^2 - 2x + 1$ (iv) $f_4(x) = e^x$

(ii) $f_2(x) = \sin 2\pi x$ (v) $f_5(x) = \ln x$

(iii) $f_3(x) = \cos 2\pi x$ (vi) $f_6(x) = 1/(1 + x)$

7.4.2. The Bernstein basis polynomials $b_{n,k}$ exhibit a wide variety of interesting behaviors including and beyond Lemma 7.4.3. Suppose $x \in [0, 1]$, $n \in \mathbb{N} \cup \{0\}$, and $k = 0, 1, \dots, n$. Prove the following statements.

(i) $b_{n,k}(x) \geq 0$ (*nonnegativity*).

(ii) $\displaystyle\sum_{k=0}^{n} b_{n,k}(x) = \sum_{k=0}^{n} \binom{n}{k} x^k (1-x)^{n-k} = 1$ (*probability, partition of unity*).
Hint: Apply the Binomial Theorem 1.2.24 to $(x + (1 - x))$.

(iii) $\displaystyle\sum_{k=0}^{n} \frac{k}{n} b_{n,k}(x) = \sum_{k=0}^{n} \frac{k}{n} \binom{n}{k} x^k (1-x)^{n-k} = x$ (*mean*).
Hint: Take the derivative of $(x + y)^n$ with respect to x while treating y as a constant, then let $y = 1 - x$.

(iv) $\displaystyle\sum_{k=0}^{n} \left(x - \frac{k}{n}\right)^2 b_{n,k}(x) = \sum_{k=0}^{n} \left(x - \frac{k}{n}\right)^2 \binom{n}{k} x^k (1-x)^{n-k} = \frac{x(1-x)}{n}$ (*variance*).
Hint: Take the second derivative of $(x + y)^n$ with respect to x while treating y as a constant, then let $y = 1 - x$.

(v) $b_{n,k}(1 - x) = b_{n,n-k}(x)$.

(vi) $b_{n,k}(x)$ where $0 < k \leq n$ has a root at $x = 0$ with multiplicity k.

(vii) $b_{n,k}(x)$ where $0 \leq k < n$ has a root at $x = 1$ with multiplicity $n - k$.

(viii) Antiderivatives: $\displaystyle\int b_{n,k}(x)\,dx = \frac{1}{n+1}\sum_{j=k+1}^{n+1} b_{n+1,j}(x).$

(ix) Definite integrals are constant for each $n \in \mathbb{N}$:

$$\int_0^1 b_{n,k}(x)\,dx = \frac{1}{n+1}. \tag{7.4.46}$$

(x) The derivative $b'_{n,k}$ is a linear combination of Bernstein polynomials of lower degree:

$$b'_{n,k}(x) = n(b_{n-1,k-1}(x) - b_{n-1,k}(x)). \tag{7.4.47}$$

(xi) The jth derivatives at 0 and 1 are nice:

$$b_{n,k}^{(j)}(0) = \frac{n!}{(n-k)!}\binom{j}{k}(-1)^{k+j} \qquad \text{and} \qquad b_{n,k}^{(j)}(1) = (-1)^j b_{n,n-k}^{(j)}(0). \tag{7.4.48}$$

(xii) For $n > 0$, $b_{n,k}$ has a unique maximum attained at k/n and

$$\max b_{n,k}([0,1]) = b_{n,k}\left(\frac{k}{n}\right) = \frac{k^k}{n^n}(n-k)^{n-k}\binom{n}{k}. \tag{7.4.49}$$

Chapter 8

Series

Series formalize the notion of infinite linear combinations and sums with infinitely many terms by taking the terms of a sequence and adding them together one term at a time. In turn, series define a notion of infinite polynomials through *power series*. The chapter brings the main content of the book to a conclusion with an exploration of Taylor polynomials and series.

8.1 Series of real numbers

First, let's define series of real numbers. Series of functions are defined later. Although series of vectors in Euclidean spaces make for a meaningful and interesting topic, they are not explored in this book.

Definition 8.1.1: Series

A *series* of real numbers is an object of the form

$$\sum_{n=1}^{\infty} x_n = x_1 + x_2 + x_3 + \cdots \tag{8.1.1}$$

where (x_n) is a sequence of real numbers called the *terms* of the series. For each index $k \in \mathbb{N}$, the *kth partial sum* of the above series is the linear combination s_k given by

$$s_k = \sum_{n=1}^{k} x_n = x_1 + x_2 + \cdots + x_k. \tag{8.1.2}$$

Also, the *k-tail of a series* is the series defined by the sequential k-tail $(x_{n \geq k})$ where we have

$$\sum_{n=k}^{\infty} x_n = x_k + x_{k+1} + x_{k+2} + \cdots. \tag{8.1.3}$$

Remark 8.1.2: Indexing series

In practice, series can take more general forms than (8.1.1) such as

$$\sum_{n=0}^{\infty} x_n = x_0 + x_1 + x_2 + \cdots \tag{8.1.4}$$

where the indices begin with $n = 0$. That is, the indices can begin at any integer. For instance, we have

$$\sum_{n=-3}^{\infty} \frac{1}{2^n} = 8 + 4 + 2 + 1 + \frac{1}{2} + \frac{1}{4} + \frac{1}{8} + \cdots \tag{8.1.5}$$

where $n \in \mathbb{N} \cup \{-3, -2, -1, 0\}$. In this more general setting, every k-tail of a series as in (8.1.3) is itself a series whose indices begin at an integer k.

Series converge when their partial sums converge as a sequence. Also, the *sum* of a series is the limit of its sequence of partial sums.

Definition 8.1.3: Convergence and sum of a series

A series of real numbers

$$\sum_{n=1}^{\infty} x_n = x_1 + x_2 + x_3 + \cdots \tag{8.1.6}$$

converges when its sequence of partial sums (s_k) converges as in Definition 2.2.1. In this case, the limit of (s_k) is a real number s called the *sum* of the series and we write

$$\sum_{n=1}^{\infty} x_n = \lim_{k \to \infty} \left(\sum_{n=1}^{k} x_n \right) = \lim_{k \to \infty} s_k = s. \tag{8.1.7}$$

Also, a series *diverges* if its sequence of partial sums diverges.

Remark 8.1.4: Series and sequences

Convergence of series is determined by the sequence of partial sums. As a result, the approach to proofs on the convergence of series often involves the manipulation of partial sums followed by properties of sequential limits from throughout Chapter 2 impact the results we obtain on series. The first example of the section exhibits this idea.

Example 8.1.5: A telescoping series

Consider the series of positive numbers given by

$$\sum_{n=1}^{\infty} \frac{1}{n(n+1)} = \frac{1}{1 \cdot 2} + \frac{1}{2 \cdot 3} + \frac{1}{3 \cdot 4} + \cdots. \tag{8.1.8}$$

This series converges and its sum is 1.

Scratch Work 8.1.6: The partial sums telescope

The trick to this proof is to recognize that the partial sums telescope. Note that for each $n \in \mathbb{N}$, we can find a common denominator by multiplying by a couple of nice versions of 1 to get

$$\frac{1}{n} - \frac{1}{n+1} = \frac{1}{n}\left(\frac{n+1}{n+1}\right) - \frac{1}{n+1}\left(\frac{n}{n}\right) \tag{8.1.9}$$

$$= \frac{n+1}{n(n+1)} - \frac{n}{n(n+1)} \tag{8.1.10}$$

$$= \frac{(n+1) - n}{n(n+1)} \tag{8.1.11}$$

$$= \frac{1}{n(n+1)}. \tag{8.1.12}$$

Reversing this process by adding a nice version of zero to the numerator leads to a sequence of partial sums, each of which telescope.

Proof for Example 8.1.5. Consider the series

$$\sum_{n=1}^{\infty} \frac{1}{n(n+1)} = \frac{1}{1\cdot2} + \frac{1}{2\cdot3} + \frac{1}{3\cdot4} + \cdots. \tag{8.1.13}$$

For each $n \in \mathbb{N}$, by adding a nice version of zero to the numerator and splitting the fraction in two, we have

$$\frac{1}{n(n+1)} = \frac{(n+1) - n}{n(n+1)} = \frac{n+1}{n(n+1)} - \frac{n}{n(n+1)} = \frac{1}{n} - \frac{1}{n+1}. \tag{8.1.14}$$

Thus, for each $k \in \mathbb{N}$, the partial sum s_k telescopes in that

$$s_k = \sum_{n=1}^{k} \frac{1}{n(n+1)} \tag{8.1.15}$$

$$= \sum_{n=1}^{k} \left(\frac{1}{n} - \frac{1}{n+1}\right) \tag{8.1.16}$$

$$= \left(\frac{1}{1} - \frac{1}{2}\right) + \left(\frac{1}{2} - \frac{1}{3}\right) + \frac{1}{3} - \cdots - \frac{1}{k} + \left(\frac{1}{k} - \frac{1}{k+1}\right) \tag{8.1.17}$$

$$= 1 - \frac{1}{k+1} \tag{8.1.18}$$

By the linearity of convergent sequences (Theorem 2.3.9), we have

$$\lim_{k\to\infty} s_k = \lim_{k\to\infty}\left(1 - \frac{1}{k+1}\right) = \lim_{k\to\infty} 1 - \lim_{k\to\infty} \frac{1}{k+1} = 1 - 0 = 1. \tag{8.1.19}$$

Therefore, by the definition of convergence for series (Definition 8.1.3), the series given by $\sum_{n=1}^{\infty} 1/(n(n+1))$ converges to the sum 1 and

$$\sum_{n=1}^{\infty} \frac{1}{n(n+1)} = \lim_{k\to\infty} s_k = 1. \tag{8.1.20}$$

$\square$

The geometric sums and decimal expansions explored in Section 2.7 series. In fact, infinite decimal expansions (Definition 2.7.7) are the sums of convergent series thanks to Theorem 2.7.8.

Corollary 8.1.7: Decimals are series

Every infinite decimal expansion is the sum of a convergent series. Moreover,

$$0.x_1x_2\ldots = \lim_{k\to\infty}\left(\sum_{n=1}^{k}\frac{x_n}{10^n}\right) = \sum_{n=1}^{\infty}\frac{x_n}{10^n} \tag{8.1.21}$$

where (x_n) is a sequence of digits satisfying

$$x_n \in \{0,1,2,\ldots,9\} \quad \text{for each} \quad n \in \mathbb{N}. \tag{8.1.22}$$

Scratch Work 8.1.8: Interpret previous results

The proof follows from interpreting the results of Theorem 2.7.8 on infinite decimal expansions as series. See Definitions 2.7.7, 8.1.1, and 8.1.3.

Proof of Lemma 7.2.6. Suppose (x_n) is a sequence of digits given by

$$x_n \in \{0,1,2,\ldots,9\} \quad \text{for each} \quad n \in \mathbb{N}. \tag{8.1.23}$$

Consider the infinite decimal expansion $0.x_1x_2\ldots$ and the modified sequence $(x_n/10^n)$ along with the series

$$\sum_{n=1}^{\infty}\frac{x_n}{10^n}. \tag{8.1.24}$$

By Definitions 2.7.7, 8.1.1, and 8.1.3 along with Theorem 2.7.8, this series converges to the infinite decimal expansion $0.x_1x_2\ldots$ since it is also the limit of the partial sums of the series. That is, we have

$$0.x_1x_2\ldots = \lim_{k\to\infty}\left(\sum_{n=1}^{k}\frac{x_n}{10^n}\right) = \sum_{n=1}^{\infty}\frac{x_n}{10^n}. \tag{8.1.25}$$

$\square$

Geometric sums from Definition 2.7.1 lead to a vital class of series: *geometric series.*

Definition 8.1.9: Geometric series

A *geometric series* is a series of the form

$$\sum_{n=0}^{\infty} ar^n = a + ar + ar^2 + \cdots \tag{8.1.26}$$

where $a, r \in \mathbb{R}$. The real number a is called the *initial term* and r is called the *common ratio*. Also, we use the convention $r^0 = 1$.

Geometric series provide an important class of series, especially since we know exactly when they converge *and* what they converge to. Make sure you are comfortable with this result; geometric series pop up in many proofs and exercises.

Theorem 8.1.10: Closed form for geometric series

A geometric series of the form

$$\sum_{n=0}^{\infty} ar^n = a + ar + ar^2 + \cdots \tag{8.1.27}$$

converges if and only if $a = 0$ or $|r| < 1$. In this case, the sum of the series is given by

$$\sum_{n=0}^{\infty} ar^n = \frac{a}{1 - r}. \tag{8.1.28}$$

Scratch Work 8.1.11: Consider the partial sums

The convergence or divergence of a series stems from the behavior of its partial sums. In the case of geometric series, the partial sums are geometric sums as in Definition 2.7.1 (though in the current context the role of the variables n and k are reversed). Specifically, thanks to the Geometric Sum Formula 2.7.2, a geometric series of the form (8.1.27) has partial sums given by

$$s_k = \sum_{n=0}^{k} ar^n = a + ar + ar^2 + \cdots + ar^k = \frac{a(1 - r^{k+1})}{1 - r} \tag{8.1.29}$$

for each index $k \in \mathbb{N}$ and $r \neq 1$. Note how the k-th partial sum s_k very much depends on the behavior of the expression r^{k+1}. The proof breaks down the argument into several cases largely based on how the sequence (r^{k+1}) behaves.

Proof of Theorem 8.1.10. Throughout the proof, consider geometric series of the form

$$\sum_{n=0}^{\infty} ar^n = a + ar + ar^2 + \cdots \tag{8.1.30}$$

where $a, r \in \mathbb{R}$. When $r \neq 1$, the partial sums are given by

$$s_k = \sum_{n=0}^{k} ar^n = a + ar + ar^2 + \cdots + ar^k = \frac{a(1 - r^{k+1})}{1 - r}, \tag{8.1.31}$$

which follows from the Geometric Sum Formula 2.7.2.

Case (i), $a = 0$: Suppose $a = 0$. Then the partial sums of (8.1.30) are all zero. That is, for each index $k \in \mathbb{N}$ we have

$$s_k = \frac{0(1 - r^{k+1})}{1 - r} = 0. \tag{8.1.32}$$

Hence, the corresponding geometric series converges to 0 since

$$\sum_{n=0}^{\infty} 0 r^n = \lim_{k \to \infty} s_k = \lim_{n \to \infty} 0 = 0. \tag{8.1.33}$$

The remaining cases consider nonzero values of a.

Case (ii), $|r| < 1$: Suppose $a \neq 0$ and $|r| < 1$. By Corollary 2.4.19, we have

$$\lim_{k \to \infty} r^{k+1} = 0. \tag{8.1.34}$$

Hence, by the linearity of limits of sequences (Theorem 2.3.9) we have

$$\sum_{n=0}^{\infty} ar^n = \lim_{k \to \infty} s_k = \lim_{k \to \infty} \left(\frac{a(1 - r^{k+1})}{1 - r} \right) = \frac{a}{1 - r}. \tag{8.1.35}$$

Case (iii), $r = -1$: Suppose $a \neq 0$ and $r = -1$. Then $r^{k+1} = (-1)^{k+1}$ for each index $k \in \mathbb{N}$, meaning $r^{k+1} = 1$ when k is odd and $r^{k+1} = -1$ when k is even. Hence, the partial sums alternate between a and 0 since

$$s_k = \frac{a(1 - (-1)^{k+1})}{1 - (-1)} = \begin{cases} 0, & \text{when } k \text{ is odd,} \\ a, & \text{when } k \text{ is even.} \end{cases} \tag{8.1.36}$$

The subsequences of (s_k) determined by odd and even indices, respectively (s_{2j-1}) and (s_{2j}), converge to different limits since

$$\lim_{j \to \infty} s_{2j-1} = 0 \neq a = \lim_{j \to \infty} s_{2j}. \tag{8.1.37}$$

Therefore, by the Divergence Criteria for Sequences 2.6.9, the sequence of partial sums (s_k) and the geometric series $\sum_{n=0}^{\infty} ar^n$ diverge.

Case (iv), $r = 1$: Suppose $a \neq 0$ and $r = 1$. Then $r^{k+1} = 1$ for every index $k \in \mathbb{N}$ and so the sequence of partial sums is unbounded since

$$s_k = \sum_{n=0}^{k} a 1^{k+1} = a(k + 1). \tag{8.1.38}$$

Therefore, by the Divergence Criteria for Sequences 2.6.9, the sequence of partial sums $(s_k) = (a(k+1))$ and the geometric series $\sum_{n=0}^{\infty} ar^n$ diverge.

Case (v), $|r| > 1$: Suppose $a \neq 0$ and $|r| > 1$. Then (r^{k+1}) is unbounded, and so the sequence of partial sums is unbounded as well since

$$s_k = \frac{a(1 - r^{k+1})}{1 - r} = \frac{a}{1 - r} - r^k \left(\frac{r}{1 - r} \right). \tag{8.1.39}$$

Therefore, by the Divergence Criteria for Sequences 2.6.9, the sequence of partial sums (s_k) and the geometric series $\sum_{n=0}^{\infty} ar^n$ diverge. $\quad\square$

Every real number can be thought of as the sum of lots of different series. Also, we can always reindex a series to start with $n = 0, 1$, or any integer we like.

Example 8.1.12: Sum to 1

Geometric series of various forms sum to 1. For instance,

$$\sum_{n=1}^{\infty} \frac{1}{2^n} = \sum_{n=0}^{\infty} \frac{1/2}{2^n} = \frac{1}{2} + \frac{1}{4} + \frac{1}{8} + \cdots = 1 \qquad \text{and} \tag{8.1.40}$$

$$\sum_{n=1}^{\infty} \frac{9}{10^n} = \sum_{n=0}^{\infty} \frac{9/10}{10^n} = \frac{9}{10} + \frac{9}{100} + \cdots = 0.999\ldots = 1. \tag{8.1.41}$$

Since both series are geometric series (Definition 8.1.9), both sums follow from Theorem 8.1.10 by setting $a = r = 1/2$ and $a = 9/10$ with $r = 1/10$, respectively. We have

$$\sum_{n=0}^{\infty} \frac{1}{2^{n+1}} = \sum_{n=0}^{\infty} \frac{1/2}{2^n} = \frac{1/2}{1 - (1/2)} = 1 \qquad \text{and} \tag{8.1.42}$$

$$\sum_{n=1}^{\infty} \frac{9}{10^n} = \sum_{n=0}^{\infty} \frac{9/10}{10^n} = \frac{9/10}{1 - (1/10)} = 1. \tag{8.1.43}$$

Also, see Example 2.7.10 where it is shown that $0.999\ldots = 1$ from the perspective of infinite decimal expansions as limits of sequences of finite decimal expansions.

As a first application of geometric series, we can prove that the set of rational numbers $\mathbb{Q}$ is, in some sense, arbitrarily small. Specifically, even though the rationals are unbounded and dense in the real line, $\mathbb{Q}$ is contained in the union of a countable number of open intervals whose total length can be made as small as we like. Try to draw it yourself to get a sense of what this might mean.

Theorem 8.1.13: The set of rational numbers is arbitrarily small

For every $\varepsilon > 0$, there is a sequence of open intervals (I_n) where $I_n = (a_n, b_n)$ for each index $n \in \mathbb{N} \cup \{0\}$,

$$\mathbb{Q} \subseteq \bigcup_{n=0}^{\infty} I_n \qquad \text{and} \qquad \sum_{n=0}^{\infty} (b_n - a_n) = \varepsilon. \tag{8.1.44}$$

Scratch Work 8.1.14: Break ε into countably many pieces

The key idea is to take the positive number ε and break it into countably many pieces whose sum is ε. Resetting the initial term to $a = \varepsilon/2$ for the first geometric series in Example 8.1.12 yields

$$\sum_{n=0}^{\infty} \frac{\varepsilon}{2^{n+1}} = \sum_{n=1}^{\infty} \frac{\varepsilon/2}{2^n} = \frac{\varepsilon/2}{1 - (1/2)} = \varepsilon. \tag{8.1.45}$$

From here, the open intervals I_n stem from any enumeration of the rationals (r_n) where, for each index $n \in \mathbb{N} \cup \{0\}$, we pair the rational number r_n with the open interval I_n centered at r_n whose *length*—the distance between its endpoints—is exactly $\varepsilon/2^{n+1}$. Setting the endpoints a_n and b_n to be the real numbers at exactly $\varepsilon/2^{n+2}$ away from r_n gets the job done.

Proof of Theorem 8.1.13. Let (r_n) be an enumeration of the rationals $\mathbb{Q}$ starting with index $n = 0$ and let $\varepsilon > 0$. For each index $n \in \mathbb{N} \cup \{0\}$, define

$$a_n = r_n - \frac{\varepsilon}{2^{n+2}}, \qquad b_n = r_n + \frac{\varepsilon}{2^{n+2}}, \tag{8.1.46}$$

and

$$I_n = (a_n, b_n) = \left(r_n - \frac{\varepsilon}{2^{n+2}}, r_n + \frac{\varepsilon}{2^{n+2}} \right). \tag{8.1.47}$$

Since (r_n) accounts for every rational number and $r_n \in I_n = (a_n, b_n)$ for each index $n \in \mathbb{N} \cup \{0\}$, we have

$$\mathbb{Q} \subseteq \bigcup_{n=0}^{\infty} I_n = \bigcup_{n=0}^{\infty} (a_n, b_n). \tag{8.1.48}$$

Furthermore, we have

$$b_n - a_n = \left(r_n + \frac{\varepsilon}{2^{n+2}} \right) - \left(r_n - \frac{\varepsilon}{2^{n+2}} \right) = \frac{2\varepsilon}{2^{n+2}} = \frac{\varepsilon}{2^{n+1}}. \tag{8.1.49}$$

Therefore,

$$\sum_{n=0}^{\infty} (b_n - a_n) = \sum_{n=0}^{\infty} \frac{\varepsilon}{2^{n+1}} = \sum_{n=1}^{\infty} \frac{\varepsilon/2}{2^n} = \frac{\varepsilon/2}{1 - (1/2)} = \varepsilon. \tag{8.1.50}$$

$\square$

**Remark 8.1.15: Countable subsets of a Euclidean space
are arbitrarily small**

Scratch Work 8.1.14 and the proof of Theorem 8.1.13 can be modified to prove a more general statement when S is a countable subset of a Euclidean space $\mathbb{R}^m$. In this case, S is arbitrarily small in the following sense: For every $\varepsilon > 0$, there is a sequence of open neighborhoods (V_n) where for each index $n \in \mathbb{N} \cup \{0\}$ the neighborhood V_n is centered at some point in S and has radius ε_n where

$$S \subseteq \bigcup_{n=0}^{\infty} V_n \qquad \text{and} \qquad \sum_{n=0}^{\infty} \varepsilon_n = \varepsilon. \tag{8.1.51}$$

The details of such a proof are left as an exercise.

To close out the section, consider an important pair of examples.

Example 8.1.16: Harmonic series

The series

$$\sum_{n=1}^{\infty} \frac{1}{n} = 1 + \frac{1}{2} + \frac{1}{3} + \cdots \tag{8.1.52}$$

is known as the *harmonic series*. Despite the fact that its sequence of terms converges to zero, the harmonic series diverges. That is,

$$\lim_{n \to \infty} \frac{1}{n} = 0 \qquad \text{and yet} \qquad \sum_{n=1}^{\infty} \frac{1}{n} \text{ diverges.} \tag{8.1.53}$$

Scratch Work 8.1.17: Consider the partial sums

As with geometric series, the convergence or divergence of the harmonic series can be determined by the behavior of the partial sums. Here, they are shown to be arbitrarily large in that every time we skip from one partial sum to another by doubling the number of terms considered, we add at least $1/2$. For instance:

$$s_1 = 1, \tag{8.1.54}$$

$$s_2 = 1 + \frac{1}{2}, \tag{8.1.55}$$

$$s_4 = 1 + \frac{1}{2} + \left(\frac{1}{3} + \frac{1}{4} \right) \tag{8.1.56}$$

$$> 1 + \frac{1}{2} + \left(\frac{1}{4} + \frac{1}{4} \right) = 1 + \frac{2}{2}, \qquad \text{and} \tag{8.1.57}$$

$$s_8 = 1 + \frac{1}{2} + \left(\frac{1}{3} + \frac{1}{4} \right) + \left(\frac{1}{5} + \frac{1}{6} + \frac{1}{7} + \frac{1}{8} \right) \tag{8.1.58}$$

$$> 1 + \frac{1}{2} + \left(\frac{1}{4} + \frac{1}{4} \right) + \left(\frac{1}{8} + \frac{1}{8} + \frac{1}{8} + \frac{1}{8} \right) = 1 + \frac{3}{2}. \tag{8.1.59}$$

In general, an induction argument shows that for each index $k \in \mathbb{N}$ we have

$$s_{2^k} = \sum_{n=1}^{2^k} \frac{1}{n} > 1 + \frac{k}{2}. \tag{8.1.60}$$

Thus, the sequence of partial sums is unbounded and so, by the Divergence Criteria for Sequences 2.6.9, the harmonic series diverges.

The details of the induction argument and the proof are left as an exercise.

The number e from calculus is known as *Euler's number*, and one way to define it is as the sum of a special convergent series of positive numbers.

Example 8.1.18: Euler's number e

Euler's number e is the sum of the series whose terms are $1/(n!)$ for each index $n \in \mathbb{N} \cup \{0\}$. That is, this series converges and we define e as the sum. Hence,

$$e = \sum_{n=0}^{\infty} \frac{1}{n!}. \tag{8.1.61}$$

Scratch Work 8.1.19: Comparing partial sums

The goals is to show Euler's number e is well-defined by showing its series converges. As with all the proofs in the section up to this point, the result is obtained by considering partial sums and taking advantage of properties of sequential limits. Here, the convergence of the given series follows from showing the partial sums form an increasing sequence which is bounded above and, therefore, converges by the Monotone and Bounded Convergence Theorem 2.4.9.

Proof for Example 8.1.18. Consider the series whose terms are $1/n!$ for each $n \in \mathbb{N} \cup \{0\}$. Since $1/n \le 1/2$ when $n \ge 2$, for each $n \in \mathbb{N} \cup \{0\}$ we have

$$0 \le \frac{1}{n!} = \frac{1}{1 \cdot 2 \cdot 3 \cdots (n-1) \cdot n} \le \frac{1}{\underbrace{1 \cdot 2 \cdot 2 \cdots 2 \cdot 2}_{n \text{ factors}}} = \frac{1}{2^n}. \tag{8.1.62}$$

The sequence of partial sums (s_k) is bounded above by 2 since, for each $k \in \mathbb{N} \cup \{0\}$,

$$s_k = \sum_{n=0}^{k} \frac{1}{n!} \le \sum_{n=0}^{k} \frac{1}{2^n} = \frac{1 - (1/2)^{n+1}}{1 - (1/2)} \le \frac{1}{1/2} = 2, \tag{8.1.63}$$

where the sum on the right is a geometric sum and the Geometric Sum Formula 2.7.2 applies. Also, the sequence of partial sums (s_k) is increasing since

$$s_k = \sum_{n=0}^{k} \frac{1}{n!} \le \left(\sum_{n=0}^{k} \frac{1}{n!} \right) + \frac{1}{(k+1)!} = \sum_{n=0}^{k+1} \frac{1}{2^n} = s_{k+1}. \tag{8.1.64}$$

Hence, by the Monotone and Bounded Convergence Theorem 2.4.9, the sequence of partial sums (s_k) converges. Therefore, the number e is well-defined by

$$e = \sum_{n=0}^{\infty} \frac{1}{n!} \tag{8.1.65}$$

since this series converges by Definition 8.1.3. $\qquad \square$

The next section develops some key properties of series of real numbers.

Exercises

8.1.1. Prove convergent series are linear: Suppose $\sum_{n=1}^{\infty} x_n$ and $\sum_{n=1}^{\infty} y_n$ are convergent series of real numbers which converge to x and y, respectively, and suppose $\alpha \in \mathbb{R}$. Then $\sum_{n=1}^{\infty}(x_n + y_n)$ converges to $x + y$ and $\sum_{n=1}^{\infty}(\alpha x_n)$ converges αx. That is,

(i) $\displaystyle\sum_{n=1}^{\infty}(x_n + y_n) = \sum_{n=1}^{\infty} x_n + \sum_{n=1}^{\infty} y_n = x + y$ (*additivity*);   and

(ii) $\displaystyle\sum_{n=1}^{\infty}(\alpha x_n) = \alpha \sum_{n=1}^{\infty} x_n = \alpha x$ (*homogeneity*).

Hint: The convergence of a series is defined by the convergence of the sequence of partial sums (Definition 8.1.3), so apply the linearity of sequential limits (Theorem 2.3.9) to the partial sums.

8.1.2. Assume convergent series are linear as in the previous exercise. Prove the following corollary: Suppose $q \in \mathbb{N}$ and for each $j = 1, \dots, q$ we have $\alpha_j \in \mathbb{R}$ and the series of real numbers $\sum_{n=1}^{\infty} x_{j,n}$ converges to x_j. Then the series of linear combinations $\sum_{n=1}^{\infty} \sum_{j=1}^{q}(\alpha_j x_{j,n})$ converges to the linear combination of series $\sum_{j=1}^{q} \sum_{n=1}^{\infty}(\alpha_j x_{j,n})$, and we have

$$\sum_{n=1}^{\infty}\sum_{j=1}^{q}(\alpha_j x_{j,n}) = \sum_{j=1}^{q}\left(\alpha_j \sum_{n=1}^{\infty} x_{j,n}\right) = \sum_{j=1}^{q}(\alpha_j x_j). \tag{8.1.66}$$

8.1.3. Prove the harmonic series $\displaystyle\sum_{n=1}^{\infty} \frac{1}{n}$ diverges by completing the argument started in Scratch Work 8.1.17.

8.1.4. Determine whether the series $\displaystyle\sum_{n=1}^{\infty} \frac{1}{n + 3^n}$ converges or diverges.

8.1.5. Prove $\displaystyle\lim_{n\to\infty}\left(1 + \frac{1}{n}\right)^n = e$ by completing the following steps.

(i) Use the Binomial Theorem 1.2.24 to show for each $n \in \mathbb{N}$ we have

$$y_n = \left(1 + \frac{1}{n}\right)^n = 1 + 1 + \frac{1}{2!}\left(1 - \frac{1}{n}\right) + \frac{1}{3!}\left(1 - \frac{1}{n}\right)\left(1 - \frac{2}{n}\right) + \cdots$$
$$+ \frac{1}{n!}\left(1 - \frac{1}{n}\right)\left(1 - \frac{2}{n}\right)\cdots\left(1 - \frac{n-1}{n}\right). \tag{8.1.67}$$

(ii) Use (i) to show (y_n) is increasing and bounded above by e. Why does this show $\displaystyle\lim_{n\to\infty} y_n = \lim_{n\to\infty}\left(1 + \frac{1}{n}\right)^n$ converges?

(iii) Let $n, k \in N$ where $n \geq k$ and s_k is the kth partial sum of the series that defines e in Example 8.1.18. Show that $s_k \leq y_n$.

(iv) Combine the above results to complete the proof.

8.1.6. This exercise shows Euler's number e is irrational.

(i) Let $k \in N$ and let s_k denote the kth partial sum of the series that defines e in Example 8.1.18. Prove

$$0 < e - s_n < \frac{1}{k!k}. \tag{8.1.68}$$

Note that this inequality shows how rapidly the series $\sum_{n=1}^{\infty} 1/n!$ converges to its sum e.

(ii) Use (i) and a contradiction argument to prove e is irrational.

8.1.7. Construct a function $h : \mathbb{R} \to \mathbb{R}$ satisfying the following properties and prove your result:

(i) h is increasing.

(ii) h is continuous at every $c \in \mathbb{R}\backslash\mathbb{N}$.

(iii) h is discontinuous at every $c \in \mathbb{N}$.

(iv) h is bounded. Hint: Split a positive bound b into countably many pieces as done with ε in Scratch Work 8.1.14.

8.1.8. Prove that if an infinite decimal expansion is repeating, then the infinite decimal expansion converges to a rational number.

8.1.9. Prove countable sets are arbitrarily small (as mentioned in Remark 8.1.15): If S is a countable subset of a Euclidean space $\mathbb{R}^m$, then for every $\varepsilon > 0$ there is a sequence of open neighborhoods (V_n) where for each index $n \in \mathbb{N} \cup \{0\}$ the neighborhood V_n is centered at some point in S and has radius ε_n where

$$S \subseteq \bigcup_{n=0}^{\infty} V_n \quad \text{and} \quad \sum_{n=0}^{\infty} \varepsilon_n = \varepsilon. \tag{8.1.69}$$

8.2   Properties of series

The primary motivation for this section is determining conditions under which a series converges or diverges.

First up, the proof of Example 8.1.18 can be modified to yield a characterization of convergent series when the terms are nonnegative.

Theorem 8.2.1: Nonnegative and bounded partial sums

Suppose $\sum_{n=1}^{\infty} a_n$ is a series of nonnegative real numbers (specifically, $a_n \geq 0$ for all $n \in \mathbb{N}$). Then $\sum_{n=1}^{\infty} a_n$ converges if and only if the sequence of partial sums (s_k) is bounded.

Proof of Theorem 8.2.1. Suppose $\sum_{n=1}^{\infty} a_n$ is a series of real numbers where, throughout the proof, $a_n \geq 0$ for every $n \in \mathbb{N}$.

For the forward implication, assume $\sum_{n=1}^{\infty} a_n$ converges. Since a series converges when its sequence of partial sums converge (Definition 8.1.3), and convergent sequences are bounded (Theorem 2.3.15), it follows that the sequence of partial sums (s_k) is bounded.

For the backward implication, assume the sequence of partial sums (s_k) is bounded. For each for each $k \in \mathbb{N}$, $a_{k+1} \geq 0$ implies

$$s_k = \sum_{n=0}^{k} a_n \leq \left(\sum_{n=0}^{k} a_n \right) + a_{k+1} = \sum_{n=0}^{k+1} a_n = s_{k+1}. \tag{8.2.1}$$

Hence, (s_k) is increasing as well. So by the Monotone and Bounded Convergence Theorem 2.4.9, the sequence of partial sums (s_k) converges. Therefore, the series $\sum_{n=1}^{\infty} a_n$ converges by Definition 8.1.3. □

Convergent series are characterized by a Cauchy criterion. Like the other Cauchy criteria, convergence is assured even without a candidate for the limit/derivative/integral/sum in mind. However, the Cauchy criterion for series below looks a bit different from the others. We can concentrate on the difference of partial sums, leading to a finite sum of terms starting at some index past a threshold.

Theorem 8.2.2: Cauchy criterion for series

Suppose $\sum_{n=1}^{\infty} x_n$ is a series of real numbers. Then $\sum_{n=1}^{\infty} x_n$ converges if and only if, for every $\varepsilon > 0$, there is a threshold $n_\varepsilon \in \mathbb{N}$ such that for all $j, k \in \mathbb{N}$ where $j > k \geq n_\varepsilon$ we have

$$\left| \sum_{n=k+1}^{j} x_n \right| = |x_{k+1} + x_{k+2} + \cdots + x_{j-1} + x_j| < \varepsilon. \tag{8.2.2}$$

Scratch Work 8.2.3: Apply the Cauchy criterion for sequences to the partial sums

The threshold follows from the Cauchy criterion for sequences 2.6.5 when the partial sums of the series are taken into consideration. The sum in inequality (8.2.2) comes from canceling the first k terms from the jth partial sum of the series, like this:

$$s_j - s_k = \sum_{n=1}^{j} x_n - \sum_{n=1}^{k} x_n \tag{8.2.3}$$

$$= (x_1 + \cdots + x_k + x_{k+1} + \cdots + x_{j-1} + x_j) \tag{8.2.4}$$

$$- (x_1 + \cdots + x_k) \tag{8.2.5}$$

$$= x_{k+1} + \cdots + x_{j-1} + x_j \tag{8.2.6}$$

$$= \sum_{n=k+1}^{j} x_n. \tag{8.2.7}$$

On to the proof.

Proof of Theorem 8.2.2. Suppose $\sum_{n=1}^{\infty} x_n$ is a series of real numbers and $j, k \in \mathbb{N}$ such that $j > k$. By taking the difference of the jth and kth partial sums as in Scratch Work 8.2.3, we have

$$\left| \sum_{n=1}^{j} x_n - \sum_{n=1}^{k} x_n \right| = \left| \sum_{n=k+1}^{j} x_n \right| = |x_{k+1} + x_{k+2} + \cdots + x_{j-1} + x_j|. \tag{8.2.8}$$

So, the hypothesis for the backward implication is identical to the definition of a Cauchy sequence (Definition 2.6.1) when applied to the sequence of partial sums, meaning it is also equivalent to the convergence of the series. That is, by the definition of convergence for series as the converge of the sequence of partial sums (Definition 8.1.3) and the Cauchy criterion for sequences 2.6.5 applied to the partial sums, we have the desired result: The series $\sum_{n=1}^{\infty} x_n$ converges if and only if, for every $\varepsilon > 0$, there is a threshold $n_\varepsilon \in \mathbb{N}$ such that for all $j, k \in \mathbb{N}$ where $j > k \geq n_\varepsilon$ we have

$$\left| \sum_{n=1}^{j} x_n - \sum_{n=1}^{k} x_n \right| = \left| \sum_{n=k+1}^{j} x_n \right| = |x_{k+1} + x_{k+2} + \cdots + x_{j-1} + x_j| < \varepsilon. \tag{8.2.9}$$

$\square$

The harmonic series $\sum_{n=1}^{\infty}(1/n)$ from Example 8.1.16 shows us that even if the terms of a series converge to zero, we cannot conclude the series itself converges. On the other hand, a corollary of the Cauchy criterion for series 8.2.2 tells us that when a series of real numbers converges, its sequence of terms must converge to zero. This result is also known as the "nth term test".

Corollary 8.2.4: Convergent series have terms converging to zero

If $\sum_{n=1}^{\infty} x_n$ converges, then (x_n) converges to zero. That is,

$$\sum_{n=1}^{\infty} x_n \text{ converges} \quad \Longrightarrow \quad \lim_{n \to \infty} x_n = 0. \tag{8.2.10}$$

Equivalently, the contraposition states

$$\lim_{n \to \infty} x_n \neq 0 \quad \Longrightarrow \quad \sum_{n=1}^{\infty} x_n \text{ diverges}. \tag{8.2.11}$$

Proof of Corollary 8.2.4. Suppose $\sum_{n=1}^{\infty} x_n$ converges. Then by Cauchy criterion for series 8.2.2 applies to the difference between consecutive partial sums. Thus, for every $\varepsilon > 0$ there is a threshold $n_\varepsilon \in \mathbb{N}$ such that for all $k \in \mathbb{N}$ where $k + 1 > k \geq n_\varepsilon$, we have

$$\left| \sum_{n=1}^{k+1} x_n - \sum_{n=1}^{k} x_n \right| = |x_{k+1}| = |x_{k+1} - 0| < \varepsilon. \tag{8.2.12}$$

Therefore, the index $n_\varepsilon + 1$ serves as a threshold for the convergence of the sequence of terms to zero (Definition 2.2.1) and

$$\lim_{n \to \infty} x_n = 0. \tag{8.2.13}$$

$\square$

The convergence of a series comes with an interesting dichotomy when we consider the corresponding series of absolute values.

Definition 8.2.5: Absolute and conditional convergence

Consider a series of real numbers $\sum_{n=1}^{\infty} a_n$ along with the series of absolute values $\sum_{n=1}^{\infty} |a_n|$.

(i) If $\sum_{n=1}^{\infty} |a_n|$ converges, then $\sum_{n=1}^{\infty} a_n$ *converges absolutely.*

(ii) If $\sum_{n=1}^{\infty} |a_n|$ diverges but $\sum_{n=1}^{\infty} a_n$ converges, then $\sum_{n=1}^{\infty} a_n$ *converges conditionally.*

The lack of symmetry between the definitions of absolute and conditional convergence can be explained by the following example and another corollary of the Cauchy criterion for series 8.2.2.

Example 8.2.6: Alternating harmonic series

The series

$$\sum_{n=1}^{\infty} \frac{(-1)^{n+1}}{n} = 1 - \frac{1}{2} + \frac{1}{3} - \frac{1}{4} \pm \cdots \tag{8.2.14}$$

is known as the *alternating harmonic series.* This series converges, as shown in a bit once we have more tools at our disposal (see Example 8.2.10). On the other hand, its series of absolute values is the divergent harmonic series from Example 8.1.16. That is,

$$\sum_{n=1}^{\infty} \left| \frac{(-1)^{n+1}}{n} \right| = \sum_{n=1}^{\infty} \frac{1}{n}. \tag{8.2.15}$$

Hence, the alternating harmonic series is conditionally convergent by Definition 8.2.5.

Corollary 8.2.7: Absolute convergence implies convergence

If $\sum_{n=1}^{\infty} |a_n|$ converges, then $\sum_{n=1}^{\infty} a_n$ converges.

Proof of Corollary 8.2.7. Suppose $\sum_{n=1}^{\infty} a_n$ converges absolutely so that $\sum_{n=1}^{\infty} |a_n|$ converges. Then by induction on the triangle inequality (1.2.35) and the Cauchy criterion for series 8.2.2 applied to $\sum_{n=1}^{\infty} |a_n|$, for every $\varepsilon > 0$ there is a threshold $n_\varepsilon \in \mathbb{N}$ such that for all $j, k \in \mathbb{N}$ where $j > k \geq n_\varepsilon$ we have

$$\left| \sum_{n=k+1}^{j} a_n \right| = |a_{k+1} + a_{k+2} + \cdots + a_j| \leq |a_{k+1}| + |a_{k+1}| + \cdots + |a_j| = \left| \sum_{n=k+1}^{j} |a_n| \right| < \varepsilon. \quad (8.2.16)$$

Therefore, by the Cauchy criterion for series 8.2.2 applied to $\sum_{n=1}^{\infty} a_n$ this time, the series $\sum_{n=1}^{\infty} a_n$ converges. $\quad\square$

Remark 8.2.8: Summary of absolute and conditional convergence

Now that Corollary 8.2.7 has been established:

A series $\sum_{n=1}^{\infty} a_n$ converges absolutely when both the series of absolute values $\sum_{n=1}^{\infty} |a_n|$ and the original series $\sum_{n=1}^{\infty} a_n$ converge.

On the other hand:

A series $\sum_{n=1}^{\infty} a_n$ converges conditionally if and only if the series of absolute values $\sum_{n=1}^{\infty} |a_n|$ diverges while the original series $\sum_{n=1}^{\infty} a_n$ converges.

The remainder of the section develops tests that determine the convergence or divergence of series.

There are many ways for us to determine whether a given series of real numbers converges or not. However, in general it can be very difficult to determine the value of the sum when a series converges. The focus of this section is to develop tests for convergence or divergence of series, but *not* the values of sums.

Theorem 8.2.9: Alternating Series Test

Suppose (a_n) is decreasing and converges to zero. Then the *alternating series* given by

$$\sum_{n=1}^{\infty} (-1)^{n+1} a_n = a_1 - a_2 + a_3 - a_4 + \cdots \quad (8.2.17)$$

converges.

Proof of the Alternating Series Test 8.2.9. Suppose (a_n) is decreasing and converges to zero. So, $\lim_{n\to\infty} a_n = 0$ and for each $n \in \mathbb{N}$

$$a_n \geq a_{n+1} \quad \Longleftrightarrow \quad a_n - a_{n+1} \geq 0. \quad (8.2.18)$$

Multiplying by -1 and rearranging a bit yields we

$$-(a_n - a_{n+1}) \leq 0. \quad \text{and} \quad -(a_{n+1} - a_n) \geq 0. \quad (8.2.19)$$

Also, since convergent series are bounded, the Mononotone and Bounded Convergence Theorem 2.4.9 tells us

$$\inf(a_n) = \lim_{n \to \infty} a_n = 0. \tag{8.2.20}$$

An infimum is a lower bound by Definition 1.1.14, so for each $n \in \mathbb{N}$ we have

$$a_n \geq 0 = \inf(a_n). \tag{8.2.21}$$

That is, a_n is nonnegative, a fact used throughout the proof.

To establish the hypothesis of the Cauchy criterion for series 8.2.2, let $j, k \in \mathbb{N}$ where $j > k$ and consider the difference of the partial sums

$$s_j - s_k = \sum_{n=k+1}^{j} (-1)^{n+1} a_n. \tag{8.2.22}$$

From here, the argument depends on the parity of k and j as either even or odd and its effect on the sum in (8.2.22).

First, suppose k is odd. Then by regrouping the terms starting with $k + 1$, we have

$$s_j - s_k = \sum_{n=k+1}^{j} (-1)^{n+1} a_n \tag{8.2.23}$$

$$= a_{k+1} - a_{k+2} + a_{k+3} - a_{k+4} + a_{k+5} + \cdots + (-1)^{j+1} a_j \tag{8.2.24}$$

$$= a_{k+1} \underbrace{-(a_{k+2} - a_{k+3})}_{\leq 0} \underbrace{-(a_{k+4} - a_{k+5})}_{\leq 0} + \cdots + \underbrace{\begin{cases} -(a_{j-1} - a_j), & \text{if } j \text{ is odd,} \\ -a_j, & \text{if } j \text{ is even.} \end{cases}}_{\leq 0} \tag{8.2.25}$$

The grouped summands that follow a_{k+1} are all nonpositive (≤ 0) thanks to the inequalities from the beginning of the proof. Hence,

$$s_j - s_k \leq a_{k+1} \leq a_k. \tag{8.2.26}$$

Now suppose k is even. Then by regrouping the terms starting with $k + 1$, we have

$$s_j - s_k = \sum_{n=k+1}^{j} (-1)^{n+1} a_n \tag{8.2.27}$$

$$= -a_{k+1} + a_{k+2} - a_{k+3} + a_{k+4} - a_{k+5} + \cdots + (-1)^{j+1} a_j \tag{8.2.28}$$

$$= -a_{k+1} \underbrace{-(a_{k+3} - a_{k+2})}_{\geq 0} \underbrace{-(a_{k+5} - a_{k+4})}_{\geq 0} + \cdots + \underbrace{\begin{cases} +a_j, & \text{if } j \text{ is odd,} \\ -(a_j - a_{j-1}), & \text{if } j \text{ is even.} \end{cases}}_{\geq 0} \tag{8.2.29}$$

This time, the grouped summands that follow a_{k+1} are all nonnegative (≥ 0). Hence,

$$s_j - s_k \geq -a_{k+1} \geq -a_k. \tag{8.2.30}$$

So, whether k is even or odd, inequalities (8.2.26) and (8.2.30) both incorporate even and odd via k and $k+1$. As a result, for every $j, k \in \mathbb{N}$ where $j > k$ we have

$$-a_k \leq s_j - s_k \leq a_k \qquad \Longleftrightarrow \qquad |s_j - s_k| \leq a_k. \tag{8.2.31}$$

Finally, since (a_n) converges to zero, by the definition of sequential limit (Definition 2.2.1) there is a threshold $n_\varepsilon \in \mathbb{N}$ such that for all $j > k \geq n_\varepsilon$ we have

$$|s_j - s_k| \leq a_k = |a_k - 0| < \varepsilon. \tag{8.2.32}$$

Therefore, by the Cauchy criterion for series, $\sum_{n=1}^{\infty} (-1)^{n+1} a_n$ converges. $\square$

Example 8.2.10: Revisiting the alternating harmonic series

The alternating harmonic series

$$\sum_{n=1}^{\infty} \frac{(-1)^{n+1}}{n} = 1 - \frac{1}{2} + \frac{1}{3} - \frac{1}{4} \pm \cdots \tag{8.2.33}$$

is indeed conditionally convergent (Definition 8.2.5). As noted in Example 8.2.6, its series of absolute values is the divergent harmonic series from Example 8.1.16. That is,

$$\sum_{n=1}^{\infty} \left| \frac{(-1)^{n+1}}{n} \right| = \sum_{n=1}^{\infty} \frac{1}{n}. \tag{8.2.34}$$

On the other hand the alternating harmonic series itself converges thanks to the Alternating Series Test 8.2.9 where we set $a_n = 1/n$ for each $n \in \mathbb{N}$. Hence,

$$\sum_{n=1}^{\infty} (-1)^{n+1} a_n = \sum_{n=1}^{\infty} \frac{(-1)^{n+1}}{n} = 1 - \frac{1}{2} + \frac{1}{3} - \frac{1}{4} \pm \cdots \tag{8.2.35}$$

converges conditionally (Definition 8.2.5).

The *Comparison Test* is a classic result from calculus which we are now in position to prove. Essentially, part (i) tells us that when the absolute values of terms are eventually bounded by the terms of a convergent series, the original series converges absolutely. Part (ii) says that if when the absolute values of terms are eventually bounded below by the terms of a divergent series of nonnegative terms, the series of absolutely values diverges.

Note that the conclusion of part (ii) of only addresses the series of absolute values $\sum_{n=1}^{\infty} |a_n|$, not the original series $\sum_{n=1}^{\infty} a_n$.

Theorem 8.2.11: Comparison Test

Suppose $\sum_{n=1}^{\infty} a_n$ and $\sum_{n=1}^{\infty} b_n$ are series of real numbers where $b_n \geq 0$ for all $n \in \mathbb{N}$.

(i) If $\sum_{n=1}^{\infty} b_n$ converges and there is an index $j_0 \in \mathbb{N}$ such that

$$n \in \mathbb{N} \text{ with } n \geq j_0 \qquad \Longrightarrow \qquad |a_n| \leq b_n, \tag{8.2.36}$$

then $\sum_{n=1}^{\infty} a_n$ converges absolutely.

(ii) If $\sum_{n=1}^{\infty} b_n$ diverges and there is an index $k_0 \in \mathbb{N}$ such that

$$n \in \mathbb{N} \text{ with } n \geq k_0 \quad \Longrightarrow \quad b_n \leq |a_n|, \tag{8.2.37}$$

then $\sum_{n=1}^{\infty} |a_n|$ diverges.

Scratch Work 8.2.12: Use the Cauchy criterion for series once again

Part (i) of the Comparison Test follows from yet another manipulation of partial sums with an application of the Cauchy criterion for series 8.2.2. This time, we must be careful to choose an index that satisfies both the Cauchy criterion and the implication (8.2.36). By the way, the implication (8.2.36) can be taken to mean the terms $|a_n|$ are eventually less than or equal to b_n.

Part (ii) follows from part (i) via contradiction and switching the roles of the series.

Proof of the Comparison Test 8.2.11. Throughout, suppose $\sum_{n=1}^{\infty} a_n$ and $\sum_{n=1}^{\infty} b_n$ are series of real numbers where $b_n \geq 0$ for all $n \in \mathbb{N}$.

  Part (i): Assume $\sum_{n=1}^{\infty} b_n$ converges and there is an index $j_0 \in \mathbb{N}$ such that

$$n \in \mathbb{N} \text{ with } n \geq j_0 \quad \Longrightarrow \quad |a_n| \leq b_n. \tag{8.2.38}$$

Since $\sum_{n=1}^{\infty} b_n$ converges, by the Cauchy criterion for series 8.2.2, there is a threshold $n_\varepsilon \in \mathbb{N}$ such that for all $j, k \in \mathbb{N}$ where $j > k \geq n_\varepsilon$ we have

$$\left| \sum_{n=k+1}^{j} b_n \right| = b_{k+1} + b_{k+2} + \cdots + b_j < \varepsilon. \tag{8.2.39}$$

Now define $q_\varepsilon = \max\{j_0, n_\varepsilon\} \in \mathbb{N}$ and suppose $j > k \geq q_\varepsilon$. Then both (8.2.38) and (8.2.39) hold and we have

$$\sum_{n=k+1}^{j} |a_n| = |a_{k+1}| + |a_{k+2}| + \cdots + |a_j| = b_{k+1} + b_{k+2} + \cdots + b_j < \varepsilon. \tag{8.2.40}$$

Therefore, $\sum_{n=1}^{\infty} |a_n|$ converges by Cauchy criterion for series 8.2.2, so $\sum_{n=1}^{\infty} a_n$ converges absolutely.

  Part (ii): Assume $\sum_{n=1}^{\infty} b_n$ diverges and there is an index $k_0 \in \mathbb{N}$ such that

$$n \in \mathbb{N} \text{ with } n \geq k_0 \quad \Longrightarrow \quad b_n \leq |a_n|. \tag{8.2.41}$$

To argue via contradiction, assume $\sum_{n=1}^{\infty} |a_n|$ converges and so $\sum_{n=1}^{\infty} a_n$ converges absolutely (Definition 8.2.5). Since $b_n \geq 0$ implies $b_n = |b_n| \leq |a_n|$ for large enough n, part (i) of the Comparison Test 8.2.11 (with the roles of $|a_n|$ and b_n reversed) implies we have $\sum_{n=1}^{\infty} |b_n| = \sum_{n=1}^{\infty} b_n$ converges. This contradicts the assumption that $\sum_{n=1}^{\infty} b_n$ diverges. $\qquad \square$

Next up is a result due to Augustin-Louis Cauchy, the same person whose name is attached to so many of the Cauchy criteria found throughout the book.

Theorem 8.2.13: Cauchy Condensation Test

Suppose (a_n) is a decreasing sequence of nonnegative terms ($0 \le a_{n+1} \le a_n$ for all $n \in \mathbb{N}$). Then

$$\sum_{n=1}^{\infty} a_n \text{ converges} \quad \Longleftrightarrow \quad \sum_{k=0}^{\infty} 2^k a_k \text{ converges.} \tag{8.2.42}$$

When these series converge we have

$$\frac{1}{2}\sum_{k=0}^{\infty} 2^k a_k \le \sum_{n=1}^{\infty} a_n \le \sum_{k=0}^{\infty} 2^k a_k. \tag{8.2.43}$$

Scratch Work 8.2.14: A subtle comparison of terms

The powers of 2 have shown up from time to time, like with one of the geometric series that sums to 1 in Example 8.1.12 and in Scratch Work 8.1.16 which leads to a proof of the divergence of the harmonic series in Example 8.1.16. Here, they give us a way to compare the partial sums of the related series in the statement of Theorem 8.2.13 that allow us to take advantage of the Comparison Test 8.2.11 to get the convergence of one of the series to yield the other. In particular, by the Geometric Sum Formula 2.7.2, the geometric sum of 2^n from $n = 0$ to k simplifies nicely to a positive integer:

$$\sum_{n=0}^{k} 2^n = 1 + 2 + 4 + \cdots + 2^k = \frac{1 - 2^{k+1}}{1 - 2} = 2^{k+1} - 1 \in \mathbb{N}. \tag{8.2.44}$$

This fortuitous result allows us to compare not just the terms but the *indices* of the series in question, facilitating the proof.

Also, since the terms in both series are nonnegative, their partial sums define increasing sequences which, if they converge, they converge to suprema as in the Monotone and Bounded Convergence Theorem 2.4.9.

Proof of Theorem 8.2.13. Suppose $(a_n) \subseteq \mathbb{R}$ where $0 \le a_{n+1} \le a_n$ for all $n \in \mathbb{N}$, and consider the pair of series

$$\sum_{n=1}^{\infty} a_n \quad \text{and} \quad \sum_{k=0}^{\infty} 2^k a_k. \tag{8.2.45}$$

For each $j \in \mathbb{N}$ and $q \in \mathbb{N} \cup \{0\}$, let s_j denote the jth partial sum of $\sum_{n=1}^{\infty} a_n$ and let t_q denote the qth partial sum of $\sum_{k=0}^{\infty} 2^k a_k$ so that

$$s_j = a_1 + a_2 + a_3 + \cdots + a_j \quad \text{and} \tag{8.2.46}$$

$$t_q = a_1 + 2a_2 + 4a_4 + \cdots + 2^q a_{2^q}. \tag{8.2.47}$$

Then both sequences of partial sums (s_j) and (t_q) are increasing. So, by the

From here, we split the argument into two parts depending on which series is assumed to converge and how the index j compares to the index 2^q.

To prove the forward implication, assume $\sum_{n=1}^{\infty} a_n$ converges and $j \geq 2^q$. So as in Scratch Work 8.2.14, we have

$$\sum_{n=0}^{q-1} 2^n = 1 + 2 + 4 + \cdots + 2^{q-1} = \frac{1 - 2^q}{1 - 2} \leq 2^q - 1 \leq 2^q \leq j. \tag{8.2.48}$$

Since $0 \leq a_{n+1} \leq a_n$ for every $n \in \mathbb{N}$, we have

$$\frac{1}{2}t_q = \frac{1}{2}\left(a_1 + 2a_2 + 4a_4 + \cdots + 2^q a_{2^q}\right) \tag{8.2.49}$$

$$= \frac{1}{2}a_1 + a_2 + 2a_4 + \cdots + 2^{q-1}a_{2^q} \tag{8.2.50}$$

$$= \frac{1}{2}a_1 + a_2 + \underbrace{(a_4 + a_4)}_{2 \text{ terms}} + \cdots + \underbrace{(a_{2^q} + \cdots + a_{2^q})}_{2^{q-1} \text{ terms}} \tag{8.2.51}$$

$$\leq a_1 + a_2 + \underbrace{(a_3 + a_4)}_{2 \text{ terms}} + \cdots + \underbrace{(a_{2^{q-1}+1} + \cdots + a_{2^q})}_{2^{q-1} \text{ terms}} \tag{8.2.52}$$

$$= a_1 + a_2 + a_3 + a_4 \cdots + a_j \tag{8.2.53}$$

$$= s_j. \tag{8.2.54}$$

Now, since $\sum_{n=1}^{\infty} a_n$ converges and (s_j) is increasing, the Monotone and Bounded Convergence Theorem 2.4.9 implies

$$\frac{1}{2}t_q \leq s_j \leq \sup\{s_j : j \in \mathbb{N}\} = \lim_{j \to \infty} s_j = \sum_{n=1}^{\infty} a_n. \tag{8.2.55}$$

Thus, $\sum_{n=1}^{\infty} a_n$ is an upper bound for $(t_q/2)$. Therefore, by another application of the Monotone and Bounded Convergence Theorem 2.4.9 along with the fact that a supremum is the *least* upper bound (Theorem 1.3.10), the series $\sum_{k=0}^{\infty} 2^k a_k$ converges and

$$\frac{1}{2}\sup\{t_q : q \in \mathbb{N}\} = \frac{1}{2}\lim_{q \to \infty} t_q = \frac{1}{2}\sum_{k=0}^{\infty} 2^k a_{2^k} \leq \sum_{n=1}^{\infty} a_n. \tag{8.2.56}$$

Therefore, the forward implication holds.

To prove the backward implication, assume $\sum_{k=0}^{\infty} 2^k a_k$ converges and $j \leq 2^q$. Again, as in is Scratch Work 8.2.14, we have

$$j \leq 2^q \leq \sum_{n=0}^{q} 2^n = 1 + 2 + 4 + \cdots + 2^q = \frac{1 - 2^{q+1}}{1 - 2} = 2^{q+1} - 1 \in \mathbb{N}. \tag{8.2.57}$$

Then, as done in Scratch Work 8.1.16 on the divergence harmonic series, by grouping the terms of the partial sum s_j by taking indices in successive chunks of powers of two, and keeping in mind

$0 \le a_{n+1} \le a_n$ for every $n \in \mathbb{N}$, we get

$$s_j = a_1 + a_2 + a_3 + \cdots + a_j \tag{8.2.58}$$
$$\le a_1 + (a_2 + a_3) + (a_4 + a_5 + a_6 + a_7) + \cdots + (a_{2^q} + \cdots + a_{2^{q+1}-1}) \tag{8.2.59}$$
$$\le a_1 + \underbrace{(a_2 + a_2)}_{2 \text{ terms}} + \underbrace{(a_4 + a_4 + a_4 + a_4)}_{4 \text{ terms}} + \cdots + \underbrace{(a_{2^q} + \cdots + a_{2^q})}_{2^q \text{ terms}} \tag{8.2.60}$$
$$= a_1 + 2a_2 + 4a_4 + \cdots + 2^q a_{2^q} \tag{8.2.61}$$
$$= t_q. \tag{8.2.62}$$

Now, since $\sum_{k=0}^{\infty} 2^k a_k$ converges converges and (t_q) is increasing, the Monotone and Bounded Convergence Theorem 2.4.9 implies

$$s_j \le t_q \le \sup\{t_q : q \in \mathbb{N} \cup \{0\}\} = \lim_{q \to \infty} t_q = \sum_{k=0}^{\infty} 2^k a_{2^k}. \tag{8.2.63}$$

Thus, $\sum_{k=0}^{\infty} 2^k a_{2^k}$ is an upper bound for (s_j). Therefore, by another application of the Monotone and Bounded Convergence Theorem 2.4.9 and noting a supremum is the *least* upper bound (Theorem 1.3.10), $\sum_{n=1}^{\infty} a_n$ converges and

$$\sup\{s_j : j \in \mathbb{N}\} = \lim_{j \to \infty} s_j = \sum_{n=1}^{\infty} a_n \le \sum_{k=0}^{\infty} 2^k a_k. \tag{8.2.64}$$

Therefore, the backward implication holds.

Overall, if either $\sum_{k=0}^{\infty} 2^k a_k$ or $\sum_{k=0}^{\infty} 2^k a_{2^k}$ converges, then both converge and

$$\frac{1}{2} \sum_{k=0}^{\infty} 2^k a_k \le \sum_{n=1}^{\infty} a_n \le \sum_{k=0}^{\infty} 2^k a_k. \tag{8.2.65}$$

$\square$

One payoff of the Cauchy Condensation Test 8.2.13 is the *p*-series test from calculus.

Theorem 8.2.15: *p*-series

Suppose $p \in \mathbb{R}$. Then the so-called *p*-series given by

$$\sum_{n=1}^{\infty} \frac{1}{n^p} \tag{8.2.66}$$

converges if and only if $p > 1$.

Proof of Theorem 8.2.15. First, suppose $p \le 0$. Then $-p = \alpha \ge 0$ and for each $n \in \mathbb{N}$,

$$\frac{1}{n^p} = n^{-p} = n^\alpha \ge 1. \tag{8.2.67}$$

So $(1/n^p) = (n^\alpha)$ does not converge to zero. Hence, the series

$$\sum_{n=1}^{\infty} \frac{1}{n^p} = \sum_{n=1}^{\infty} n^\alpha. \tag{8.2.68}$$

diverges by Corollary 8.2.4.

Now suppose $p > 0$. The Cauchy Condensation Test 8.2.13 tells us

$$\sum_{n=1}^{\infty}(1/n^p) \text{ converges} \qquad \Longleftrightarrow \qquad \sum_{k=0}^{\infty}(2^k/2^{kp}) \text{ converges}. \qquad (8.2.69)$$

Note that for each $k \in \mathbb{N} \cup \{0\}$ we have

$$2^k \cdot \frac{1}{2^{kp}} = \frac{2^k}{2^{kp}} = 2^{(1-p)k}. \qquad (8.2.70)$$

So taking series over k yields

$$\sum_{k=0}^{\infty} 2^k \cdot \frac{1}{2^{kp}} = \sum_{k=0}^{\infty} 2^{(1-p)k}. \qquad (8.2.71)$$

The series on the right is a geometric series with initial term $a = 1$ and common ratio $r = 2^{(1-p)}$. So by Theorem 8.1.10, the series converges if and only if $|r| = 2^{(1-p)} < 1$. Since

$$2^{(1-p)} < 1 \qquad \Longleftrightarrow \qquad 1-p < 0 \qquad \Longleftrightarrow \qquad p > 1, \qquad (8.2.72)$$

both $\sum_{n=1}^{\infty}(1/n^p)$ and $\sum_{k=0}^{\infty}(2^k/2^{kp})$ converge if and only if $p > 1$. $\qquad \square$

Geometric sums and series have already appeared multiple times so far. They provide a vital basis for comparison in the scratch work and proofs even more results from calculus, such as the *Ratio* and *Root Tests*, two classic results from calculus.

The Ratio Test tells us that a series $\sum_{n=1}^{\infty} a_n$ converges when the ratios $|a_{n+1}|/|a_n|$ are eventually less than some constant strictly less than one.

Theorem 8.2.16: Ratio Test

Suppose $\sum_{n=1}^{\infty} a_n$ is a series of nonzero real numbers.

(i) If there is a bound $0 < b < 1$ and an index $j_0 \in \mathbb{N}$ such that

$$n \in \mathbb{N} \text{ with } n \geq j_0 \qquad \Longrightarrow \qquad \left|\frac{a_{n+1}}{a_n}\right| \leq b, \qquad (8.2.73)$$

then $\sum_{n=1}^{\infty} a_n$ converges absolutely.

(ii) If there is an index $k_0 \in \mathbb{N}$ such that

$$n \in \mathbb{N} \text{ with } n \geq k_0 \qquad \Longrightarrow \qquad \left|\frac{a_{n+1}}{a_n}\right| \geq 1, \qquad (8.2.74)$$

then $\sum_{n=1}^{\infty} a_n$ diverges.

Scratch Work 8.2.17: Compare to a geometric series

Part (i) of the Ratio Test 8.2.16 follows from a comparison with the geometric series defined by the bound $0 < b < 1$. Part (ii) holds since the terms do not converge to zero.

Proof of the Ratio Test 8.2.16. Throughout the proof, suppose $\sum_{n=1}^{\infty} a_n$ is a series of nonzero real numbers.

Part (i): Assume there is a bound $0 < b < 1$ and an index $j_0 \in \mathbb{N}$ such that

$$n \in \mathbb{N} \text{ with } n \geq j_0 \quad \Longrightarrow \quad \left| \frac{a_{n+1}}{a_n} \right| \leq b. \tag{8.2.75}$$

Then we have

$$|a_{j_0+1}| \leq b|a_{j_0}|, \tag{8.2.76}$$
$$|a_{j_0+2}| \leq b|a_{j_0+1}| \leq b^2|a_{j_0}|, \tag{8.2.77}$$
$$\vdots \tag{8.2.78}$$
$$|a_{j_0+q}| \leq b^q|a_{j_0}|, \tag{8.2.79}$$

for all $q \in \mathbb{N}$. Hence, for any $n \in \mathbb{N}$ where $n \geq j_0$ it follows that

$$|a_n| \leq b^{n-j_0}|a_{j_0}| = b^{-j_0}|a_{j_0}|b^n. \tag{8.2.80}$$

Since $0 < b < 1$ and after a bit of reindexing, $\sum_{n=1}^{\infty} b^{-j_0}|a_{j_0}|b^n$ is a convergent geometric series by Theorem 8.1.10. Therefore, by part (i) of the Comparison Test 8.2.11, $\sum_{n=1}^{\infty} a_n$ converges absolutely.

Part (ii): Assume there is an index $k_0 \in \mathbb{N}$ such that

$$n \in \mathbb{N} \text{ with } n \geq k_0 \quad \Longrightarrow \quad \left| \frac{a_{n+1}}{a_n} \right| \geq 1. \tag{8.2.81}$$

Then for every $q \in \mathbb{N}$ we have

$$1 \leq |a_{k_0}| \leq |a_{k_0+2}| \leq \cdots \leq |a_{k_0+q}|. \tag{8.2.82}$$

Equivalently, for all $n \in \mathbb{N}$ with $n \geq k_0$ we have

$$|a_n| = |a_n - 0| \geq 1 \tag{8.2.83}$$

Hence, the sequence of terms (a_n) does not converge to 0. More specifically, the k_0-tail $(a_{n \geq k_0})$ is away from 0. So as in Remark 2.2.2, no neighborhood of 0 contains a tail of (a_n), thus (a_n) does not converge to 0. Therefore, by Corollary 8.2.4, $\sum_{n=1}^{\infty} a_n$ diverges. $\square$

The Root Test says a series $\sum_{n=1}^{\infty} a_n$ converges when the nth roots $\sqrt[n]{|a_n|}$ are eventually less than some constant strictly less than one.

Theorem 8.2.18: Root Test

Suppose $\sum_{n=1}^{\infty} a_n$ is a series of real numbers.

(i) If there is a bound $0 < b < 1$ and an index $j_0 \in \mathbb{N}$ such that

$$n \in \mathbb{N} \text{ with } n \geq j_0 \quad \Longrightarrow \quad \sqrt{|a_n|} \leq b, \tag{8.2.84}$$

then $\sum_{n=1}^{\infty} a_n$ converges absolutely.

(ii) If there is an index $k_0 \in \mathbb{N}$ such that

$$n \in \mathbb{N} \text{ with } n \geq k_0 \quad \Longrightarrow \quad \sqrt{|a_n|} \geq 1, \tag{8.2.85}$$

then $\sum_{n=1}^{\infty} a_n$ diverges.

Scratch Work 8.2.19: Again, compare to a geometric series

The proof of the Root Test 8.2.18 follows from an argument similar to the proof of of the Ratio Test 8.2.16. This time, for part (i) compare the given series to a convergent geometric series defined by the bound $0 < b^2 < 1$. Part (ii) holds since the terms do not converge to zero.

Proof of the Root Test 8.2.18. Throughout the proof, suppose $\sum_{n=1}^{\infty} a_n$ is a series of nonzero real numbers.

Part (i): Assume there is a bound $0 < b < 1$ and an index $j_0 \in \mathbb{N}$ such that

$$n \in \mathbb{N} \text{ with } n \geq j_0 \quad \Longrightarrow \quad \sqrt{|a_n|} \leq b. \tag{8.2.86}$$

Then for all $q \in \mathbb{N}$ we have

$$\sqrt{|a_{j_0+1}|} \leq b\sqrt{|a_{j_0}|}, \tag{8.2.87}$$

$$\sqrt{|a_{j_0+2}|} \leq b\sqrt{|a_{j_1}|} \leq b^2\sqrt{|a_{j_0}|}, \tag{8.2.88}$$

$$\vdots \tag{8.2.89}$$

$$\sqrt{|a_{j_0+q}|} \leq b^q\sqrt{|a_{j_0}|}. \tag{8.2.90}$$

Hence, for any $n \in \mathbb{N}$ where $n \geq j_0$ it follows that

$$\sqrt{|a_n|} \leq b^{n-j_0}\sqrt{|a_{j_0}|} \quad \Longrightarrow \quad |a_n| \leq b^{2(n-j_0)}|a_{j_0}|. \tag{8.2.91}$$

Since $0 < b < 1$ implies $0 < b^2 < b < 1$ and after a bit of reindexing, $\sum_{n=1}^{\infty} b^{-2j_0}|a_{j_0}|b^{2n}$ is a convergent geometric series by Theorem 8.1.10. Therefore, by part (i) of the Comparison Test 8.2.11, $\sum_{n=1}^{\infty} a_n$ converges absolutely.

Part (ii): Assume there is an index $k_0 \in \mathbb{N}$ such that

$$n \in \mathbb{N} \text{ with } n \geq k_0 \quad \Longrightarrow \quad \sqrt{|a_n|} \geq 1 \quad \Longrightarrow \quad |a_n| \geq 1. \tag{8.2.92}$$

Then for all $n \in \mathbb{N}$ with $n \geq k_0$ we have

$$|a_n| = |a_n - 0| \geq 1 \tag{8.2.93}$$

Hence, and as in the proof of part (ii) of the Ratio Test 8.2.16, the sequence of terms (a_n) does not converge to 0. Therefore, by Corollary 8.2.4, $\sum_{n=1}^{\infty} a_n$ diverges. $\square$

The next section investigates the last big topic of the book: *series of functions*.

Exercises

8.2.1. Determine whether each series converges or diverges.

(i) $\displaystyle\sum_{n=1}^{\infty} \ln\left(\frac{n+1}{n}\right)$

(ii) $\displaystyle\sum_{n=1}^{\infty} \frac{(-1)^{n+1}}{\sqrt{n}}$

(iii) $\displaystyle\sum_{n=1}^{\infty} \frac{(-1)^{n+1}(n^2-3)}{n^2+10}$

(iv) $\displaystyle\sum_{n=0}^{\infty} e^{-2n}$

(v) $\displaystyle\sum_{n=1}^{\infty} \left(\frac{3}{2n}+\frac{3}{2^n}\right)$

(vi) $\displaystyle\sum_{n=1}^{\infty} \frac{1}{(n+2)(3n-2)}$

(vii) $\displaystyle\sum_{n=1}^{\infty} \frac{\sqrt{n+5}}{(2n-1)(3n+4)}$

(viii) $\displaystyle\sum_{n=1}^{\infty} \frac{10}{\sqrt{n^2+5}+\sqrt{n^2+6n+9}}$

(ix) $\displaystyle\sum_{n=1}^{\infty} (\sqrt{n+1}-\sqrt{n})$

(x) $\displaystyle\sum_{n=1}^{\infty} \frac{\sqrt{n+1}-\sqrt{n}}{n}$

8.2.2. Consider what happens when a finite number of the terms of a sequence or series of real numbers are changed.

(i) Prove that if (a_n) converges to ℓ and $b_n = a_n$ for all but finitely many $n \in \mathbb{N}$, then (b_n) converges to ℓ.

(ii) Prove that if $\sum_{n=1}^{\infty} x_n$ converges to s and $y_n = x_n$ for all but finitely many $n \in \mathbb{N}$, then $\sum_{n=1}^{\infty} y_n$ converges but perhaps not to s.

8.2.3. Consider the series $\sum_{n=1}^{\infty} c_n$ where

$$c_n = \begin{cases} n+10, & \text{if } n \leq 100, \\ \dfrac{2}{3^n}, & \text{if } n > 100. \end{cases} \tag{8.2.94}$$

Prove $\sum_{n=1}^{\infty} c_n$ converges and determine the sum. Hint: Carl Gauss' formula for the sum of the first n consecutive positive integers gives us a nice closed form. We have

$$\sum_{k=1}^{n} k = 1+2+\cdots+n = \frac{n(n+1)}{2}. \tag{8.2.95}$$

8.2.4. Consider a series $\sum_{n=1}^{\infty} x_n$ along with the series of squares $\sum_{n=1}^{\infty} x_n^2$.

   (i) Prove that if $\sum_{n=1}^{\infty} x_n$ and converges and $x_n \geq 0$ for all $n \in \mathbb{N}$, then $\sum_{n=1}^{\infty} x_n^2$ converges.

   (ii) Find an example where $\sum_{n=1}^{\infty} x_n$ diverges but $\sum_{n=1}^{\infty} x_n^2$ converges.

   (iii) Find an example where $\sum_{n=1}^{\infty} x_n$ converges but $\sum_{n=1}^{\infty} x_n^2$ diverges.

8.2.5. Suppose $x_n > 0$ and $\lim_{n \to \infty}(n x_n) = \ell \neq 0$. Prove $\sum_{n=1}^{\infty} x_n$ diverges.

8.2.6. Suppose $x_n > 0$ and $(n^2 x_n)$ converges. Prove $\sum_{n=1}^{\infty} x_n$ converges.

8.2.7. Given a series $\sum_{n=1}^{\infty} a_n$, for each $n \in \mathbb{N}$ define

$$a_n^+ = \begin{cases} a_n, & \text{if } a_n \geq 0, \\ 0, & \text{if } a_n < 0, \end{cases} \quad \text{and} \quad a_n^- = \begin{cases} 0, & \text{if } a_n \geq 0, \\ a_n, & \text{if } a_n < 0. \end{cases} \tag{8.2.96}$$

Prove $\sum_{n=1}^{\infty} a_n$ converges absolutely if and only if both $\sum_{n=1}^{\infty} a_n^-$ and $\sum_{n=1}^{\infty} a_n^+$ converge. Also, in this case we have

$$\sum_{n=1}^{\infty} a_n = \sum_{n=1}^{\infty} a_n^+ - \sum_{n=1}^{\infty} a_n^- \quad \text{and} \quad \left| \sum_{n=1}^{\infty} a_n \right| = \sum_{n=1}^{\infty} a_n^+ + \sum_{n=1}^{\infty} a_n^-. \tag{8.2.97}$$

8.2.8. Prove the *Summation by Parts* formula: Consider sequences (x_n) and (y_n). Let

$$s_n = \sum_{j=1}^{n} x_j = x_1 + x_2 + \cdots + x_n \tag{8.2.98}$$

and set $s_0 = 0$. Show that

$$s_n = \sum_{j=k}^{n} x_j y_j = s_n y_{n+1} - s_{k-1} y_k + \sum_{j=k}^{n} s_j(y_j - y_{j+1}). \tag{8.2.99}$$

8.2.9. Use the Summation by Parts formula above to prove *Abel's Test*: If $\sum_{n=1}^{\infty} x_n$ converges and (y_n) is a decreasing sequence of nonnegative terms, i.e.,

$$y_1 \geq y_2 \geq y_3 \geq \cdots \geq 0, \tag{8.2.100}$$

then $\sum_{n=1}^{\infty} x_n y_n$ converges. Hint: Use the Comparison Test on $\sum_{n=1}^{\infty} s_n(y_n - y_{n+1})$.

8.2.10. Use the Summation by Parts formula or Abel's Test above to prove *Dirichlet's Test*: If the partial sums of $\sum_{n=1}^{\infty} x_n$ converges and (y_n) is a decreasing sequence of nonnegative terms that converges to zero, i.e.,

$$y_1 \geq y_2 \geq y_3 \geq \cdots \geq 0 \quad \text{and} \quad \lim_{n \to \infty} y_n = 0, \tag{8.2.101}$$

then $\sum_{n=1}^{\infty} x_n y_n$ converges.

8.2.11. Convergent series have an interesting dichotomy when the terms are rearranged: All rearrangements of an absolutely convergent series converge to the same sum; conditionally convergent series can be rearranged to converge to given any extended real number in $[-\infty, \infty]$.

A *rearrangement of* of $\sum_{n=1}^{\infty} x_n$ is a series $\sum_{n=1}^{\infty} x_{f(n)}$ where $f : \mathbb{N} \to \mathbb{N}$ is a bijection (meaning the same terms are used but reordered).

(i) Prove that if $\sum_{n=1}^{\infty} x_n$ converges absolutely to the sum s, then any rearrangement $\sum_{n=1}^{\infty} x_{f(n)}$ converges absolutely to the same sum s.

(ii) Prove that if $r \in \mathbb{R}$ and $\sum_{n=1}^{\infty} x_n$ converges conditionally, then there is a rearrangement $\sum_{n=1}^{\infty} x_{f(n)}$ which converges r.

(iii) Prove that if $\sum_{n=1}^{\infty} x_n$ converges conditionally, then for any $-\infty \leq \alpha < \beta \leq \infty$ there is a rearrangement $\sum_{n=1}^{\infty} x_{f(n)}$ whose partial sums $(s_{f(n)})$ satisfy

$$-\infty \leq \alpha = \liminf_{n \to \infty} s_{f(n)} < \limsup_{n \to \infty} s_{f(n)} = \beta \leq \infty. \tag{8.2.102}$$

(iv) Prove that if $\sum_{n=1}^{\infty} x_n$ converges conditionally, then for any $-\infty \leq \alpha < \beta$ there is a rearrangement $\sum_{n=1}^{\infty} x_{f(n)}$ whose partial sums $(s_{f(n)})$ satisfy

$$\mathrm{Slim}(s_{f(n)}) = \mathbb{R}. \tag{8.2.103}$$

8.3   Series of functions

The final broad topic of topic of the book is the notion of an infinite sum of functions made precise by *series of functions*. The topic is massive, our coverage of it is just the tip of the iceberg.

As in Remark 8.1.2 for series of real numbers, the indices of a series of functions can start at any integer. For the sake of simplicity, the indices start at $n = 1$.

Definition 8.3.1: Series of functions

A *series of functions* is an object of the form

$$\sum_{n=1}^{\infty} f_n(x) = f_1(x) + f_2(x) + f_3(x) + \cdots \tag{8.3.1}$$

where $D \subseteq \mathbb{R}$, $x \in D$, and (f_n) is a sequence of real-valued functions defined on the common domain D. For each index $k \in \mathbb{N}$, the *kth partial sum* is the linear combination $s_k : D \to \mathbb{R}$ defined for each $x \in D$ by

$$s_k(x) = \sum_{n=1}^{k} f_n(x) = f_1(x) + f_2(x) + \cdots + f_k(x). \tag{8.3.2}$$

Also, the *k-tail of a series* is the series of functions defined by the sequential *k*-tail $(f_{n \geq k})$

---

for each $x \in D$ where

$$\sum_{n=k}^{\infty} f_n(x) = f_k(x) + f_{k+1}(x) + f_{k+2}(x) + \cdots . \tag{8.3.3}$$

Sometimes, the variable x is suppressed and we respectively write the series and its partial sums as

$$\sum_{n=1}^{\infty} f_n = f_1 + f_2 + f_3 + \cdots \qquad \text{and} \qquad s_k = \sum_{n=1}^{k} f_n = f_1 + f_2 + \cdots + f_k. \tag{8.3.4}$$

The convergence of sequences of functions in Chapter 7 is explored by considering *pointwise* and *uniform* versions. Pointwise convergence is when each output sequence converges and has its own pointwise threshold (Definition 7.1.5). Uniform convergence is when all the output sequences converge and respect a single uniform threshold (Sauron's Nani, see Definition 7.2.1). A similar pair is defined for convergence of series of functions through partial sums in Definition 8.3.1.

Definition 8.3.2: Pointwise and uniform convergence of series of functions

A series of functions $\sum_{n=1}^{\infty} f_n$ *converges pointwise at* c if the sequence of partial sums $(s_k) = (\sum_{n=1}^{k} f_n)$ converges pointwise at c. Also, $\sum_{n=1}^{\infty} f_n$ *converges pointwise* when it converges pointwise at every point of the common domain. In this case, the *pointwise limit* of $\sum_{n=1}^{\infty} f_n$ is the pointwise limit f of the sequence of partial sums (s_k). See Definition 7.1.5.

Similarly, $\sum_{n=1}^{\infty} f_n$ *converges uniformly* if the sequence of partial sums (s_k) converges uniformly. In this case, the *uniform limit* of $\sum_{n=1}^{\infty} f_n$ is the uniform limit f of the sequence of partial sums (s_k). See Definition 7.2.1.

The definitions for pointwise and uniform convergence for series are so dense that some unpacking is order.

Remark 8.3.3: Unpacked pointwise and uniform convergence of series of functions

Suppose $D \subseteq \mathbb{R}$, $c \in D$, $k \in \mathbb{N}$, and $f_n : D \to \mathbb{R}$ for every $n \in \mathbb{N}$. Another way to think of the pointwise convergence at c of a series of functions $\sum_{n=1}^{\infty} f_n$ is that the *output series* $\sum_{n=1}^{\infty} f_n(c)$ converges as a series of real numbers (Definition 8.1.3). In this case,

$$f(c) = \sum_{n=1}^{\infty} f_n(c) = \lim_{k \to \infty} \sum_{n=1}^{k} f_n(c) = \lim_{k \to \infty} s_k(c) \tag{8.3.5}$$

where $s_k(c)$ is the kth partial sum of the output series $\sum_{n=1}^{\infty} f_n(c)$.

So, considering each $c \in D$ separately, for every $\varepsilon > 0$, by the definition of sequential limit

(Definition 2.2.1) there is a *pointwise threshold* $k_\varepsilon(c) \in \mathbb{N}$ such that

$$k \in \mathbb{N} \text{ with } k \geq k_\varepsilon(c) \quad \Longrightarrow \tag{8.3.6}$$

$$|s_k(c) - f(c)| = \left| \sum_{n=1}^{k} f_n(c) - \sum_{n=1}^{\infty} f_n(c) \right| = \left| \sum_{n=k+1}^{\infty} f_n(c) \right| < \varepsilon. \tag{8.3.7}$$

NOTE: The pointwise threshold $k_\varepsilon(c) \in \mathbb{N}$ depends on the specific input c, and the inequality in (8.3.6) only holds for the outputs of c.

On the other hand, for the uniform convergence a series of functions $\sum_{n=1}^{\infty} f_n$ to its uniform limit f means for every $\varepsilon > 0$ *and for all* $x \in D$, by the definition of uniform convergence (Definition 7.2.1) there is a *uniform threshold* $k_\varepsilon \in \mathbb{N}$ such that—for all $x \in D$—we have

$$k \in \mathbb{N} \text{ with } k \geq k_\varepsilon \quad \Longrightarrow \tag{8.3.8}$$

$$|s_k(x) - f(x)| = \left| \sum_{n=1}^{k} f_n(x) - \sum_{n=1}^{\infty} f_n(x) \right| = \left| \sum_{n=k+1}^{\infty} f_n(x) \right| < \varepsilon. \tag{8.3.9}$$

Here, the uniform threshold $k_\varepsilon \in \mathbb{N}$ *is independent of the inputs* and, as a result, serves as the pointwise threshold $k_\varepsilon(c)$ for every input $c \in D$ at the same time.

As a first example, let's revisit geometric series (Definition 8.1.9).

Example 8.3.4: Geometric series of functions

Let $a \in \mathbb{R}$ and consider the series of functions on $(1, 1)$ defined by

$$\sum_{n=0}^{\infty} a x^n, \qquad \text{for every } x \in (-1, 1). \tag{8.3.10}$$

For each $c \in (-1, 1)$, the output series is a convergent geometric series with initial term a and common ratio c by Theorem 8.1.10. Moreover, this series of functions converges pointwise with pointwise limit $g : (-1, 1) \to \mathbb{R}$ given by

$$g(x) = \sum_{n=0}^{\infty} a x^n = \frac{a}{1 - x}, \qquad \text{for each } x \in (-1, 1). \tag{8.3.11}$$

However, the series does not converge uniformly for $a \neq 0$. The proof is left as an exercise.

A result due to Weierstrass gives us a convenient way to check for the uniform convergence of a series of functions.

Theorem 8.3.5: Weierstrass M-Test

Suppose (f_n) is a sequence of functions on D and (M_n) is a sequence of nonnegative uniform bounds where

$$|f_n(x)| \leq M_n \quad \text{for all } x \in D \text{ and } n \in \mathbb{N}. \tag{8.3.12}$$

If $\sum_{n=1}^{\infty} M_n$ converges, then $\sum_{n=1}^{\infty} f_n$ converges uniformly on D.

Scratch Work 8.3.6: Cauchy criteria yet again

The proof follows from Cauchy criteria for sequences 2.6.5 and uniform convergence 7.2.11 by considering the sequence of partial sums along with some induction on the triangle inequality (1.2.35).

Proof of the Weierstrass M-Test 8.3.5. Suppose the hypotheses hold and let $\varepsilon > 0$. By the Cauchy criterion for sequences 2.6.5, the sequence of partial sums $(\sum_{n=1}^{k} M_n)$ is Cauchy. So by Definitions 2.6.1 and 8.3.1 along with the fact each M_n is a uniform bound for $|f_n|$, there is a threshold $n_\varepsilon \in \mathbb{N}$ such that for all inputs $x \in D$ and all indices $j, k \in \mathbb{N}$ where $j > k \geq n_\varepsilon$ we have

$$|s_j(x) - s_k(x)| = \left| \sum_{n=k+1}^{j} f_n(x) \right| \leq \sum_{n=k+1}^{j} |f_n(x)| \leq \sum_{n=k+1}^{j} M_n < \varepsilon, \tag{8.3.13}$$

where the equation holds thanks to Scratch Work 8.2.3 and first inequality holds by induction on the triangle inequality (1.2.35). Hence, the sequence of partial sums (s_k) satisfies the Cauchy criterion for uniform convergence 7.2.11 with uniform threshold n_ε. Therefore, $\sum_{n=1}^{\infty} f_n$ converges uniformly on D. $\square$

Example 8.3.7: Weierstrass functions

Suppose $|b| < 1$ and consider the series of functions defined on the real line by

$$\sum_{n=0}^{\infty} b^n \cos nx \quad \text{for every } x \in \mathbb{R}. \tag{8.3.14}$$

This series converges uniformly on $\mathbb{R}$ and its uniform limit is an example of a *Weierstrass function*. The original Weierstrass functions are similar and explored in the exercises.

Proof for Example 8.3.7. Let $M_n = |b|^n$ for each $n \in \mathbb{N} \cup \{0\}$. Note that for every $x \in \mathbb{R}$ and we have

$$|b^n \cos nx| \leq |b|^n = M_n. \tag{8.3.15}$$

Since $|b| < 1$, Theorem 8.1.10 tells us $\sum_{n=0}^{\infty} M_n$ is a convergent geometric series with

$$\sum_{n=0}^{\infty} M_n = \sum_{n=0}^{\infty} |b|^n = \frac{1}{1 - |b|}. \tag{8.3.16}$$

Therefore, by the Weierstrass M-Test 8.3.5 where we set $f_n(x) = b^n \cos nx$,

$$\sum_{n=0}^{\infty} b^n \cos nx \tag{8.3.17}$$

converges uniformly on $\mathbb{R}$.   □

The Weierstrass functions in Example 8.3.7 are continuous, integrable over compact intervals, and differentiable. This string of results follows from properties of uniform convergence applied to partial sums.

First up: A uniformly convergent series of continuous functions converges to a continuous limit.

> ### Corollary 8.3.8: Continuity and uniformly convergent series
>
> If $\sum_{n=1}^{\infty} f_n$ converges uniformly to f on D and f_n is continuous on D for each $n \in \mathbb{N}$, then f is continuous on D.

Proof of Corollary 8.3.8. Suppose $\sum_{n=1}^{\infty} f_n$ converges uniformly to f on D and f_n is continuous on D for each $n \in \mathbb{N}$. By Definition 8.3.2, the sequence of partial sums (s_k) converges uniformly to f on D. Since each partial sum s_k is a linear combination of continuous functions, the uniform limit f is continuous by Theorem 7.3.1.   □

Next, a uniformly convergent series of integrable functions converges to an integrable limit, and the integral of the uniform limit is the series of integrals.

> ### Corollary 8.3.9: Integration and uniformly convergent series
>
> If $\sum_{n=1}^{\infty} f_n$ converges uniformly to f on $[a,b]$ and f_n is continuous on $[a,b]$ for each $n \in \mathbb{N}$, then the uniform limit f is integrable over $[a,b]$ and we have
>
> $$\int_a^b f = \int_a^b \left(\sum_{n=1}^{\infty} f_n \right) = \sum_{n=1}^{\infty} \left(\int_a^b f_n \right). \tag{8.3.18}$$

Proof of Corollary 8.3.9. Suppose $\sum_{n=1}^{\infty} f_n$ converges uniformly to f on $[a,b]$ and f_n is continuous on $[a,b]$ for each $n \in \mathbb{N}$. By Definition 8.3.2, the sequence of partial sums (s_k) converges uniformly to f on $[a,b]$. Since a partial sum is a linear combination and linear combinations of integrable functions are integrable by Corollary 6.3.9, s_k is integrable for each $k \in \mathbb{N}$. Therefore, the uniform limit f is integrable by Theorem 7.3.1. Additionally,

$$\int_a^b f = \int_a^b \left(\sum_{n=1}^{\infty} f_n \right) = \int_a^b \left(\lim_{k \to \infty} s_k \right) = \lim_{k \to \infty} \left(\int_a^b s_k \right) = \lim_{k \to \infty} \left(\sum_{n=1}^{k} \int_a^b f_n \right) = \sum_{n=1}^{\infty} \left(\int_a^b f_n \right). \tag{8.3.19}$$

□

Finally, the relationship between differentiation and uniform convergence of series has a different flavor. This relationship is codified by the following corollary of Theorem 7.3.8 by applying it to partial sums of derivatives. The proof is omitted.

Corollary 8.3.10: Differentiation and uniformly convergent series of derivatives

Suppose (f_n) is a sequence of differentiable functions on $[a, b]$ where the series of derivatives $\sum_{n=1}^{\infty} f_n'$ converges uniformly to a function g on $[a, b]$. If there is some $x_0 \in [a, b]$ whose output series $\sum_{n=1}^{\infty} f_n(x_0)$ converges, then there is a function $f : [a, b] \to \mathbb{R}$ such that

(i) $\sum_{n=1}^{\infty} f_n$ converges uniformly to f on $[a, b]$ with

$$f(x) = \sum_{n=1}^{\infty} f_n(x) \tag{8.3.20}$$

for every $x \in [a, b]$; and

(ii) f is differentiable on $[a, b]$ and f is an antiderivative of g with

$$f'(x) = \sum_{n=1}^{\infty} f_n'(x) = g(x) \tag{8.3.21}$$

for every $x \in [a, b]$.

The next section explores the idea of "infinite polynomials" defined as *power series*.

Exercises

8.3.1. Use geometric series and the equation $x = 1 - (1 - x)$ to find a series of functions that converges pointwise on $(0, 2)$ to $f(x) = 1/x$.

8.3.2. Use geometric series and the equation $1 + x^2 = 1 - (-x^2)$ to find a series of functions that converges pointwise on $(-1, 1)$ to $g(x) = 1/(1 + x^2)$.

8.3.3. Prove the series

$$\sum_{n=1}^{\infty} \frac{x^6 + 4^n}{x^4 + 6^n} \tag{8.3.22}$$

converges to a continuous function on the real line $\mathbb{R}$.

8.3.4. Prove the series

$$\sum_{n=1}^{\infty} \frac{20 \sin\left(\pi x^n - \sqrt{20n}\right)}{n^2} \tag{8.3.23}$$

converges to a continuous function on the real line $\mathbb{R}$.

8.3.5. Give an example of a series of functions which converges uniformly but where the Weierstrass M-Test 8.3.5 does not apply.

8.3.6. Karl Weierstrass proved there are continuous which are *nowhere differentiable*, meaning their derivatives do not exist at any input. This fact was derided by contemporaries who called such functions *monsters*. The functions, now called *Weierstrass functions*, are defined by power series of the form

$$\sum_{n=0}^{\infty} a^n \cos\left(b^n \pi x\right) \tag{8.3.24}$$

where $0 < a < 1$ and $b \in \mathbb{N}$ such that

$$ab > 1 + \frac{3\pi}{2}. \tag{8.3.25}$$

(i) Prove every Weierstrass function defines a continuous function on the real line $\mathbb{R}$.

(ii) Find a closed form for the integral integral of a Weierstrass function over $[0, 1/2]$:

$$\int_0^{1/2} \sum_{n=0}^{\infty} a^n \cos\left(b^n \pi x\right) dx. \tag{8.3.26}$$

Proofs of the nowhere differentiability of these Weierstrass functions are labor-intensive, so only those interested in how that works should attempt a proof themselves or search the literature. (Maybe a proof will be included in a future edition of this book.)

8.4   Power series

Power series are a special case of series of functions that give meaning to the notion of polynomials with infinite degree. The key motivation for the section is the determination of how and when power series converge.

Definition 8.4.1: Power series

Given $c \in \mathbb{R}$, a *power series centered at* $c = 0$ is a series of functions of the form

$$\sum_{n=0}^{\infty} a_n (x - c)^n \tag{8.4.1}$$

where (a_n) is a sequence of real numbers called the *coefficients* of the power series.

Power series give us a way to define the exponential function e^x.

Example 8.4.2: The exponential function e^x

The exponential function e^x is defined by the power series

$$e^x = \sum_{n=0}^{\infty} \frac{x^n}{n!}. \tag{8.4.2}$$

An application of the Ratio Test 8.2.16 shows that this power series converges pointwise on the real line, meaning e^x is well-defined. To that end, fix $x \in \mathbb{R}$ and note that for every $n \in \mathbb{N}$ we have

$$\left| \frac{x^{n+1}/(n+1)!}{x^n/n!} \right| = \frac{|x|}{n+1}. \tag{8.4.3}$$

This ratio of terms is eventually less than 1 by the Archimedean Property 1.4.6 which says there is some $n_{|x|} \in \mathbb{N}$ such that $|x| < n_{|x|}$ where

$$n \in \mathbb{N} \text{ with } n \geq n_{|x|} \quad \implies \quad \left| \frac{x^{n+1}/(n+1)!}{x^n/n!} \right| = \frac{|x|}{n+1} \leq \frac{|x|}{n_{|x|}} < 1. \tag{8.4.4}$$

(We also could have argued using the definition of sequential limit in Definition 2.2.1 since $\lim_{n \to \infty} |x|/(n+1)! = 0$.) Therefore, the Ratio Test 8.2.16 tells us the power series (8.4.2) converges pointwise on $\mathbb{R}$.

It is possible for a power series to converge only at its center.

Example 8.4.3: Divergence except at the center

Consider the power series centered at $c = 0$ given by

$$\sum_{n=0}^{\infty} n! x^n. \tag{8.4.5}$$

Another application of the Ratio Test 8.2.16 shows that this power series diverges at every $x \neq 0$. To see this, fix $x \in \mathbb{R} \setminus \{0\}$ and note that for every $n \in \mathbb{N}$ we have

$$\left| \frac{(n+1)! x^{n+1}}{n! x^n} \right| = (n+1)|x|. \tag{8.4.6}$$

This ratio of terms is eventually greater than 1 by the Corollary 1.4.8. That is, for some $n_{|x|} \in \mathbb{N}$ we have

$$\frac{1}{n_{|x|}} < |x| \quad \implies \quad 1 < n_{|x|}|x|. \tag{8.4.7}$$

As a result, we have

$$n \in \mathbb{N} \text{ with } n \geq n_{|x|} \quad \implies \quad \left| \frac{(n+1)! x^{n+1}}{n! x^n} \right| = (n+1)|x| > n_{|x|}|x| > 1. \tag{8.4.8}$$

Therefore, the Ratio Test 8.2.16 tells us the power series (8.4.5) diverges except at $x = 0$.

In general, a power series converges on the whole real line, neighborhoods of its center c (and maybe endpoints), or only at the center.

> **Theorem 8.4.4: Convergence of power series at an off-center point implies convergence on an interval**
>
> If a power series $\sum_{n=0}^{\infty} a_n(x - c)^n$ converges at some $x_0 \in \mathbb{R}$, then for any r such that $0 < r < |x_0 - c|$, this power series converges uniformly on the compact interval
>
> $$[c - r, c + r] = \{x \in \mathbb{R} : |x - c| \le r\}. \tag{8.4.9}$$
>
> In particular, if $|y - c| < |x_0 - c|$, then $\sum_{n=0}^{\infty} a_n(y - c)^n$ converges absolutely. Additionally, when $x_0 \ne c$, $\sum_{n=0}^{\infty} a_n(x - c)^n$ converges pointwise on the open interval $(c - |x_0|, c + |x_0|)$.

> **Scratch Work 8.4.5: Use a multitude of results on series**
>
> The proof follows from a combination of results on series found in this chapter. Yet again, a comparison with geometric series plays a key role.

Proof of Theorem 8.4.4. Suppose the series or real numbers $\sum_{n=0}^{\infty} a_n(x_0 - c)^n$ converges for some $x_0 \in \mathbb{R}$. Since the terms of convergent series converge to zero (Corollary 8.2.4), and since convergent sequences are bounded (Theorem 2.3.15), there is a bound $b > 0$ such that for all $n \in \mathbb{N}$ we have

$$|a_n(x_0 - c)^n| \le b. \tag{8.4.10}$$

Next, suppose r satisfies $0 < r < |x_0 - c|$. Then for every $x \in \mathbb{R}$ where

$$|x - c| \le r < |x_0 - c|, \tag{8.4.11}$$

it follows that for each index $n \in \mathbb{N}$ we have

$$|a_n(x - c)^n| = |a_n(x - c)^n| \frac{|x_0 - c|^n}{|x_0 - c|^n} \le b \frac{|x - c|^n}{|x_0 - c|^n} < b \left(\frac{r}{|x_0 - c|} \right)^n. \tag{8.4.12}$$

Since $0 < r < |x_0 - c|$ implies $0 < r/|x_0 - c| < 1$, the series

$$\sum_{n=0}^{\infty} b \left(\frac{r}{|x_0 - c|} \right)^n \tag{8.4.13}$$

is a convergent geometric series by Theorem 8.1.10. Therefore, by the Weierstrass M-Test 8.3.5, the power series $\sum_{n=0}^{\infty} a_n(x - c)^n$ converges uniformly on

$$[c - r, c + r] = \{x \in \mathbb{R} : |x - c| \le r\}. \tag{8.4.14}$$

Also, we conclude $\sum_{n=0}^{\infty} a_n(y - c)^n$ converges absolutely when $|y - c| < |x_0 - c|$.

To show $\sum_{n=0}^{\infty} a_n(x-c)^n$ converges pointwise on the open interval $(c-|x_0|, c+|x_0|)$ when $x_0 \neq c$, suppose $t \in (c-|x_0|, c+|x_0|)$. If $t = c$, then $t - c = 0$ and the power series converges absolutely at t with

$$\sum_{n=0}^{\infty} a_n(t-c)^n = a_0. \tag{8.4.15}$$

If $t \neq c$, choose $r_t = |t-c|$ so that $0 < r_t < |x_0 - c|$. The above argument for uniform convergence applies and tells us

$$\sum_{n=0}^{\infty} a_n(t-c)^n \tag{8.4.16}$$

converges. Therefore, $\sum_{n=0}^{\infty} a_n(x-c)^n$ converges pointwise on $(c-|x_0|, c+|x_0|)$. $\qquad\square$

Unlike the linear combinations of monomials that define polynomials (Definition 4.5.1), it is not necessarily the case that a power series defines a function whose domain is the whole real line. It can be a challenge to determine when a power series converges from scratch.

> **Example 8.4.6: An alternating power series**
>
> Consider the power series centered at $c = 0$ given by
>
> $$\sum_{n=1}^{\infty} \frac{x^{n+1}}{n} = x^2 - \frac{x^3}{2} + \frac{x^4}{3} - \frac{x^5}{4} + \cdots \tag{8.4.17}$$
>
> This power series converges pointwise on $[-1, 1)$ and diverges on $\mathbb{R} \backslash [-1, 1)$.

Proof for Example 8.4.6. To kick things off, suppose $x = -1$. We get

$$\sum_{n=1}^{\infty} \frac{(-1)^{n+1}}{n} = 1 - \frac{1}{2} + \frac{1}{3} - \frac{1}{4} + \cdots, \tag{8.4.18}$$

which is the convergent alternating harmonic series from Example 8.2.6. So by Theorem 8.4.4, $\sum_{n=1}^{\infty} (x^{n+1}/n)$ converges uniformly on $[-r, r]$ for any r where $0 < r < |-1-0| = 1$. Since r is arbitrary, $\sum_{n=1}^{\infty} (x^{n+1}/n)$ converges pointwise on $[-1, 1)$.

To see that $\sum_{n=1}^{\infty} (x^{n+1}/n)$ diverges on $\mathbb{R} \backslash [-1, 1)$, first note that when $x = 1$ we get

$$\sum_{n=1}^{\infty} \frac{1}{n} = 1 + \frac{1}{2} + \frac{1}{3} + \frac{1}{4} + \cdots, \tag{8.4.19}$$

which is the divergent harmonic series from Example 8.1.16.

Now suppose $|x| > 1$. After a bit of effort, we can show divergence with the Ratio Test 8.2.16. For each index $n \in \mathbb{N}$ we have

$$\frac{|x|^{n+2}/(n+1)}{|x|^{n+1}/n} = |x|\left(\frac{n}{n+1}\right) = |x|\left(1 - \frac{1}{n+1}\right). \tag{8.4.20}$$

By the homogeneity of sequential limits (Theorem 2.3.9),

$$\lim_{n\to\infty} |x| \left(\frac{n}{n+1}\right) = |x|. \tag{8.4.21}$$

Since $|x| > 1$ implies $|x| - 1 > 0$ (which we can treat as an error ε) and noting $n/(n+1) < 1$, by the definition of sequential limit (Definition 2.2.1) there is a threshold n_x such that

$$n \in \mathbb{N} \text{ with } n \geq n_x \quad \Longrightarrow \quad \left| |x| \left(\frac{n}{n+1}\right) - |x| \right| = |x| - |x| \left(\frac{n}{n+1}\right) < |x| - 1 \tag{8.4.22}$$

$$\Longrightarrow \quad -|x| \left(\frac{n}{n+1}\right) < -1 \tag{8.4.23}$$

$$\Longrightarrow \quad |x| \left(\frac{n}{n+1}\right) > 1. \tag{8.4.24}$$

Therefore, by part (ii) of the Ratio Test 8.2.16, $\sum_{n=1}^{\infty}(x^{n+1}/n)$ diverges on $\mathbb{R}\backslash[-1,1)$. $\qquad\square$

Thanks to Theorem 8.4.4, we can split the domains on which a power series converges into three types: The singleton $\{c\}$ comprising the center of the power series; the whole real line $\mathbb{R}$; or a bounded interval centered at c.

> ### Corollary 8.4.7: Classifying sets of convergence
>
> Given a power series $\sum_{n=0}^{\infty} a_n(x-c)^n$, the set of points S where the series converges is either
>
> (i) the singleton $\{c\}$,
>
> (ii) the real line $\mathbb{R}$, or
>
> (iii) one of following bounded intervals:
> $(c-u, c+u)$, $(c-u, c+u]$, $[c-u, c+u)$, or $[c-u, c+u]$,
>   where $u = \sup\{|x-c| : x \in S\} > 0$.

Proof of Corollary 8.4.7. Let S denote the set of points where $\sum_{n=0}^{\infty} a_n(x-c)^n$ converges. We have $c \in S$ since $\sum_{n=0}^{\infty} a_n(c-c)^n$ converges to the first term a_0.

If no other point is in S, then $S = \{c\}$ and (i) holds.

Now, suppose S contains c at least one other element $y \neq c$, and consider the set

$$T = \{|x - c| : x \in S\}. \tag{8.4.25}$$

Then T is nonempty with at least one positive element given by $|y - c| > 0$. From here, the set T is either bounded above or not.

Suppose T is not bounded above. Then for any $x \in \mathbb{R}$ there is some $t \in T$ where

$$|x - c| < t. \tag{8.4.26}$$

By Theorem 8.4.4, we have $x \in S$. Therefore, $S = \mathbb{R}$ and (ii) holds.

Finally, suppose T defined the same way but is now bounded above. Then T has a supremum by the Axiom of Completeness 1.3.8. Since a supremum is an upper bound, we have

$$0 < |y - c| \leq \sup T = u. \tag{8.4.27}$$

Now, if $x \in \mathbb{R}$ where $|x - c| > u$, then $|x - c|$ is too large to be in T, and so $x \notin S$. Hence,

$$S \subseteq \{x : |x - c| \leq u\} = [c - u, c + u]. \tag{8.4.28}$$

But if $x \in \mathbb{R}$ where $|x - c| < u$, then $u - |x - c| > 0$. Since a supremum is arbitrarily close to its set (Definition 1.1.14), there is some $x_0 \in S$ with $|x_0 - c| \in T$ such that

$$0 < u - |x_0 - c| < u - |x - c| \quad \implies \quad |x - c| < |x_0 - c|. \tag{8.4.29}$$

Since $x_0 \in S$, Theorem 8.4.4 implies $x \in S$ as well. And so by (8.4.28) we have

$$(c - u, c + u) \subseteq S \subseteq [c - u, c + u]. \tag{8.4.30}$$

Therefore, (iii) holds since S must be either $(c-u, c+u)$, $(c-u, c+u]$, $[c-u, c+u)$, or $[c-u, c+u]$, depending on whether S contains either, neither, or both $c - u$ and $c + u$. $\qquad \square$

Corollary 8.4.7 justifies the following trio of options to define the *radius of convergence*.

Definition 8.4.8: Radius of convergence

Given a power series $\sum_{n=0}^{\infty} a_n (x - c)^n$ centered at c with S denoting the set of points where this series converges as in Corollary 8.4.7, the *radius of convergence* is the extended real number R determined as follows:

(i) $R = 0$ if $S = \{0\}$;

(ii) $R = \infty$ if $S = \mathbb{R}$; or

(iii) $R = u = \sup\{|x - c| : x \in S\}$ otherwise.

Also, the set S is called the *interval of convergence*.

Remark 8.4.9: Within, beyond, or at the radius of convergence

When a power series $\sum_{n=0}^{\infty} a_n (x - c)^n$ centered at c has radius of convergence R, Corollary 8.4.7 yields the following breakdown:

(i) If $0 < R \leq \infty$ and $y \in (c - R, c + R)$, or equivalently $|y - c| < R$, then the series $\sum_{n=0}^{\infty} a_n (y - c)^n$ converges.

(ii) If $0 \leq R < \infty$ and $|z - c| > R$, then $\sum_{n=0}^{\infty} a_n (z - c)^n$ diverges.

(iii) If $0 < R < \infty$ and $|w - c| = R$, then $\sum_{n=0}^{\infty} a_n (w - c)^n$ may converge or it may diverge.

An immediate consequence of Theorem 8.4.4 is the uniform convergence of a power series on compact subsets of the interval of convergence.

Corollary 8.4.10: Uniform convergence of power series

Suppose $\sum_{n=0}^{\infty} a_n(x-c)^n$ is a power series centered at c with radius of convergence $R > 0$. If K is a compact set where

$$K \subseteq V_c(R) = (c - R, c + R), \qquad (8.4.31)$$

then $\sum_{n=0}^{\infty} a_n(x-c)^n$ converges uniformly on K.

Proof of Corollary 8.4.10. Suppose $\sum_{n=0}^{\infty} a_n(x-c)^n$ has radius of convergence $R > 0$ and K is compact with $K \subseteq (c - R, c + R)$. Since K is compact, both $\min K$ and $\max K$ exist and are elements of K. Also, there is some $q \in (c - R, c + R)$ such that, by part (vi) of Theorem 1.3.2,

$$c - R < q \leq \min K \leq y \leq \max K \leq q < c + R \quad \implies \quad |y - c| \leq |q - c| < R \qquad (8.4.32)$$

for every $y \in K$ we have. So, as in Remark 8.4.9, $\sum_{n=0}^{\infty} a_n(q-c)^n$ converges. Therefore, by Theorem 8.4.4 and since $K \subseteq [c - q, c + q]$, the power series $\sum_{n=0}^{\infty} a_n(x-c)^n$ converges uniformly on both K and $[c - q, c + q]$. $\qquad \square$

The connection between compact sets and the uniform convergence of power series established in Corollary 8.4.10 leads to a string of results on continuity, integration, and differentiation of power series thanks to results developed in Chapter 7.

Theorem 8.4.11: Continuity of power series

Suppose $\sum_{n=0}^{\infty} a_n(x-c)^n$ is a power series centered at c with radius of convergence $R > 0$. If K is a compact set where

$$K \subseteq V_c(R) = (c - R, c + R), \qquad (8.4.33)$$

then the uniform limit $f : K \to \mathbb{R}$ defined by

$$f(x) = \sum_{n=0}^{\infty} a_n(x-c)^n \qquad \text{for all } x \in K \qquad (8.4.34)$$

is uniformly continuous on K.

Proof of Theorem 8.4.11. Suppose $\sum_{n=0}^{\infty} a_n(x-c)^n$ is a power series centered at c with radius of convergence $R > 0$ and K is a compact set where

$$K \subseteq V_c(R) = (c - R, c + R). \qquad (8.4.35)$$

By Corollary 8.4.10, the power series converges uniformly on K. Hence, the uniform limit $f : K \to \mathbb{R}$ given by

$$f(x) = \sum_{n=0}^{\infty} a_n(x-c)^n \qquad \text{for all } x \in K \qquad (8.4.36)$$

is well-defined. Let (s_k) denote the sequence of partials sums given by

$$s_k(x) = \sum_{n=0}^{k} a_n(x - c)^n \qquad \text{for all } x \in K \quad \text{and} \quad k \in \mathbb{N}. \tag{8.4.37}$$

Each s_k is a linear combination of continuous polynomials, and so each s_k is continuous by Corollary 4.5.7. Moreover, by Theorem 7.3.1 applied to (s_k), f is continuous on K since it is the uniform limit of (s_k) and

$$f(x) = \sum_{n=0}^{\infty} a_n(x - c)^n = \lim_{k \to \infty} s_k(x) \qquad \text{for all } x \in K. \tag{8.4.38}$$

Since K is compact, Theorem 4.7.13 tells us f is uniformly continuous on K. $\qquad\square$

The next theorem tells us the integral of a power series is a series of integrals. Its proof is similar to that of Theorem 8.4.11.

Theorem 8.4.12: Integral of a power series

Suppose $\sum_{n=0}^{\infty} a_n(x - c)^n$ is a power series centered at c with radius of convergence $R > 0$. If a compact interval $[a, b]$ satisfies

$$[a, b] \subseteq V_c(R) = (c - R, c + R), \tag{8.4.39}$$

then the uniform limit $f : [a, b] \to \mathbb{R}$ defined by

$$f(x) = \sum_{n=0}^{\infty} a_n(x - c)^n \tag{8.4.40}$$

is integrable over $[a, b]$ and we have

$$\int_a^b f(x)\, dx = \int_a^b \left(\sum_{n=0}^{\infty} a_n(x - c)^n \right) dx = \sum_{n=0}^{\infty} \left(\int_a^b a_n(x - c)^n\, dx \right). \tag{8.4.41}$$

Proof of Theorem 8.4.12. Suppose $\sum_{n=0}^{\infty} a_n(x - c)^n$ is a power series centered at c with radius of convergence $R > 0$ and $[a, b]$ is a compact interval satisfying

$$[a, b] \subseteq V_c(R) = (c - R, c + R). \tag{8.4.42}$$

By Corollary 8.4.10, the power series converges uniformly on $[a, b]$. Hence, the uniform limit $f : [a, b] \to \mathbb{R}$ given by

$$f(x) = \sum_{n=0}^{\infty} a_n(x - c)^n \qquad \text{for all } x \in K \tag{8.4.43}$$

is well-defined. Let (s_k) denote the sequence of partials sums given by

$$s_k(x) = \sum_{n=0}^{k} a_n(x - c)^n \qquad \text{for all } x \in [a, b] \quad \text{and} \quad k \in \mathbb{N}. \tag{8.4.44}$$

Each s_k is a linear combination of integrable polynomials, and so each s_k is integrable over $[a,b]$ by Corollary 6.3.9. Moreover, by Theorem 7.3.4 applied to (s_k), f is integrable over $[a,b]$ and we have

$$\int_a^b f(x)\,dx = \int_a^b \left(\sum_{n=0}^\infty a_n(x-c)^n\right) dx \tag{8.4.45}$$

$$= \int_a^b \left(\lim_{k\to\infty} s_k(x)\right) dx \tag{8.4.46}$$

$$= \lim_{k\to\infty}\left(\int_a^b s_k(x)\,dx\right) \tag{8.4.47}$$

$$= \lim_{k\to\infty}\left(\int_a^b \sum_{n=0}^k a_n(x-c)^n\,dx\right) \tag{8.4.48}$$

$$= \lim_{k\to\infty}\left(\sum_{n=0}^k \int_a^b a_n(x-c)^n\,dx\right) \tag{8.4.49}$$

$$= \sum_{n=0}^\infty\left(\int_a^b a_n(x-c)^n\,dx\right). \tag{8.4.50}$$

$\square$

To conclude the section, an analogous result holds for derivatives of power series. The omitted proof is comparable to those of Theorems 8.4.11 and 8.4.12, yet different enough to serve as a nice exercise. See Theorem 7.3.8 and Remark 7.3.9.

Theorem 8.4.13: Derivative of a power series

Suppose $\sum_{n=0}^\infty a_n(x-c)^n$ is a power series centered at c with radius of convergence $R>0$. If a compact interval $[a,b]$ satisfies

$$[a,b] \subseteq V_c(R) = (c-R, c+R), \tag{8.4.51}$$

then the uniform limit $f:[a,b]\to\mathbb{R}$ defined by

$$f(x) = \sum_{n=0}^\infty a_n(x-c)^n \qquad \text{for all } x\in[a,b] \tag{8.4.52}$$

is differentiable over $[a,b]$ and we have

$$f'(x) = \sum_{n=0}^\infty (a_n(x-c)^n)' = \sum_{n=1}^\infty na_n(x-c)^{n-1}. \tag{8.4.53}$$

Additionally, the radius of convergence of the series of derivatives $\sum_{n=1}^\infty na_n(x-c)^{n-1}$ is also R.

Example 8.4.14: The derivative of e^x is itself

Consider the exponential function e^x from Example 8.4.2 where it is defined by the power series centered at $c = 0$ given by

$$f(x) = e^x = \sum_{n=0}^{\infty} \frac{x^n}{n!}, \tag{8.4.54}$$

and note that the radius of convergence is $R = \infty$. By taking large enough compact intervals, Theorem 8.4.13 applies and tells us $f(x) = e^x$ is differentiable at every $x \in \mathbb{R}$. Moreover, by the Power Rule, the linearity of derivatives (Theorem 5.4.1), and some reindexing we have

$$f'(x) = \sum_{n=0}^{\infty} \frac{nx^{n-1}}{n!} = \sum_{n=1}^{\infty} \frac{x^{n-1}}{(n-1)!} = \sum_{n=0}^{\infty} \frac{x^n}{n!} = f(x) = e^x. \tag{8.4.55}$$

That is, e^x is its own derivative.

The next section concludes the main content of the book with *Taylor polynomials* and *Taylor series*.

Exercises

8.4.1. Prove Theorem 8.4.13.

8.4.2. Suppose $\sum_{n=0}^{\infty} a_n$ converges conditionally. Prove that for every $c \in \mathbb{R}$, the power series $\sum_{n=0}^{\infty} a_n(x-c)^n$ has radius of convergence $R = 1$.

8.4.3. Suppose (a_n) is a sequence of real numbers where

$$0 < \ell \le |a_n| \le u \tag{8.4.56}$$

for some $\ell, u \in (0, \infty)$ and all $n \in \mathbb{N}$. Prove the power series $\sum_{n=1}^{\infty} a_n x^n$ has radius of convergence $R = 1$.

8.4.4. This exercise leads to a series expansion for π.

   (i) Use geometric series to find a series representation for

$$g(x) = \frac{1}{1+x^2} \tag{8.4.57}$$

   which converges pointwise on $(-1, 1)$.

   (ii) Use part (i) and the antiderivative formula

$$\arctan x = \int \frac{1}{1+x^2} \, dx \tag{8.4.58}$$

to show

$$\arctan x = \sum_{n=1}^{\infty} \frac{(-1)^{n-1}}{2n-1} x^{2n-1}. \tag{8.4.59}$$

(iii) Use (ii) to find a series expansion for π.

8.4.5. Suppose the power series $\sum_{n=0}^{\infty} a_n x^n$ is a power series whose terms (a_n) satisfy

$$a_n \neq 0 \text{ for each } n \in \mathbb{N} \quad \text{and} \quad \lim_{n\to\infty} \left| \frac{a_{n+1}}{a_n} \right| = \ell. \tag{8.4.60}$$

Prove the *Ratio Test for power series* from calculus:

(i) If $\ell = 0$, then $\sum_{n=0}^{\infty} a_n x^n$ converges pointwise for all $x \in \mathbb{R}$.

(ii) If $\ell > 0$, then $1/\ell$ is the radius of convergence of $\sum_{n=0}^{\infty} a_n x^n$.

8.4.6. Use the Ratio Test 8.2.16 to prove a stronger result than the previous exercise where the condition

$$\lim_{n\to\infty} \left| \frac{a_{n+1}}{a_n} \right| = \ell \tag{8.4.61}$$

is replaced. Specifically,

(i) Suppose $a_n \neq 0$ for each $n \in \mathbb{N}$ and there exist $q > 0$ and $n_q \in \mathbb{N}$ such that

$$n \in \mathbb{N} \text{ with } n \geq n_q \quad \Longrightarrow \quad \left| \frac{a_{n+1}}{a_n} \right| \leq q. \tag{8.4.62}$$

Prove $\sum_{n=0}^{\infty} a_n x^n$ converges absolutely for $|x| < 1/q$.

(ii) Suppose $a_n \neq 0$ for each $n \in \mathbb{N}$ and there exist $p > 0$ and $n_p \in \mathbb{N}$ such that

$$n \in \mathbb{N} \text{ with } n \geq n_p \quad \Longrightarrow \quad \left| \frac{a_{n+1}}{a_n} \right| \geq p > 0. \tag{8.4.63}$$

Prove $\sum_{n=0}^{\infty} a_n x^n$ diverges $|x| > 1/p$.

8.4.7. Suppose the power series $\sum_{n=0}^{\infty} a_n x^n$ is a power series whose sequence of terms (a_n) satisfies

$$\lim_{n\to\infty} \sqrt[n]{|a_n|} = \ell. \tag{8.4.64}$$

Prove the *Root Test for power series* from calculus:

(i) If $\ell = 0$, then $\sum_{n=0}^{\infty} a_n x^n$ converges pointwise for all $x \in \mathbb{R}$.

(ii) If $\ell > 0$, then $1/\ell$ is the radius of convergence of $\sum_{n=0}^{\infty} a_n x^n$.

8.4.8. A stronger version of the above Root Test for power series holds. Given a power series $\sum_{n=0}^{\infty} a_n x^n$, set

$$\alpha = \limsup_{n\to\infty} \sqrt[n]{|a_n|} \leq \infty. \tag{8.4.65}$$

Prove the radius of convergence R of the power series $\sum_{n=0}^{\infty} a_n x^n$ is given by

$$R = \begin{cases} \infty, & \text{if } \alpha = 0, \\ 0, & \text{if } \alpha = \infty, \\ \dfrac{1}{\alpha} & \text{if } 0 < \alpha < \infty. \end{cases} \tag{8.4.66}$$

8.4.9. Fix $c \in \mathbb{R}$ and determine the radii and intervals of convergence of the power series

$$\sum_{n=0}^{\infty} n^2 (x - c)^n \qquad \text{and} \qquad \sum_{n=0}^{\infty} \frac{1}{n^2} (x - c)^n. \tag{8.4.67}$$

8.4.10. Let $p \in \mathbb{R}$ and consider the power series $\sum_{n=0}^{\infty} \frac{1}{n^p} x^n$. Use p-series (Theorem 8.2.15) to breakdown the cases of possible radii of convergence and prove your result.

8.4.11. Determine the radius and interval of convergence of the power series $\sum_{n=0}^{\infty} \frac{2^n n!}{n^n} x^n$.

8.5 Taylor polynomials and Taylor series

Polynomials have played a key role throughout the book. They are extremely nice to work with since they are continuous, differentiable, and integrable, and computers can evaluate them. They are also arbitrarily close to all continuous functions on compact intervals in the uniform sense laid out in the Weierstrass Approximation Theorem 7.4.7.

In this final section of the chapter, polynomials, along with their infinite extensions given by power series, are shown to have deep relationships with differentiable functions of varying degree established in a local sense through *Taylor polynomials* and *Taylor series*.

Notation 8.5.1: kth derivative

For an index $k \in \mathbb{N} \cup \{0\}$ and a function real-valued function f which is k-times differentiable, the kth derivative of f is denoted by $f^{(k)}$. In particular, the kth derivative is defined recursively in that $f^0 = f$ and $f^{(k+1)} = (f^{(k)})'$.

Taylor polynomials pop out from playing around with integration by parts. A version of this theorem and a similar proof can be found in [8].

Theorem 8.5.2: Taylor polynomial and remainder via row integration by parts (RIP)

Suppose $I \subseteq \mathbb{R}$ is an interval, $f : I \to \mathbb{R}$, $n \in \mathbb{N} \cup \{0\}$, f is $(n+1)$-times differentiable at $c \in I$, and $f^{(k)}$ is continuous on I for $k = 1, \ldots, n+1$. Then for every $x \in I$ we have

$$f(x) = \sum_{k=0}^{n} \frac{f^{(k)}(c)}{k!}(x-c)^k + \int_c^x \frac{f^{(n+1)}(t)}{n!}(x-t)^n \, dt \tag{8.5.1}$$

$$= f(c) + f'(c)(x-c) + \frac{f''(c)}{2}(x-c)^2 + \cdots + \frac{f^{(n)}(c)}{n!}(x-c)^n \tag{8.5.2}$$

$$+ \int_c^x \frac{f^{(n+1)}(t)}{n!}(x-t)^n \, dt. \tag{8.5.3}$$

Scratch Work 8.5.3: Play around with RIP and the Fundamental Theorem of Calculus

By thinking of f as an antiderivative of f', the Fundamental Theorem of Calculus I 6.1.15 tells us

$$f(x) - f(c) = \int_c^x f'(t) \, dt = \int_c^x -f'(t)(-1) \, dt. \tag{8.5.4}$$

Here's the sneaky part: To take advantage of row integration by parts (RIP), choose $u = -f'(t) = -f^{(1)}(t)$ and, hence, $dv = -1 \, dt$. In the table below, each pair of rows represents an iteration of integration by parts and the fourth column encodes the new integral from each iteration. Differentiation (diff.) and integration (int., that is, finding antiderivatives) are performed with respect to t. The antiderivative of -1 is taken to be $(x-t)$, where x is treated as a constant. Other antiderivatives are chosen in a similar fashion. Ultimately, for $n \in \mathbb{N}$ we have:

(alt.) $\pm$		(diff.) u	(int.) dv	$(\pm \int)$ $\int u \, dv$
$+$	$\to$	$-f^{(1)}(t)$	-1	$\int f'(t) \, dt$
$-$	$\to$	$-f^{(2)}(t)$	$(x-t)$	$\int f''(t)(x-t) \, dt$
$+$	$\to$	$-f^{(3)}(t)$	$-\dfrac{(x-t)^2}{2!}$	$\int \dfrac{f'''(t)}{2}(x-t)^2 \, dt$
$\vdots$		$\vdots$	$\vdots$	$\vdots$
$(-1)^{n-1}$	$\to$	$-f^{(n)}(t)$	$(-1)^n \dfrac{(x-t)^{n-1}}{(n-1)!}$	$\int \dfrac{f^{(n)}(t)}{(n-1)!}(x-t)^{(n-1)} \, dt$
$(-1)^n$	$\to$	$-f^{(n+1)}(t)$	$\to$ $(-1)^{n+1} \dfrac{(x-t)^n}{n!}$	$\int \dfrac{f^{(n+1)}(t)}{n!}(x-t)^n \, dt$

Therefore, by gathering succesive "uv" terms by following the arrows and noting the right-most column contains the new integral that comes from each iteration of row integration by parts, we have

$$f(x) - f(c) = \int_c^x -f^{(1)}(t)(-1)\, dt \tag{8.5.5}$$

$$= \left[-f^{(1)}(t)(x-t) - \frac{f^{(2)}(t)}{2!}(x-t)^2 - \cdots - \frac{f^{(n)}(t)}{n!}(x-t)^n \right]_c^x \tag{8.5.6}$$

$$+ \int_c^x \frac{f^{(n+1)}(t)}{n!}(x-t)^n\, dt \tag{8.5.7}$$

$$= \sum_{k=1}^n \frac{f^{(k)}(c)}{n!}(x-c)^k + \int_c^x \frac{f^{(n+1)}(t)}{n!}(x-t)^n\, dt, \tag{8.5.8}$$

where the notation $[\ldots]_c^x$ means we evaluate the expression $[\ldots]$ at both x and c, then take the difference.

The result follows from adding $f(c)$ across the above equations. The proof amounts to making sure the steps taken throughout this scratch work are valid.

Proof of Theorem 8.5.2. Suppose $I \subseteq \mathbb{R}$ is an interval, $f : I \to \mathbb{R}$, $n \in \mathbb{N} \cup \{0\}$, f is $(n+1)$-times differentiable at $c \in I$, and $f^{(k)}$ is continuous on I for $k = 1, \ldots, n+1$. Fix $x \in I$ and treat $t \in I$ as a variable. Since polynomials are continuous (Theorem 4.5.2) and products of continuous functions are continuous (Theorem 4.5.8), the functions

$$\frac{f^{(k+1)}(t)}{k!}(x-t)^k \tag{8.5.9}$$

are continuous on I for each $k = 0, \ldots, n$. Hence, each of these functions is integrable over the compact interval with endpoints x and c by Theorem 6.2.17. From there, thinking of f as an antiderivative of f' and following Scratch Work 8.5.3, the Fundamental Theorem of Calculus I 6.1.15 and iterations of integration by parts (via RIP) yield

$$f(x) - f(c) = \int_c^x -f^{(1)}(t)(-1)\, dt \tag{8.5.10}$$

$$= \left[-f^{(1)}(t)(x-t) - \frac{f^{(2)}(t)}{2!}(x-t)^2 - \cdots - \frac{f^{(n)}(t)}{n!}(x-t)^n \right]_c^x \tag{8.5.11}$$

$$+ \int_c^x \frac{f^{(n+1)}(t)}{n!}(x-t)^n\, dt \tag{8.5.12}$$

$$= \sum_{k=1}^n \frac{f^{(k)}(c)}{n!}(x-c)^k + \int_c^x \frac{f^{(n+1)}(t)}{n!}(x-t)^n\, dt, \tag{8.5.13}$$

where the notation $[\ldots]_c^x$ means we evaluate the expression $[\ldots]$ at both x and c, then take the difference. Adding $f(c)$ produces the desired result

$$f(x) = \sum_{k=0}^{n} \frac{f^{(k)}(c)}{k!}(x - c)^k + \int_c^x \frac{f^{(n+1)}(t)}{n!}(x - t)^n \, dt \tag{8.5.14}$$

$$= f(c) + f'(c)(x - c) + \frac{f''(c)}{2}(x - c)^2 + \cdots + \frac{f^{(n)}(c)}{n!}(x - c)^n \tag{8.5.15}$$

$$+ \int_c^x \frac{f^{(n+1)}(t)}{n!}(x - t)^n \, dt. \tag{8.5.16}$$

$\square$

It's time to formally define the key objects of the section: *Taylor polynomials.*

Definition 8.5.4: Taylor polynomial

Suppose I is an interval, $f : I \to \mathbb{R}$, and f is n-times differentiable at $c \in I$. The *nth Taylor polynomial of f at c* is the function $T_n(f, c) : I \to \mathbb{R}$ given by

$$T_n(f, c)(x) = \sum_{k=0}^{n} \frac{f^{(k)}(c)}{n!}(x - c)^k \tag{8.5.17}$$

$$= f(c) + f'(c)(x - c) + \frac{f''(c)}{2}(x - c)^2 + \cdots + \frac{f^{(n)}(c)}{n!}(x - c)^n. \tag{8.5.18}$$

$T_n(f, c)$ is also called the *Taylor polynomial of f of degree n at c.*

Example 8.5.5: Taylor polynomials of $\sin x$

Consider the function $g : \mathbb{R} \to \mathbb{R}$ given by $g(x) = \sin x$. For each $j \in \mathbb{N} \cup \{0\}$ and every $x \in \mathbb{R}$ we have

$$g(x) = \sin x \qquad\qquad\qquad g^{(4j)}(x) = \sin x$$

$$g'(x) = \cos x \qquad\qquad\qquad g^{(4j+1)}(x) = \cos x$$

$$g''(x) = -\sin x \qquad\qquad\qquad g^{(4j+2)}(x) = -\sin x$$

$$g^{(3)}(x) = -\cos x \qquad\qquad\qquad g^{(4j+3)}(x) = -\cos x$$

$$g^{(4)}(x) = \sin x \qquad\qquad\qquad g^{(4j+4)}(x) = \sin x$$

$$\vdots \qquad\qquad\qquad\qquad\qquad\qquad \vdots$$

So with $n = 4$ and $c = 0$ we have

$$T_4(g, 0)(x) = g(0) + g'(0)x + \frac{g''(0)}{2}x^2 + \frac{g^{(3)}(0)}{3!}x^3 + \frac{g^{(4)}(0)}{4!}x^4 \tag{8.5.19}$$

$$= x - \frac{x^3}{6}. \tag{8.5.20}$$

For $n = 4$ and $c = \pi/2$ instead, we have

$$T_4\left(g, \pi/2\right)(x) = g\left(\frac{\pi}{2}\right) + g'\left(\frac{\pi}{2}\right)\left(x - \frac{\pi}{2}\right) + \frac{g''\left(\frac{\pi}{2}\right)}{2}\left(x - \frac{\pi}{2}\right)^2 \tag{8.5.21}$$

$$+ \frac{g^{(3)}\left(\frac{\pi}{2}\right)}{3!}\left(x - \frac{\pi}{2}\right)^3 + \frac{g^{(4)}\left(\frac{\pi}{2}\right)}{4!}\left(x - \frac{\pi}{2}\right)^4 \tag{8.5.22}$$

$$= 1 - \frac{1}{2}\left(x - \frac{\pi}{2}\right)^2 + \frac{1}{24}\left(x - \frac{\pi}{2}\right)^4. \tag{8.5.23}$$

Try plotting $g(x) = \sin x$, $T_4(g, 0)(x)$, and $T_4\left(g, \pi/2\right)(x)$ with some mathematical software. $T_4(g, 0)(x)$ approximates $\sin x$ pretty well near $c = 0$ while $T_4\left(g, \pi/2\right)(x)$ approximates $\sin x$ near $c = \pi/2$, but neither is good approximation of $\sin x$ over larger intervals.

Remark 8.5.6: Differentiable functions are locally polynomial

As mentioned at the start of the section, continuous functions on compact intervals are *uniformly approximated* by polynomials across their whole domain. Specifically, the Weierstrass Approximation Theorem on $[0, 1]$ 7.4.5 says that when $f : [0, 1] \to \mathbb{R}$ is continuous, then for every $\varepsilon > 0$ there is an index $n_\varepsilon \in \mathbb{N}$ whose Bernstein polynomial $B_{n_\varepsilon}(f)$ yields

$$x \in [0, 1] \qquad \Longrightarrow \qquad |B_{n_\varepsilon}(f)(x) - f(x)| < \varepsilon. \tag{8.5.24}$$

(The more general Weierstrass Approximation Theorem 7.4.7 extends this result to continuous functions on arbitrary compact intervals.)

In the current setting, n-times differentiable functions are *locally approximated* by their nth Taylor polynomials. In fact, we have already seen a specific instance of the local linearity of differentiable functions. By merging Lemma 5.3.10 with the 1st Taylor polynomial $T_1(f, c)$ from Definition 8.5.4, we have for every $\varepsilon > 0$ there is a threshold $\delta > 0$ where

$$|x - c| < \delta \qquad \Longrightarrow \qquad |f(x) - (f'(c)(x - c) + f(c))| < \varepsilon \tag{8.5.25}$$
$$\Longleftrightarrow \qquad |f(x) - T_1(f, c)(x)| < \varepsilon. \tag{8.5.26}$$

The next chunk of the section leads to a generalization of Lemma 5.3.10 for n-times differentiable functions where

$$|f(x) - T_n(f, c)(x)| < \varepsilon \tag{8.5.27}$$

on some neighborhood of c.

To get at (8.5.27) and see how well Taylor polynomials approximate their function f, consider the differences known as the *remainders* or *error functions*.

Definition 8.5.7: Remainder of a Taylor polynomial

Suppose I is an interval, $f : I \to \mathbb{R}$, and f is n-times differentiable at $c \in I$. The nth *remainder of f at c* is the function $R_n(f, c) : I \to \mathbb{R}$ given by

$$R_n(f, c)(x) = f(x) - T_n(f, c)(x) \tag{8.5.28}$$

where $T_n(f, c)$ is the nth Taylor polynomial of f at c.

A corollary of Theorem 8.5.2 identifies an integral form for the remainder of a Taylor polynomial. The proof is omitted since it amounts to checking Definition 8.5.7 against the conclusion of Theorem 8.5.2.

Corollary 8.5.8: Integral form of the remainder

Suppose I is an interval, $f : I \to \mathbb{R}$, $n \in \mathbb{N} \cup \{0\}$, f is $(n + 1)$-times differentiable at $c \in I$, and $f^{(k)}$ is continuous on I for $k = 1, \ldots, n + 1$. Then for every $x \in I$, the remainder $R_n(f, c) : I \to \mathbb{R}$ is given by

$$R_n(f, c)(x) = f(x) - T_n(f, c)(x) = \int_c^x \frac{f^{(n+1)}(t)}{n!}(x - t)^n \, dt. \tag{8.5.29}$$

The remainder of a Taylor polynomial is also given by derivatives.

Theorem 8.5.9: Lagrange form of the remainder

Suppose I is an interval, $f : I \to \mathbb{R}$, $n \in \mathbb{N} \cup \{0\}$, f is $(n + 1)$-times differentiable at $c \in I$, and $f^{(k)}$ is continuous on I for $k = 1, \ldots, n + 1$. Then for each $x_0 \in I$, there is an input y_n between x_0 and c where the value of the remainder $R_n(f, c)(x_0)$ is given by

$$R_n(f, c)(x_0) = \frac{f^{(n+1)}(y_n)}{(n + 1)!}(x_0 - c)^{n+1}. \tag{8.5.30}$$

Scratch Work 8.5.10: The key input is from the Intermediate Value Theorem

Continuous functions attain all values between any two outputs as described by the Intermediate Value Theorem 4.6.5, essentially guaranteeing each output has an input. So, to get the key input y_n, the approach is to show the function

$$\frac{f^{(n+1)}(t)}{(n + 1)!}(x_0 - c)^{n+1} \tag{8.5.31}$$

is continuous on the compact interval between x_0 and c, and then show the value of the remainder $R_n(f, c)(x_0)$ is between outputs of this function. A wide variety of results come into play for this proof!

Proof of Theorem 8.5.9. Suppose the hypotheses of Theorem 8.5.9 hold and assume, without loss of generality, that $c < x_0$. Since $f^{(n+1)}$ is continuous on the compact interval $[c, x_0]$, it attains its extreme values by the Extreme Value Theorem 4.6.9. So, let

$$\ell_n = \min f^{(n+1)}([c, x_0]) \qquad \text{and} \qquad u_n = \max f^{(n+1)}([c, x_0]). \qquad (8.5.32)$$

For all for all $t \in [c, x_0]$ we have $x_0 - t \geq 0$, so the bounds $\ell_n \leq f^{(n+1)}(t) \leq u_n$ imply

$$\frac{\ell_n}{n!}(x_0 - t)^n \leq \frac{f^{(n+1)}(t)}{n!}(x_0 - t)^n \leq \frac{u_n}{n!}(x_0 - t)^n. \qquad (8.5.33)$$

Since polynomials are continuous (Theorem 4.5.2) and products of continuous functions are continuous (Theorem 4.5.8), the three functions in (8.5.33) are continuous on $[c, x_0]$. Hence, each is also integrable over $[c, x_0]$ by Theorem 6.2.17. So by the order property of integrals (Corollary 6.3.10) and the linearity of integration (Theorem 6.3.6) together with the integral form of the remainder in Corollary 8.5.8, we have

$$\frac{\ell_n}{n!}\int_c^{x_0}(x_0 - t)^n\, dt \leq R_n(f, c)(x_0) = \int_c^{x_0}\frac{f^{(n+1)}(t)}{n!}(x_0 - t)^n\, dt \leq \frac{u_n}{n!}\int_c^{x_0}(x_0 - t)^n\, dt. \qquad (8.5.34)$$

Evaluating the integrals on the left and right using the Fundamental Theorem of Calculus I 6.1.15 via antiderivative $-(x_0 - t)^{n+1}/(n + 1)$ yields

$$\frac{\ell_n(x_0 - c)^{n+1}}{(n + 1)!} \leq R_n(f, c)(x_0) \leq \frac{u_n(x_0 - c)^{n+1}}{(n + 1)!}. \qquad (8.5.35)$$

To take advantage of the Intermediate Value Theorem 4.6.5, note that $(x_0 - c)^{n+1}/(n + 1)$ is a positive constant. So scaling $\ell_n \leq f^{(n+1)}(t) \leq u_n$ by this constant gives us

$$\frac{\ell_n(x_0 - c)^{n+1}}{(n + 1)!} \leq \frac{f^{(n+1)}(t)(x_0 - c)^{n+1}}{(n + 1)!} \leq \frac{u_n(x_0 - c)^{n+1}}{(n + 1)!} \qquad (8.5.36)$$

for all $t \in [c, x_0]$. Also, by (8.5.32), these new bounds are attained and we have

$$\frac{\ell_n(x_0 - c)^{n+1}}{(n + 1)!} = \min \frac{(x_0 - c)^{n+1}}{(n + 1)!}f^{(n+1)}([c, x_0]) \qquad \text{and} \qquad (8.5.37)$$

$$\frac{u_n(x_0 - c)^{n+1}}{(n + 1)!} = \max \frac{(x_0 - c)^{n+1}}{(n + 1)!}f^{(n+1)}([c, x_0]). \qquad (8.5.38)$$

By the linearity of continuity (Theorem 4.5.5), the scaled function $\frac{(x_0-c)^{n+1}}{(n+1)!}f^{(n+1)}$ is continuous on $[c, x_0]$. Note (8.5.35) and (8.5.37) combine to say the remainder $R_n(f, c)(x_0)$ is between outputs of this scaled function, so by the Intermediate Value Theorem 4.6.5, there is some input $y_n \in [c, x_0]$ where

$$R_n(f, c)(x_0) = \frac{f^{(n+1)}(y_n)}{(n + 1)!}(x_0 - c)^{n+1}. \qquad (8.5.39)$$

$\square$

Taylor series provide the setting for the final results of the chapter.

Definition 8.5.11: Taylor series

Suppose I is an interval, $c \in I$, and $f : I \to \mathbb{R}$ where $f^{(n)}(c)$ exists for all $n \in \mathbb{N}$. The *Taylor series of f at c* is the power series

$$\sum_{n=0}^{\infty} \frac{f^{(n)}(c)}{n!}(x - c)^n. \tag{8.5.40}$$

The special case where $c = 0$ is called the *Maclaurin series of f*. Note that kth partial sum of a Taylor series is $T_k(f, c)$, the kth Taylor polynomial of f at c as in Definition 8.5.4.

The Taylor series of a function is only meaningful when the derivatives of all orders exist at a specified value c. When a function's derivatives exist on a common interval, the function is *smooth*

Definition 8.5.12: Smooth functions

Suppose I is an interval. A function $f : I \to \mathbb{R}$ is *smooth* or *infinitely differentiable I* if the kth derivative $f^{(k)}$ exists on I for every $k \in \mathbb{N}$.

Smooth functions equal their Taylor series when the remainders tend to zero pointwise. The same result can be found at the end of [6].

Theorem 8.5.13: Smooth functions and Taylor series

Suppose I is an interval, $c \in I$, and f is smooth on I. Then for a given $x_0 \in I$,

$$f(x_0) = \sum_{n=0}^{\infty} \frac{f^{(n)}(c)}{n!}(x_0 - c)^n \qquad \Longleftrightarrow \qquad \lim_{k \to \infty} R_k(f, c)(x_0) = 0. \tag{8.5.41}$$

Scratch Work 8.5.14: Limits of partial sums and linearity of sequential limits

The proof relies on fundamentals of convergent series and sequences of real numbers: The sum of a series is the limit of partial sums (Definition 8.1.3) and sequential limits are linear (Theorem 2.3.9). However, the particular form of the remainder $R_k(f, c)$ does not matter.

Proof of Theorem 8.5.13. Suppose f is smooth on interval I and $x_0, c \in I$. Note that $f(x_0)$ is constant with respect to $k \in \mathbb{N}$, so

$$f(x_0) = \lim_{k \to \infty} f(x_0). \tag{8.5.42}$$

Both implications follow from the same string of various equivalent statements, though followed in opposite directions:

$$f(x_0) = \sum_{n=0}^{\infty} \frac{f^{(n)}(c)}{n!}(x_0 - c)^n \quad \Longleftrightarrow \quad \lim_{k\to\infty} f(x_0) = \lim_{k\to\infty} T_k(f, c)(x_0) \tag{8.5.43}$$

$$\Longleftrightarrow \quad \lim_{k\to\infty} f(x_0) - \lim_{k\to\infty} T_k(f, c)(x_0) = 0 \tag{8.5.44}$$

$$\Longleftrightarrow \quad \lim_{k\to\infty} (f(x_0) - T_k(f, c)(x_0)) = 0 \tag{8.5.45}$$

$$\Longleftrightarrow \quad \lim_{k\to\infty} R_k(f, c)(x_0) = 0. \tag{8.5.46}$$

The first equivalence above follows from (8.5.42) and identifying the Taylor series as the limit of the sequence of Taylor polynomials (Definitions 8.1.3, 8.5.4, and 8.5.11). The second follows from subtracting/adding $\lim_{k\to\infty} T_k(f, c)(x_0)$. The third is the linearity of sequential limits (Theorem 2.3.9). And the last equivalence follows from the definition of the remainder as the difference between the function and a Taylor polynomial (Definition 8.5.7). $\qquad\square$

A fundamental example of Taylor series closes out the main content of the book. First, a lemma will help us get a hold of remainders. It also supplies a proof for Exercise 2.4.6.

Lemma 8.5.15: Factorials dominate monomials

For every $x \in \mathbb{R}$ we have

$$\lim_{n\to\infty} \frac{x^n}{n!} = 0. \tag{8.5.47}$$

Scratch Work 8.5.16: Ratios are eventually small

For each $n \in \mathbb{N}$, the numerator and denominator of the ratio $x^n/n!$ each has n factors. So we can write

$$\frac{x^n}{n!} = \left(\frac{x}{1}\right)\left(\frac{x}{2}\right)\cdots\left(\frac{x}{n-1}\right)\left(\frac{x}{n}\right). \tag{8.5.48}$$

The Archimedean Property 1.4.6 tells us there is an index $n_0 \in \mathbb{N}$ large enough so that

$$|x| < n_0 \quad \Longrightarrow \quad 0 \le \frac{|x|}{n_0} < 1. \tag{8.5.49}$$

For $n > n_0$, we can split (8.5.48) at the index $n_0 - 1$, like this

$$\frac{x^n}{n!} = \left(\frac{x}{1}\right)\left(\frac{x}{2}\right)\cdots\left(\frac{x}{n-1}\right)\left(\frac{x}{n}\right) \tag{8.5.50}$$

$$= \left(\frac{x}{1}\right)\left(\frac{x}{2}\right)\cdots\left(\frac{x}{n_0-1}\right)\left(\frac{x}{n_0}\right)\cdots\left(\frac{x}{n-1}\right)\left(\frac{x}{n}\right). \tag{8.5.51}$$

The factors to the left of the split stay constant as n continues to grow. On the other hand, each factor to the right of the split has an absolute value less than some $0 < c_0 < 1$, and we get more and more of them as n grows. Hence, with a little more effort, we can apply Corollary 2.4.19 to complete the proof.

Proof of Lemma 8.5.15. Fix $x \in \mathbb{R}$. By the Archimedean Property 1.4.6, there is an index $n_0 \in \mathbb{N}$ large enough so that for all $n \in \mathbb{N}$ where $n \geq n_0 \in \mathbb{N}$ we have

$$|x| < n_0 \leq n \qquad \Longrightarrow \qquad 0 \leq \frac{|x|}{n} \leq \frac{|x|}{n_0} < 1. \tag{8.5.52}$$

Now set $c_0 = |x|/n_0$. Splitting the ratio $|x|^n/n!$ as in Scratch Work 8.5.16 yields

$$\frac{|x|^n}{n!} = \left(\frac{|x|}{1}\right)\left(\frac{|x|}{2}\right)\cdots\left(\frac{|x|}{n-1}\right)\left(\frac{|x|}{n}\right) \tag{8.5.53}$$

$$= \underbrace{\left(\frac{|x|}{1}\right)\left(\frac{|x|}{2}\right)\cdots\left(\frac{|x|}{n_0-1}\right)}_{n_0-1 \text{ factors}} \cdot \underbrace{\left(\frac{|x|}{n_0}\right)\cdots\left(\frac{|x|}{n-1}\right)\left(\frac{|x|}{n}\right)}_{n-n_0 \text{ factors}} \tag{8.5.54}$$

$$\leq \left(\frac{|x|^{n_0-1}}{(n_0-1)!}\right)\left(\frac{|x|}{n_0}\right)^{n-n_0} \tag{8.5.55}$$

$$= \left(\frac{|x|^{n_0-1}}{(n_0-1)!}\right) c_0^{n-n_0} \tag{8.5.56}$$

$$= \underbrace{\left(\frac{|x|^{n_0-1}}{c_0^{n_0}(n_0-1)!}\right)}_{\text{constant w.r.t. } n} c_0^n. \tag{8.5.57}$$

Since $0 \leq c_0 = |x|/n_0 < 1$, Corollary 2.4.19 applies and pairs with the linearity of sequential limits (Theorem 2.3.9) to give us

$$\lim_{n\to\infty} \frac{|x|^n}{n!} = 0. \tag{8.5.58}$$

Finally, for every $x \in \mathbb{R}$ and $n \in \mathbb{N}$ we have

$$-\frac{|x|^n}{n!} \leq \frac{x^n}{n!} \leq \frac{|x|^n}{n!}. \tag{8.5.59}$$

Therefore, by the linearity of and Squeeze Theorem for sequential limits (Theorems 2.3.9 and 2.4.3), for every $x \in \mathbb{R}$ we have

$$\lim_{n\to\infty} \frac{x^n}{n!} = 0. \tag{8.5.60}$$

$\square$

One final example to conclude the book.

Example 8.5.17: Taylor series of e^x

The Maclaurin series of $f(x) = e^x$ is exactly the power series that defines e^x in Example 8.4.2. Moreover, $f(x) = e^x$ equals the sum of its Taylor series centered at *any* $c \in \mathbb{R}$. That is, for every $x, c \in \mathbb{R}$ we have

$$f(x) = e^x = \sum_{n=0}^{\infty} \frac{x^n}{n!} = \sum_{n=0}^{\infty} \frac{f^{(n)}(c)}{n!}(x - c)^n = \sum_{n=0}^{\infty} \frac{e^c}{n!}(x - c)^n. \tag{8.5.61}$$

Scratch Work 8.5.18: Show the remainders converge to zero

To show a function equals its Taylor series at a given input x_0, Theorem 8.5.13 tells us it suffices to show the sequence of remainders $(R_n(f, c)(x_0))$ converges to 0. Example 8.4.14 tells us the exponential function e^x is equal to all of its derivatives. Paired with some assumptions about e^x from calculus and the Lagrange form of the remainder as a derivative from Theorem 8.5.9, we have enough to prove e^x equals its Taylor series centered at any $c \in \mathbb{R}$.

Proof for Example 8.5.17. Consider the exponential function $f(x) = e^x$ given by

$$f(x) = e^x = \sum_{n=0}^{\infty} \frac{x^n}{n!} \tag{8.5.62}$$

as in Example 8.4.2. Fix $c, x_0 \in \mathbb{R}$ with $x_0 \neq c$ and let $n \in \mathbb{N} \cup \{0\}$. Since e^x equals its derivatives of all orders (Example 8.4.14), the nth remainder $R_n(f, c)(x_0)$ is given by

$$R_n(f, c)(x_0) = \frac{f^{(n+1)}(y_n)}{(n+1)!}(x_0 - c)^{n+1} = \frac{e^{y_n}}{(n+1)!}(x_0 - c)^{n+1} \tag{8.5.63}$$

for some y_n between x_0 and c. Also, since e^x is increasing (a fact from calculus that is assumed but not prove here) and $-(|x_0| + |c|) \leq y_n \leq |x_0| + |c|$, we have

$$e^{-(|x_0|+|c|)} \leq e^{y_n} \leq e^{|x_0|+|c|}. \tag{8.5.64}$$

Multiplying across by a key nonnegative number yields

$$\frac{e^{-(|x_0|+|c|)}}{(n+1)!}(x_0 - c)^{n+1} \leq R_n(f, c)(x_0) = \frac{e^{y_n}}{(n+1)!}(x_0 - c)^{n+1} \leq \frac{e^{|x_0|+|c|}}{(n+1)!}(x_0 - c)^{n+1} \tag{8.5.65}$$

By the linearity of and Squeeze Theorem for sequential limits (Theorems 2.3.9 and 2.4.3) along with Lemma 8.5.15 we have

$$\lim_{n \to \infty} R_n(f, c)(x_0) = \lim_{n \to \infty} \frac{e^{y_n}}{(n+1)!}(x_0 - c)^{n+1} = 0. \tag{8.5.66}$$

Therefore, by Theorem 8.5.13, for every $x_0, c \in \mathbb{R}$ we have

$$f(x) = e^x = \sum_{n=0}^{\infty} \frac{x^n}{n!} = \sum_{n=0}^{\infty} \frac{f^{(n)}(c)}{n!}(x - c)^n = \sum_{n=0}^{\infty} \frac{e^c}{n!}(x - c)^n. \tag{8.5.67}$$

$\square$

Exercises

8.5.1. Determine Maclaurin series for $g(x) = \sqrt{1-x}$ (i.e., Taylor series centered at $c = 0$). Prove the series converges to $g(x) = \sqrt{1-x}$ when $-1 < x \leq 0$.

8.5.2. Determine Taylor series for $h(x) = \ln x$ centered at $c = 1$. Where does this series to $h(x) = \ln x$?

8.5.3. Suppose $p : \mathbb{R} \to \mathbb{R}$ is a polynomial of degree $n \in \mathbb{N} \cup \{0\}$ given by

$$p(x) = \sum_{k=0}^{n} a_k x^k = a_0 + a_1 x + a_2 x^2 + \cdots + a_n x^n. \tag{8.5.68}$$

(i) Prove that for each $c \in \mathbb{R}$, the Taylor series of p centered at c is a finite sum (all terms with index $j > n$ are zero) and

$$p(x) = \sum_{k=0}^{n} \frac{f^{(k)}(c)}{k!} (x - c)^k. \tag{8.5.69}$$

(ii) Consider the polynomial $p(x) = 4x^3 - 2x^2 + x - 1$. Use part (i) to determine other polynomials equal to p derived from the Taylor series of p at $c = -1, 1$, and 2, respectively.

8.5.4. This exercise builds on the Taylor polynomials of $\sin x$ in Example 8.5.5 and the proof that e^x equals its Taylor series centered at any point (Example 8.5.17).

(i) Verify the formulas for the Taylor series centered at $c = 0$ (so, the Maclaurin series) of $\sin x$ and $\cos x$ using the relationship between their derivatives from calculus:

(a) $\sin x = \displaystyle\sum_{n=0}^{\infty} \frac{(-1)^n}{(2n+1)!} x^{2n+1} = x - \frac{x^3}{3!} + \frac{x^5}{5!} - \cdots$.

(b) $\cos x = \displaystyle\sum_{n=0}^{\infty} \frac{(-1)^n}{(2n)!} x^{2n} = 1 - \frac{x^2}{2!} + \frac{x^4}{4!} - \cdots$.

(ii) Prove $\sin x$ and $\cos x$ equal their Maclaurin series at every $x \in \mathbb{R}$.

(iii) Prove $\sin x$ is an odd function and $\cos x$ is an even function.

8.5.5. Suppose $f : (-1, \infty) \to \mathbb{R}$ is given by $f(x) = \ln(1 + x)$. Use the fact from calculus that

$$f'(x) = \frac{1}{1+x} \qquad \text{for all } x \in (-1, \infty) \tag{8.5.70}$$

to establish the following results.

(i) Derive the formula for the Taylor series of $f(x) = \ln(1 + x)$ centered at $c = 0$ and determine the radius of convergence.

(ii) Prove

$$\ln 2 = \sum_{n=1}^{\infty} \frac{(-1)^{n+1}}{n} = 1 - \frac{1}{2} + \frac{1}{3} - \frac{1}{4} + \frac{1}{5} - \cdots . \tag{8.5.71}$$

(iii) Derive the formula for the Taylor series of $g(x) = \ln\left(1 + x^2\right)$ centered at $c = 0$ and determine the radius of convergence.

8.5.6. Use previously established results and exercises to find Taylor series expansions of the following functions and determine precisely for which values of x the original function equals the sum of its Taylor series.

(i) $x \sin x^3$

(iii) $\dfrac{e^x - 1}{x}$

(ii) $x \cos x^2$

(iv) $\ln\left(1 + x^2\right)$

8.5.7. Prove that if f is smooth on an interval I and there is a uniform bound $b \geq 0$ on the derivatives where

$$|f^{(n)}(x)| \leq b \tag{8.5.72}$$

for all $x \in I$ and $n \in \mathbb{N} \cup \{0\}$, then f equals its Taylor series on I.

8.5.8. Prove the following weaker form of Theorem 8.5.9: Suppose $n \in \mathbb{N} \cup \{0\}$, f is $(n+1)$-times differentiable at $c \in [0, x]$, $f^{(k)}$ is continuous on $[0, x]$ for $k = 1, \ldots, n + 1$, and there is a bound $b \geq 0$ on the derivatives where $|f^{(n+1)}(x)| \leq b$ for all $x \in [0, x]$. Prove

$$R_n(f, 0)(x) \leq \frac{bx^{n+1}}{(n+1)!}. \tag{8.5.73}$$

8.5.9. This exercise shows the extent to which a convergent Taylor series can differ from the function that defined it, which is to say, completely. The derivatives make use of the ∞/∞ case of *L'Hospital's Rule*, loosely stated and not proven here: Suppose $\lim_{x \to c} |f(x)| = \infty$ and $\lim_{x \to c} |g(x)| = \infty$. Then

$$\lim_{x \to c} \frac{f'(x)}{g'(x)} = \ell \quad \implies \quad \lim_{x \to c} \frac{f(x)}{g(x)} = \ell. \tag{8.5.74}$$

Consider the function $f : \mathbb{R} \to \mathbb{R}$ given by

$$f(x) = \begin{cases} e^{-1/x^2}, & \text{if } x \neq 0, \\ 0, & \text{if } x = 0. \end{cases} \tag{8.5.75}$$

(i) Prove f is uniformly continuous on $\mathbb{R}$.

(ii) Prove f is smooth on $\mathbb{R}$ and $f^{(n)}(0) = 0$ for every $n \in \mathbb{N} \cup \{0\}$.

(iii) Prove f is not equal to its Taylor series centered at $c = 0$ (i.e., its Maclaurin series) at any $x \neq 0$. That is, despite the fact that the series converges at every point of the real line, prove

$$f(x) \neq \sum_{n=0}^{\infty} \frac{f^{(n)}(0)}{n!} x^n \tag{8.5.76}$$

for all $x \neq 0$.

Bibliography

[1] S. Abbott, *Understanding Analysis*, 2nd edition, Springer, New York, NY, 2015.

[2] C. Adams and R. Franzosa, *Introduction to Topology: Pure and Applied*, Pearson, NJ, 2008.

[3] J. Barnes, *Teaching Real Analysis in the Land of Make Believe*, PRIMUS, vol. 14, Taylor and Francis, UK, 2007, pp. 366–372.

[4] J. E. Borzellino, *WHOSE LIMIT IS IT ANYWAY?*, PRIMUS, vol. 11, Taylor and Francis, UK, 2001, pp. 265–274.

[5] B. Cornu, *Limits*, Advanced Mathematical Thinking, Kluwer Academic Publishers, Dordrecht, Netherlands, 1991, pp. 153–166.

[6] J. Cummings, *Real Analysis: A Long-Form Mathematics Textbook*, 2nd edn., self-published.

[7] E. D. Gaughan, *Introduction to Analysis*, 5th edn., AMS, Providence, RI, 1998.

[8] D. Horowitz, *Tabular integration by parts*, College Math. J., **21**, no. 4, 1990, pp. 307–311.

[9] S. Larsen and C. Swinyard, *Coming to Understand the Formal Definition of Limit: Insights Gained From Engaging Students in Reinvention*, Journal for Research in Mathematics Education, vol. 43 (4), 2012, pp. 465-493.

[10] W. Rudin, *Principles of Mathematical Analysis*, 3rd edn., McGraw Hill, 1976.

[11] S. Seager, *Analysis Boot Camp: An Alternative Path to Epsilon-Delta Proofs in Real Analysis*, PRIMUS, vol. 30 (1), 2020.

[12] F. Su, *Mathematics for Human Flourshing*, Yale, 2020.

Index